Evolution of Man

CONTRIBUTORS:

E. N. da C. Andrada
Robert Ardrey
Isaac Asimov
Marston Bates
Henri Bergson
Ludwig von Bertalanffy
W. I. B. Beveridge
Harrison Brown
Colin Clark
Charles Darwin
Kingsley Davis
Theodosius Dobzhansky
René Dubos
Loren Eiseley
Francis Galton
Ernst Haeckel
J. B. S. Haldane
Edward T. Hall
A. V. Hill
Adolf Hitler
Hudson Hoagland
Aldous Huxley
Julian S. Huxley
Thomas H. Huxley
H. B. D. Kettlewell
David Krech
Joseph Wood Krutch
Jean Baptiste Lamarck
L. S. B. Leakey
Joshua Lederberg
Martin Lüscher
Kevin Lynch
Archibald MacLeish

Floyd Matson
James McConnell
P. B. Medawar
Ashley Montagu
Ruth Moore
Desmond Morris
H. J. Muller
Friedrich Nietzsche
Fairfield Osborn
William Paley
Donald C. Peattie
Lloyd Rodwin
David Sarnoff
Amram Scheinfeld
Hans Selye
George Bernard Shaw
Charles Sherrington
George G. Simpson
Edmund Sinnott
B. F. Skinner
Roger W. Sperry
Saul Steinberg
John H. Storer
Maurice Sussman
Pierre Teilhard de Chardin
William Vogt
Alfred R. Wallace
G. P. Wells
H. G. Wells
Lancelot Law Whyte
Norbert Wiener
John Rowan Wilson

EVOLUTION OF
MAN

Prepared by AMERICAN FOUNDATION FOR
CONTINUING EDUCATION

Edited by LOUISE B. YOUNG

New York   OXFORD UNIVERSITY PRESS   1970

PREFACE

Evolution of Man is a book of readings focusing on the nature of man and his role in the evolutionary process. It is designed to provide a background of knowledge about the issues concerning man's control of man and his environment through science—problems such as eugenics, birth control, pollution, ecology, and city planning, as well as the broader issue of the total direction of human development.

The most important link between science and the humanities lies in biology and more specifically in the science of man. Through the evolutionary concept, man's position in the world of nature is seen in perspective as the latest phase of a continuing process that started with the first living molecule, perhaps a billion years ago. Against the background of this evolutionary history, man's problems, his significance, and his potential can best be understood. Biology tells us that the success of each species depends upon a harmonious relationship between the organism and the environment. Man is the only organism that has learned to significantly modify its environment. This ability confers great evolutionary opportunities, but it also creates grave dangers. The direction of change is no longer determined by natural law. The future has become man's responsibility, and therefore an understanding of the scientific view of man should enter into the decisions that will mold that future.

This volume of readings and two companion volumes have grown out of the conviction that an understanding of science has become increasingly important to the responsible citizen in a modern democratic society. More and more matters of public policy involve judgments concerning the potential of scientific research, the limits of scientific expertise, and the role of science in our society. At the same time it is evident that today an increasing number of responsible and otherwise well-educated adults are content to leave science out of their education, unlike their counterparts of several centuries ago who knew and discussed the latest scientific ideas of their day. Galileo's discoveries and Newton's laws were understood by poets, artists, and philosophers as well as by scientists. Courses and demonstrations were offered and interest ran high even among the ladies of the court. But today many people consider science a subject that should be left to the experts. They believe it has become so complicated that the latest and most interesting aspects cannot be understood without years of preparation and, furthermore, that although the ideas may be intellectually intriguing, they are not relevant to the problems of everyday life.

The traditional science courses in schools and colleges have reinforced these impressions. Beginning science courses are designed primarily as a foundation for a particular scientific discipline. But science for the non-scientist requires a less specialized approach with emphasis on general principles rather than detailed facts. It should provide an introduction to some of the exciting new concepts, and an appreciation of the ways these ideas relate to the rest of our culture.

In 1960 the American Foundation for Continuing Education began work under a grant from the National Science Foundation to develop a new kind of program for adult education in science. After several years of experimentation and testing, a study-discussion format was chosen centering on three books of readings. Each book was designed to deal with one broad area of scientific research and to illuminate one important new aspect of modern life which has developed out of scientific discovery.

The first book, *Exploring the Universe,* published in 1963, was planned to provide the layman with a background of understanding of the principles on which the Space Age has been built, to give him a glimpse of the mystery and the majesty of the universe which man has begun to explore and, above all, to suggest the methods and nature of the search itself. *Exploring the Universe* was used as the basis for a series of shows by the same name prepared by National Educational Television and shown over educational channels during 1963.

The readings in the second book, *The Mystery of Matter,* deal with the very smallest features of the physical world and show how scientific concepts were developed to account for the nature of both living and non-living matter. The readings explore many of the implications of the Atomic Age—political issues such as the pros and cons of nuclear testing, philosophical implications such as the rise of materialism and the new non-mechanical view of reality that is beginning to emerge from modern physics.

This third book of readings, *Evolution of Man,* is planned to prepare the citizen for the age of increasing control over our environment and the direction of human development. Are we entering a Humanistic Era or an Age of Automation applied to human beings? By understanding the evolutionary process as a whole and man's place within this process the responsible citizen can take a more constructive part in making the decisions that will determine the future of man.

These three volumes are designed to complement each other and can most profitably be read together. Throughout the series every effort has been made to avoid repeating explanations of the same scientific material. For instance, the theories concerning the evolution of the universe are covered in *Exploring the Universe*, the evolution of inorganic matter is treated in *The Mystery of Matter*, and neither of these subjects is included in *Evolution of Man*, which begins with the evolution of biological species. Students who want to appreciate the full sweep of the evolutionary process, which can be traced with remarkable continuity from a primeval nebula to a modern man, should study all three volumes.

However, each individual book is also designed to stand alone. Any information necessary for comprehension is included at that point. The readings have been chosen to describe scientific concepts in a way that can be understood by an intelligent adult without any previous knowledge in science or mathematics. The technical terms are explained in context as they first appear. For later reference, these words are listed in the Index with the page number on which the explanation occurs. Many of the selections are excerpts from the writings of great scientists. The clarity of their exposition demonstrates dramatically that an idea which is thoroughly understood by the author can be explained in a meaningful way to a person who is not an expert in the field. It is hoped that when the reader finds that he is able to read these selections with pleasure and understanding he will continue to keep himself up to date on the latest and most exciting developments by taking advantage of the many excellent books and articles that are written for laymen.

As mentioned earlier, the format of these books was dictated by the desire to demonstrate that the intelligent citizen can understand interesting new scientific concepts and, perhaps more importantly, to demonstrate the relevance of these concepts to the rest of our culture. By showing historically how ideas develop rather than simply explaining the present state of scientific knowledge, these selections assist the reader to gain insight into how scientists work, under what conditions important advances are made, and why scientific theories change from time to time. Moreover, by the inclusion of selections from novelists, philosophers, and public figures, the reader is able to appreciate how scientific theories reflect the general intellectual climate of the time in which they were conceived and how they, in turn, influence the artistic, social, even the political attitudes of their day.

The readings in this book are divided into ten parts, each part organized around one broad general question. These questions are not just rhetorical. They are matters on which no final answer has yet been reached. The fact that these and many other questions do exist illustrates an important aspect of scientific research. Science is not a closed book of accepted facts but a never-ending search, achieving at best a nearer and nearer approximation to an understanding of the physical world.

The first articles in each part are usually descriptive and expository of a scientific concept. Later articles turn to an exploration of the human response to these ideas or the religious, philosophical, and social implications. Throughout the selections many questions are posed which are not answered. Often different points of view are presented with no editorial decision made about which is the right point of view because the aim of these readings is to stimulate thought about these issues. Confronted by differences in points of view, the reader will be encouraged to think through his own ideas and to deepen his understanding of the part science plays in his world.

Since articles are selected to complement one another and are arranged to bring into focus a fundamental aspect of science, they should be read in the order in which they appear. Many of the selections have been excerpted. Titles have sometimes been changed to indicate more clearly the nature of the selection's function in this book. In some cases footnotes and illustrations have been omitted because in this context they do not contribute to the central issue; editorial notes and illustrations have been added where they were deemed useful. There are also a comprehensive index and biographies of the authors (page 623). A list of suggestions for further reading appears at the end of each part. The titles listed are just a sample of the fine books available, more of which are being published each month.

As in other programs that the AFCE has prepared and published, we do not represent or support any particular viewpoint with respect to the public policy questions or the religious or social issues contained in these readings. The AFCE is a non-profit educational organization devoted to the development of study programs in the liberal disciplines for adults concerned with their own continuing education.

We are grateful to the many scientists and educators throughout the country who assisted us in developing this series of books for adult education in science. We should especially like to thank Dr. Earl Evans, Professor of Biochemistry at the University of Chicago; Dr. Joseph Kaplan, Professor of Physics at the University of California at Los Angeles; Dr. Irving Klotz, Professor of Chemistry at Northwestern; and

Dr. Paul B. Weisz, Professor Emeritus of Biology at Rockefeller University. Final responsibility, however, for the selections in this volume rests with the editor.

L.B.Y.

Winnetka, Ill.
August 1969

TABLE OF CONTENTS

Part 4 WHAT IS THE ORIGIN OF MAN?

Part 5 DID MIND EVOLVE BY NATURAL SELECTION?

1

Is There Sufficient Evidence for Evolution?

The selections in Part 1 describe the facts and observations that have led to the general acceptance of the theory of evolution. How did these facts discredit the older theory of static forms that were separately created? Is the proof final and irrefutable? What constitutes scientific proof and what is the value of a scientific theory?

Is There Sufficient Evidence for Evolution?

Introduction

THE THEORY OF EVOLUTION has probably had a greater impact upon the thought and temper of our times than any other scientific theory. Teilhard de Chardin says: "What makes and classifies a 'modern' man (and a whole host of our contemporaries is not yet 'modern' in this sense) is having become capable of seeing in terms not of space and time alone, but also of duration . . . and above all having become incapable of seeing anything otherwise—anything—not even himself." [1]

Contrary to common opinion, evolution was not discovered by Charles Darwin. What Darwin (and also simultaneously Alfred Wallace) discovered was an explanation of how evolution could have happened by natural forces, and thereby gained scientific acceptance for the idea of evolutionary change which had been suggested by a number of scientists during the preceding century. In this first series of readings we are going to examine the idea of evolution, omitting, as far as possible, any discussion of the causes that bring it about. We will look at evolution as a way of seeing the world, and examine a few of the consequences of this modern view.

Ever since men first began to reflect upon the nature of the mysterious world into which they had been born, they have been struck by two aspects that seem to be contradictory. There is obviously change; we see movement and activity going on about us all the time. "We never step twice into the same rivers," said Heraclitus, "for fresh waters are ever flowing in upon you." But underlying the small fluctuations there are also longer rhythms and cycles which appear to recur perpetually, bringing back the same repetition of day and night, the same procession of constellations across the evening sky, the same recurrence of life in the spring. Is change only a surface phenomenon like ripples within a fixed sphere of permanence? Or could it be that both change and cyclic rhythms are real manifestations of a cosmic process that is really "going somewhere," like a spiral staircase that does turn back on itself at each winding but that never returns to exactly the same place?

Various explanations of the relationship of change and permanence have been in the ascendancy at different times in history. During ancient and medieval times the Platonic theory of eternal forms—of change being

1. Pierre Teilhard de Chardin, *The Phenomenon of Man*, Harper & Row, Publishers, 1965, New York, p. 218.

only a distorted reflection of a fixed and permanent reality—was the dominant belief of religious and philosophical, as well as scientific thought.

Even today the desire for a return to some sort of absolute reality has not entirely lost its sway. The theory of the universe as a giant pulsating atom undergoing regular expansions and contractions is one scientific example of the theory of cyclic return. However, in general it seems fair to say that belief in progress has become a basic characteristic of modern thought and that the biological theory of evolution has been partly responsible for this new world view.

Belief in progress is a dynamic force in our civilization, sparking an era of unprecedented change and a continuing drive for improvement in the human condition. Would the aims of our society be different if we did not believe that mankind could improve? If we did not believe that the universe was in fact evolving and mankind was developing and maturing along with it? Unfortunately, the belief in progress is also used to justify a number of evils. These dangers and distortions we will explore in later readings.

When a scientific theory has caused such a profound alteration in the philosophy of our times it is important that people understand the nature of the truth that is revealed by scientific theory. Science is one way of getting at the truth—some people believe, the only valid way—but the truth which it reveals is of a tentative nature. It does not provide ultimate or absolute truth such as that claimed by religious revelations or philosophical systems. On the other hand, mankind yearns for certainties and the non-scientist, looking at the miracles which science has achieved, is apt to attribute to scientific discoveries a finality which they do not possess. Sometimes this confusion is compounded by the enthusiastic scientist who presents his favorite theory or discovery in such terms that it sounds like an irrefutable fact. Other scientists, of course, understand the spirit in which these declarations should be taken. As George and Muriel Beadle say: "They know—often to their despair as human beings—that science is an endless opening of sealed boxes which turn out to have more sealed boxes inside. The more one learns, the more there is to learn. There is never a last word." [2]

A scientific theory is primarily a working hypothesis whose object is to coordinate apparently diverse phenomena. It acts as an aid to investigation by creating an orderly relationship among known facts and by suggesting experiments that may lead to new knowledge. Sometimes the new facts are not in harmony with the old theory which must then evolve in order to accommodate them. "A theory," says J. J. Thomson, "is a policy

2. George and Muriel Beadle, *The Language of Life, An Introduction to the Science of Genetics,* Doubleday & Company, Garden City, N.Y., p. 53.

rather than a creed." [3] A million pieces of evidence supporting it will not be sufficient to achieve a final proof. On the other hand, one piece of evidence that cannot be explained within the context of the theory may be enough to disprove it. As Wells, Huxley, and Wells point out: "A single· human tooth *in situ* in a coal seam would demolish the entire fabric of modern biology." [4]

Nevertheless, such a piece of evidence has not been found and millions of facts have been found supporting modern evolutionary theory. Under these circumstances most biologists have become convinced "beyond reasonable doubt" of the truth of the theory. Is this legal criterion as close as science can bring us to absolute certainty? Is it sufficient evidence on which to base a whole philosophy of life? Perhaps, in a scientific age, we must learn to live with some uncertainty, to be flexible and adaptable in our thinking, just as organisms must constantly adapt to a changing physical world in order to evolve.

3. J. J. Thomson, *The Corpuscular Theory of Matter*, Charles Scribner's Sons, New York, 1907.
4. See page 31.

The Idea of Evolution

Cosmic Cycles and Biological Progress [*]

by René Dubos (1962)

THE MYTH OF ETERNAL RETURN

The very idea of progress is far from evident and universal. Indeed, belief in progress represents such a recent attitude, that, according to historians, the idea did not become widespread until the eighteenth century.

It is obvious that men who live close to nature—as did most men until not so long ago—have little occasion to experience continuous progress toward something new and better. What impresses them, rather, is the recurrence of the old, the endless repetition of similar events. Ecclesiastes had good reason to lament more than two thousand years ago that "there is nothing new under the sun"—either in the ways of nature or in the ways of men. Every day the sun rises and sets; every month the moon displays its regular phases; every year the seasons come and go, bringing in their train a more or less predictable order of changes in the weather, in the vegetation, in animal life, and in human moods. Social phenomena also exhibit a distressing predictability. Civilizations are born, flourish, and decay; hope follows despair; good and evil seem forever to compete and to engage in a tug of war on the stage of the human comedy.

This endless recurrence of natural events, and these monotonous repetitions in the human drama, so profoundly influenced the primitive mind that they shaped many of its beliefs and practices. All people in the past became emotionally involved in the natural cycles, and they marked the turning points in recurrent phenomena by religious festivals and rites, most of which persist even today. The most obvious, of course, are the pagan celebrations in all parts of the world to mark the reawakening of life in the spring, and the gathering of crops in the fall—celebrations which have become so deeply integrated in human life that they still stir deep currents in the blood of modern city-dwellers. Just as the sap starts

[*] From René Dubos, *The Torch of Life, Continuity in Living Experience,* the Credo series edited by Ruth Nanda Anshen, Simon and Schuster, Inc. Copyright © 1962 by Pocket Books, Inc. Reprinted by permission.

flowing in the trees, and the birds begin nesting when the sun becomes warmer, so do men and women experience the need to expand their increased vital energy—either in the form of Mardi Gras, of pole dances, or of marriage ceremonies. And everywhere in the fall, harvest festivals mark the ripening of the grain, of the pumpkin, and of the grapes.

Very early in the human epic, earlier perhaps than the emergence of festivities dedicated to food and to the animal senses, the rhythms of nature inspired celebrations of a more abstract quality. The worship of the sun at a special moment of the year was practiced among the Chaldeans and the Natchez in much the same way as it had been in prehistoric times at Stonehenge and Carnac. The solar events also inspired some of the most cherished and beautiful myths of the ancient Mediterranean world. Daphne, the dawn, sprang from the waters at the first blush of morning light; then, as the tints of early day faded gradually in the light of the rising orb, she fled from Apollo who tried to win her. . . .

There is no doubt that the many legends, processions, mysteries, and sacrifices which ruled ancient life according to the regular pattern of natural events, imposed upon the human mind the concept that the world repeated itself, endlessly, from year to year. The awareness of this eternal return is still, of course, a powerful force among us. Witness the vigor of the New Year tradition when every person, except the most disenchanted, entertains for a few moments the illusion that once more everything starts anew.

As shown by Mircea Eliade in *Cosmos and History*, the myth of eternal return has had a profound significance in the development of human thought. Ancient man generally believed that there are cyles in social history, just as there are recurrent phenomena in nature. There were Golden Ages in ancient times, and there will be Golden Ages in the future; the trials of the moment are but the winter months between the times of plenty. Ancient man could not conceive of endless progress, because he believed that history must repeat itself indefinitely.

THE TORCH OF LIFE

The deep reality of the endless cycle between life and matter is made evident to the senses by looking at the earth, handling it, smelling it, and observing its changes from season to season. After death, plants and animals fall to the ground; there they rot and putrefy; and in the process they are transmuted into the humus of the soil out of which new life is produced. Under a thousand symbols, men of all religions and philosophies have sung and portrayed this repeated return of living things to inanimate matter, and the ever-repeated emergence of new life. There is, indeed, a fascination and perhaps truth in the ancient creed that life is always arising anew from matter, as Aphrodite came out of the foam of the sea. Men have also long believed that living things are endowed with a special kind of force—in itself eternal—capable after death of entering into all sorts of new, diverse combinations to re-create life. In the ancient Latin world these problems found their most passionate expression in Lucretius' poem "De Rerum Natura."

Untiringly, Lucretius reiterated in his poem that nothing arises save by the death of something else, that Nature remains always young and whole in spite of death at work everywhere; that all the living forms that we observe about us are but so many passing forms of a permanent substance; that all things come from dust, and to dust return—but a dust eternally fertile. "What came from earth goes back into the earth," Lucretius wrote. "What was sent down from the ethereal vault is readmitted to the precincts of heaven. Death does not put an end to things by annihilating the component particles; it only breaks up their conjunction. Then it links them in new combination." Or elsewhere: "Whatever earth contributes to feed the growth of others is restored to it. The universal mother is also the common grave. . . . Sea and river and springs are perennially replenished, and the flow of fluid is unending."

There is, indeed, scientific truth and philosophical depth, as well as poetic beauty, in the famous lines of Lucretius' "De Rerum Natura," . . . "Mortals live by mutual interchange. One breed increases by another's decrease. The generations of living things pass in swift succession, and *like runners in a race they hand on the torch of life.*"

It is, in fact, a universal rule of nature that the tissues of dead plants and animals undergo decomposition and are thus returned to the envelope of soil, water, and atmosphere at the surface of the globe. Should any component of organic life fail to decompose and be allowed to accumulate, it would soon cover the world and imprison in its inert mass chemical constituents essential to the continuation of life. Of this, however, there seems to be no danger. Experience shows that substances of animal or plant origin do not accumulate because they undergo, sooner or later, a chain of chemical alterations which break them down stepwise, into simpler and simpler compounds. It is in this fashion that after death the chemical elements are returned to nature for the support of new life. There is literal truth in the Biblical saying that "all are of the dust, and all turn to dust again."

The eternal movement from life, through organic matter, down to simple chemical molecules, and back into life again, is the biochemical expression of the myth of eternal return. Scientists have long been preoccupied with this complex cycle and have known that its continuation is essential to the maintenance of life on earth. . . .

Even elementary textbooks illustrate, with elaborate, pictorial images, the various chemical steps through which carbon, phosphorus, and nitrogen pass from plant and animal to microbe and then back to soil, as in a cyclic dance. In a curious way, modern scientific findings are thus giving substance to some of the ancient myths of eternal return. Much more exciting, however, is the fact that scientific discoveries have revealed an aspect of life which had hardly been perceived in the past.

Living things do not behave as passive messengers when they transmit matter from one generation to the next or from one form of life to another. Passing the torch of life is truly a continuous act of creation. In fact, the torch of life is not passed as in a race. New torches are endlessly lighted from it, to be carried over all parts of the earth. And, mysteriously, during this process, in the course of aeons, living things have progressively become more varied and diversified through the operation of mutational and other evolutionary changes. There are, obviously, recurrent cycles and eternal returns, but there is also forward motion. The future does more than repeat the past. Awareness of this forward motion of life—called progress when applied to man—is perhaps the most distinctive feature of modern philosophy and science. It is what differentiates most profoundly the temper of modern times from that of all past ages, even the most enlightened.

BIOLOGICAL PROGRESS

Whereas the belief in recurrent cycles—in a closed *circular* course of everything in the cosmos—fits the practical experience of men living in intimate contact with nature, the belief in progress as an expanding process endlessly generating new creations is of the nature of an intellectual *tour de force*. It demands a very long-range view of natural and human history, an awareness of the creative power of rational knowledge, and an almost blind faith in the wisdom of man. Belief in progress apparently began timidly in classical Greece, but it remained only a vague, unconvincing idea until the seventeenth century. Even Newton held to the traditional view of the stability and perfect design in nature and, therefore, could hardly conceive of biological and social progress. It was only after him that the sense of change and mutability began to grow. The belief in progress became a kind of religious faith among the philosophers of the Enlightenment with scientific inclinations; it has spread more and more vigorously ever since, weakening and almost destroying in the hearts and the minds of men the emotional power of belief in eternal return.

The Dawn of Evolutionary Theory *
by Loren Eiseley (1958)

THE SCALE OF NATURE

As we survey the course of scientific history it would appear inevitable that the present world would have given man his first clues to the history of life. Yet it is interesting to observe that only the existence in the West of a certain type of theological philosophy caused men to look upon the world around them in a way, or in a frame, that would prepare the Western mind for the final acceptance of evolution. Strange though it may sound, it was a combination of Judeo-Greek ideas, amalgamated within the medieval church itself, which were to form part of the foundation out of which finally arose, in the eighteenth and nineteenth centuries, one of the greatest scientific achievements of all time: the recovery of the lost history of life, and the demonstration of its total interrelatedness. This achievement, however, waited upon the transformation of a static conception of nature into a dynamic one. It was just this leaven which the voyagers supplied with their unheard-of animals, and apes that were scarcely distinguishable from savage men.

Widespread in the literature of the seventeenth and eighteenth centuries, and easily traceable into earlier periods, is the theological doctrine known variously as the *Scala Naturae*, Chain of Being, *echelle des être*, Ladder of Perfection, and by other similar titles. . . . There can be little doubt that the rise of comparative anatomy is inextricably linked to the history of the Chain of Being concept with its gradations of complexity in living forms. In making this observation, however, we have to keep in mind one salient fact. Strange though it may sound to a modern evolutionist this gradation of organisms implied nothing in the way of phylogenetic relationship. Equally it implied nothing in the way of evolutionary transformations and it specifically denied the possibility that any organism could become extinct. The whole scheme was as rigidly fixed as the medieval social world itself. Indeed it is to some degree a powerful mental projection of that world.

"There is in this Universe a Stair," [says Thomas] Browne, "rising not disorderly, or in confusion, but with a comely method and proportion."

* From Loren Eiseley, *Darwin's Century*. Doubleday & Company, Inc. Garden City, New York. Copyright © 1958 by Loren Eiseley. Reprinted by permission.

And since the Scale of Nature runs from minerals by insensible degrees upward through the lower forms of life to man, and beyond him to purely spiritual existences like the angels, ourselves, compounded of both dust and spirit, become "that great and true Amphibium, whose nature is disposed to live . . . in divided and distinguished worlds." We exist, in short, in both the material and spiritual universes. In this respect *Homo duplex,* as he was sometimes called, occupies a place on the scale of life as a link betweeen the animal and spiritual natures. Man suffers from this division and it contributes to his ofttimes confused and contradictory behavior.

If, however, this serried array of living forms does not denote a physical phylogenetic connection, what can it be said to represent? It is just here that we enter upon the very real differences which exist between the recognition of grades of complexity in nature and the assumption that a lower level of complexity evolves physically into that of a higher—that an ape, by way of illustration, may evolve into a man. The scholars of the eighteenth century recognized quite well that the ape stood next to man on the Scale of Nature, but they did not find this spectacle as appalling as a nineteenth-century audience listening to Thomas Huxley. There was a very simple reason for this: The *Scala Naturae* in its pure form asserts the immutability of species. The entire chain of life is assumed to have been created in its present order when God by creative fiat brought the universe out of chaos.

As we remarked earlier the scale is static. Creation is not considered as still in progress. Thus the resemblances between living things are not the result of descent with modification but rather are the product of the uniformity and continuity of the divine act. Since the world was assumed by theologians and scientists alike to be only a few thousand years old, the question of evolution could arise only with the greatest difficulty. There was literally not time enough for such a creation. Theologically it was also held that animal species could not become extinct. By and large, men eyed askance the notion that a whole order of life could disappear. Such piecemeal disappearances from the Scale of Nature seemed to threaten the confidence reposed in divine providence.

As time went on, evidences for the past existence of organisms no longer to be found on the planet began to be brought forward but were received with obvious reluctance. Few wished to believe the reports and their reception was not encouraging. Since knowledge of some parts of the world was scant, even well into the eighteenth century, one favorite resort was to accept the disappearance of certain forms of life in Europe but with the proviso that the creatures probably still survived in remote areas of the earth.

It was a convenient evasion of a question which had theological over-
tones. For just this reason, however, the often mentioned reports of mammoths surviving into colonial times in interior North America have to be
viewed with great skepticism. The intellectual climate of the times promoted and encouraged such accounts. Always the creatures lay just a little
farther on, first in the Virginia woods, or in Labrador, then deeper into
the interior or "across the lakes." They were heard bellowing in the
woods, or seen grazing on the plains of South America. In no case, however, is the documentation satisfactory, nor were hides or tusks from recent beasts shipped home to adorn the cabinets of eager scientists. With
the acceptance of the idea of total and successive extinctions of past
faunas at the dawn of the nineteenth century the sporadic reports of living mammoths or mastodons began to fade. It must also be remembered
in this connection that until the great age of the world and its successive
strata were grasped, no great antiquity could be attributed even to fossil
bones.

It remains a curious episode in the history of science that the Scale of
Nature doctrine which denied extinction should at the same time have encouraged the comparative anatomical observations which would eventually lead to the discovery of extinction. Even more important, the idea of
phylogenetic relationship along the scale of life would emerge almost
simultaneously. The attention which perfectly orthodox thinkers were encouraged to give to the ascending ladder of being, their eagerness to trace
every degree of continuous relationship in the productions of the divine
being, their zealous efforts to show that the apparent missing links in the
scale could be found, enormously stimulated the study of taxonomy and
variation.

All that the Chain of Being actually needed to become a full-fledged
evolutionary theory was the introduction into it of the conception of time
in vast quantities added to mutability of form. It demanded, in other
words, a universe not made but being made continuously. It is ironic and
intriguing that the fixed hierarchical order in biology began to pass almost
contemporaneously with the disappearance of the feudal social scale in
the storms of the French Revolution. It was France, whose social system
was dissolving, that produced the first modern evolutionists. As we look
back upon the long reign of the Scale of Being, whose effects, as we shall
later see, persisted well into the nineteenth century, we may observe that
the seed of evolution lay buried in this traditional metaphysic which indeed prepared the Western mind for its acceptance. "Thus disguised and
protected," writes Lois Whitney, "did the hypothesis of evolution have, as
it were, a happy seed time, a period in which to germinate and take root,
before the orthodox world scented the danger." . . .[1]

1. *Primitivism and the Idea of Progress,* Baltimore, 1934, p. 158.

LINNAEUS

An orderly and classified arrangement of life was an absolute necessity before the investigation of evolution, or even its recognition, could take place. Before life and its changes and transmutations can be pursued into the past, the orders of complexity in the living world must be thoroughly grasped. Comparative anatomy must have reached a point of development sufficient to permit the scientist to distinguish a living animal from one no longer in existence. Moreover, the naturalist must be able to recognize affinity and relationship in the midst of difference. He must be able to observe the likeness which reveals the interrelatedness of life across the gulf of time and yet, equally, pointing to distinctions of detail, the student must be able to say "here change is evident." Knowledge of this degree of sophistication could not come in a day. As the great Swedish taxonomist [Carolus] Linnaeus was to remark later, "The first step of science is to know one thing from another. This knowledge consists in their specific distinctions; but in order that it may be fixed and permanent distinct names must be given to different things, and those names must be recorded and remembered." [2] John Ray [born in 1627] was a modern in his search for a natural system of classification based upon clear structural affinities.

In this respect Ray anticipated and influenced Linnaeus. Moreover, in his emphasis upon "natural system," in his concern with behavior, he had perhaps a more far-ranging philosophic mind than his successor. He not only helped make possible the *Systema Naturae* of Linnaeus: he was also the forerunner of Gilbert White, of Paley's *Natural Theology* and finally of the *Origin of Species*. . . .

In 1707, two years after the aged and infirm John Ray had died at Black Notley, Linnaeus was born in southern Sweden. It was a time of marked English influence in Sweden. Many young men of family journeyed to London, and English philosophy and science exerted great influence upon Swedish culture. Linnaeus, being in modest circumstances, took his medical degree in Holland, where he came in contact with the great Dutch scientist Hermann Boerhaave and launched the first edition of his best-known work, the *Systema Naturae*, in 1735. In 1736 he visited England and made a solid acquaintanceship in learned circles. From that time onward his prestige in English science was enormous—a genuine mass phenomenon. . . .

That pure naming and systems of classification got a little out of hand and took on a one-sided emphasis which persisted well into the nine-

2. Sir James Edward Smith, *A Selection of the Correspondence of Linnaeus and Other Naturalists from the Original Manuscripts*, London, 1821, Vol. 2, p. 460.

teenth century need not be attributed solely to Linnaeus. He rose to fame in a period of great wonder and eagerness to explore and catalogue the products of far lands. New words were pouring into European speech. The name was all and Linnaeus, with his gift for precise definition, with his exquisite taste for order, was providing the framework necessary to science before science could proceed to other things.

Further, if Linnaeus pursued the name, the name in its own way led to things no man could foresee. It was in his time, and owing greatly to his influence, that naturalists began to be apportioned posts on voyages of exploration. Cook's voyage on the *Endeavor* in 1768, to which Sir Joseph Banks contributed so heavily, is a case in point. It set the pattern which led eventually to Darwin's voyage on the *Beagle*. . . .

THE FIXITY OF SPECIES

Scientists have long accused the church of holding back, by its preconceived beliefs, the progress of the evolutionary philosophy. The matter is actually more complicated than this. Science, in the establishment of species as a fixed point from which to examine the organic world, gave to the concept a precision and fixity which it did not originally possess. Categories of plant and animal life . . . did not, in earlier centuries, possess the clarity that they began to take on at the hands of Ray and Linnaeus. As one astute but anonymous writer observed over fifty years ago: "Until the scientific idea of 'species' [3] acquired form and distinctness there could be no dogma of 'special' creation in the modern sense. This form and distinctness it did not possess until the naturalists of the seventeenth century began to substitute exactness of definition for the previous vague characterizations of the objects of nature." [4]

As scientific delight and enthusiasm over the naming of new species grew with the expanding world of the voyagers, the conviction of the stability and permanence of the living world increased. Strict definition, so necessary to scientifically accurate analysis, led in the end to the total crystallization of the idea of order. It is true, as we have observed earlier, that the notion of the fixed Scale of Being and the Christian conception of time, as well as the Biblical account of Creation, all tended to discount the evolutionary hypothesis, but, ironically enough, it was Linnaeus with his proclamation that species were absolutely fixed since the beginning who intensified the theological trend. His vast prestige in both scientific

3. For animals and plants that reproduce sexually, with cross fertilization, a species is defined as a group of individuals capable of breeding with one another but not with members of other species. Other organisms are classified into species on the basis of structural similarities. Ed.

4. "Lamarck, Darwin and Weismann," *The Living Age*, 1902, Vol. 235, p. 519.

and cultivated circles made it assured that his remarks would be heeded. Henceforth the church would take the fixity of species for granted. Science, in its desire for classification and order, had found itself satisfactorily allied with a Christian dogma whose refinements it had contributed to produce.

Yet no sooner had Linnaeus proclaimed his dogma than, while working in the botanical gardens of his patron Clifford at Hartecamp, he grew aware of the modern "sportiveness" of nature. He saw varieties appear spontaneously, he saw "abnormal" plants derived from normal ones. Like Ray before him, but perhaps more clearly, he was forced to distinguish between the true species of the Creator and the varietal confusion and disorder of the moment, which might be artificially manipulated by the skill of gardeners. In this way he attempted to cling to his original thesis. He had assumed that all species come from original pairs created on a small island which, in the beginning, had constituted the only dry land, the original Eden of the world.

As one pursues this subject through his multitudinous writings and the ever mounting editions of the *Systema Naturae* one can trace a growing uncertainty and doubt. He sees the possibility of new species arising through crossbreeding. He confesses that he dare not decide "whether all these species are the children of time, or whether the Creator from the very beginning of the world had restricted this course of development to a definite number of species." [5] He cautiously removes from later editions of the *Systema* the statement that no new species can arise. The fixity of species, the precise definition of the term, is no longer secure. "*Nullae species novae*" had been accepted by the world, but to the master taxonomist who had drawn the lines of relationship with geometric precision all was now wavering toward mutability and formlessness. Only the natural orders now seemed stable. What this actually might have meant to Linnaeus who had placed man along with monkeys in his order of the Primates, it is far too late to determine satisfactorily, but that he toyed with ideas of strange animal mixtures and permutations we know [Figures 1–1 and 1–2].

THE COMTE DE BUFFON

Linnaeus had one great rival in the public affection; this was the Comte de Buffon (1707–88). In 1749 he published the first volume of his huge *Histoire naturelle*, a set of studies of the living world destined to have a wide circulation and to be translated into many languages. It was grace-

5. Cited by Hagberg, *Carl Linnaeus*, Jonathan Cape, London, 1952, p. 202.

Figure 1–1. The tenth edition of Systema Naturae *by Linnaeus included the species* Homo Monstrosus *(as well as the species* Homo Sapiens*). The belief in semihuman creatures living in remote parts of the world had persisted for thousands of years and these legends were kept alive by the tall tales brought home by explorers during the middle ages. Some examples of Homo Monstrosus described by authors such as Homer, Herodotus and Megasthenes are shown above: the headless people, the one-eyed Cyclops, the Sciapodes with one enormous foot which they used as an umbrella, the big-lipped people, the long-eared Phanesians, the goat-footed people (later called satyrs). (These woodcuts are from the* Liber Chronicarum *by Hartmann Schedel, Nuremburg, 1493. Courtesy Rare Book Division, The New York Public Library, Astor, Lenox and Tilden Foundations. Figure added by the editor.)*

Figure 1-2. Etchings showing imaginative specimens of ape-like creatures that were believed to be subhuman species. (From A History of Four-footed Beasts by Edward Topsell, published in London, 1607. Courtesy Rare Book Division, The New York Public Library, Astor, Lenox and Tilden Foundations. Figure added by the editor.)

fully, even entrancingly written, and here and there the author managed, not too conspicuously, to touch upon a number of forbidden topics.

The book was written to appeal to two sorts of readers: those interested in the simple description of animals and those intellectuals who might wish to think about what they saw. We need not expect complete candor on the part of a man writing a century before Darwin. Buffon had doubts, hesitations, and fears. He wrote at times cryptically and ironically. He brought forward an impressive array of facts suggesting evolutionary changes and then arbitrarily denied what he had just been at such pains to propose. It is not always possible to determine when he was exercising an honest doubt of his own and when he was playing a game. In any case he could not leave this dangerous subject alone. It fascinated him as, a century later, it was to fascinate Darwin. He had devised a theory of "degeneration." The word sounds odd and a trifle morbid today, because we are in the habit of thinking of life as "evolving," "progressing" from one thing to another. Nevertheless, Buffon's "degeneration" is nothing more than a rough sketch of evolution. He implied by this term simply change, a falling away from some earlier type of animal into a new mold. Curiously enough, as his work proceeded, Buffon managed, albeit in a somewhat scattered fashion, at least to mention *every significant ingredient which was to be incorporated into Darwin's great synthesis of 1859*. He did not, however, quite manage to put these factors together. Specifically they may be analyzed as follows:

1. Buffon observed a tendency for life to multiply faster than its food supply and thus to promote a struggle for existence on the part of living things. "Nature," he said, "turns upon two steady pivots, unlimited fecundity which she has given to all species; and those innumerable causes of destruction which reduce the product of this fecundity. . . ."[6]

2. He recognized that within a single species there were variations in form. In domestic plants and animals these variations were often heritable, so that by careful selection the stock could be improved and the direction of the improvement controlled. "There is," he wrote, "a strange variety in the appearance of individuals, and at the same time a constant resemblance in the whole species."[7] He recognized "our peaches, our apricots, our pears" to be "new productions with ancient names. . . . It was only by sowing and rearing an infinite number of vegetables of the same species, that some individuals were recognized to bear better and more succulent fruit than others. . . ."[8] Similarly he noted that in the case of the domestic hen and pigeon "a great number of races have been

6. *Buffon's Natural History*, London, 1812, Vol. 5, p. 88.
7. Ibid., pp. 128–29.
8. Ibid., Vol. 2, p. 346.

lately produced, all of which propagate their kinds." "In order to improve Nature," he commented in another volume, "we must advance by gradual steps." [9]

3. Buffon was impressed by the underlying similarity of structure among quite different animals, an observation which is a necessary prelude to tracing out ancestral relationships in the fossil past. "There exists," he said, "a primitive and general design, which may be traced to a great distance, and whose degradations are still slower than those of figure or other external relations. . . ." [10]

He philosophized warily that among the numerous families brought into existence by the Almighty "there are lesser families *conceived by Nature and produced by Time*." [11] Such remarks, woven into the web of orthodoxy, at times grow bolder, as when he suggested "that each family, as well in animals as in vegetables, comes from the same origin, and even that all animals are come from one species, which, in the succession of time, by improving and degenerating, has produced all the races of animals which now exist." [12] "Improvement and degeneration," he had earlier remarked, "are the same thing; for they both imply an alteration of original constitution." Though Buffon was quick to add something to satisfy the ecclesiastical authorities after such a remark, it is interesting to observe that in repeating all creatures have really been specially created, he says "*we ought to believe* that they were then *nearly* such as they appear at present." [13] It is obviously a most grudging concession.

4. . . . . Buffon anticipated the need of a greatly lengthened time scale in order to account for the stratification of the planet and the history of life upon it. "Nature's great workman," he said, "is time." By modern standards, of course, his estimates of the antiquity of the globe are very constricted but in his own time they were unorthodox. He thought that it had taken some seventy-two thousand years for the globe to cool from an incandescent state sufficiently to allow for the appearance of life. He assumed that the heat of the globe was imperceptibly diminishing. Further, he calculated that roughly another seventy thousand years would elapse before the planet was so chilled as to be unable to sustain life on its surface.[14]

5. He accepted the fact that some of the animal life of the earth had become extinct. This he ascribed to the cooling of the earth which had eliminated the warmth-loving fauna of an earlier day. He thought that

9. Ibid., Vol. 4, p. 102.
10. Ibid., pp. 160–61.
11. Ibid., p. 162. (Italics mine. L.E.)
12. Ibid., Vol. 5, pp. 184–85.
13. Ibid., p. 185. (Italics mine. L.E.)
14. Ibid., Vol. 2, p. 337.

many existing species would, in time, perish for the same reasons. He recognized the bones of mammoth as being those of extinct elephants and foresaw the value of paleontology. "To know all the petrifactions of which there are no living representatives," he remarked, "would require long study and an exact comparison of the various species of petrified bodies, which have been found in the bowels of the earth. This science is still in its infancy." [15] By this means, however, man, through the use of comparative anatomy, might be enabled to "remount the different ages of nature." It will eventually be possible, Buffon thought, to place milestones "on the eternal route of time."

6. Buffon also recognized the value of an experimental approach to evolutionary problems. The relations between species, he contended, could never be unraveled without long continued and difficult breeding experiments. "At what distance from man," he hinted slyly, "shall we place the large apes, who resemble him so perfectly in conformation of body . . . ? Have not the feeble species been destroyed by the stronger, or by the tyranny of man . . . ?" [16] Although in some passages he was careful to maintain the distinctive qualities of man, . . .

The count died in 1788, ten years after Linnaeus, although both had been born in the same year. It was a good time to go. The next year his son was to perish with other aristocrats in the fury of the Terror, proudly and reproachfully saying as he waited on the scaffold, "Citizens, my name is Buffon." It was the end of an age. Buffon like Linnaeus had had a world reputation, had had his specimens passed graciously through warring fleets, had corresponded with Franklin, had been one of the leading figures in a great country at the height of its intellectual powers.

It is a great pity that his ideas were scattered and diffused throughout the vast body of his *Natural History* with its accounts of individual animals. Not only did this concealment make his interpretation difficult, but it lessened the impact of his evolutionary ideas. If he had been able to present his thesis in a single organized volume, it is possible that he himself might have argued his points more cogently and perhaps seen more fully the direction of his thought. . . .

ERASMUS DARWIN AND LAMARCK

There has been considerable difference of opinion among students of evolutionary thought upon the origin of the views of Erasmus Darwin (1731–1802) [grandfather of Charles Darwin] and Jean Lamarck (1744–

15. Ibid., p. 250.
16. Ibid., Vol. 4, p. 218.

1829). Some have contended that Lamarck was stimulated by Erasmus Darwin's work, which was published prior to his own. Others claim both arrived at their ideas independently. Still another view would derive both men's ideas, in essence at least, from Buffon. This latter position is the most plausible. The contention for complete originality of thought on the part of both authors can be least sustained, for by the end of the eighteenth century the *idea* of unlimited organic change had been spread far and wide. It certainly was not a popular doctrine, but it had long been known in intellectual circles, largely through the popularity of Buffon.

Increasing interest in scientific breeding had also intensified public interest in the alteration of animal and plant forms. There were many who might not have been willing to say that all life arose from a single organic corpuscle, but who were vaguely and uncertainly aware that living forms might vary within limits. If pressed to name those limits with precision, they would have evinced discomfort. Lamarck and the earlier Darwin should be seen simply as continuing and enhancing a little stream of evolutionary thought which, beginning with ideas of purely specific or generic change—alteration, in other words, within narrow limits—was growing steadily bolder in the range of its thinking.

Of the two men, Lamarck was the more complete and systematic thinker. Erasmus Darwin's importance lies less in his scientific achievement than in his relationship to Charles Darwin and in his indirect influence upon Charles. (He died seven years before Charles Darwin was born.) Nevertheless, the priority of Erasmus over Lamarck is clear. His *Zoonomia* was published in 1794, but there is correspondence extant which indicates that its author was at work upon it as early as 1771.[17] He had as insatiable a passion for the odd facts of natural history as his grandson. . . . He was aware that life, because of its ever changing aspect, is not always perfectly adjusted to its surrounding environment. He is the undoubted source from which his grandson drew the idea of sexual selection, and in Canto IV of *The Temple of Nature* he sketched a ghastly picture of the struggle for existence. He estimated the antiquity of the earth in terms of "millions of ages." In spite of the diversity of life he recognized through "a certain similitude on the features of nature . . . that *the whole is one family of one parent.*"[18] Though similar quotations expressing Erasmus Darwin's grasp of comparative morphology and stores of odd learning could be multiplied from his works, we may come directly to the point. He himself remarks in *The Botanic Garden:* "As all the families both of plants and animals appear in a state of perpetual im-

17. Bashford Dean, "Two Letters of Dr. Darwin: the Early Date of His Evolutional Writings," *Science,* 1906, n.s. Vol. 23, pp. 986–87.
18. Preface to the *Zoonomia.*

provement or degeneracy,[19] it becomes a subject of importance to detect the causes of these mutations."

What, then, is Erasmus Darwin's explanation of the mechanics of evolution? It lies essentially in "the power of acquiring new parts, attended with new propensities, directed by irritations, sensations, volitions, and associations; and thus possessing the faculty of continuing to improve by its own inherent activity, and of delivering down those improvements by generation to its posterity, world without end!" [20] The key here lies in the words "irritations," "sensations," and "volitions." Erasmus Darwin, in partial but not complete contrast to his grandson, believed in the inheritance of acquired characteristics. Lamarck's philosophy was markedly similar.

Jean Baptiste Lamarck was intimately acquainted in his earlier years with Buffon, but it was not until his late fifties, in 1802, that he expressed himself as favoring the evolutionary hypothesis. . . .

Lamarck believed in a constant, spontaneous generation, so far as low forms of life were concerned, and he assumed a living scale of life which, in some respects, is reminiscent of the old *Scala Naturae*, although he broke partially away from the simple ladder arrangement. He believed in alteration rather than extinction. Any missing taxonomical links simply remained to be discovered. Thus, in so far as he studied man, he would have derived him from a living primate—probably the ever serviceable orang. As the world alters, as geographic and climatic areas change, new influences are brought to bear upon plant and animal life. In the course of long ages transformations in this life occur. . . .

Lamarck's name has by historical chance become so heavily associated with the doctrine of acquired characteristics that it is often assumed he invented it. Yet after an exhaustive treatment of the subject, running back through several centuries, [Professor] Zirkle remarks: "It is interesting for us to note how many of Lamarck's contemporaries stated that such characters were inherited and to note how completely these statements have been overlooked by modern biologists." [21] Zirkle goes on to establish the presence of the idea in medical, biological, and travel books. It was the commonly accepted doctrine of the time and, indeed, the first apt to be explored in the advance of biology. What Erasmus Darwin and Lamarck both did was to apply a very ancient hypothesis, one might almost say a folk-belief, to the explanation of continuing organic change and modification. Lamarck, whose work is the most thoroughgoing, saw clearly the cumulative advantages of such change in the creation of the higher orga-

19. Compare with Buffon's phraseology cited on p. 18.
20. *Zoonomia*, Vol. 1, p. 572.
21. Conway Zirkle, "The Early History of the Idea of Acquired Characters and of Pangenesis," *Proceedings of the American Philosophical Society*, 1946, n.s. Vol. 35, p. 111.

nisms. Right or wrong, there was nothing startlingly new about this—all the originality lay in its application to evolution.

As Professor Gillispie has pointed out, Lamarck was a late eighteenth-century Deist. Evolution, in his eyes, "was the accomplishment of an immanent purpose to perfect the creation." [22] Thus in his thought the old fixed ladder of being had been transformed into an "escalator." Life, in simple forms, is constantly emerging, and, through its own inner perfecting principle or drive, it begins to achieve complexity and to ascend toward higher levels.

THE INFLUENCE OF ERASMUS DARWIN

There were two separate channels by which Charles Darwin was familiarized with the general idea of evolution in his youth. Though the little autobiography which he wrote at the urging of his children in his declining years is not particularly explicit upon such points as this, one such channel can be documented, and the other, though not extensively discussed by Darwin, can scarcely be ignored as an almost certain source of information.

One, the more certain channel, lies in the poetry and prose of grandfather Erasmus Darwin, which achieved sufficient world renown as to make it very certain that the ideas of Erasmus would be discussed in family circles. Moreover, the schoolboy who boasted of a fondness for Shakespeare would surely have tried the wares of a poet within the immediate confines of the family. In fact, Darwin himself tells us that he had read his grandfather's prose work, the *Zoonomia,* and though he maintains it had no effect on him, it is not without interest that Darwin's first trial essay on the road to the *Origin* he entitled *Zoonomia.* Furthermore, in one of the unconsciously revelatory statements of which Darwin was sometimes capable he tells us, right after disclaiming that the *Zoonomia* had affected him, that "at this time I admired greatly the *Zoonomia.*" [23]

We need not at this point, however, raise the question of what the youth believed—quite possibly he did not know himself. It is sufficient to establish the fact that such ideas were likely to have been assimilated early enough as to have had a familiar ring. The theory of evolution would thus have lost the shocking and heretical implications that it had for the uninitiated.

Darwin himself at one point confesses, albeit a little reluctantly, that "it

22. Charles C. Gillispie, "The Formation of Lamarck's Evolutionary Theory," *Archives Internationales d'Histoire de Science,* 1957, Vol. 9, pp. 323–38.
23. *Life and Letters of Charles Darwin,* 3 vols., edited by Francis Darwin, London: John Murray, 1888. Vol. 1, p. 38.

is probable that the hearing rather early in life such views maintained and praised may have favored my upholding them under a different form in my *Origin of Species.*" Believing or unbelieving, young Charles had been raised in a family of somewhat unconventional and free-thinking traditions. We know further that he had a passionate attachment to nature and an equal revulsion against the conventional classical education of the time. At length he was packed off to Edinburgh in the hope that he would follow a medical career as his father and grandfather had done before him. Luckily for science, the sensitive youth could not endure the more ghastly aspects of medical practice and his stay at Edinburgh was short. In that brief period, however, he made the acquaintance of Dr. Robert Grant (1793–1874). Their relation was for a short time pleasant, but subsequently a coolness arose which persisted throughout their later years.[24] Grant was something of an anomaly in the Scotland of that day. In Paris he had picked up an acquaintance with Lamarckian evolution and was immensely enthusiastic about it. One day, walking with Darwin, he expounded the Lamarckian philosophy. Once more young Darwin listened, so he says, "in silent astonishment," again disclaiming any effect on his mind. Yet he listened, and listened well enough apparently to remember the episode into remote old age. He was then but a youth of sixteen; the voyage of the *Beagle* was still six years away. . . .

THE VOYAGE OF THE "BEAGLE"

Tiring of the medical round, Darwin drifted to Cambridge with the thought of entering the ministry, but he continued to cultivate naturalists, to dabble in geology, and to fear the wrath of his exasperated father who had grown weary of his eternal hunting and his lack of scholarly application. One thing, however, is significant: the boy attracted and held the attention of distinguished older men. They sensed something unexpressed within him. Through the good offices of the botanist Henslow and the winning over of his father by his uncle, Josiah Wedgwood, he was permitted to go as naturalist on the voyage of the *Beagle.* The ship sailed in 1831.

One can, in a sense, regard the voyage of the *Beagle* as a romantic interlude. One can point out that every idea Darwin developed was lying fallow in England before he sailed. One can show that sufficient data had been accumulated to enable a man of great insight to have demonstrated the fact of evolution and the theory of natural selection by sheer deduction in a well-equipped library. All of this is doubtless true. Yet it is sig-

24. P. H. Jesperson, "Charles Darwin and Dr. Grant," *Lychnos,* 1948–49, Vol. 1, pp. 159–67.

nificant that the two men who actually fully developed the principle of natural selection, Charles Darwin and Alfred Russel Wallace, were both travelers to the earth's farthest reaches, and both had been profoundly impressed by what they had seen with their own naked eyes and with the long thoughts that come with weeks at sea.

Evidence of Evolution

Galápagos Archipelago *
by Charles Darwin (1835)

The natural history of these islands is eminently curious, and well deserves attention. Most of the organic productions are aboriginal creations, found nowhere else; there is even a difference between the inhabitants of the different islands; yet all show a marked relationship with those of America, though separated from that continent by an open space of ocean, between 500 and 600 miles in width. The archipelago is a little world within itself, or rather a satellite attached to America, whence it has derived a few stray colonists, and has received the general character of its indigenous productions. Considering the small size of these islands, we feel the more astonished at the number of their aboriginal beings, and at their confined range. Seeing every height crowned with its crater, and the boundaries of most of the lava-streams still distinct, we are led to believe that within a period, geologically recent, the unbroken ocean was here spread out. Hence, both in space and time, we seem to be brought somewhat near to that great fact—that mystery of mysteries—the first appearance of new beings on this earth.

 * * *

It was most striking to be surrounded by new birds, new reptiles, new shells, new insects, new plants, and yet by innumerable trifling details of structure, and even by the tones of voice and plumage of the birds, to have the temperate plains of Patagonia, or the hot dry deserts of Northern Chile, vividly brought before my eyes. Why, on these small points of land, which within a late geological period must have been covered by the ocean, which are formed of basaltic lava, and therefore differ in geological character from the American continent, and which are placed under a peculiar climate,—why were their aboriginal inhabitants, associated, I may add, in different proportions both in kind and number from those on the continent, and therefore acting on each other in a different manner— why were they created on American types of organization? It is probable

* From Charles Darwin, *The Voyage of the Beagle*.

26

that the islands of the Cape de Verd group resemble, in all their physical conditions, far more closely the Galápagos Islands than these latter physically resemble the coast of America; yet the aboriginal inhabitants of the two groups are totally unlike; those of the Cape de Verd Islands bearing the impress of Africa, as the inhabitants of the Galápagos Archipelago are stamped with that of America.

✻   ✻   ✻

[There is] a most singular group of finches, related to each other in the structure of their beaks, short tails, form of body, and plumage: there are thirteen species, which Mr. Gould has divided into four sub-groups. All these species are peculiar to this archipelago; and so is the whole group,

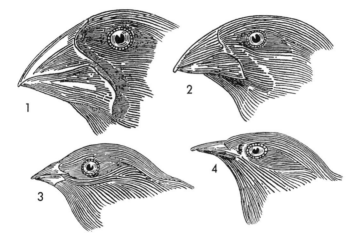

Figure 1–3. (1) *Geospiza magnirostris.* (2) *Geospiza fortis.* (3) *Geospiza parvula.* (4) *Certhidea olivacea.*

with the exception of one species of the sub-group Cactornis, lately brought from Bow Island, in the Low Archipelago. Of Cactornis, the two species may be often seen climbing about the flowers of the great cactus-trees; but all the other species of this group of finches, mingled together in flocks, feed on the dry and sterile ground of the lower districts. The males of all, or certainly of the greater number, are jet black; and the females (with perhaps one or two exceptions) are brown. The most curious fact is the perfect gradation in the size of the beaks in the different species of Geospiza, from one as large as that of a hawfinch to that of

chaffinch, and (if Mr. Gould is right in including his sub-group, Certhidea, in the main group), even to that of a warbler. [The largest beak in
the genus Geospiza is shown in Figure 1–3, a, and the smallest in c; but
instead of there being only one intermediate species, with a beak of the
size shown in b, there are no less than six species with insensibly graduated beaks. The beak of the sub-group Certhidea, is shown in d.] The
beak of Cactornis is somewhat like that of a starling; and that of the
fourth sub-group, Camarhynchus, is slightly parrot-shaped. Seeing this
gradation and diversity of structure in one small, intimately related group
of birds, one might really fancy that from an original paucity of birds in

Figure 1–4. The large ground-finch Geospiza magnirostris *has evolved a thick,
powerful beak for breaking open hard seeds, but it also eats flowers, fruits, and
some insects. It can eat harder seeds than other ground-finches and so it is not
in direct competition with the smaller-billed birds. (Photograph by Robert I.
Bowman, San Francisco State College. Reprinted by permission. Figure added
by the editor.)*

this archipelago, one species had been taken and modified for different ends.

❀   ❀   ❀

Dr. Hooker has furnished me with several other most striking illustrations of the difference of the species on the different islands. He remarks that this law of distribution holds good both with those genera confined to the archipelago, and those distributed in other quarters of the world: in like manner we have seen that the different islands have their proper spe-

Figure 1–5. The tool-using finch is one of nature's rarest phenomena; a bird which uses an implement to get its food. Since there are no woodpeckers in the Galápagos, this finch has been able to fill the woodpecker's normal niche, but it lacks that bird's long tongue for extracting insects from the holes it drills in trees. So the finch has learned to use a cactus spine to do the job. (Photograph by Irenäus Eibl-Eibesfeldt, Max-Planck-Institut. Reprinted by permission. Figure added by the editor.)

cies of the mundane genus of tortoise, and of the widely distributed American genus of the mocking-thrush, as well as of two of the Galápageian sub-groups of finches, and almost certainly of the Galápageian genus Amblyrhynchus.

The distribution of the tenants of this archipelago would not be nearly so wonderful, if, for instance, one island had a mocking-thrush, and a second island some other quite distinct genus;—if one island had its genus of lizard, and a second island another distinct genus, or none whatever;—or if the different islands were inhabited, not by representative species of the

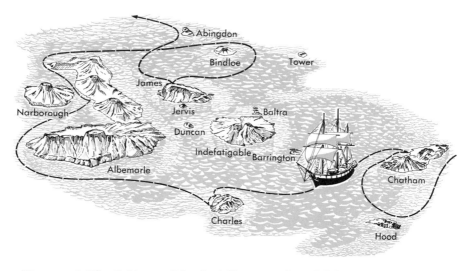

Figure 1–6. The Galápagos Islands. (Illustration by Adolph Brotman. Reprinted by permission. Figure added by the editor.)

same genera of plants, but by totally different genera, as does to a certain extent hold good; for, to give one instance, a large berry-bearing tree at James Island has no representative species in Charles Island. But it is the circumstance, that several of the islands possess their own species of the tortoise, mocking-thrush, finches, and numerous plants, these species having the same general habits, occupying analogous situations, and obviously filling the same place in the natural economy of this archipelago, that strikes me with wonder. It may be suspected that some of these representative species, at least in the case of the tortoise and of some of the birds, may hereafter prove to be only well-marked races; but this would be of equally great interest to the philosophical naturalist. I have said that most of the islands are in sight of each other: I may specify that Charles Island is fifty miles from the nearest part of Chatham Island, and thirty-three miles from the nearest part of Albemarle Island. Chatham Island is sixty miles from the nearest part of James Island, but there are two intermediate islands between them which were not visited by me. . . . I never dreamed that islands, about fifty or sixty miles apart, and most of them in sight of each other, formed of precisely the same rocks, placed under a quite similar climate, rising to a nearly equal height, would have been differently tenanted; but . . . this is the case. It is the fate of most voyagers, no sooner to discover what is most interesting in any locality, than they are hurried from it; but I ought, perhaps, to be thankful that I obtained sufficient material to establish this most remarkable fact in the distribution of organic beings.

The Facts Supporting Evolution *

by H. G. Wells, Julian S. Huxley, and G. P. Wells (1934)

THE NATURE OF THE PROOF

Before we go on to the evidence . . . let us consider what our evidence must show if Evolution is to be accepted as the general process of life.

First, then, all things living, or once living, must fall into a branching plan. Everything in the past must be reasonably shown to be either ancestral to a living thing or else without descendants; there must be no renewal of the process, nothing in the past must be plainly derived from some later form. Every mammal, for example, is held to be descended from a reptilian ancestor. Suppose in the early Coal Measures, before ever a reptile existed, we found the skull of a horse or a lion. Then the whole vision of Evolution would vanish. A single human tooth *in situ* in a coal seam would demolish the entire fabric of modern biology. But never do we find any such anachronisms. The order of descent is always observed.

Next there must be an orderly sequence in fossil forms, so far as they are found. We must see very distinctly that form passes into form. In the days of Darwin such sequences were hard to find. In those days there were probably not a hundredth part of the present multitude of fossils that are now collected and arranged. The Creationists pointed trium-

* From H. G. Wells, J. Huxley, and G. P. Wells, *The Science of Life*, Doubleday and Co., New York. Copyright 1929 by H. G. Wells, J. Huxley, and G. P. Wells. Reprinted by permission of Collins-Knowlton-Wing, Inc.

phantly to a gapped and fragmentary story, sustained by hypothesis, broken up by "missing links." Darwin was challenged to show anywhere in the fossil record the steps by which one species has passed into another. It was then quite a difficult challenge. To-day we have an answer, a score of answers, to that challenge, beyond Darwin's utmost hopes.

Then if animals have been specially created just as they are to fit special conditions, it is reasonable to suppose they are perfectly and completely adjusted to those conditions. There is no reason why any animal should fail to have any structure that might be helpful in its way of life, or possess any structure it has no need for. If a cat lives on birds and a tiger on ground game, is there any reason why a cat should not have wings because a tiger has not? But if the diverse species have been evolved step by step, a certain disharmony is to be expected between inherited structure and reactions, and the full possibilities of the life a creature leads. The second section of our evidence then will be an examination of plant and animal structure to see how far animal and vegetable organs are special to their needs, and how far they have the air of being primarily an inheritance merely fitted to those needs and limited in that fitting by conditions of descent.

And then the way animals and plants are scattered over the world will not be haphazard if Evolution is really the truth of life. If we found a region where an animal might live abundantly and that animal is not there, but somewhere else in the world, then if we are to believe in Creation we have to find Creation very remiss upon the distributive side; but if we believe in Evolution, then it is quite reasonable to suppose that an animal evolved in one part of our planet may never get to another for all the fitness of conditions there. All that also we will illustrate and weigh. . . .

THE EVIDENCE OF THE ROCKS

The Nature and Scale of the Record of the Rocks

. . . Right up to the end of the eighteenth century the comparatively few fossils then known were almost universally regarded as mere curiosities; many dismissed them as sports of Nature, freaks of the earth, and not really the remains of flesh-and-blood, while at most they got credit for being witnesses to the universal biblical Deluge. Nothing more could be expected until geology had made her profound advance of introducing time, and time on a vast scale, into our ideas about the earth's crust.

Let us recall how that extension of time dawned upon the human intelligence.

Everybody knows that flowing waters bring down sediment and deposit it in layers on the floor of seas and lakes, or on flood-plains. Sometimes the deposit reaches the surface, as in deltas. Consider the great tongue of the Mississippi delta, built out into the sea for sixty miles and more, the Mississippi brings down every year over four hundred million tons of sediment. At other times the deposit spreads over the bottom, as when a pond is gradually filled up, or an alluvial meadow built, layer upon layer, from the silt laid down by successive floods. Layers of material may be laid down in other ways; the great spit of Dungeness, on England's south coast, grows out to sea at the rate of over five feet a year, from shingle brought along the coast by the waves and currents. In moor country, deep layers of peat are formed by the successive death of the bottom parts of the bog plants. Currents and waves deposit stretches and banks of sand in quiet bays. After a volcanic eruption vast quantities of dust and pumice and rock fragments fall in the neighbouring sea and sink to the bottom. And from the surface layers of the ocean a constant rain of billions of skeletons of animals and plants, many of them microscopic, but . . . incredibly abundant, is always falling softly towards the depths.

In these and other ways new materials are to-day being accumulated in the form of sheets or layers of varying extent in innumerable regions of the globe, and they must obviously have been accumulating in the same sort of way through all the ages since liquid water has existed on our planet. These accumulations of slowly deposited layers are what we call sedimentary or *stratified rocks* [1] (as opposed to the *igneous rocks*, forced in among the other layers of the crust in a molten state from below, or belched out over the surface by volcanoes), and the sheets themselves are technically called by the Latin word for layers—*strata*.

In these stratified rocks fossils are often found entombed—the remains and traces of dead animals and plants which were often strikingly different from any creatures alive to-day. We need give but a couple of examples. If you happen to spend your holiday on the Dorset coast near Lyme Regis and search the crumbling cliffs near by (or, indeed, if you explore any clay quarry across from Dorset to Peterborough), you will be pretty sure to find some bi-concave bone discs, the vertebrae of some large animal. Persistent and lucky hunters have found whole skeletons containing such vertebrae. They are of giant reptiles, christened Ichthyosaurs, or "fish-lizards," wholly unlike anything existing to-day, obviously aquatic, for their limbs are converted into paddles.

In the same clay you will be likely to find other fossils, the spiral am-

---

1. By the geologist, all constituents of the earth's crust, except the actual soil, are called *rocks*, whether they are hard granite or limestone, friable chalk or sandstone, or soft clay or loess.

monites, often beautifully patterned. [See Figure 3–1.] These, too, though their shells are to be found in millions embedded in various rocks, have never been discovered alive. But we know that these shells were inhabited by creatures not unlike the pearly nautilus of to-day, and more distantly resembling the cuttlefish and octopus.

It would not take much reflection, one might think, to realize that in any such sedimentary deposit, whether thick or thin, the lower layers must have been laid down before those above them. But such were our ancestors' prejudices and preconceptions, based for the most part upon the belief in the sudden creation of the world at a not very remote period, that it was not until the turn of the eighteenth century that this fundamental but elementary idea was properly put forward, to become from thenceforward the basis of geology. William Smith, an English surveyor, as his work took him from one part of the country to another, noted that a number of characteristic rocks, such as chalk, oolite limestone, red sandstone, or gault clay, occurred as layers which covered large areas of country. Moreover, wherever these layers occurred, they were always in the same order. The gault clay, for instance, was always close below the chalk, the green sand always immediately below the gault, the oolite limestone many layers below the chalk, the red sandstone several layers below the oolite, and so forth. And, a third point, each layer of rock was characterized not merely by the material of which it is made, but also by the fossils which it contains. This last was of vital importance; often two layers of clay or of sandstone may be nearly indistinguishable in their consistency and materials, but easily distinguished by their contained fossils. For instance, the London clay, over which London is built, lies above the chalk layer. It contains fossil fruits of palms and conifers, some nautilus shells, numerous characteristic sea-snails, and a few mammals. No ammonites or ichthyosaurs have ever been discovered in it. The gault clay, on the other hand, from below the chalk, has no plant fruits, but does contain ammonites, often uncoiled in a peculiar way instead of regularly spiral; while the Oxford clay, a thick layer close above the oolite limestone, has huge numbers of ammonites, almost all built as regular spirals.

Such facts as these obviously mean that we ought to be able to arrange all the sedimentary rocks of the world in a series, according to their age; and, this once accomplished, all the fossils in the earth's crust will fall into their time-sequence, too. The task has been accomplished for the great majority of layers. As a result, we can say that one kind of fossil belonged to an animal which lived and died before another kind of animal found fossilized in another layer; and the bewildering variety of life becomes more orderly through receiving an arrangement in time. To take merely the same examples we first mentioned, ammonites are found to be absent from all layers below the coal measures and from all above the chalk, but present in all congenial layers between these limits; while ichthyosaurs, though they also have the chalk as their upper limit, only extend downwards through about two-thirds of the layers in which ammonites are found.

This fundamental principle, that the different layers of the earth's crust can be arranged in a time-sequence, is the basis of that department of science known as palaeontology. These sheets of inert matter are the pages of the book of our planet's history. They lie scattered over the globe, often torn, defaced, or crumpled. But patience and reason combined have been able to reconstruct whole chapters and sections of that great book. In it we can read not only the physical changes that the world has experienced —when the Rockies were built, or the Scottish Highlands worn down to mere stumps of their former grandeur, the date of great Ice Ages, aeons before the last Ice Age, or of the appalling flow of lava which overwhelmed a quarter of a million square miles in North-Western India—but also the history of Life, printed on the pages of the book in the form of fossils, hieroglyphs which to persevering study reveal readily enough the secret of their picture-writing.

The principle is both fundamental and simple; but there are sometimes difficulties in applying it. Part of the record may have been destroyed or defaced till the life-story it contains becomes illegible; or a whole set of pages may have been crumpled or turned upside down into reverse order (as in the upthrust of some mountain ranges), so that their proper rearrangement is a matter of the greatest difficulty; or an isolated page or chapter from some out-of-the-way corner of the globe may be hard to place.

Happily such difficulties only concern parts of the record; whole chapters of it have the pages all tidily in order, the fossil-writing abundant and easy to interpret. In most cases these confusions have been analyzed and overcome, and the result is that, with rare exceptions, the fossil-bearing rocks of the world can now be assigned to their proper position in the time-scale—their right place among the pages of the Book of Earth.

As a result, the earth's history has been divided up, as this book is divided into books, chapters, sections, and so forth, into subdivisions of various grades. Two sets of terms are used, according as we are thinking in terms of geological time, or in terms of layers of rock. The main divisions of geological time are usually called Eras; within each Era a number of Periods (or Epochs) are distinguished, and they are further divided into sub-periods.

When, on the other hand, we are speaking of rock-layers, we refer to a

System as our main unit, roughly corresponding to the rocks laid down during one Period of time. . . . The Cretaceous System is thus an actual set of rock layers laid down during the time of the Cretaceous Period.

Now, the fact of primary importance in the history of life displayed by these geological Periods is the orderly succession of living forms. They *progress*. They progress from simple beginnings to more complex and versatile types. At the bottom (earliest) of our rock series come rocks with barely a trace of life and then in succession life unfolds. Comes first the ARCHEOZOIC ERA, with the dawn of life. Then the PROTEROZOIC ERA, with creatures as highly organized as worms. Then the vast PALEOZOIC ERA. There are no vertebrata at all and no evidence of land life in its opening period, the CAMBRIAN. Then in a second period, the ORDOVICIAN, is the dawn of vertebrate life. Then comes the SILURIAN, in which fishes and some land plants and invertebrata appear. Then DEVONIAN and CARBONIFEROUS, with an ever-increasing amount of land forms, and the whole of the era closes with the PERMIAN, in which reptiles first appear.

Above these comes the MESOZOIC SYSTEM of rocks, that gigantic volume which tells of the Era of mighty reptiles and coniferous and cycad-like plants. Its formations (which like the Periods of the Paleozoic derive their names either from the districts in which the rocks are well-developed or from well-defined physical characters) are the TRIASSIC, the JURASSIC, the CRETACEOUS.

Finally comes the CENOZOIC ERA, the age of modern life, of mammals, birds, grasses, flowering plants, and trees.

*   *   *

Archbishop Ussher, less than three hundred years ago, dated the creation of the world in 4004 B. C. (and gave the day and hour, too!), and his calculations still adorn the margin of the Authorized Version of the Bible. Though they are wrong, they are at least wrong on a grand scale, since the age of the earth is somewhere about . . . a million times greater than he supposed. Even to-day the average man tends to think the six-thousand-year antiquity of Babylon or Egypt enormous. But just as astronomy is teaching us to think of cosmic space on a wholly different scale from geography, to be measured in terms of "light-years" running into ten thousands of millions of miles, so geology is making it necessary to think of earth-history on a wholly different time-scale from human history, in terms of million-year periods, to which a decade bears almost the same proportion as an hour does to a century, and a century as a day does to a whole generation of human life.

To think in such magnitudes is not so difficult as many people imagine. The use of different scales is simply a matter of practice. We very soon

Geologic Time Chart

System and Period	Series and Epoch	Distinctive Records of Life	Began (Millions of Years Ago)
	CENOZOIC ERA		
Quaternary	Recent (last 11,000 years)		
	Pleistocene	Early man	2+
	Pliocene	Large carnivores	10
	Miocene	Whales, apes, grazing forms	27
Tertiary	Oligocene	Large browsing mammals	38
	Eocene	Rise of flowering plants	55
	Paleocene	First placental mammals	65–70
	MESOZOIC ERA		
Cretaceous		Extinction of dinosaurs	130
Jurassic		Dinosaurs' zenith, primitive birds, first small mammals	180
Triassic		Appearance of dinosaurs	225
	PALEOZOIC ERA		
Permian		Reptiles developed, conifers abundant	260
Carboniferous			
Upper (Pennsylvanian)		First reptiles, coal forests	300
Lower (Mississippian)		Sharks abundant	340
Devonian		Amphibians appeared, fishes abundant	405
Silurian		Earliest land plants and animals	435
Ordovician		First primitive fishes	480
Cambrian		Marine invertebrates	550–570
	PRECAMBRIAN TIME		
		Few fossils	more than 3,490

(From *Encyclopaedia Britannica*, Vol. 18, p. 44, 1968. Reprinted by permission.)

get used to maps, though they are constructed on scales down to a hundred-millionth of natural size; we are used to switching over from thinking in terms of seconds and minutes to some other problem involving years and centuries; and to grasp geological time all that is needed is to stick tight to some magnitude which shall be the unit on the new and magnified scale—a million years is probably the most convenient—to grasp its meaning once and for all by an effort of imagination, and then to think of all passage of geological time in terms of this unit.

Defects and Happy Finds in the Record

❊   ❊   ❊

We must remember that fossilization is the fate of very few animals and plants. Only one in a million makes its mark in the Book of Life. The great majority of dead things simply decay and disappear and their material is returned to the general circulation of Nature to be built up into the bodies of new organisms. But once in a while a corpse is preserved more permanently; it falls into mud or silt or some place where the bacteria of decay cannot get at it. Insects have been caught and sealed and preserved for countless years in the fossilized resin we call amber. The bodies of mammoths can be dug out of frozen mud-cliffs in Siberia with the skin and flesh still preserved. And even when the flesh decays away the bones may escape the dissolving action of rain or other waters and ultimately find their way into the palaeontologist's cabinet.

But it should be remembered that these direct preservations of the material of extinct creatures are the rarest accidents. Most fossils (for any dug-up trace of an organic being is called a fossil) are not actual surviving bits of corpses at all, but bits of corpses which have been changed into rock by a slow translation and replacement. They are copies at second hand of the original writing. As the bones or other enduring fragments lie buried they slowly dissolve away and are replaced more or less completely by mineral substances. In a word they are "petrified." In this process, of course, they undergo varying degrees of distortion, although in one or two exceptional cases the translation is astonishingly accurate, each little difference of texture in the original being faithfully reflected in the mineralized fossil. The record may be even more indirect than that; it may be a dried footprint or the hollow impress of a bit of skin or a shell. From these scattered and accidental remains the palaeontologist patiently reconstructs his picture of the world as it used to be.

Occasionally a lucky chance makes us realize how scanty our information really is.

In California, for instance, a pool of water with sticky margins impregnated with tar proved a death-trap to thousands of creatures as they came down to drink, and to hordes of carnivorous mammals, like wolves and sabre-toothed tigers, which endeavoured to catch the drinkers when they stuck fast. Now the palaeontologist finds a hoard of treasure in that one pool. [See Figure 1–7.] In France, in the drought of 1911, it was noted that all the fish in a pool burrowed into the mud when the water dried up, and were eventually baked hard in their hundreds. Unless the geologists of future ages happen to hit on such a patch of trapped life, they are not likely to find more than isolated bones of these kinds of fish.

Figure 1–7. A candidate for fossilization, a squirrel joins other animals trapped in past millennia by La Brea Tar Pits in California. (Photograph courtesy of the Los Angeles County Museum. Figure added by the editor.)

Almost the only complete skeletons of the extraordinary giant reptile, Iguanodon, are those of a whole troop, twenty-nine of them, old and young, which were found in a Belgian cave, obviously entombed by some accident.

Fossil ants, we note, are often found in amber. Amber is the fossilized gum which once exuded as resin from long decayed pine-trees. Only an insignificant fraction of the ant population of the world gets trapped in resin; and only an insignificant fraction of the resin is hardened and preserved as amber; and yet almost all our knowledge of the ants of the past is derived from specimens in amber.

The amount of detail preserved to us is also very much a matter of luck. As a rule, nothing survives but the skeleton, and even that may be distorted and sometimes partly rotted. Now and again, however, happy accidents have caused traces of the softer parts to survive to our day. For example, one or two specimens of the strange, duck-billed dinosaur, trachodon, in the Upper Cretaceous, died and fell on a patch of soft mud. They decayed and the place where they had been became covered with sand. But the mud beneath held the impression of their skins with surprising fidelity—so that now we have the form of their skins preserved, with a mould in hardened mud and a cast in hardened sand. Some of the dolphin-like ichthyosaurs left records of their flesh and fins; some even of their faeces, marked with a spiral twist by the folds of their intestine, and containing undigested remains of the fossil squid-like creatures called belemnites.

The very first known land beasts left footprints in the mud across which they lumbered [see Figure 1–8]; the bird-reptile Archaeopteryx stamped

Figure 1–8. Tracks of dinosaur. The living example of youthful Homo sapiens *was added to indicate the size of the tracks. (Photograph courtesy of the American Museum of Natural History. Figure added by the editor.)*

its feathers with astonishing detail in the fine-grained lithographic lime-stone of Solenhofen.

But there are even older happy chances than these. Some Devonian plants are so minutely petrified that we can study the precise shape of the microscopic hairs protruding from their leaves. And recently Walcott has discovered a wonderful array of invertebrates imprinted on Middle Cambrian shales, nearly as clear to view as when they were swimming about, ages before the first fish, almost five hundred million years ago. The soft mud, now pressed hard, reveals to us the outlines of jelly-fish, annelid worms, with their appendages and bristles, small crustacea, with all their soft leaf-like appendages, even the outlines of their stomachs, and arrow-worms, just like those that swim in our modern seas, with their transparent fins. Walcott has also found swarms of bacteria from still ear-lier rocks. It is doubtful whether this is to be considered the most extraor-dinary case of fossilization; it is certainly rivalled by some Paleozoic fishes (to be seen in New York) which have been so delicately petrified that thin slices of their muscles, ground down to transparency, and looked at under a high power of the microscope, show the cross-sections of the muscle-fibres as clearly as a fresh-made preparation from a modern dog-fish. How the microscopic structure of living tissue came to be thus trans-lated into stone we do not fully understand; but at least the rarity of such lucky finds brings home to us the multiplicity of what is lost for ever.

The various difficulties thus put in the way of palaeontologists are of two main kinds. There are those which lead to gaps and imperfections in the fossil record, whether through the fewness of animals and plants which became fossilized, the washing away of large sections of the crust when brought above water, for wind, rain, and frost to destroy, or the de-struction of fossils by the heat and pressure of metamorphosis. And there are those which make it difficult to arrange what fossils we have in the right order, whether the difficulty springs from the turning of layers topsy-turvy and from faulting, or from the fact that two contemporaneous layers might show quite different fossils, either because they were laid down in quite different situations, or because, though comparable in the environment they provide, they were situated far apart on the earth's sur-face.

The imperfection of the record is unfortunate, but nothing more. It makes the labours of fossil-hunters greater, and should warn us against clamouring for the immediate discovery of this or that "missing link." But patience and the exploration of more and more of the earth's surface are bringing their own reward. Every year the record becomes less scrappy, and in many groups of animals, what fifty years ago seemed impossible to hope for has been achieved, and an unbroken series discovered, leading through ages of time from simple to highly developed types. The difficulty is not one of principle.

Editor's Note. The techniques of radioactive dating which have recently been discovered have put geology and paleontology on a much more accurate basis than was possible when this selection was written. Radioactive dating measures changes in the atomic structure of certain types of matter. In each case an unstable "parent" element is converted at a very slow but constant rate into more stable "daughter" elements. By measuring the relative proportions of "parent" and "daughter" atoms in a given sample, scientists can tell how long the process has been going on. Carbon-14 dating (the first to be extensively used for this purpose) is accurate for about 50,000 years into the past. Potassium-argon dating, which is based upon a slower radioactive process, is ideal for dating objects millions of years old.

A Sample Section in the History of Life: The Evolution of Horses

And now let us take one of the better preserved sections from this vast, confused autobiography which life has written in the rocks. It is a section of which we shall give a considerable amount of detail and rather a bothersome multiplicity of generic names. The reader is under no obligation to remember these names, but they have to be "produced in court" for the purposes of our proof.

It is loudly argued by many Creationists and semi-Creationists that there is no fully-worked-out pedigree of any existing forms of life, and that there is nothing to dispose of the view that at irregular intervals creative forces intervene in the evolving process and make life take a convulsive stride forward. This, however, is not the case. So far from there being no well-worked-out pedigree, in which the successive forms in some group of animals are seen visibly modified and differentiated, there are now several such family trees in existence. We are giving here the past record of the existing horses. They have been evolved from a small, four-toed Eocene mammal and every step in the process is traceable.

It is worth noting that the earliest known three-toed fossil horse was described as recently as 1860, the year after Darwin published the *Origin of Species,* and that it was not till about 1870 that any serious attempt was or could be made to establish the horse's ancestry from fossils. Many startling finds were made in the 'seventies and 'eighties presenting the story in rough outline. But it has been the patient accumulation of specimens since then which has filled in the details and made it convincing. As the fossil-bearing rocks have been more intensively explored, link after link has been brought to light, until now we are able to reconstruct an almost unbroken chain of change extending for over forty million years.

The existing horses constitute a very distinct family of animals. No

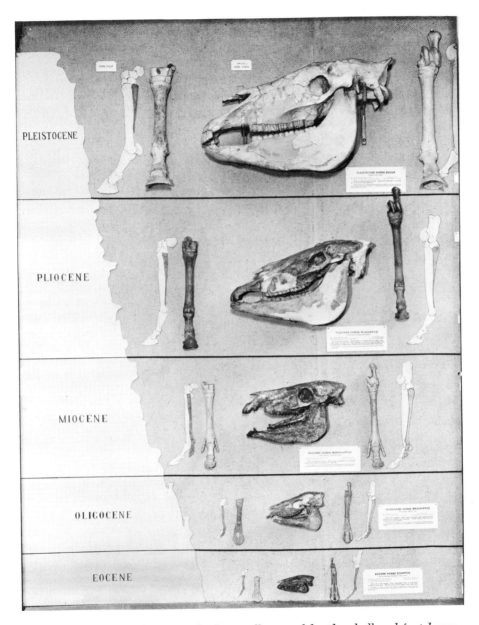

Figure 1–9. The evolution of the horse, illustrated by the skull and foot bones. (Photograph courtesy of the American Museum of Natural History. Figure added by the editor.)

other vertebrates have but one toe to each foot; and no other animals have quite similar teeth. The family now comprises one genus only, called *Equus* and including seven species—two Asiatic wild asses (the onager and kiang), one African wild ass, the little Przevalsky's horse from Asia, and three kinds of zebra from Africa. A fourth kind of zebra, the quagga, has recently been exterminated. In addition, there are, of course, the many varieties of domestic horse and donkey, brought into being by man's selective breeding. Now all these wild horses and asses and zebras live in much the same way. They run swiftly over hard, level plains (wild asses in the Mongolian desert have been timed doing their forty miles per hour), and they feed on a food that is difficult to chew—on the tough leaves and stems of grasses, which are often hardened by a certain amount of flinty matter. And corresponding with this hardness of ground and of food, we find special provisions in their feet and teeth.

The single toe of a horse's foot corresponds to the third or middle toe of the more ordinary five-toed foot. Only the last joint of the toe touches the ground; the hoof in which this last joint is encased is the exact equivalent of an overgrown toe-nail. The horse's wrist and ankle are far above the ground, forming the joints commonly called "knee" and "hock"; the true knee is what is styled the stifle. The region corresponding to our palm or sole contains the single elongated cannon-bone. But the other fingers or toes are not completely absent, for attached to the hinder angles of this cannon-bone are two little splint-bones; and that these are the remains of the second and fourth fingers and toes is amply proved not merely by their position, and by the fact that one or both of them occasionally develop the missing joints and a miniature hoof, but also, as we shall soon see, by their development in the embryo.

The horse's limb, then, is a specially modified limb; it is a limb in which one toe is enormous and strong, and in which the others have dwindled more or less completely away. It is a limb devoted to a special function. On open, grassy plains the best means of escape from enemies lies in speed; and it is for speed on comparatively level and hard ground that the legs of the horse are suited. Everything else has been sacrificed to that. The elongation of the actual foot-region gives a better leverage; the concentration of all the limb-muscles in the upper part of the legs allows for rapid swing; a limb which consists of a single pillar, joined so as to move only fore-and-aft, transmitting all the weight downwards to a single expanded hoof, is stronger than one which can be moved in all directions, or than one in which the lower arm or leg contains two bones (as in ourselves), and there is less "give" than in a limb ending in several toes. Accordingly the horse, though it can only execute a very few kinds of movements with its limbs, though it sinks in soft ground owing to lack of spreading toes, though it is not well adapted to broken country, triumphs in speed on dry and rolling plains.

Incidentally, length of leg makes length of neck a necessity; without that the horse could not reach down to its food.

The grinding teeth of a horse are beautifully and elaborately adapted for dealing with the tough grasses that he consumes. They are peculiar in three ways. First, they are all alike, instead of the true molars being more complex than the premolars as is usual. The premolars, we may remark in passing, are those grinding teeth which have predecessors in the milk dentition; the molars have not. Secondly, these quite similar molars and premolars of the horse all have an extremely complicated surface-pattern. Before they cut the jaw-bone they are covered with hard, glossy enamel, rising up in ridges round a couple of deep cavities. These cavities later get filled up with cement, a substance not quite so hard as enamel, which is secreted by special glands as the tooth breaks through the gum. Under the enamel is the softer dentine. Use soon grinds off the top of the teeth, with the result that their tops become nearly flat; but as the three materials wear down at different rates, sharp edges of enamel stand up a little beyond the cement, and a little higher above the dentine. The whole forms a remarkably effective miniature millstone, with the advantage over our millstones that it keeps its grinding-ridges sharp as it is worn down.

The third point about the teeth is that they are of remarkable depth and that during the first eight years of life they have no closed roots but go on growing from below, like a rabbit's front teeth, as they are worn away above. After this roots are formed, new growth ceases and the teeth are simply pushed up to compensate for the wear at the surface till they are all worn away, and the animal dies because it cannot chew its food. Our own teeth, and most mammalian teeth, are finished and complete as soon as they have erupted, but in a horse completion is delayed for eight years to give a longer working life.

The horse's teeth, then, are admirably adapted for chewing up grasses; and the size and peculiar shape of his head is due to the need for finding room for these powerful and deep-rooted living millstones, and for the muscles to work them. . . .

If we go back stage by stage through the rocks of the whole Cenozoic Period, we find that the horse has recorded its pedigree in fossils. There are four main stages. In the last, the fossil horses resemble the living forms in all essentials of teeth and feet, differing only in details of proportion. They are all grass-plain animals.

In the stage before this there are no one-hoofed horses. Instead we find smaller creatures, of obviously horse-like type, but with three hoofs on each foot; the two outer hoofs, however, are small and must have been useless in running, since they did not reach the ground, but hung in the air like the dew-claws of deer and other animals. The teeth had a less elaborate grinding system, and were much shorter. Fossil bones of these —shall we call them the fathers of the horses?—are not uncommon in China, and (with others) are dug up to be sold to apothecaries as "dragons' teeth," that essential ingredient in the Chinese pharmacopoeia. Professor Watson tells us that even the Chinese labourers employed to dig them up recognize the skulls as like those of their donkeys.

In the next, and still older, assemblage of forms, the grandparents, so to speak, the ancestral horses were no larger than a large dog or a small Shetland pony. They also were three-toed; but all three hoofs touched the ground, and in addition they possessed on the fore-foot the trace of a fourth toe, in the form of a little splint against the cannon-bone; the tooth-pattern again was less elaborate, the whole tooth shorter, and there was no trace of the cement which in all later forms filled up the valleys between the ridges of enamel and dentine and so ensured a flat grinding surface throughout. Yet one can easily recognize the skeletons even of this stage as those of horses—three-toed and rather lumpish horses, but horses.

Finally, in the earliest stage, the far ancestors, in which we can still definitely detect the tendencies which culminated in the modern horse, none of the animals were bigger than a medium-sized terrier; there were four little hoofs on the fore-foot and three on the hind; sometimes the hind-foot also showed two splint-bones representing the missing first and fifth digits; the teeth were very short, with only indications of the system of grinding ridges, and the premolars were not so large nor of so complicated a pattern as the molars.

Besides the teeth and toes, other characters, too, show steady parallel changes as we go back through time. The earlier forms had shorter necks and faces, less tightly fitted wrists and ankles, two separate bones instead of one in the lower arm and leg. In a word, as we go back we find horses less and less adapted efficiently for swift running and for grinding hard vegetable food and more and more like the other generalized mammals of the early Eocene. . . .

There are two complications of the horse's history. First, though the main trend is always on towards the modern horse, many of the fossil types represent side twigs which have died out sooner or later leaving only the central branch to grow on to culmination. Second, the horses, being a very mobile species which can travel great distances rapidly, have a wide arena for their development. The earliest forms so far found are from the Western United States; but from North America they soon invaded the Old World across a land connection where is now Bering

Strait; and the living tide flowed back and forth between the Old World and the New, or was dammed back, according as this bridge emerged or was sunk under the waters; it invaded Africa, and flowed down into South America, when the Central American connection came into existence, in the late Miocene Period. Thus, as climate changed and barriers were bridged, the various types of horse would move from place to place of this wide scene, so that sometimes quite new types suddenly appear in the

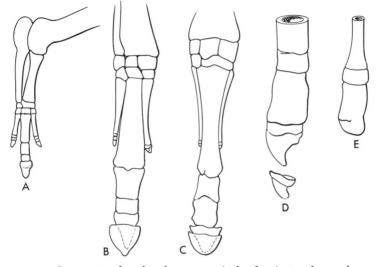

Figure 1–10. Stages in the development of the hoofs in the embryo of the modern horse. (A) The skeleton of the limb of an embryo horse six weeks old, showing three toes. (B) The same, eight weeks old showing the side toes much reduced relative to the middle toe, on which the hoof is forming. (C) The same in a five months' embryo, showing side toes reduced to splints. (D) The end of a side toe of (C), much enlarged, showing its rudimentary hoof. (E) The end of a side toe at birth; the various bones are still separate, not joined as in adult horses. (Modified from drawings by Professor Cossar Ewart.)

local record—invaders which were evolved in some other locality where perhaps fossils have not yet been discovered. But in spite of these obscuring factors the story has been clearly worked out; the fossils are so numerous that it can be construed without any doubt at all. . . .

Step by step, variety by variety, the progressive changes can be traced. One can hardly say where one species ends and another begins. Doubtless our knowledge of fossil horses will be further filled in and rounded off in the future, as new specimens turn up; but new discoveries can do no more now than fill in a little gap here, correct a minor error there. The essential facts are already before us in their fullness. In one long gallery one might assemble all these stages. We have here in a crushing multitude of steadily progressing specimens just that complete, continuous exhibition of Evolution in action the Creationist has demanded. . . .

It would be possible to give a number of other evolutionary series, almost or quite as perfect as those we have described in detail. The Titanotheres, strange, horned, extinct mammals, rival the horses in fullness of record. So do the camels. The tapir and rhinoceros branches are not far behind. . . .

THE EVIDENCE FROM PLANT AND ANIMAL STRUCTURE

Structural Plans, Visible and Invisible

We have reviewed the geological facts which constitute the direct evidence for Evolution. . . . Yet the indirect evidence must not be passed over. For one thing, it is impressive to see how each line of evidence confirms the same story; and for another, the indirect evidence throws light upon many of the facts and methods of Evolution which the direct evidence does not touch.

The first line of indirect evidence is comparative anatomy. This is often very similar to the direct evidence derived from fossil forms; in a number of cases the linking forms between groups are not wholly extinct, but a few of them linger on to the present day. . . . The mammals and reptiles are linked by the duck-billed Platypus and its allies; worms and arthropods by the grub-like Peripatus; vertebrates and sea-urchins by Balanoglossus. So also the dog-faced but tree-living lemurs link monkeys with insectivorous mammals, the tailless apes are half-way in structure between monkeys and man, the lungfish help bridge the gap between fish and terrestrial vertebrates.

But the chief evidence from comparative anatomy comes from the broad study of structural plan. Each of the great phyla [2] of animals and plants is characterized by a common plan that underlies the construction of its members. That is fact. But why should it be fact? What is the sense of flying bat, swimming whale, burrowing mole, and jumping jerboa, all being built on one plan, while a wholly different plan runs through flying butterfly, swimming water-boatman, burrowing mole-cricket and jumping grasshopper?

Long before the time of Darwin, naturalists had recognized these underlying similarities of plan; Cuvier endeavoured to explain them by asserting that each main plan, or archetype as he called it, corresponded to an idea in the mind of God, who had rung changes on it in the process

2. A phylum is a primary division of the animal or vegetable kingdom. Ed.

of creation. It is difficult to understand why only a small, definite number of archetypal ideas should have been thus divinely conceived. Why should God be limited in his ideas? And, further, as we shall see, the facts of embryology make that conception unacceptable. In any case, the idea of Evolution provides a more natural and much simpler presentation of the reality. If bats, whales, moles, jerboas, and the rest of the mammals were all descended from some common stock, then it would be *expected* that they should all show the same general plan, that they would start with that and vary from that to meet the demands of their distinctive ways of life.

What could be more different at first sight than whale's flipper, human arm, horse's foreleg, and bat's wing? Yet the skeleton of each is very

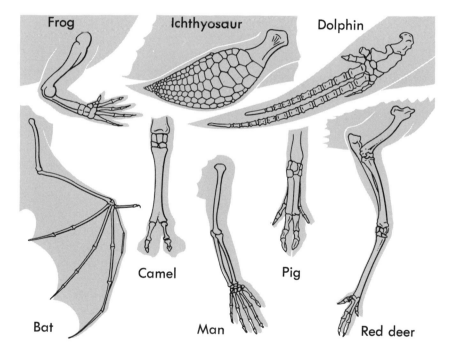

Figure 1–11. The structural plan of the vertebrate's fore-limb is exemplified by that of man. The general plan remains the same throughout the vertebrates, from amphibians up, though the details may be altered. In the Frog, the two bones of the lower arm are fused. In the Ichthyosaur, they are extremely short-ened, and extra rows of finger-bones are added. In the Dolphin, two of the fingers are elongated; in the Bat, four. In the Pig, the thumb has vanished. In the Red Deer, the second and fifth fingers are on the way to disappearance, and in the Camel only the third and fourth are left.

plainly built on the same plan—a plan originally comprising one long bone in the upper arm, two in the lower, ten little knobbly ones in the wrist, and five jointed fingers to end up with.

This original model is distorted, cut about, modified. Sometimes one or two parts are enormously enlarged, like the two bones of the horse's forearm; sometimes parts are shortened and broadened, like the humerus, radius, and ulna of whales; sometimes they shrink almost to nothing, like the second and fourth toes of the horses, or wholly disappear, like their first and fifth toes; but the general plan remains as the common point of departure. (Figure 1–11). . .

Vestiges: The Evidence of the Useless

There are certain facts of anatomy which have proved not merely difficult, but impossible to explain on any other assumption than that of Evolution. These are what are called vestiges—organs which are useless to their possessor, but resemble and correspond to useful organs in other creatures. Such organs are often loosely called rudimentary organs; however, since they seem definitely not to be the beginnings of something better, but rather to represent the ruins of past usefulness, it is better to style them vestigial.

Perhaps the most striking vestigial organs are the legs of whales. Whales have, in their flippers, well-developed fore-limbs; but externally they show no trace of hind-limbs. However, if they are dissected, one or two little bones are to be found embedded in the flesh in the region of the hind-limb. In some whales, a pair of long rods is all that remains, representing the vestige of the limb-skeleton, while the limb is altogether gone; in others, the vestige of the hip-girdle has a vestige of a thigh-bone attached. These bones are wholly useless—there is no trace of limbs for them to support, and they have not been turned to other uses. If we believe in the special creation of each kind of whale, or even of whales as a group, we must confess that these limb-vestiges spell nonsense. But if whales have evolved from land mammals, their presence is not only natural but full of significance. Leviathan we realize is not a perfect, immaculate whale, made as a whale and as nothing else, but the descendant of a land animal doing its best to swim.

Very similar vestiges of limbs are found in some snakes. No snake has any trace of a fore-limb, and most lack hind-limbs, too. But in the boas, pythons, and one or two others, vestiges of hip-girdle and hind-limbs are to be found. Sometimes these seem to be wholly useless, while in other cases, although they have no use as limbs, they protrude as two claws, which doubtless serve some new if minor function. If, as their general anatomy indicates, snakes have evolved from lizards, these vestiges make sense; without the background of Evolution, they are inexplicable. . . .

The Evidence of the Embryo

About a hundred years ago, von Baer, the great embryologist, omitted to label some specimens of embryos which he had put away in spirit. When he came to examine them later he found—but we will quote his own words:—"I am quite unable to say to what class they belong. They may be lizards, or small birds, or very young mammals, so complete is the similarity in the mode of formation . . . of all these animals." Thinking over this he came to formulate a general law—that animals resembled each other more and more the farther back we pursued them in development. This law in general holds good, and this resemblance of embryos or larvae is a very striking fact, very difficult to explain save on evolutionary lines. A child of two can tell a pig from a man, a hen from a monkey, an elephant from a snake. But these animals are only easy to tell apart in the later stages of their development. When they were early embryos, they were all so alike that not merely the average man but the average biologist would not be able to distinguish them, and even a specialist in embryology might be pardoned a mistake (Figure 1–12).

But this is by no means all. The embryos of different animals, in addition to being more like each other as development is traced backwards, show also a widening contrast with their parents and their adult destiny. They become unlike their adult selves, but at the same time and in the same respects their construction comes to resemble that of quite other types of animals. To go back to von Baer's unlabelled specimens, not only are the early embryos of man, cat, hen, and snake so alike that they are hard to tell apart, but one of the ways in which they are alike is in having their heart, main arteries, and neck-region built on the same plan as in fish. Their heart is not divided, wholly or partly, into right and left halves, but is a single series of pumping chambers, just like the heart of a fish; on the side of the neck is a series of clefts in just the position of a fish's gill-slits; there is a series of arteries running down between the clefts just as in fish; and, indeed, the whole arrangement of blood-vessels and nerves and their relation to the clefts is piscine, and not in the least indicative of the arrangement they themselves will show later. These clefts never bear gills; the resemblance is not complete; the reader must not run away with the idea that the human embryo ever has the "gill-slits of a fish." But it has a transitory rude passage through that type of structure. It is not, so to speak, a reproduction; it is an imperfect memory.

Once more this means nothing—indeed, makes nonsense—if we are to

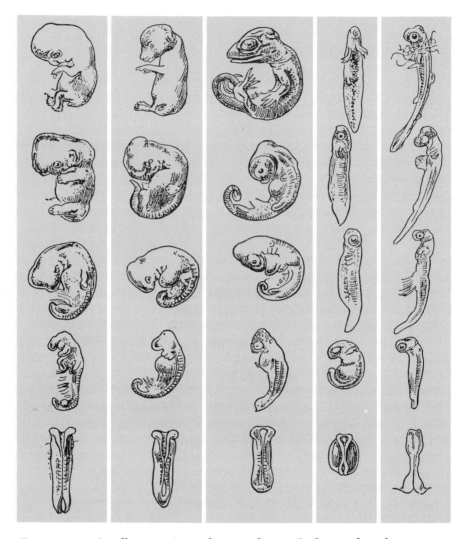

Figure 1–12. A collection of vertebrate embryos. Each upright column repre-sents the development of a single type—left to right: Man, Rabbit, Lizard, Newt, and Dogfish—the earliest embryos being below and the latest above. Note that the early stages are very like each other, and that the animals diverge as they develop. In the first stage the nerve-folds are closing in to form the brain. Then the gill-clefts appear. In the land animals these close later; in the newt and dogfish feathery gills appear. The human tail and its gradual shorten-ing are clearly seen.

believe that land animals were created as land animals. But it at once becomes pregnant with meaning if we accept the fact of Evolution, for then we can understand that snakes and hens and human beings and all other air-breathing vertebrates are fundamentally fishlike, that they start on the fishward road and turn away from it towards their higher structural achievement. When they reproduce the old disposition asserts itself; they start towards the old water-way and turn aside towards the uplands. Because of that recurrent urge each individual animal repeats within its individual cycle of life these uneffaced tendencies from the remote part of its race. In Amphibia the recapitulation is much more thoroughgoing; they have not only the clefts but the gills, and most of them actually do breathe by means of gills in their early tadpole stages, and physiologically are indeed fish.

Nearly half a century later, Haeckel, looking at the facts of embryology in the light of evolutionary ideas, broadened and reformulated and perhaps rather exaggerated von Baer's law. Haeckel's revision was this, that every animal in the course of its individual development tends to recapitulate the development of the race; and from this time onwards the facts on which the law is based have been called the facts of recapitulation. But it is a general and not a complete recapitulation. Evolution can affect every part of a life-cycle, and if a stage wastes much time or energy, Nature, who is no historian, will abbreviate it or cut it out quite ruthlessly.

Exactly how far Haeckel's law takes us—what are its limitations; whether recapitulation ever shows us an animal's adult ancestors; whether present development ever does more than recapitulate ancestral development; what is the cause of recapitulation; and why some of the characters and structures are regularly recapitulated in development, others only occasionally or not at all—all this we cannot here discuss.[3] What we are here concerned with are the positive, visible facts. Tens of thousands of animals do recapitulate the past during their development—do, without any apparent advantage in so doing, show organs and constructions which occur elsewhere in the adults of less specialized creatures: and in none of these tens of thousands of cases is this departure intelligible save on the view that in so doing they are repeating phases that were once final forms in the earlier evolution of the race. (See Fig. 1–10.)

3. Haeckel's statement that ontogeny recapitulates phylogeny is not strictly true. The individual does not run through the *adult* stages of its evolutionary ancestors. But if we use the idea of recapitulation "in a purely descriptive sense, implying no more than that ancestral plans of structure may be retained in development, and so shed light on evolution and even reveal unexpected relationships, then it is a legitimate and useful term." For more information on this subject see Julian Huxley, *Evolution in Action*, New American Library of World Literature, New York, pp. 17–19. Ed.

There is probably no single case of development among many-celled animals which does not show some recapitulatory feature. Even the origin of all sexually produced individuals in a single cell, the fertilized egg, in a certain sense recapitulates the origin of many-celled from single-celled forms. But there are plenty of more definite examples. Every human being and every other vertebrate at a certain stage of his or her development has an unjointed notochord in place of the future jointed backbone, even though in all save the lampreys and certain fish this notochord-rod vanishes entirely. And the most primitive chordate, Amphioxus, has a notochord all its life long and never develops a backbone. . . .

A much less all-pervading example, but none the less a very pretty one, we have already cited in the early horse-stock. The three-toed Merychippus had an elaborate grinding pattern on its permanent premolar teeth; but the pattern of the milk-teeth was simpler, like the pattern of the permanent teeth in geologically earlier and more primitive horse-ancestors. Here we can actually put side by side the adult ancestral form and its young descendant recapitulating it before passing on to the more highly evolved structure. . . .

The evidence of the fossils in the rocks is direct evidence for Evolution. But the evidence of embryology, though indirect, is more immediate. You can watch the individual animal indulge in these amazing reminiscences, and pass almost before your eyes from ancestral primitiveness to adult modernity. All the facts have a simple and straightforward meaning if Evolution be a fact, while a denial of Evolution leaves them unexplained and apparently inexplicable. . . .

THE DISTRIBUTION OF LIVING THINGS

Everybody knows that different animals come from different countries —the platypus from Australia and Tasmania, the zebra from Africa, the marmoset from South America, the musk ox from Greenland and Arctic Canada, and so on; and the same is, of course, true for plants. But it is not always grasped that different regions differ in respect of whole groups of their animal and plant inhabitants. Contrast the three southern continents: Africa south of the Sahara (that sea of sand which is as much a barrier to life as any sea of water), South America, and Australia. All comprise both temperate and tropical regions; all have their mountains, forests, and open plains. But their animal populations are extremely unlike. If, for the sake of brevity, we restrict ourselves to the mammal population, we find that Africa is characterized by an abundance of antelopes of many kinds, by rhinoceroses, giraffes, elephants, wart-hogs, zebras, lions, leopards, baboons, and buffaloes, and in the rain-forests, by gorillas and chimpanzees, okapi and many kinds of monkeys. Farther south the coneys or hyraxes and the extraordinary aard-varks are very characteristic. The whole giraffe family, with both giraffe and okapi, is found in no other region, nor is the aard-vark family.

Just as characteristic as the presences are the absences. There are no deer, no beavers, no field-mice or voles, no shrews, no bears, and scarcely any goats or sheep (we are, of course, speaking only of animals found wild).

Contrast this assemblage of mammals with that found in South America. Here live llamas and their relatives, edentates like the sloths, the true ant-eaters and the armadillos, primitive monkeys with prehensile tails, vampire bats, peccaries, tapirs, guinea-pigs, vizcachas and agoutis, opossums. None of these occur in Africa; and most of them are either wholly restricted to South America, or at the most penetrate a little way into Central or North America. The whole order of the true edentates is confined to this region.

Finally, Australia (with which for our present purpose we must include the neighbouring islands of Tasmania and New Guinea) is more peculiar still. Before the advent of white men, it contained none of the higher placental mammals whatever, with the exception of bats, whose wings, of course, give them facilities for spreading denied to mere land forms, with a few ubiquitous mice and the dingo dog, both probably introduced by the early human immigrants of the country. But by way of compensation Australia possesses a unique menagerie (now, alas, rapidly dwindling, with many species in danger of extermination if protective measures are not introduced) of the two lower sub-classes of mammals, the pouched marsupials and the egg-laying monotremes.

In point of fact, no egg-laying mammals occur outside this area, and no marsupials, except the American group of opossums, and one curious little creature called Coenolestes, with teeth in some ways recalling those of kangaroos, from South America. All the rest are Australian—kangaroos and wallabys, cuscuses and phalangers, wombats and bandicoots, marsupial wolf and Tasmanian devil, pouched ant-eaters and pouched moles and pouched mice—some forty genera, with hundreds of species. Add to this the platypus and its egg-laying confrères, the spiny ant-eaters, and you have indeed a strange zoo.

Now, it might naturally be supposed, and in the past often was assumed, that each species and each group lived in the region best suited to it. But such is demonstrably and obviously not the case. New Zealand has no native mammals save a bat or two and possibly one species of rat— and yet introduced mammals thrive and multiply. Rabbits, for instance, have run wild over large areas, and red deer introduced from Scotland have not only thriven, but have grown much larger than they ever do in their native land. Then the house-sparrow has spread and the starling is spreading over the whole of North America, in spite of the competition of the hundreds of kinds of native birds. The few horses introduced by the Spanish conquerors of South America multiplied and ran wild in huge herds over the pampas. Far from the native Australian birds and animals being especially well adapted to Australian conditions, they are no match for the species that have been introduced from other regions. The mere mention of rabbits will make an Australian farmer cross. And when we come to plants we find that one of the gravest problems of agriculture in various countries, notably New Zealand and Australia, is to prevent introduced species like the prickly pear and blackberry from overrunning the country and ousting not only the native plants, but man and his agricultural efforts as well.

Why then are whole groups of related animals tied down to limited regions of the world? What meaning is there in the restriction of the giraffe family to Africa, the whole of the edentate order of sloths and armadillos and ant-eaters to Southern America, all the monotreme sub-class and almost all the marsupials to Australia?

The answer is to be found in the past, in the history of evolving life in relation to the history of the seas and continents. Through fossils we are able to discover not only the past development of existing groups, but also their past distribution in each epoch. Geology, on the other hand, can tell us a great deal about the extent of sea and land in past periods. It can do this by studying where in each epoch marine deposits were laid down, where there were deserts, or evaporating inland seas which produced beds of salt, where the invading ocean had carved beaches, where ice-sheets had passed or mountain-ranges had been elevated. Through such evidence, geology is able to say definitely that the present distribution of land and water is in no way permanent. In the past, the main land-masses of the world have been connected and disjoined in many other ways; and geology can often tell us just when and where the connections and partings were made, and what was the distribution of seas and continents during a particular geological period.[4] [See Figure 4–13.]

Animals such as land-mammals can and do migrate slowly until they are spread over the whole of a land-mass. But there are barriers which they cannot cross. The sea is the most formidable of such barriers, ice-sheets are another, and broad deserts may be nearly as bad. Thus, the distribution of any group of land-animals will depend upon three factors—

4. According to a new geophysical theory the present land masses may have been part of two large continents, one in the northern and one in the southern hemisphere about 200 or 300 million years ago. See J. R. Heirtzler, "Sea-Floor Spreading," *Scientific American,* December 1968. Ed.

first, upon the region where the group happened to originate; second, upon the connections which this region then and later happened to have with other land-masses; and third, upon the fate of the group in the different regions to which it obtained access.

If mammals first evolved after New Zealand had been cut off by a barrier of ocean from all the continents, we should not expect to find any land-mammals in New Zealand. If lung-fish were once widely distributed all over the world, but later were all but extinguished in the struggle for existence, we should expect to find the few existing lung-fish scattered in isolated regions which happened to favour their survival; it is along such lines that our reasoning must run.

The key to the present distribution lies in past distribution. When palaeontology and geology are able to provide us with evidence, the distribution of animals and plants ceases to be a puzzle and becomes a simple matter of history. . . .

CONCLUSION

All these lines of evidence lead to the same conclusion. The way in which each one corroborates all the others is impressive enough, but let it not be forgotten that the actual examples we have chosen are but a fragment of the mass available. If an idea is true, it will apply in every part of its domain. The domain of the idea of organic evolution is the whole domain of life; and the final evidence for Evolution is that throughout the whole domain the idea of Evolution helps our comprehension. It explains old discoveries and leads us on to new; it draws order out of confusion; it gives meaning to what is otherwise meaningless, and brings thousands of isolated facts into a single related whole.

Does the Evidence for Evolution
Establish It as a Scientific Fact?

Evolution *

by Theodosius Dobzhansky (1956)

Evolution as a historical fact was proved beyond reasonable doubt not later than in the closing decades of the nineteenth century. No one who takes the trouble to become familiar with the pertinent evidence has at present a valid reason to disbelieve that the living world, including man, is a product of an evolutionary development. According to the classical saying, "Fata volentem ducunt, nolentem trahunt"—which can be freely translated to mean that "Necessity guides those who accept it, and drags those who resist it." To be sure, antievolutionists still exist. But it is fair to say that most of them are not well informed, while the informed exceptions display biases which make arguments futile and facts useless. Does it indeed make sense to declare oneself unconvinced by the evidence for evolution only because no one has actually seen the "fact" of the transformation of the dawn horse into a present-day horse, or of the prehuman into a human being? This is like the Midwestern gentleman who was unconvinced that the earth was round, even after he took a "round-the-world" trip. Nobody has seen atoms and molecules either, but we act as though atoms and molecules were facts and not hypotheses. Evolution took place mostly before there were observers to see it and to record their observations. Any such historical fact must be inferred on the basis of evidence which can be seen today. It is no mere matter of taste whether one rejects or accepts the inference. A hypothesis may, at a certain level of scientific knowledge, be forced upon every reasonable person.

Innumerable facts of comparative morphology, physiology, embryology, paleontology, and geographic distribution suggest that evolution has taken place. These facts make sense if the hypothesis of evolution is granted; they do not make sense otherwise.

Editor's Note. Scientists accept the theory of evolution as proven "beyond reasonable doubt." But until 1968 there were some states in this country where the teaching of evolution in the public schools was illegal, and the subject is still controversial enough in many areas to cause some publishers of elementary textbooks to omit the subject of evolution for fear of jeopardizing general adoption of the textbooks. A recent example of this kind of opposition occurred in April 1968. The members of the Village Baptist Church of Buffalo Grove, Illinois, sent a petition to School District No. 21 of Wheeling, Illinois, protesting the teaching of evolution as a fact in the public schools. Evolution, they declared, is just a *theory* and has never been proven to be a *fact*. Therefore, the teaching of it as fact is both wrong and detrimental.

Protests of this kind raise the question of what constitutes a scientific proof. To answer this question we must consider the nature of scientific theory and the difference between scientific and religious or philosophical "truth."

The Nature of Scientific Theory *
by W. I. B. Beveridge

Generalisations can never be *proved*. They can be tested by seeing whether deductions made from them are in accord with experimental and observational facts, and if the results are not as predicted, the hypothesis or generalisation may be *disproved*. But a favourable result does not prove the generalisation, because the deduction made from it may be true without its being true. Deductions, themselves correct, may be made from palpably absurd generalisations. For instance, the truth of the hypothesis that plague is due to evil spirits is not established by the correctness of the deduction that you can avoid the disease by keeping out of the reach of the evil spirits. In strict logic a generalisation is never proved and remains on probation indefinitely, but if it survives all attempts at disproof it is accepted in practice, especially if it fits well into a wider theoretical scheme.

If scientific logic shows we must be cautious in arriving at generalisations ourselves, it shows for the same reasons that we should not place excessive trust in any generalisation, even widely accepted theories or laws. Newton did not regard the laws he formulated as the ultimate truth, but probably most following him did until Einstein showed how well-founded Newton's caution had been. In less fundamental matters how often do we see widely accepted notions superseded!

* From W. I. B. Beveridge, *The Art of Scientific Investigation*, W. W. Norton & Company, Inc. Copyright W. W. Norton & Company, Inc., revised edition 1957. Reprinted by permission. All rights reserved.

Scientific Truth *

by E. N. da C. Andrada (1956)

It is the task of the philosopher to reflect upon the general nature of the happenings, material and spiritual, that make up the life of man, and to endeavour to work out some scheme which shall help to reconcile conflicting appearances and simplify, by the investigation of first principles, the complex tangle of events in which our being is involved. It is for him to try to find some kind of an answer to the eternal 'Why?' which mankind, bewildered by the problems of good and evil, of life and death, has been uttering, now in the stammer of childhood, now in the harsh voice of agony, now in the quiet tones of reflecting age, since man has been a thinking animal. The nature of appearance and reality, the meaning of truth and falsehood, the scope and implication of our knowledge, the significance of the conception of beauty—these are among the hard questions on which the philosopher must exercise his powers. They are very wide and very elusive, difficult to enunciate satisfactorily, more difficult to solve in the least particular. New points of view can be found, but how are we to judge when a real advance has been made?

The task of the man of science is more modest: it is not to answer the everlasting 'Why?' but the no less everlasting 'How?' He deals with the facts of observation, and tries to reduce them into a system, so that, if we admit certain principles to start with, things which are actually known to happen systematically can be shown to follow as necessary consequences, and a method of looking for new ones is suggested. The principles themselves are chosen to suit the facts, which for the man of science are all-important—*les principes ne se démontrent pas*. Whether the principles are in the absolute sense true is not a matter on which the man of science, as such, feels called to argue: if their consequences agree with Nature they are true enough to be useful, at any rate. The fundamental principles of science are, therefore, often called *working hypotheses*, since they are devised with the sole purpose of furnishing a basis upon which a system may be built corresponding in appearance with the behaviour of the material world, wherever we are able to make measurements or observations for comparison. We consider that an advance has been made when a wider range of the phenomena which are observed has been brought within the scope of one general principle.

It follows that a scientific theory may be abandoned when it has proved itself insufficient without in any way weakening the general worth of the scientific method. . . . We do not claim any finality for the theory [of the atom, for instance]: some new discovery may suddenly force us to modify our ideas in many particulars, some new discoverer may show that certain complications can be simplified with advantage, but the successes of the present theory show that we shall probably have to retain many of its general features. It is an excellent working hypothesis because it has shown us law where law was not hitherto discovered, and connections between different phenomena where before we knew of no connection. It has enabled us to arrange our known facts in a more convenient and logical way, and has pointed the way to the discovery of very interesting new facts. It is justified by its works, but it is not final. Science is a living thing, and living things develop.

On this view, any particular scientific theory is a provisional tool with which we carve knowledge of the material world out of the block of Nature. It may at any moment be supplanted by an improved version or by a completely new theory, but this is only to say that when we get a better tool, which does all that this one does and something more as well, we will abandon our present tool. To refuse to use a tool because some day a better one may be invented is folly: in the same way, not to make use of a theory which has been proved capable of explaining a great many facts, and of suggesting new lines of research, because it has acknowledged flaws, and is incapable of explaining other facts, would be folly. To use another metaphor, the history of science may, as has been said, be full of beautiful theories slain by ugly little facts, but those theories did not die in vain if before their death they had subdued a vast number of jarring facts into a law-abiding populace. Nor do theories generally die a final death: often they are resurrected with some new feature which gets over the old difficulty.

The difference, then, between any religious belief and a scientific theory is that the former has for the believers an element of absolute truth: it is a standard by which they stand or fall, and to abandon it is dishonour and sin. The scientific theory is, however, only true as long as it is useful. The man of science regards even his best theory as a makeshift device to help him on his way, and is always on the lookout for something better and more comprehensive.

Suggestions for Further Reading

* Carrington, Richard: *A Guide to Earth History,* Mentor Books, The New American Library, Inc., 1961. A comprehensive account of the origin of our planet and the evolution of life upon it. The style is interesting and easy-to-understand, with clear charts, illustrations, and maps.

* Darwin, Charles: *The Voyage of the Beagle,* Bantam Books, 1958. Darwin's own fascinating journal of the experiences and observations that led him to his theory of natural selection. The short selection in this Part illustrates Darwin's style which is still fresh and interesting today.

* Eiseley, Loren: *Darwin's Century,* Anchor Books, Doubleday & Co., Inc., 1960. A perceptive account of the discoveries which paved the way for Darwin's theory, the impact of the theory itself, and the reaction to it in subsequent years.

Evolution, Life Science Library, Time Inc., 1962. A lavishly illustrated book with text by Ruth Moore introducing the layman to basic ideas in evolution and genetics. Selections from this book are used in Part 4.

* Hurley, Patrick M.: *How Old Is the Earth?* Anchor Books, Doubleday & Co., Inc., 1959. An illuminating Science Study Series book that explains man's efforts to understand the origin, age, and structure of the earth, and how radioactivity has given the modern geologist a new, definitive research technique.

* Rapport, Samuel, and Helen Wright (eds.): *The Crust of the Earth,* Signet Science Library Book, The New American Library of World Literature, Inc., 1955. A good layman's introduction to geological science.

* Simpson, George Gaylord: *Life of the Past,* A Yale Paperbound, Yale University Press, 1961. A paleontologist describes the ancient animals which have been reconstructed solely by means of fossil remains. The reader obtains a clear insight into the work and aims of paleontologists.

* Paperback edition

2

What Causes Evolution?

The principal theories of the causes of evolution are presented in these readings. Does the theory of natural selection account for the facts of evolution in the light of present day knowledge?

What Causes Evolution?

Introduction

IT IS ONE thing to postulate that evolutionary change has occurred and quite another one to advance a logical reason why this change should have taken place. What caused an organism like the four-toed Eohippus, for instance, to evolve into the modern horse? What caused some organisms to undergo radical and progressive transformations while others like the king crab remained the same over millions of years?

Many theories have been advanced in answer to this question. Some of the theories invoke spiritual forces such as an inner drive or an innate destiny of the species. Other theories invoke only natural forces. The readings in this part describe the natural explanations because they come most properly under the realm of scientific investigation. In Part 3 some consideration will also be given to the philosophical consequences of modern evolutionary theory and whether the facts leave any room for religious or spiritual interpretations.

In pre-Socratic times the natural philosopher Empedocles proposed a primitive version of the theory of natural selection—that nature creates at random many different organisms and that only the best adapted of these survive.

> Hence, doubtless, Earth prodigious forms at first
> Engendered, of face and members most grotesque;
> Monsters half-man, half-woman, not from each
> Distant, yet neither total; shapes unsound,
> Footless, and handless, void of mouth or eye,
> Or from misjunction, maimed, of limb with limb:
> To act all impotent, or flee from harm,
> Or nurture take their loathsome days to extend.
> These sprang at first, and things alike uncouth;
> Yet vainly; for abhorrent Nature quick
> Checked their vile growths; . . .
> Hence, doubtless, many a tribe has sunk suppressed,
> Powerless its kind to breed. . . .
> Centaurs lived not; nor could shapes like these
> Live ever.[1]

Such early speculations, however, were not pursued by other thinkers, probably because the idea of evolution itself was not generally accepted until many years later. As we have seen, the idea of evolution was sug-

1. Empedocles, c.495–c.435 B.C. From Lucretius, *On the Nature of Things,* Book V.

gested by several philosophers in the eighteenth century. But Jean Baptiste Lamarck, writing about 1802, was the first man to present a definitive theory of evolutionary change and to suggest a reason for the process. His explanation was based on the common observation that those parts of the body which are used most become larger and more effective and that those which are neglected diminish in size and usefulness. In order for these variations in the individual to produce evolutionary change in the species these "acquired characteristics" would have to be passed on by heredity. In Lamarck's day there was no reason to believe that acquired traits could not be passed on. Nothing was known about the chromosome mechanism of heredity, and the various vague ideas about inheritance which had been proposed did not rule out the possibility of inheriting acquired characteristics. Charles Darwin also assumed some inheritance of this kind and postulated a system of "gemmules" which were present in all the tissues of the body. These gemmules, he believed, migrated into the reproductive cells and thence to the next generation, where they directed the development of the corresponding tissues of the offspring. On such a basis the inheritance of an acquired characteristic could be carried by the gemmules from the particular organ that had undergone a change in size or function.

It was more than a century after Lamarck that Mendel's work was rediscovered and the chromosome theory of heredity was developed. August Weismann, writing in the early years of the twentieth century, declared that the chromosome theory made the inheritance of acquired traits impossible. He coined the name *germ plasm* for that part of an organism which could eventually grow into the next generation. This germ plasm differentiated very early in the development of the individual (the *soma*) and thereafter, Weismann pointed out, it could not be affected by the experiences or efforts of the organism itself. Today geneticists know that some interaction does take place between the germ plasm and its environment. However, there is still no indication that changes which occur in the germ plasm are in any way influenced by the needs or the demands of the organism. The science of genetics, however, did support Darwin's theory of natural selection and answered a number of objections which had been raised to his theory.

Looking back from the summit of our present knowledge, it is easy to ridicule Lamarck's ideas. But the reader should try to place himself back in the context of the scientific knowledge of the early nineteenth century to appreciate Lamarck's contribution and to understand why some of his thoughts still have proponents today. The suggestion that the individual organism is an active force in the evolutionary process is an appealing idea, in keeping with our belief in the dignity and importance of the individual. We would *like* to believe that the strivings of individuals can have an effect on the future of the species.

Bias in favor of this view was so powerful that just a few years ago neo-Lamarckism was the only evolutionary theory acceptable to the Communist regime in Russia. Biologists who made the mistake of pointing out that scientific experiments consistently showed that acquired traits were not inherited were quickly silenced or forced to recant. And for a decade or so biological research in the sense of an open inquiry was impossible in the Soviet Union.

This extreme endorsement of Lamarckism led to a powerful reaction against it in the democratic world. As Joseph Wood Krutch says: "Lamarckism became an epithet to be hurled rather than a position to be analyzed." [2]

Emotional responses to scientific theories are especially common in biology because these theories are involved with areas that mean a great deal to us as human beings. Evolutionary theory has been considered a threat to organized religion and has been used to justify or repudiate various social and political systems. The fact that biology is not a strictly quantitative science leaves the door open to differences of opinion which are often passionately maintained and which cannot be easily resolved.

It is generally accepted, for instance, that natural selection produces evolutionary change. However, in the present state of our knowledge it is impossible to demonstrate conclusively that *no other* agency may be at work. In physics we can demonstrate that the speed at which a satellite is moving is sufficient to maintain it in a certain orbit around the earth, that no unknown agency need be evoked to explain this relative motion. But in evolutionary theory we cannot prove, for example, that the advantages of increased mobility and new sources of food are sufficient to explain the evolution of wings in the bird. Darwin, himself, did not believe that natural selection was the only cause. "I am convinced," he said, "that natural selection has been the main but not the exclusive means of modifications." And he went on to express frustration that this statement had been consistently passed over. [3]

Darwin would probably have been even more distressed by the oversimplified position taken by his followers in the late nineteenth century

2. See Joseph Wood Krutch selection, p. 115. Actually, some distinction should be drawn between Lamarckism and neo-Lamarckism. The neo-Lamarckian school that developed during the latter part of the nineteenth century (and later endorsed by the Lysenko group in Russia) did not exactly follow Lamarck's theory. It stressed a more direct action of the environment on organic structure than Lamarck had proposed and omitted the concept of a perfecting principle in which Lamarck had believed.

3. See Charles Darwin selection, p. 82.

and early twentieth century. The neo-Darwinists, particularly Weismann, rejected all of Darwin's theory except natural selection. They set about to prove that the struggle for existence, resulting in the death of the unfit and survival of the fit, was all that was needed to explain the evolutionary process.

Since Weismann's time the concept of natural selection has been considerably broadened and refined. It is understood now to be a complex process operating by differential reproduction (instead of just differential survival) and involving an interplay between environmental and inherited factors. Natural selection in this more refined sense is generally believed to be the only cause that is needed to explain evolution.

It is a basic principle in science that it is unsound to set up more than one hypothesis when one will suffice to explain a phenomenon.[4] However, it is also a scientist's prerogative to question, and it is in keeping with the ideal of openness in scientific investigation to give a fair hearing to any dissenting voices who are suggesting causes that come within the proper realm of science. New facts are constantly being discovered. Evolutionary theory today is very different from that which was first proposed by Darwin. It may well be that by the turn of the century our understanding of evolution will be quite different from that which is held today. One of the fascinations of science is that it is not just a collection of accepted facts. The ultimate solution, like the pot of gold at the end of the rainbow, is always receding before us.

4. This principle, known as "Ockam's razor", was enunciated by William of Ockam in 1318: *"Entia non sunt multiplicanda praeter necessitem"*; literally, beings are not to be multiplied without necessity.

History of Evolutionary Theory

What Is a Species? [*]

by Jean Baptiste Lamarck (1801–06)

A great many facts teach us that gradually, as the individuals of one of our species change their situation, climate, mode of life, or habits, they thus receive influences which gradually change the consistence and the proportions of their parts, their form, their faculties, even their organization; so that all of them participate eventually in the changes which they have undergone.

In the same climate very different situations and exposures at first cause simple variations in the individuals which are found exposed there; but as time goes on the continual differences of situation of individuals of which I have spoken, which live and successively reproduce in the same circumstances, give rise among them to differences which are, in some degree, essential to their being, in such a way that at the end of many successive generations these individuals, which originally belonged to another species, are at the end transformed into a new species, distinct from the other.

For example, if the seeds of a grass, or of every other plant natural to a humid field, should be transplanted, by an accident, at first to the slope of a neighboring hill, where the soil, although more elevated would yet be quite cool, so as to allow the plant to live, and then after having lived there, and passed through many generations there, it should gradually reach the poor and almost arid soil of a mountainside—if the plant should thrive and live there and perpetuate itself during a series of generations, it would then be so changed that the botanists who should find it there would describe it as a separate species.

The same thing happens to animals which circumstances have forced to

[*] From Forest Ray Moulton and Justus J. Schifferes, editors, *The Autobiography of Science,* Doubleday & Company, Inc., Garden City, New York. Copyright © 1960 by Justus J. Schifferes. Reprinted by permission of Justus Schifferes. This article is a briefer version of three essays published in *Lamarck, The Founder of Evolution,* translated by Alpheus S. Packard, Longmans, Green & Company, London, 1901.

change their climate, manner of living, and habits; but for these the influences of the causes which I have just cited need still more time than in the case of plants to produce the notable changes in the individuals, though in the long run, however, they always succeed in bringing them about. . . .

It appears, as I have already said, that *time* and *favorable conditions* are the two principal means which Nature has employed in giving existence to all her productions. We know that for her time has no limit and that consequently she has it always at her disposal.

As to the circumstances of which she has had need and of which she makes use every day in order to cause her productions to vary, we can say that they are in a manner inexhaustible.

The essential ones arise from the influence and from all the environing media, from the diversity of local causes, of habits, of movements, of action, finally of means of living, of preserving their lives, of defending themselves, of multiplying themselves, et cetera. Moreover, as the result of these different influences, the faculties, developed and strengthened by use, became diversified by the new habits maintained for long ages, and by slow degrees the structure, the consistence, in a word the nature, the condition of the parts and of the organs consequently participating in all these influences, became preserved and were propagated by generation.

The bird which necessity drives to the water to find there the prey needed for its subsistence separates the toes of its feet when it wishes to strike the water and move on its surface. The skin, which unites these toes at their base, contracts in this way the habit of extending itself. Thus in time the broad membranes which connect the toes of ducks, geese, et cetera, are formed in the way indicated.

But one accustomed to live perched on trees has necessarily the end of the toes lengthened and shaped in another way. Its claws are elongated, sharpened, and are curved and bent so as to seize the branches on which it so often rests.

Likewise we perceive that the shore bird, which does not care to swim, but which, however, is obliged to approach the water to obtain its prey, will be continually in danger of sinking in the mud, but wishing to act so that its body shall not fall into the liquid, it will contract the habit of extending and lengthening its feet. Hence it will result in the generations of these birds which continue to live in this manner that the individuals will find themselves raised as if on stilts, on long naked feet; namely, denuded of feathers up to and often above the thighs.

I could here pass in review all the classes, all the orders, all the genera and species of animals which exist, and make it apparent that the conformation of individuals and of their parts, their organs, their faculties, et cetera, is entirely the result of circumstances to which the race of each species has been subjected by nature.

I could prove that it is not the form either of the body or of its parts which gives rise to habits, to the mode of life of animals, but, on the contrary, it is the habits, the mode of life, and all the influential circumstances which have, with time, made up the form of the body and of the parts of animals. With the new forms new faculties have been acquired and gradually Nature has reached the state in which we actually see her.

Indeed, we know that all the time that an organ, or a system of organs is rigorously exercised throughout a long time, not only its power, and the parts which form it, grow and strengthen themselves, but there are proofs that this organ, or system of organs, at that time attracts to itself the principal active forces of the life of the individual, because it becomes the cause which, under these conditions, makes the functions of other organs to be diminished in power.

Thus not only every organ or every part of the body, whether of man or of animals, being for a long period and more vigorously exercised than the others, has acquired a power and facility of action that the same organ could not have had before, and that it has never had in individuals which have exercised less, but also we consequently remark that the excessive employment of this organ diminishes the functions of the others and proportionately enfeebles them.

The man who habitually and vigorously exercises the organ of his intelligence develops and acquires a great facility of attention, of aptitude for thought, et cetera, but he has a feeble stomach and strongly limited muscular powers. He, on the contrary, who thinks little does not easily, and then only momentarily fixes his attention, while habitually giving much exercise to his muscular organs, has much vigor, possesses an excellent digestion, and is not given to the abstemiousness of the savant and man of letters.

Moreover, when one exercises long and vigorously an organ or system of organs, the active forces of life (in my opinion, the nervous fluid) have taken such a habit of acting toward this organ that they have formed in the individual an inclination to continue to exercise which it is difficult for it to overcome.

Hence it happens that the more we exercise an organ, the more we use it with facility, the more does it result that we perceive the need of continuing to use it at the times when it is placed in action. So we remark that the habit of study, of application, of work, or of any other exercise of our organs or of any one of our organs, becomes with time an indispensable need to the individual, and often a passion which it does not know how to overcome. . . .

Thus we are assured that that which is taken for *species* among living bodies, and that all the specific differences which distinguish these natural productions, have no absolute *stability*, but that they enjoy only a relative *stability*; which it is very important to consider in order to fix the limits which we must establish in the determination of that which we must call *species*.

———————

Editor's Note. The theory of natural selection was presented to the world simultaneously by two Englishmen who were also world travelers. Charles Darwin had been working on a detailed and carefully reasoned presentation of his theory for over twenty years, ever since his return from the voyage around the world on the *Beagle.* Then, in 1858, he was startled to receive in the post an essay written by a young naturalist, Alfred Russel Wallace, who had been studying the flora and fauna of the Malay Archipelago. This essay stated in a very concise and convincing manner the central concept of the theory of natural selection. Both men reported later that the germ of the idea had been suggested to them by Thomas Malthus's essay on population.

> In 1838, [writes Darwin in his *Autobiography*] that is, fifteen months after I had begun my systematic enquiry, I happened to read for amusement "Malthus on Population," and being well prepared to appreciate the struggle for existence which everywhere goes on from long-continued observation of the habits of animals and plants, it at once struck me that under these circumstances favourable variations would tend to be preserved, and unfavourable ones to be destroyed. The result of this would be the formation of new species. Here then I had at last got a theory by which to work.[1]

Wallace's unexpected competition spurred Darwin into rushing his own work to completion but, at the same time, he was scrupulously fair in respecting Wallace's rights to credit for the discovery. Through his friends Charles Lyell and Joseph Hooker he arranged for a simultaneous presentation of his theory and Wallace's at a meeting of the Linnaean Society in August 1858. The following selection is from the paper Wallace sent to Darwin and which was published in the *Journal of the Proceedings of the Linnaean Society.*

1. From Charles Darwin, *Life and Letters*, 1887.

On the Tendency of Varieties To Depart Indefinitely
from the Original Type *
by Alfred Russel Wallace (1858)

One of the strongest arguments which have been adduced to prove the
original and permanent distinctness of species is that varieties produced
in a state of domesticity are more or less unstable and often have a tendency, if left to themselves, to return to the normal form of the parent
species; and this instability is considered to be a distinctive peculiarity of
all varieties, even of those occurring among wild animals in a state of nature, and to constitute a provision for preserving unchanged the originally
created distinct species. . . .

But it is the object of the present paper to show that this assumption is
altogether false, that there is a general principle in nature which will
cause many varieties to survive the parent species, and to give rise to successive variations departing further and further from the original type,
and which also produces, in domesticated animals, the tendency of varieties to return to the parent form.

THE STRUGGLE FOR EXISTENCE

The life of wild animals is a struggle for existence. The full exertion of
all their faculties and all their energies is required to preserve their own
existence and provide for that of their infant offspring. The possibility of
procuring food during the least favorable seasons and of escaping the attacks of their most dangerous enemies are the primary conditions which
determine the existence both of individuals and of entire species. These
conditions will also determine the population of a species; and by a careful consideration of all the circumstances we may be enabled to compre-
hend, and in some degree to explain, what at first sight appears so inexplicable—the excessive abundance of some species, while others closely
allied to them are very rare. . . .

Wildcats are prolific and have few enemies; why, then, are they never
as abundant as rabbits? The only intelligible answer is that their supply of

* From Alfred Russel Wallace, "On the Tendency of Varieties To Depart Indefinitely from the Original Type," *Journal of the Proceedings of the Linnaean
Society*, August 1858.

food is more precarious. It appears evident, therefore, that so long as a country remains physically unchanged the numbers of its animal population cannot materially increase. If one species does so, some others requiring the same kind of food must diminish in proportion. The numbers that die annually must be immense; and as the individual existence of each animal depends upon itself, those that die must be the weakest—the very young, the aged, and the diseased—while those that prolong their existence can only be the most perfect in health and vigor—those who are best able to obtain food regularly and avoid their numerous enemies. It is, as we commenced by remarking, "a struggle for existence," in which the weakest and least perfectly organized must always succumb. . . .

If now we have succeeded in establishing these two points—first, that the animal population of a country is generally stationary, being kept down by a periodical deficiency of food and other checks; and, second, that the comparative abundance or scarcity of the individuals of the several species is entirely due to their organization and resulting habits, which, rendering it more difficult to procure a regular supply of food and to provide for their personal safety in some cases than in others, can only be balanced by a difference in the population which have to exist in a given area—we shall be in a condition to proceed to the consideration of varieties, to which the preceding remarks have a direct and very important application.

USEFUL VARIATIONS WILL TEND TO INCREASE; USELESS OR HURTFUL
VARIATIONS TO DIMINISH

Most or perhaps all the variations from the typical form of a species must have some definite effect, however slight, on the habits or capacities of the individuals. Even a change of color might, by rendering them more or less distinguishable, affect their safety; a greater or less development of hair might modify their habits. More important changes, such as an increase in the power or dimensions of the limbs or any of the external organs, would more or less affect their mode of procuring food or the range of country which they could inhabit. It is also evident that most changes would affect, either favorably or adversely, the powers of prolonging existence. An antelope with shorter or weaker legs must necessarily suffer more from the attacks of the feline carnivora; the passenger pigeon with less powerful wings would sooner or later be affected in its powers of procuring a regular supply of food; and in both cases the result must necessarily be a diminution of the population of the modified species.

If, on the other hand, any species should produce a variety having slightly increased powers of preserving existence, that variety must inevitably in time acquire a superiority in numbers. These results must follow as surely as old age, intemperance, or scarcity of food produce an increased mortality. In both cases there may be many individual exceptions; but on the average the rule will invariably be found to hold good. All varieties will therefore fall into two classes—those which under the same conditions would never reach the population of the parent species, and those which would in time obtain and keep a numerical superiority. Now let some alteration of physical conditions occur in the district—a long period of drought, a destruction of vegetation by locusts, the irruption of some new carnivorous animal seeking "pastures new"—any change, in fact, tending to render existence more difficult to the species in question, and tasking its utmost powers to avoid complete extermination; it is evident that, of all the individuals composing the species, those forming the least numerous and most feebly organized variety would suffer first, and, were the pressure severe, must soon become extinct. The same causes continuing in action, the parent species would next suffer, would gradually diminish in numbers, and with a recurrence of similar unfavorable conditions might also become extinct. The superior variety would then alone remain, and on a return to favorable circumstances would rapidly increase in numbers and occupy the place of the extinct species and variety.

SUPERIOR VARIETIES WILL ULTIMATELY EXTIRPATE THE ORIGINAL SPECIES

The variety would now have replaced the species, of which it would be a more perfectly developed and more highly organized form. It would be in all respects better adapted to secure its safety and to prolong its individual existence and that of the race. Such a variety could not return to the original form; for that form is an inferior one and could never compete with it for existence. Granted, therefore, a "tendency" to reproduce the original type of the species, still the variety must ever remain preponderant in numbers, and under adverse physical conditions again alone survive. But this new, improved, and populous race might itself in course of time give rise to new varieties, exhibiting several diverging modifications of form, any of which, tending to increase the facilities for preserving existence, must, by the same general law, in their turn become predominant. Here, then, we have progression and continued divergence deduced from the general laws which regulate the existence of animals in a state of nature and from the undisputed fact that varieties do frequently occur. . . .

The hypothesis of Lamarck—that progressive changes in species have been produced by the attempts of animals to increase the development of their own organs and thus modify their structure and habits—has been repeatedly and easily refuted by all writers on the subject of varieties and species, and it seems to have been considered that when this was done the whole question had been finally settled; but the view here developed renders such hypothesis quite unnecessary, by showing that similar results must be produced by the action of principles constantly at work in nature. The powerful retractile talons of the falcon and the cat tribes have not been produced or increased by the volition of those animals; but among the different varieties which occurred in the earlier and less highly organized forms of these groups, those always survived longest which had the greatest facilities for seizing their prey. Neither did the giraffe acquire its long neck by desiring to reach the foliage of the more lofty shrubs and constantly stretching its neck for the purpose, but because any varieties which occurred among its antetypes with a longer neck than usual at once secured a fresh range of pasture over the same ground as their shorter-necked companions, and on the first scarcity of food were thereby enabled to outlive them. Even the peculiar colors of many animals, more especially of insects, so closely resembling the soil or leaves or bark on which they habitually reside, are explained on the same principle; for though in the course of ages varieties of many tints may have occurred, yet those races having colors best adapted to concealment from their enemies would inevitably survive the longest.

The Origin of Species *
by Charles Darwin (1872)

INTRODUCTION

When on board H.M.S. "Beagle," as naturalist, I was much struck with certain facts in the distribution of the organic beings inhabiting South America, and in the geological relations of the present to the past inhabitants of that continent. These facts, as will be seen in the latter

* From Charles Darwin, *The Origin of Species*, first edition, 1859, sixth edition, 1872.

chapters of this volume, seemed to throw some light on the origin of species—that mystery of mysteries, as it has been called by one of our greatest philosophers. On my return home, it occurred to me, in 1837, that something might perhaps be made out on this question by patiently accumulating and reflecting on all sorts of facts which could possibly have any bearing on it. After five years' work I allowed myself to speculate on the subject, and drew up some short notes; these I enlarged in 1844 into a sketch of the conclusions, which then seemed to me probable: from that period to the present day I have steadily pursued the same object. I hope that I may be excused for entering on these personal details, as I give them to show that I have not been hasty in coming to a decision.

My work is now (1859) nearly finished; but as it will take me many more years to complete it, and as my health is far from strong, I have been urged to publish this Abstract. I have more especially been induced to do this, as Mr. Wallace, who is now studying the natural history of the Malay archipelago, has arrived at almost exactly the same general conclusions that I have on the origin of species. In 1858 he sent me a memoir on this subject, with a request that I would forward it to Sir Charles Lyell, who sent it to the Linnean Society, and it is published in the third volume of the Journal of that society. Sir C. Lyell and Dr. Hooker, who both knew of my work—the latter having read my sketch of 1844—honoured me by thinking it advisable to publish, with Mr. Wallace's excellent memoir, some brief extracts from my manuscripts. . . .

In considering the Origin of Species, it is quite conceivable that a naturalist, reflecting on the mutual affinities of organic beings, on their embryological relations, their geographical distribution, geological succession, and other such facts, might come to the conclusion that species had not been independently created, but had descended, like varieties, from other species. Nevertheless, such a conclusion, even if well founded, would be unsatisfactory, until it could be shown how the innumerable species inhabiting this world have been modified, so as to acquire that perfection of structure and coadaptation which justly excites our admiration. Naturalists continually refer to external conditions, such as climate, food, &c., as the only possible source of variation. In one limited sense, as we shall hereafter see, this may be true; but it is preposterous to attribute to mere external conditions, the structure, for instance, of the woodpecker, with its feet, tail, beak, and tongue, so admirably adapted to catch insects under the bark of trees. In the case of the mistletoe, which draws its nourishment from certain trees, which has seeds that must be transported by certain birds, and which has flowers with separate sexes absolutely requiring the agency of certain insects to bring pollen from one flower to the other, it is equally preposterous to account for the structure of this parasite, with its relations to several distinct organic beings, by the effects of external conditions, or of habit, or of the volition of the plant itself.

It is, therefore, of the highest importance to gain a clear insight into the means of modification and coadaptation. At the commencement of my observations it seemed to me probable that a careful study of domesticated animals and of cultivated plants would offer the best chance of making out this obscure problem. Nor have I been disappointed; in this and in all other perplexing cases I have invariably found that our knowledge, imperfect though it be, of variation under domestication, afforded the best and safest clue. I may venture to express my conviction of the high value of such studies, although they have been very commonly neglected by naturalists.

From these considerations, I shall devote the first chapter of this Abstract to Variation under Domestication. We shall thus see that a large amount of hereditary modification is at least possible; and, what is equally or more important, we shall see how great is the power of man in accumulating by his Selection successive slight variations. I will then pass on to the variability of species in a state of nature; but I shall, unfortunately, be compelled to treat this subject far too briefly, as it can be treated properly only by giving long catalogues of facts. We shall, however, be enabled to discuss what circumstances are most favourable to variation. In the next chapter the Struggle for Existence amongst all organic beings throughout the world, which inevitably follows from the high geometrical ratio of their increase, will be considered. This is the doctrine of Malthus, applied to the whole animal and vegetable kingdoms. As many more individuals of each species are born than can possibly survive; and as, consequently, there is a frequently recurring struggle for existence, it follows that any being, if it vary however slightly in any manner profitable to itself, under the complex and sometimes varying conditions of life, will have a better chance of surviving, and thus be *naturally selected*. From the strong principle of inheritance, any selected variety will tend to propagate its new and modified form. . . .

Although much remains obscure, and will long remain obscure, I can entertain no doubt, after the most deliberate study and dispassionate judgment of which I am capable, that the view which most naturalists until recently entertained, and which I formerly entertained—namely, that each species has been independently created—is erroneous. I am fully convinced that species are not immutable; but that those belonging to what are called the same genera are lineal descendants of some other and generally extinct species, in the same manner as the acknowledged varieties of any one species are the descendants of that species. Furthermore, I am convinced that Natural Selection has been the most important, but not the exclusive, means of modification.

RECAPITULATION AND CONCLUSION

As this whole volume is one long argument, it may be convenient to the reader to have the leading facts and inferences briefly recapitulated.

That many and serious objections may be advanced against the theory of descent with modification through variation and natural selection, I do not deny. I have endeavoured to give to them their full force. Nothing at first can appear more difficult to believe than that the more complex organs and instincts have been perfected, not by means superior to, though analogous with, human reason, but by the accumulation of innumerable slight variations, each good for the individual possessor. Nevertheless, this difficulty, though appearing to our imagination insuperably great, cannot be considered real if we admit the following propositions, namely, that all parts of the organisation and instincts offer, at least, individual differences—that there is a struggle for existence leading to the preservation of profitable deviations of structure or instinct —and, lastly, that gradations in the state of perfection of each organ may have existed, each good of its kind. The truth of these propositions cannot, I think, be disputed.

It is, no doubt, extremely difficult even to conjecture by what gradations many structures have been perfected, more especially amongst broken and failing groups of organic beings, which have suffered much extinction, but we see so many strange gradations in nature, that we ought to be extremely cautious in saying that any organ or instinct, or any whole structure, could not have arrived at its present state by many graduated steps. There are, it must be admitted, cases of special difficulty opposed to the theory of natural selection; and one of the most curious of these is the existence in the same community of two or three defined castes of workers or sterile female ants; but I have attempted to show how these difficulties can be mastered. . . .

The struggle for existence inevitably follows from the high geometrical ratio of increase which is common to all organic beings. This high rate of increase is proved by calculation,—by the rapid increase of many animals and plants during a succession of peculiar seasons, and when naturalised in new countries. More individuals are born than can possibly survive. A grain in the balance may determine which individuals shall live and which shall die,—which variety or species shall increase in number, and which shall decrease, or finally become extinct. As the individuals of the same species come in all respects into the closest competition with each other, the struggle will generally be most severe between them; it will be almost equally severe between the varieties of the same species, and next in severity between the species of the same genus. On the other hand the struggle will often be severe between beings remote in the scale of nature. The slightest advantage in certain individuals, at any age or during any season, over those with which they come into competition, or better adaptation in however slight a degree to the surrounding physical conditions, will, in the long run, turn the balance.

With animals having separated sexes, there will be in most cases a struggle between the males for the possession of the females. The most vigorous males, or those which have most successfully struggled with their conditions of life, will generally leave most progeny. But success will often depend on the males having special weapons, or means of defence, or charms; and a slight advantage will lead to victory. . . .

We can to a certain extent understand how it is that there is so much beauty throughout nature; for this may be largely attributed to the agency of selection. That beauty, according to our sense of it, is not universal, must be admitted by every one who will look at some venomous snakes, at some fishes, and at certain hideous bats with a distorted resemblance to the human face. Sexual selection has given the most brilliant colours, elegant patterns, and other ornaments to the males, and sometimes to both sexes of many birds, butterflies, and other animals. With birds it has often rendered the voice of the male musical to the female, as well as to our ears. Flowers and fruit have been rendered conspicuous by brilliant colours in contrast with the green foliage, in order that the flowers may be readily seen, visited and fertilised by insects, and the seeds disseminated by birds. How it comes that certain colours, sounds, and forms should give pleasure to man and the lower animals,—that is, how the sense of beauty in its simplest form was first acquired,—we do not know any more than how certain odours and flavours were first rendered agreeable.

As natural selection acts by competition, it adapts and improves the inhabitants of each country only in relation to their co-inhabitants; so that we need feel no surprise at the species of any one country, although on the ordinary view supposed to have been created and specially adapted for that country, being beaten and supplanted by the naturalised productions from another land. Nor ought we to marvel if all the contrivances in nature be not, as far as we can judge, absolutely perfect, as in the case even of the human eye; or if some of them be abhorrent to our ideas of fitness. We need not marvel at the sting of the bee, when used against an enemy, causing the bee's own death; at drones being produced in such great numbers for one single act, and being then slaughtered by their ster-

ile sisters; at the astonishing waste of pollen by our fir-trees; at the instinctive hatred of the queen-bee for her own fertile daughters; at the ichneu-
monidae feeding within the living bodies of caterpillars; or at other such
cases. The wonder indeed is, on the theory of natural selection, that more
cases of the want of absolute perfection have not been detected.

The complex and little known laws governing the production of varieties are the same, as far as we can judge, with the laws which have governed the production of distinct species. In both cases physical conditions
seem to have produced some direct and definite effect, but how much we
cannot say. Thus, when varieties enter any new station, they occasionally
assume some of the characters proper to the species of that station. With
both varieties and species, use and disuse seem to have produced a considerable effect; for it is impossible to resist this conclusion when we look,
for instance, at the logger-headed duck, which has wings incapable of
flight, in nearly the same condition as in the domestic duck; or when we
look at the burrowing tucu-tucu, which is occasionally blind, and then at
certain moles, which are habitually blind and have their eyes covered
with skin; or when we look at the blind animals inhabiting the dark caves
of America and Europe. With varieties and species, correlated variation
seems to have played an important part, so that when one part has been
modified other parts have been necessarily modified. With both varieties
and species, reversions to long-lost characters occasionally occur. How inexplicable on the theory of creation is the occasional appearance of stripes
on the shoulders and legs of the several species of the horse-genus and of
their hybrids! How simply is this fact explained if we believe that these
species are all descended from a striped progenitor, in the same manner
as the several domestic breeds of the pigeon are descended from the blue
and barred rock-pigeon! . . .

Disuse, aided sometimes by natural selection, will often have reduced
organs when rendered useless under changed habits or conditions of life;
and we can understand on this view the meaning of rudimentary organs.
But disuse and selection will generally act on each creature, when it has
come to maturity and has to play its full part in the struggle for existence,
and will thus have little power on an organ during early life; hence the
organ will not be reduced or rendered rudimentary at this early age. The
calf, for instance, has inherited teeth, which never cut through the gums
of the upper jaw, from an early progenitor having well-developed teeth;
and we may believe, that the teeth in the mature animal were formerly
reduced by disuse, owing to the tongue and palate, or lips, having become
excellently fitted through natural selection to browse without their aid;
whereas in the calf, the teeth have been left unaffected, and on the principle of inheritance at corresponding ages have been inherited from a re-

mote period to the present day. On the view of each organism with all its separate parts having been specially created, how utterly inexplicable is it that organs bearing the plain stamp of inutility, such as the teeth in the embryonic calf or the shrivelled wings under the soldered wing-covers of many beetles, should so frequently occur. Nature may be said to have taken pains to reveal her scheme of modification, by means of rudimentary organs, of embryological and homologous structures, but we are too blind to understand her meaning.

I have now recapitulated the facts and considerations which have thoroughly convinced me that species have been modified, during a long course of descent. This has been effected chiefly through the natural selection of numerous successive, slight, favourable variations; aided in an important manner by the inherited effects of the use and disuse of parts; and in an unimportant manner, that is in relation to adaptive structures, whether past or present, by the direct action of external conditions, and by variations which seem to us in our ignorance to arise spontaneously. It appears that I formerly underrated the frequency and value of these latter forms of variation, as leading to permanent modifications of structure independently of natural selection. But as my conclusions have lately been much misrepresented, and it has been stated that I attribute the modification of species exclusively to natural selection, I may be permitted to remark that in the first edition of this work, and subsequently, I placed in a most conspicuous position—namely, at the close of the Introduction—the following words: "I am convinced that natural selection has been the main but not the exclusive means of modification." This has been of no avail. Great is the power of steady misrepresentation; but the history of science shows that fortunately this power does not long endure.

It can hardly be supposed that a false theory would explain, in so satisfactory a manner as does the theory of natural selection, the several large classes of facts above specified. It has recently been objected that this is an unsafe method of arguing; but it is a method used in judging of the common events of life, and has often been used by the greatest natural philosophers. The undulatory theory of light has thus been arrived at; and the belief in the revolution of the earth on its own axis was until lately supported by hardly any direct evidence. It is no valid objection that science as yet throws no light on the far higher problem of the essence or origin of life. Who can explain what is the essence of the attraction of gravity? No one now objects to following out the results consequent on this unknown element of attraction; notwithstanding that Leibnitz formerly accused Newton of introducing "occult qualities and miracles into philosophy." . . .

Why, it may be asked, until recently did nearly all the most eminent living naturalists and geologists disbelieve in the mutability of species? It cannot be asserted that organic beings in a state of nature are subject to no variation; it cannot be proved that the amount of variation in the course of long ages is a limited quality; no clear distinction has been, or can be, drawn between species and well-marked varieties. It cannot be maintained that species when intercrossed are invariably sterile, and varieties invariably fertile; or that sterility is a special endowment and sign of creation. The belief that species were immutable productions was almost unavoidable as long as the history of the world was thought to be of short duration; and now that we have acquired some idea of the lapse of time, we are too apt to assume, without proof, that the geological record is so perfect that it would have afforded us plain evidence of the mutation of species, if they had undergone mutation.

But the chief cause of our natural unwillingness to admit that one species has given birth to clear and distinct species, is that we are always slow in admitting great changes of which we do not see the steps. The difficulty is the same as that felt by so many geologists, when Lyell first insisted that long lines of inland cliffs had been formed, the great valleys excavated, by the agencies which we see still at work. The mind cannot possibly grasp the full meaning of the term of even a million years; it cannot add up and perceive the full effects of many slight variations, accumulated during an almost infinite number of generations.

Although I am fully convinced of the truth of the views given in this volume under the form of an abstract, I by no means expect to convince experienced naturalists whose minds are stocked with a multitude of facts all viewed, during a long course of years, from a point of view directly opposite to mine. It is so easy to hide our ignorance under such expressions as the "plan of creation," "unity of design," &c., and to think that we give an explanation when we only re-state a fact. Any one whose disposition leads him to attach more weight to unexplained difficulties than to the explanation of a certain number of facts will certainly reject the theory. A few naturalists, endowed with much flexibility of mind, and who have already begun to doubt the immutability of species, may be influenced by this volume; but I look with confidence to the future,—to young and rising naturalists, who will be able to view both sides of the question with impartiality. Whoever is led to believe that species are mutable will do good service by conscientiously expressing his conviction; for thus only can the load of prejudice by which this subject is overwhelmed be removed.

The Evolution of Evolution *

by Isaac Asimov (1960)

Darwin's book and his theory of evolution by natural selection broke on the world (and not just the scientific world) like a thunderbolt. It set up a controversy that has not entirely died even now, a hundred years later.

Much of the furore over Darwin's theory arose over its application to man. Lyell, whose geological views had so influenced Darwin, now returned the compliment. In a book entitled *The Antiquity of Man*, published in 1863, he came out strongly in favor of Darwin's theory and discussed the hundreds of thousands of years during which man (or manlike creatures) must have existed on the earth. He used as his evidence stone tools found in ancient strata.

Darwin himself published in 1871 a second book called *The Descent of Man*, in which he discussed evidence showing man to have descended from subhuman forms of life. For one thing, man contains many vestigial organs. There are traces of points on the incurved flaps of the outer ear, dating back to a time when the ear was upright and pointed, and there are tiny, useless muscles still present that are designed to move those ears (some people can even today use them to wiggle their ears). There are four bones at the bottom of the spine which are the remnants of a tail, a sign that man's ancestors did have tails. In short, man and the manlike apes had a common ancestor several millions of years ago, and the entire ape and monkey tribe (the order of *primates,* from a Latin word meaning "first") had a common ancestor even longer ago.

(The antievolutionists seized upon this to declare over and over again that Darwin claimed that man had descended from monkeys, which was, of course, a distortion. No living monkey and no living species, for that matter, are ancestral to man, nor were any claimed to be by Darwin or any other reputable evolutionist.)

Scientists other than Lyell came to Darwin's side early in the game. In Germany, the biologist Ernst Heinrich Haeckel was a powerful proponent of evolution, and in the United States the botanist Asa Gray (of Harvard) carried the ball. In France, progress was slower because of the influence

---

* From Isaac Asimov, *The Wellsprings of Life.* Copyright © 1960 by Isaac Asimov. Reprinted by permission of Abelard-Schuman, Limited, and the author. Also from *The New Intelligent Man's Guide to Science.* Copyright © 1965 by Basic Books, Inc. Reprinted by permission of Basic Books, Inc., and the author.

of Cuvier's memory (Cuvier himself had died in 1832), but even there its victory could not be long delayed. By 1880, the scientific fraternity had been mostly won over, and the doctrine of the immutability of species was just about dead.

However, the battle continued, for this was one theory in which ordinary people, who were not scientists, were also deeply involved. If Darwinism won out, what would be left of the biblical story of the Creation? The book of Genesis could be interpreted allegorically, perhaps, and made to fit Darwin, but this didn't satisfy many people who would not compromise but who insisted on a literal interpretation of every word in the Bible. (These were called "fundamentalists.") Controversy, therefore, was bitter.

The man who did more than anyone else to win the battle for evolution among educated nonscientists was the English biologist Thomas Henry Huxley. Throughout the 1850's he had been a firm believer in the immutability of species and had even argued with Darwin about it. However, when *The Origin of Species* appeared, Huxley found himself swept away by it and was at once converted from an opponent to an ardent advocate. In 1863, he, like Lyell, wrote a book on the evolution of man. It was entitled *Man's Place in Nature*. Thereafter, his writings and his lectures were read not only by scientists but also by laymen, and his views were expressed so forcefully and well that more and more were won over by him.

The notion of evolution had to contend not only with the clamor of opponents, but also with the distortions of certain proponents. For instance, one of the leading evolutionists in Darwin's time was Herbert Spencer, an English philosopher. He it was, in fact, who made the word "evolution" popular. (Darwin himself rarely used the word.)

Spencer was primarily interested in the development of human societies and was the founder of the modern science of *sociology*. When Darwin's book came out, he saw at once that the notions of evolution could be applied to sociology. If species could be formed by the forces of natural selection, why not human societies as well? Thus, he founded *social evolution*.

But Spencer invented a phrase in the process that caught on at once. That was "the survival of the fittest." Others seized upon it to justify all that was evil and unpleasant in the society of the times.

Was there unrestrained competition in business with no holds barred? Why, that only led to the "survival of the fittest." Was there unemployment? The "less fit" would starve to death and the laborers who survived would be a stronger breed. Unemployment was good for them. In the same way, war weeded out the "unfit" and allowed better and stronger nations to survive. And, of course, there were also those who used evolutionary reasoning to show that one particular group of mankind (invariably the group to which the reasoner belonged) was superior to others.

The cruelties of unrestrained competition, of militarism and of racism all existed before Spencer. They were not invented by evolution. However, the last half of the nineteenth century saw people begin to justify these ancient evils by the use of "modern science." This distortion of Darwinism made the whole notion of evolution seem unpleasant to people and strengthened the hand of those who claimed evolution to be immoral and sinful.

Because there are still many people who will use the notion of "the survival of the fittest" to justify ways of life that seem to most of us to be evil, I would like to spend some time discussing the matter.

In the first place, the phrase "the survival of the fittest" is not an illuminating one. It implies that those who survive are the "fittest," but what is meant by "fittest"? Why, those are "fittest" who survive. This is an argument in a circle.

Actually, what does one mean by the "fittest"? Suppose you were asked the question: "Which is the 'fitter,' a man or an oyster?"

Obviously a man is a much more highly organized creature, with a more efficient set of body machinery, and with tremendously greater versatility and potentialities. Who would deny that man was fitter?

But if every bit of land were suddenly placed under shallow water, then men would die and oysters would not. If mere survival is the measure of the "fittest," then under that new condition oysters would prove "fitter" than man.

In other words, "fitness" is a relative term, and has no meaning unless you mention the environmental niche you are considering. A great many species have become extinct, yet have left behind close relatives who still survive. Not only does man exist, but also rabbits, sharks, earthworms and jellyfish. The most primitive creatures who ever existed are still represented today and are even flourishing. In fact, if mere survival is the criterion of "fitness," then the king crab is many times "fitter" than man, for it has existed as a species much longer.

Of course, each species exists within its own niche, and within that niche it would have competed with other (now extinct) species. It would have showed itself "fitter" by surviving.

If we are to try to apply the notion of "fitness" to the development of human society, let us consider not just man, but also his environmental niche.

If two men and a woman were stranded on a desert island, the environmental niche would be the desert island and all it contained. If one man killed the other (by superior force or by superior guile), he would inherit the woman, so to speak, and possibly leave descendants, while the victim would not. The murderer would be "fitter" by Spencer's test, for he had survived; and by Darwin's test, too, if he left descendants.

If the same two men and a woman were in New York City, however, their environmental niche would be not only the buildings of Manhattan and the air above it and rocks below it. Their environmental niche would include all the machinery of a human society by which they would be affected as profoundly as by the inanimate environment.

This society reacts upon the individual. For instance, murderers, as a class, are a danger to society and not just to their individual victims. As long as some individuals feel free to kill, all other individuals must feel unsafe. Therefore, in all societies, even in the most primitive, murderers have, in one way or another, been hunted down and killed.

The result is that, when a human society forms part of the environmental niche, a murderer might be "fitter" than his victim, but he will be "less fit" than nonmurderers as a group, by the Spencerian test of survival.

One can also argue, in similar fashion, that within a society the dishonest businessman is "less fit" than the ethical one; that war is "less fit" than peace; that slavery is "less fit" than brotherhood.

To reach the same conclusion by another path, I might point out that "competition" must be understood in a broader sense than that of a fist fight. Competition among the individuals of a species may be a competition of comparative cooperations. One factor in the survival of a species has often been its ability to live in packs; to have one individual of a pack act as watchman while the rest graze; to be in the habit of defending themselves as a pack against an attacking enemy when individually no defense might be possible. (Predators also can hunt in packs and hunt more successfully than if each animal went off on its own.)

Any improvement in the "pack habit" increases the chances of that species' survival. Moreover, if there is variation among the species so that some groups have more of the "pack habit" than others, it is those with the better "pack habit" that will survive.

The same is true within the human species. History is full of examples of peoples who, unable to cooperate among themselves, fell under the onslaughts of others, who were individually perhaps less admirable and advanced, but who had the virtue of cooperative action. The fate of the ancient Greeks is the most tragic case in point. By temporarily uniting, they beat off the Persians; by being unable to unite later, they fell to the Macedonians.

Any society which indulges too extensively in the Spencerian notions of "the survival of the fittest" will break down through internal dissension and fall prey to other societies which are less Spencerian. We have now reached the point where it seems that unrestrained competition among nations may do us all in as a species and that some sort of non-Spencerian cooperation is essential or we will have finally proved our "unfitness" by not surviving.

If Darwin's theory managed to survive opposition and distortion, that did not mean it was flawless. As a matter of fact, it had a serious weak point, which Darwin himself recognized.

As mentioned before, Darwin had no real explanation for variation in physical characteristics. It happened, that was certain, but why?

Worse still, Darwin thought that variations consisted of infinitely small differences and that when two parents varied in some respect, the young-sters were intermediate in that same respect. But if that were so, then when mating occurred generation after generation, should not the varia-tions average out? Should not intermediacy become universal?

To put it as simply as possible, how did variation come about in the first place, and what made variation persist long enough for natural selec-tion to get in its work?

Some of the Darwinians felt the lack and tried to supply reasons. Sev-eral thought that variations did not proceed by infinitesimal steps. They suggested that evolution proceeded by jumps, so to speak. Every once in a while there might be a large variation, one too large to average out be-fore natural selection had established it.

Although this seems a daring speculation, there was actually consider-able evidence in favor of exactly this happening. Over and over again herdsmen and farmers noticed the birth of strange varieties among their livestock and crops. These anomalies were viewed with suspicion and mis-trust and were generally considered warnings of divine displeasure. (In fact, the strange varieties are often called "monsters," from a Latin word meaning "to warn." A less emotional term is "sport.")

It was not until comparatively modern times that superstition and gen-eral uneasiness gave way to the thought that sports could be made useful. In 1791, a male lamb belonging to the flock of Seth Wright, a Massachu-setts farmer, was born with unusually short legs. When it grew to matu-rity, it was bred to ewes, and the lambs that resulted were likewise short-legged. Eventually, a whole herd of short-legged sheep resulted. The advantage of the short legs was simply this: the sheep could not jump over the low fences surrounding the pasture and were, therefore, less trou-blesome to keep. This early breed appeared again, this time in Norway, and the short-legged breed was re-established.

Since 1791 many other useful sports have been discovered and bred. It seems certain, furthermore, that long before 1791, even back in prehistoric times, sports must have been preserved and bred, and this accounts for the numerous breeds of dogs and other domestic animals that have existed through the centuries.

Yet all this material in connection with sports took considerable time to penetrate science itself. After all, scientists knew little about the mechanics of herding animals and cultivating plants, while herdsmen and farmers, for their part, did not write papers describing their discoveries.

So it was not until 1884 that a book was written which systematically presented evolution as occurring by jumps. This book was by a Swiss botanist named Karl Wilhelm Von Nägeli. Even the existence of jumps did not seem enough to account for all the facts. Why did the jumps not average out? So Nägeli went on to suggest that there was some drive within the species that kept it varying in the same direction. Once a species started jumping, say, in the direction of increased size, it continued jumping in that direction faster than ordinary mating could level out the size again. In this way, the species would grow larger and larger, as the primitive horses of past ages grew from the size of a dog to the giant animals of today. A species might even grow larger than was desirable (as if it were going too fast to stop) and might become extinct through "overlargeness." This kind of "biological inertia" was called *orthogenesis.*

Though Nägeli's theory of orthogenesis was not accepted, his notion of discontinuous evolution persisted. A Dutch botanist, Hugo De Vries, set out to find as much evidence as he could for the actual occurrence of sudden large variations in species.

In 1886, De Vries came across a wild colony of the American evening primrose in which some of the individual plants were quite different from the rest. If they were crossed, they produced a new generation like themselves and not like the ordinary primrose. With continued investigations, he found new sudden changes. He called these *mutations,* from a Latin word meaning "to change."

His experiments in crossing plants also taught him a few things about the manner in which physical characteristics are inherited. By 1900, he had enough experimental evidence to feel himself ready to publish a complete theory on inheritance.

Although De Vries was not aware of it, two other botanists were making much the same observations as he was, and were getting ready to publish essentially the same theory. These were the Austrian, Erich Tschermak, and the German, Carl Erich Correns.

All three, working independently, went through previously published material on the subject, once they had worked out their theories. And all three found they should have done this first, because all three, looking through an obscure journal, *The Proceedings of the Natural History Society of Brünn,* found a paper by someone they had never heard of, someone with no scientific reputation, someone who was merely an amateur gardener.

That paper, however, that piece of work by an amateur gardener (an Augustinian monk named Gregor Mendel), detailed in full the theory which De Vries, Tschermak and Correns had each worked out independently. What is more, that original paper had appeared in 1866, thirty-four years earlier.

Each of the three scientists was true to the ideals of science. Each called the attention of the world to Mendel's paper. Each gave Mendel all the credit, and to this day the rules that govern the inheritance of physical characteristics are referred to as Mendel's Laws.

Here is the story of Mendel, as rediscovered by the three botanists.

During the 1860's, Mendel taught natural history in the monastery school at Brünn (now Brno, Czechoslovakia) and also tended the garden. He amused himself by carefully crossing plants and observing the exact results.

He worked with pea plants which existed in his garden in a number of sharply marked-off varieties, which, however, were all the same species since they could all be "crossed."

For instance, there was a variety of pea plant with a purplish-red flower, and a variety with a white flower. Peas of the red variety, crossed among themselves, produced seeds that grew into plants with red flowers only. Those of the white variety, crossed among themselves, produced seed that grew into plants with white flowers only.

But what if the red variety were crossed with the white variety? Mendel did this and observed that all the seeds developed in this cross grew into plants with red flowers. Not one white flower in the bunch! The white flower characteristic had disappeared!

Or had it? Mendel next crossed the peas of this new generation of red flowers among themselves, waited for seeds and planted them. Then, in the third generation, some of the seeds grew into plants with white flowers, pure white. To be sure, these were in the minority. Of all the third-generation seeds, almost exactly one-quarter gave plants with white flowers. The rest grew into plants with red flowers.

Next he tried another generation. Suppose the whites of the third generation were fertilized among themselves or, better yet, were fertilized with their own pollen (self-pollination)? Only white-flowered pea plants resulted.

Suppose the third-generation red-flowered plants were self-pollinated?

Here two different results showed up. There were some reds that produced only red-flowered offspring. There were others that produced both red-flowered and white-flowered ones in the ratio of three to one.

In other words, white-flowered pea plants always "bred true." Red-flowered pea plants sometimes bred true, and sometimes did not.

To explain the results of his experiments, Mendel devised a theory. He supposed that each plant contained two factors which controlled a particular characteristic, such as the color of its flowers. Today we call such factors *genes,* from a Greek word meaning "to produce."

Thus, a plant with red flowers might have two genes, each tending to produce red flowers. If we symbolize such a gene as *R* (for "red"), then the red-flowered plant containing two such genes, could be called an *RR* plant. Similarly, a white-flowered plant would contain two genes tending to produce white flowers and could be called a *WW* plant.

Mendel next supposed that each plant transmitted only one gene apiece to the ovule or to the pollen grain. The combination of the two in the process of *pollination* gave the offspring a total of two genes again.

The *RR* plant could only transmit an *R* gene to either ovule or pollen, so that when an *RR* plant is self-pollinated or cross-pollinated with another *RR* plant, only *RR* offspring result. Similarly, *WW* plants can produce only *WW* offspring.

If an *RR* plant is crossed with a *WW* plant, however, the pollen of the first carries only an *R* gene, the ovule of the latter a *W* gene and the offspring have one of each. The second generation of such a cross consists exclusively of *RW* plants. (If it is the pollen of the *WW* plant that pollinates the ovule of an *RR* plant, the result is the same. The combination of *W* and *R* still gives an *RW* plant.)

But all such *RW* plants produce red flowers only. Apparently, the presence of the *R* gene drowns out the presence of the *W* gene.

Nowadays, we call two genes which both govern the same characteristic in different ways *alleles.* Thus the *R* gene and the *W* gene are alleles because both govern flower color. Since the *R* gene produces its effect even in the presence of a *W* gene, the *R* gene is said to be the *dominant* allele, while the *W* gene is *recessive.*

However, what happens when an *RW* plant is self-pollinated? It can pass on only one gene to its pollen grains and it shows no bias in favor of either. Half the pollen grains are *R* and half are *W*. In the same way, half the ovules are *R* and half are *W*. In the pollination process, the resultant offspring can be produced in the following manner: (1) by the pollination of an *R* ovule by an *R* pollen grain to produce an *RR* individual; (2) by the pollination of an *R* ovule by a *W* pollen grain to produce an *RW* individual; (3) by the pollination of a *W* ovule by an *R* pollen grain to produce an *RW* individual (*RW* and *WR* prove to be the same); (4) by the pollination of a *W* ovule by a *W* pollen grain to produce a *WW* individual.

All four alternatives are equally probable and happen in roughly equal numbers. Three-quarters of the plants (*RR*, *RW* and *RW* by alternatives 1, 2 and 3) have red flowers. The remaining quarter (which were the *WW* plants produced by alternative 4) have white flowers.

Suppose, next, that a hybrid red-flowered plant, *RW*, is crossed with a white-flowered plant, *WW*. The hybrid plant would produce *R* pollen grains and *W* pollen grains in equal proportion and *R* ovules and *W* ovules in equal proportion. The white-flowered plants would produce only *W* pollen grains and *W* ovules. The only two possibilities when crossed would be that (1) a *W* ovule combined with a *W* pollen grain or (2) a *W* ovule combined with an *R* pollen grain. This means that if the pollen of the *RW* plant were used, the offspring would be half *WW* and half *RW*.

If the pollen of the *WW* plant were used, then there would be two possibilities again: either (1) a *W* ovule combined with a *W* pollen grain, or (2) an *R* ovule combined with a *W* pollen grain. Here, too, the offspring would be half *WW* and half *RW*.

In either case, Mendel's theory would predict that half the plants that resulted from the cross would produce red flowers and half would produce white flowers. When tried experimentally, this turned out to be so.

Mendel did not test flower color only in crossing his pea plants. Actually, he chose seven different characteristics that varied from plant to plant. There were plants with yellow seeds and others with green seeds; plants with round seeds and others with wrinkled seeds; plants with tall stems and plants with short ones and so on.

In each case, when he crossed one variety with another, one turned out to be dominant. Round seeds were dominant over wrinkled seeds, yellow seeds were dominant over green seeds, and so on. The hybrids all produced a third generation in which the recessive form showed up again in a quarter of the total.

Furthermore, each of the seven characteristics was inherited independently. For instance, a particular plant might inherit red flowers and long stems, red flowers and short stems, white flowers and long stems, or white flowers and short stems. And any of these combinations might go along with yellow, wrinkled seeds or green, wrinkled seeds; yellow, smooth seeds or green, smooth seeds. All possible combinations of the seven characteristics could develop, so that by proper crossing one could end with 128 different varieties of pea plants.

Mendel's results explained some of the difficulties that Darwin had run into. For one thing, variations among offspring did not run along an entire gamut of infinitely small steps. There were large, discrete differences. An individual plant might have red flowers or white flowers; there need be no in between.

Secondly, there was no "blending" of inheritance. Crossing red-flowered plants and white-flowered plants in these experiments resulted in red-flowered plants and not varieties of pink. What is more, even when a recessive gene seemed lost and a particular characteristic seemed to have disappeared, it was still there and would appear, unharmed and unchanged, in a later generation. Let two *W* genes come together, no matter how many generations had elapsed during which they were in the constant, overpowering presence of *R* genes; let them but come together and a white-flowered plant is the result.

The objections of Darwin's opponents that, with random mating, variations would average out and yield one long, dull mediocrity were, therefore, not valid.

Mendel wrote up his observations and sent them to Nägeli. Nägeli was not impressed, because he thought Mendel was just blindly counting plants instead of working up some fundamental new scheme like his own orthogenesis.

This was a bad break, for Mendel's theory was of fundamental importance, while Nägeli's theory was worthless. However, it was Nägeli who had the reputation and not Mendel, so the poor monk published his paper in the obscure journal and did not try to follow it up. His work remained unknown, and he himself unhonored.

Darwin died in 1882, never knowing that the greatest weakness in his theory had been patched up. Mendel died in 1884, never suspecting that he was destined for fame. Nägeli died in 1891, never dreaming what a terrible mistake he had made.

✾   ✾   ✾

CHROMOSOMES AND GENES

By the time Mendel was rediscovered a considerable amount of scientific information had also been accumulating about the process of reproduction. . . . It had become clear that every living organism, even the largest, began life as a single cell. One of the earliest microscopists, Johann Ham, an assistant of Leeuwenhoek, had discovered in seminal fluid tiny bodies that were later named "spermatozoa" (from Greek words meaning "animal seed"). Much later, in 1827, the German physiologist Karl Ernst von Baer had identified the ovum, or egg cell, of mammals. Biologists came to realize that the union of an egg and a spermatozoon formed a fertilized ovum from which the animal eventually developed by repeated divisions and redivisions.

The big question was: How did cells divide? The answer lay in a small globule of comparatively dense material within the cell first reported by Robert Brown in 1831 and named the "nucleus."

If a one-celled organism was divided into two parts, one of which contained the intact cell nucleus, the part containing the cell nucleus was able to grow and divide, but the other part could not.

Unfortunately, further study of the cell nucleus and the mechanism of division was thwarted for a long time by the fact that the cell was more or less transparent, so that its substructures could not be seen. Then the situation was improved by the discovery that certain dyes would stain parts of the cell and not others. A dye called "hematoxylin" (obtained from logwood) stained the cell nucleus black and brought it out prominently against the background of the cell. After Perkin and other chemists began to produce synthetic dyes, biologists found themselves with a variety of dyes from which to choose.

In 1879, the German biologist Walther Flemming found that with certain red dyes he could stain a particular material in the cell nucleus which was distributed through it as small granules. He called this material "chromatin" (from the Greek word for "color"). By examining this material, Flemming was able to follow some of the changes in the process of cell division. To be sure, the stain killed the cell, but in a slice of tissue he would catch various cells at different stages of cell division. They served as still pictures, which he put together to form a kind of "moving picture" of the progress of cell division.

In 1882, Flemming published an important book in which he described the process in detail. At the start of cell division, the chromatin material gathered itself together in the form of threads. The thin membrane enclosing the cell nucleus seemed to dissolve, and at the same time a tiny object just outside it divided in two. Flemming called this object the "aster," from a Greek word for "star," because radiating threads gave it a starlike appearance. After dividing, the two parts of the aster traveled to opposite sides of the cell. Its trailing threads apparently entangled the threads of chromatin, which had meanwhile lined up in the center of the cell, and the aster pulled half the chromatin threads to one side of the cell, half to the other. As a result, the cell pinched in at the middle and split into two cells. A cell nucleus developed in each, and the chromatin material that the nuclear membrane enclosed broke up into granules again.

Flemming called the process of cell division "mitosis," from the Greek word for "thread," because of the prominent part played in it by the chromatin threads. In 1888, the German anatomist Wilhelm von Waldeyer gave the chromatin thread the name "chromosome" (from the Greek for "colored body"), and that name has stuck. It should be mentioned, though, that chromosomes, despite their name, are colorless in their unstained natural state.

Continued observation of stained cells showed that the cells of each species of plant or animal had a fixed and characteristic number of chromosomes. Before a cell divides in two during mitosis, the number of chromosomes is doubled, so that each of the two daughter cells after the division has the same number as the original mother cell.

The Belgian embryologist Eduard van Beneden discovered in 1885 that the chromosomes did *not* double in number when egg and sperm cells were being formed. Consequently each egg and each sperm cell had only half the number of chromosomes that ordinary cells of the organism possessed. (The cell division that produces sperm and egg cells therefore is called "meiosis," from a Greek word meaning "to make less.") When an egg and a sperm cell combined, however, the combination (the fertilized ovum) had a complete set of chromosomes, half contributed by the mother through the egg cell and half by the father through the sperm cell. This complete set was passed on by ordinary mitosis to all the cells that made up the body of the organism developing from the fertilized egg.[1]

A number of biologists immediately saw a connection between Mendel's genes and the chromosomes that could be seen under the microscope. The first to draw a parallel was an American cytologist named Walter S. Sutton, in 1904. He pointed out that chromosomes, like genes, came in pairs, one of which was inherited from the father and one from the mother. The only trouble with this analogy was that the number of chromosomes in the cells of any organism was far smaller than the number of inherited characteristics. Man, for instance, has only 23 pairs of chromosomes. (For many years biologists thought that the human cell had 24 pairs, but careful inspection with a new type of microscope in 1957 showed that there were 23. Anyone who has ever looked at a photograph of chromosomes in a dividing human cell can understand the miscount: they look like a tangle of short spaghetti strands or of stubby worms.) Since the number of characteristics inherited by human beings certainly runs into many thousands, biologists had to conclude that chromosomes were not genes. Each must be a collection of genes.

In short order, biologists discovered an excellent tool for studying specific genes. It was not a physical instrument but a new kind of laboratory animal. In 1906, the Columbia University zoologist Thomas Hunt Morgan

1. For further information on genetics see *The Mystery of Matter,* edited by Louise B. Young, Oxford University Press, 1965, Parts 7 & 9. Ed.

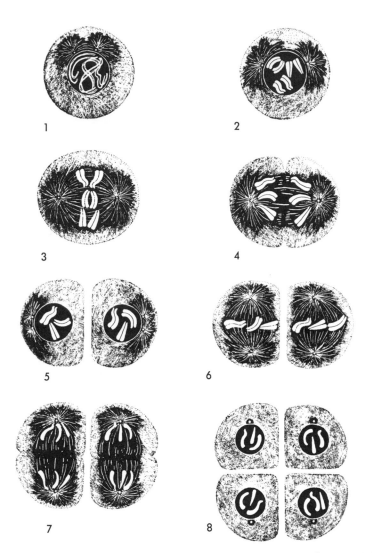

Figure 2–1. Meiosis: Since gametes from both parents join in the act of fertilization, the number of chromosomes would double every generation unless a compensatory halving occurred at some point in the maturation of the gametes. The process of cell division allowing for this is meiosis. In the numbered sequence above, which begins with paired chromosomes, a first cell division at 4–5 is followed by a second one at 7–8. Since no pairing off or longitudinal division of the chromosomes has occurred in the meantime, the end result of the original duplication and two subsequent halvings is one halving. As a result, each parent will contribute a single ("haploid") set of chromosomes through its gamete to give a complete ("diploid") set to the resulting "zygote." (After Young.) (From The Harper Encyclopedia of Science, *James R. Newman*, Ed., Harper & Row, Publishers, p. 739. *Figure added by editor.*)

conceived the idea of using fruit flies (*Drosophila melanogaster*) for research in genetics. (The term "genetics" was invented about this time by the British biologist William Bateson.)

Fruit flies had considerable advantages over pea plants (or any ordinary laboratory animal) for studying the inheritance of genes. They bred quickly and prolifically, could easily be raised by the hundreds on very little food, had scores of inheritable characteristics which could be observed readily, and had a comparatively simple chromosomal setup—only four pairs of chromosomes per cell.

With the fruit fly, Morgan and his co-workers discovered an important fact about the mechanism of inheritance of sex. They found that the female fruit fly has four perfectly matched pairs of chromosomes so that all the egg cells, receiving one of each pair, are identical so far as chromosome make-up is concerned. However, in the male one of each of the four pairs consists of a normal chromosome, called the "X chromosome," and a stunted one, which was named the "Y chromosome." Therefore when sperm cells are formed, half have an X chromosome and half a Y chromosome. When a sperm cell with the X chromosome fertilizes an egg cell, the fertilized egg, with four matched pairs, naturally becomes a female. On the other hand, a sperm cell with a Y chromosome produces a male. Since both alternatives are equally probable, the number of males and females in the typical species of living things is roughly equal. (In some creatures, notably various birds, it is the female that has a Y chromosome.)

This chromosomal difference explains why some disorders or mutations show up only in the male. If a defective gene occurs on one of a pair of X chromosomes, the other member of the pair is still likely to be normal and can salvage the situation. But in the male, a defect on the X chromosome paired with the Y chromosome generally cannot be compensated for, because the latter carries very few genes. Therefore the defect shows up.

Research on fruit flies showed that traits were not necessarily inherited independently, as Mendel had thought. It happened that the seven characteristics of pea plants that he had studied were governed by genes on separate chromosomes. Morgan found that where two genes governing two different characteristics were located on the same chromosome, those characteristics were generally inherited together (just as a passenger in the front seat of a car and one in the back seat travel together).

This genetic linkage is not, however, unchangeable. Just as a passenger can change cars, so a piece of one chromosome occasionally switches to another, swapping places with a piece from the other. Such "crossing over" may occur during the division of a cell. As a result, linked traits are separated and reshuffled in a new linkage. For instance, there is a variety

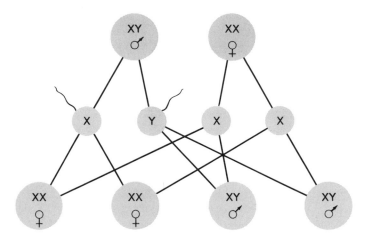

Figure 2–2. Combinations of X and Y chromosomes.

of fruit fly with scarlet eyes and curly wings. When it is mated with a white-eyed, miniature-winged fruit fly, the offspring will generally be either red-eyed and curly-winged or white-eyed and miniature-winged. But the mating may sometimes produce a white-eyed, curly-winged fly or a red-eyed, miniature-winged one as a result of crossing over. The new form will persist in succeeding generations unless another crossing over takes place.

Now picture a chromosome with a gene for red eyes at one end and a gene for curly wings at the other end. Let us say that in the middle of the chromosome's length there are two adjacent genes governing two other characteristics. Obviously, the probability of a break occurring at that particular point, separating these two genes, is smaller than the probability of a break coming at one of the many points along the length of the chromosome that would separate the genes at the opposite ends. By noting the frequency of separation of given pairs of linked characteristics by crossing over, Morgan and his co-workers were able to deduce the relative locations of the genes in question, and in this way they worked out chromosome "maps" of gene locations for the fruit fly. The location, so determined, is the "locus" of a gene.

From such chromosome maps and from a study of giant chromosomes many times the ordinary size, found in the salivary glands of the fruit fly, it has been established that the insect has a minimum of 10,000 genes in a chromosome pair. This means the individual gene must have a molecular weight of 60 million. Working from this, man's somewhat larger chromosomes may contain from 20,000 to 90,000 genes per chromosome pair, or up to 2 million altogether.

✻   ✻   ✻

It is apparent that the gene theory of inheritance is not as simple as it seemed from Mendel's experiments on the seven variations in the pea plant, and the science of inheritance (or *genetics*, as it is called) has become a complicated one, indeed.

For instance, genes cannot always be clearly differentiated into dominant and recessive. As an example, consider the human blood types. The most common types, those which are important in transfusions, are under the control of three alleles, which can be symbolized as an *A* gene, a *B* gene and an *O* gene.

A human being with a pair of *A* genes (an *AA* individual) belongs to blood type A. Similarly, a *BB* individual would belong to blood type B, and an *OO* individual to blood type O. If an *AA* mates with an *AA* (or a *BB* with a *BB*, or an *OO* with an *OO*), all the children are *AA* (or *BB* or *OO*) and are of blood type A (or B or O). These individuals who inherit the same allele from both parents are said to be *homozygous* for this particular gene.

But suppose there is crossmating. Suppose an *AA* individual mates with an *OO* individual. The *AA* contributes one gene to the offspring and this one gene can only be an *A*. The *OO* contributes one gene which can only be an *O*. The offspring, combining the two genes, is an *AO* individual. If his blood is tested, it proves to be of blood type A. Apparently the presence of the *A* gene completely masks the presence of the *O*. The *A* gene is dominant over the *O* gene. The individual who inherits different alleles is *heterozygous* for this gene.

Exactly the same thing happens in a mating between a *BB* individual and an *OO* individual. The offspring are all *BO*, and the blood type is always B. The *B* gene is dominant over the *O* gene. So far, the situation is just the same as with Mendel's pea plants. (Part of the value of Mendel's law is its perfect generality. It applies to all species, from pea plants to humans.)

Furthermore, if an *AO* individual is mated with another *AO* individual, each may contribute an *O* gene to a particular offspring. An *OO* individual of blood type O would have arisen, so that that recessive characteristic will have reappeared unharmed after having skipped a generation. The mating of a *BO* with a *BO* produces similar consequences. Again, there is no difference here between men and pea plants.

Now, however, suppose an *AA* individual is mated with a *BB* individual. The *AA* individual contributes an *A* gene, the *BB* individual a *B* gene.

All the offspring are, therefore, *AB*. But what of the blood type? Is *A* dominant over *B* or *B* over *A*? The blood, when tested, shows that neither is dominant. The blood reacts as both type *A* and type *B*, and such a case is classified as blood type *AB*. Each gene shows its full effect despite the presence of the other.

Incomplete dominance need not result in a display of the effects produced by both genes; it can show an effect produced by neither. There are plants that exist in red-flowered varieties and white-flowered varieties which, when the two varieties are crossed, produce plants with pink flowers.

The red and white colors seem to have blended, but they have not. If pink-flowered plants are crossed with pink-flowered plants (or if they are self-pollinated), the red-flowered genes and white-flowered genes sort out in all possible combinations. The offspring produce red flowers (*RR*) in one quarter of the cases, white flowers (*WW*) in another quarter, and pink flowers (*RW* or *WR*) in the remaining half.

Genes possessing two alleles (red flowers and white flowers or smooth seeds and wrinkled seeds) or even three alleles (*A*, *B* and *O* blood types) are relatively easy to work with. However, genes may possess large numbers of alleles. The gene governing the *Rh* blood types in man possesses at least eight alleles of varying types of dominance with respect to each other. The inheritance of the *Rh* genes is, therefore, so complicated that controversy concerning it continues and will probably go on for quite a while yet.

Moreover, a particular physical characteristic may be governed by the interaction of more than one set of alleles, and this further complicates the problems of inheritance. Human skin color is probably a case in point here. It is not even agreed as yet how many different sets of alleles there are contributing to this physical characteristic. The presence of more than one set makes for a large number of gradations between the extremes (and we know from experience that there are a great many color levels in the skin intermediate between that of the Swede and that of the Nigerian). It is the complicated gene pattern of such easily noticeable physical characteristics that gives the casual observer (and even one not so casual, such as Charles Darwin) the illusion that the characteristics of parents mix and blend in the offspring.

Sometimes the effect of a gene will be altered by circumstance. For instance, there is a gene, or possibly a combination of genes, that produces early baldness in a human being. There is reason to think, however, that if the circulating male sex hormone is below a certain level of concentration in the blood, baldness does not develop even in the presence of the baldness gene or genes.

If the level of circulating male sex hormones is controlled by another gene (or genes), as it well may be, this would be the case of one gene affecting the action of another.

Then, too, the action of a gene may be affected by the environment itself. A person may possess both the gene for baldness and the gene governing a high concentration of male sex hormone. Ordinarily, this would ensure early baldness. If he is castrated in childhood, however (a purely environmental effect), the male sex hormone supply is cut off at the source, and he will not go bald. (This is a baldness preventative that is unlikely to attain great popularity.)

Similarly, a person with a gene, or combination of genes, which would produce, under ideal circumstances, a height of better than six feet, will, nevertheless, not attain that height if he has been chronically undernourished as a child.

The pattern of inheritance of physical characteristics among the higher animals is studded with pitfalls like this, and it is no accident that most of the advances in genetics have been made as a result of experiments on plants, bacteria and insects. There we have the happy combination of many offspring in a short time so that statistical rules will apply well, plus the existence of relatively few and uncomplicated physical characteristics.

Human genetics is a particularly poorly studied field. Not only does the individual human being have few children only at long intervals, but, worse still, controlled matings are difficult or impossible to arrange among human beings. The human being is one of the poorest laboratory animals in existence, and the scientist studying human genetics must take what he can get in the way of mating combinations.

A particularly important factor introduced by the gene theory, as far as evolution is concerned, is that of randomness. No matter how complicated the study of genes and their interrelations gets, one thing always remains. Individual genes sort themselves out from generation to generation in a completely random fashion. Furthermore, the mutations that occur appear to be adaptively random, that is, they do not show any positive correlation with the needs of the phenotype.

Yet before the gene theory was established, there was considerable difficulty in believing that the development of species could be entirely a random affair. Could an ameba become a man through the operation of blind and random forces? It seemed there had to be some sort of purpose behind it. One way or another, there had to be direction. For instance, Nägeli's orthogenesis was a kind of purposefulness superimposed upon evolution. Species did not grow bigger or smaller at random, according to orthogenesis; once they started growing bigger, they continued willy-nilly. In the same way, an ameba that started out in the direction of the far-distant goal of man would tend to keep on the road and not stray from it.

It may well have been the implication of randomness in Mendel's theories that set Nägeli's teeth on edge and caused him to pooh-pooh Mendel's paper.

And yet to account for the facts of evolution, randomness of gene shuffling as proposed by Mendel is all that is needed, provided it is combined with the natural selection postulated by Darwin, the mutation theory suggested by De Vries and the eons of time put forward by Hutton. Using all these, we can build up a simplified scheme of the evolution of the large cats—simply as an idealized example.

Imagine members of a species of large cat living throughout Africa, southern Asia and northern South America. We can suppose them all to have identical genes and each gene to be made up of a single allele. (Such an ideal case would never actually be met with in nature.)

In accordance with the De Vries mutation theory, as genes are passed on from generation to generation, new alleles are developed. Eventually, every gene would be made up of a number of such alleles. According to Mendel's laws, these alleles would be shuffled into new and random combinations in each generation. (In fact, the evolutionary significance of sex is that it supplies a method for the shuffling of genes by having offspring collect half their genes from each of two parents. At which thought one might cry, "*Vive la signification évolutionnaire!*")

No two cats are very likely to develop the same new allele, but that makes no difference. Through unrestricted mating, an allele developed by one cat spreads gradually through the entire species. If, however, certain portions of the species are geographically isolated, the alleles developed in one portion will not spread to the others. Thus, there will be special alleles developed by the Asian cats which neither African nor South American cats will ever receive (unless they produce them independently, as, by the operation of chance, they may). The same argument holds true for special alleles developed by the African or South American cats.

The three groups will, with time, begin to show consistent differences in physical appearance, and these differences will increase with further time.

The new alleles, as they are formed, will be subject to the forces of natural selection. An allele producing a characteristic that is somewhat undesirable will be discriminated against slowly. Eventually, it will die out, except that it may be regenerated every once in a while by a new mutation. The tendency to die out will be balanced by the tendency for regeneration so that the allele will persist at a certain equilibrium level. The more undesirable the gene, the faster it will be discriminated against, and the lower the equilibrium level.

On the other hand, an allele that produces a desirable characteristic is naturally selected, since cats possessing it will live longer and breed more. If the advantage is sufficient, that allele may replace all others.

The comparative worth of a particular allele depends upon the environment to which the animal is subjected, and one that is beneficial in one environment may be detrimental in another. The differences among the Asian, African and South American cats which are initiated by chance mutations would then be accentuated by the action of natural selection (which is not random by any means).

Thus, if the African cats are creatures of the plains, a tawny coat, produced by a particular line of mutations, may be of particular use to them. Such a coat would render them unnoticeable against the background and increase the efficiency with which they can stalk game. (On the other hand, if it had a striped or spotted coat pattern, it could be seen by a half-asleep giraffe at a distance of two miles, and the cat would probably starve to death.) The tawny coat produced slowly by the random action of mutations and gene assortment would then be established as a universal characteristic among the African cats by the comparatively rapid action of natural selection.

The Asian and South American cats, however, might live in a jungle where the general background is one of sunlight filtering through leaves to make a splotched pattern of light and dark. There, a striped or spotted coat pattern would render them particularly unnoticeable. The Asian cats might develop, by random mutation, a striped coat, and natural selection would fix that as a universal characteristic. The South American cat might develop, by equally random mutation, a spotted coat, and natural selection would fix that. Had the South American cats developed stripes and the Asian cats spots, matters might have been fixed that way just as easily, but that alternative did not happen, for no better reason than that which causes a coin to fall heads, and not tails, at some particular throw, or for a pair of dice to turn up snake-eyes instead of boxcars, or a poker dealer to turn up the ace of hearts instead of the jack of diamonds.

In this imaginary development of lions, tigers and jaguars, we have the germ of a treatment for all evolution.

The over-all picture is that of groups of organisms which, by mutation, random assortment, and natural selection, and without need of any directing force at all, continually improve the manner in which each fits the particular environmental niche in which it lives.

Darwin's Missing Evidence *

by H. B. D. Kettlewell (1959)

Charles Darwin's *Origin of Species* . . . was the fruit of 26 years of laborious accumulation of facts from nature. Others before Darwin had believed in evolution, but he alone produced a cataclysm of data in support of it. Yet there were two fundamental gaps in his chain of evidence. First, Darwin had no knowledge of the mechanism of heredity. Second, he had no visible example of evolution at work in nature.

It is a curious fact that both of these gaps could have been filled during Darwin's lifetime. Although Gregor Mendel's laws of inheritance were not discovered by the community of biologists until 1900, they had first been published in 1866. And before Darwin died in 1882, the most striking evolutionary change ever witnessed by man was taking place around him in his own country.

The change was simply this. Less than a century ago moths of certain species were characterized by their light coloration, which matched such backgrounds as light tree trunks and lichen-covered rocks, on which the moths passed the daylight hours sitting motionless. Today in many areas the same species are predominantly dark! We now call this reversal "industrial melanism."

It happens that Darwin's lifetime coincided with the first great manmade change of environment on earth. Ever since the Industrial Revolution commenced in the latter half of the 18th century, large areas of the earth's surface have been contaminated by an insidious and largely unrecognized fallout of smoke particles. In and around industrial areas the fallout is measured in tons per square mile per month; in places like Sheffield in England it may reach 50 tons or more. It is only recently that we have begun to realize how widely the lighter smoke particles are dispersed, and to what extent they affect the flora and fauna of the countryside.

In the case of the flora the smoke particles not only pollute foliage but also kill vegetative lichens on the trunks and boughs of trees. Rain washes the pollutants down the boughs and trunks until they are bare and black. In heavily polluted districts rocks and the very ground itself are darkened.

Now in England there are some 760 species of larger moths. Of these more than 70 have exchanged their light color and pattern for dark or even all-black coloration. Similar changes have occurred in the moths of industrial areas of other countries: France, Germany, Poland, Czechoslovakia, Canada and the U. S. So far, however, such changes have not been observed anywhere in the tropics. It is important to note here that industrial melanism has occurred only among those moths that fly at night and spend the day resting against a background such as a tree trunk.

These, then, are the facts. A profound change of color has occurred among hundreds of species of moths in industrial areas in different parts of the world. How has the change come about? What underlying laws of nature have produced it? Has it any connection with one of the normal mechanisms by which one species evolves into another?

In 1926 the British biologist Heslop Harrison reported that the industrial melanism of moths was caused by a special substance which he alleged was present in polluted air. He called this substance a "melanogen," and suggested that it was manganous sulfate or lead nitrate. Harrison claimed that when he fed foliage impregnated with these salts to the larvae of certain species of light-colored moths, a proportion of their offspring were black. He also stated that this "induced melanism" was inherited according to the laws of Mendel.

Darwin, always searching for missing evidence, might well have accepted Harrison's Lamarckian interpretation, but in 1926 biologists were skeptical. Although the rate of mutation of a hereditary characteristic can be increased in the laboratory by many methods, Harrison's figures inferred a mutation rate of 8 per cent. One of the most frequent mutations in nature is that which causes the disease hemophilia in man; its rate is in the region of .0005 per cent, that is, the mutation occurs about once in 50,000 births. It is, in fact, unlikely that an increased mutation rate has played any part in industrial melanism.

At the University of Oxford during the past seven years we have been attempting to analyze the phenomenon of industrial melanism. We have used many different approaches. We are in the process of making a survey of the present frequency of light and dark forms of each species of moth in Britain that exhibits industrial melanism. We are critically examining each of the two forms to see if between them there are any differences in behavior. We have fed large numbers of larvae of both forms on foliage impregnated with substances in polluted air. We have observed under various conditions the mating preferences and relative mortality of the two forms. Finally we have accumulated much information about the melanism of moths in parts of the world that are far removed from industrial centers, and we have sought to link industrial melanism with the melanics of the past.

Our main guinea pig, both in the field and in the laboratory, has been the peppered moth *Biston betularia* and its melanic form *carbonaria.* This species occurs throughout Europe, and is probably identical with the North American *Amphidasis cognataria.* It has a one-year life cycle; the moth appears from May to August. The moth flies at night and passes the day resting on the trunks or on the underside of the boughs of rough-barked deciduous trees such as the oak. Its larvae feed on the foliage of such trees from June to late October; its pupae pass the winter in the soil.

The dark form of the peppered moth was first recorded in 1848 at Manchester in England. Both the light and dark forms appear in each of the photographs (Figures 2–3 and 2–4). The background of each photograph is noteworthy. In Figure 2–4 the background is a lichen-encrusted oak trunk of the sort that today is found only in unpolluted rural districts. Against this background the light form is almost invisible and the dark form is conspicuous. In Figure 2–3 the background is a bare and blackened oak trunk in the heavily polluted area of Birmingham. Here it is the dark form which is almost invisible, and the light form which is conspicuous. Of 621 wild moths caught in these Birmingham woods in 1953, 90 per cent were the dark form and only 10 per cent the light. Today this same ratio applies in nearly all British industrial areas and far outside them.

We decided to test the rate of survival of the two forms in the contrasting types of woodland. We did this by releasing known numbers of moths of both forms. Each moth was marked on its underside with a spot of quick-drying cellulose paint; a different color was used for each day. Thus when we subsequently trapped large numbers of moths we could identify those we had released and established the length of time they had been exposed to predators in nature.

In an unpolluted forest we released 984 moths: 488 dark and 496 light. We recaptured 34 dark and 62 light, indicating that in these woods the light form had a clear advantage over the dark. We then repeated the experiment in the polluted Birmingham woods, releasing 630 moths: 493 dark and 137 light. The result of the first experiment was completely reversed; we recaptured proportionately twice as many of the dark form as of the light.

For the first time, moreover, we had witnessed birds in the act of taking moths from the trunks. Although Britain has more ornithologists and bird watchers than any other country, there had been absolutely no record of

Figure 2–3. Dark and light forms of the peppered moth were photographed on the trunk of an oak blacked by the polluted air of the English industrial city of Birmingham. The light form (Biston betularia) is clearly visible; the dark form (carbonaria) is well camouflaged.

Figure 2–4. Same two forms of the peppered moth were photographed against the lichen-encrusted trunk of an oak in an unpolluted area. Here it is the dark form which may be clearly seen. The light form, almost invisible, is just below and to the right of the dark form. (Photographs by H. B. D. Kettlewell, University of Oxford. Reprinted by permission.)

birds actually capturing resting moths. Indeed, many ornithologists doubted that this happened on any large scale.

The reason for the oversight soon became obvious. The bird usually seizes the insect and carries it away so rapidly that the observer sees nothing unless he is keeping a constant watch on the insect. This is just what we were doing in the course of some of our experiments. When I first published our findings, the editor of a certain journal was sufficiently rash as to question whether birds took resting moths at all. There was only one thing to do, and in 1955 Niko Tinbergen of the University of Oxford filmed a repeat of my experiments. The film not only shows that birds capture and eat resting moths, but also that they do so selectively.

These experiments lead to the following conclusions. First, when the environment of a moth such as *Biston betularia* changes so that the moth cannot hide by day, the moth is ruthlessly eliminated by predators unless it mutates to a form that is better suited to its new environment. Second, we now have visible proof that, once a mutation has occurred, natural selection alone can be responsible for its rapid spread. Third, the very fact that one form of moth has replaced another in a comparatively short span of years indicates that this evolutionary mechanism is remarkably flexible. . . .

Now in order for the dark form of a moth to spread, a mutation from the light form must first occur. It appears that the frequency with which this happens—that is, the mutation rate—varies according to the species. The rate at which the light form of the peppered moth mutates to the dark form seems to be fairly high; the rate at which the mutation occurs in other species may be very low. For example, the light form of the moth *Procus literosa* disappeared from the Sheffield area many years ago, but it has now reappeared in its dark form. It would seem that a belated mutation has permitted the species to regain lost territory. Another significant example is provided by the moth *Tethea ocularis*. Prior to 1947 the dark form of this species was unknown in England. In that year, however, many specimens of the dark form were for the first time collected in various parts of Britain; in some districts today the dark form now comprises more than 50 per cent of the species. There is little doubt that this melanic arrived in Britain not by mutation but by migration. It had been known for a considerable time in the industrial areas of northern Europe, where presumably the original mutation occurred.

The mutation that is responsible for industrial melanism in moths is in the majority of cases controlled by a single gene. A moth, like any other organism that reproduces sexually, has two genes for each of its hereditary characteristics: one gene from each parent. The mutant gene of a melanic moth is inherited as a Mendelian dominant; that is, the effect of the mutant gene is expressed and the effect of the other gene in the pair is not. Thus a moth that inherits the mutant gene from only one of its parents is melanic.

The mutant gene, however, does more than simply control the coloration of the moth. The same gene (or others closely linked with it in the hereditary material) also gives rise to physiological and even behavioral traits. For example, it appears that in some species of moths the caterpillars of the dark form are hardier than the caterpillars of the light form. . . .

Another difference between the behavior of *B. betularia* and that of its dark form *carbonaria* is suggested by our experiments on the question of whether each form can choose the "correct" background on which to rest during the day. We offered light and dark backgrounds of equal area to moths of both forms, and discovered that a significantly large proportion of each form rested on the correct background. Before these results can be accepted as proven, the experiments must be repeated on a larger scale.[1] If they are proven, the behavior of both forms could be explained by the single mechanism of "contrast appreciation." This mechanism assumes that one segment of the eye of a moth senses the color of the background and that another segment senses the moth's own color; thus the two colors could be compared. Presumably if they were the same, the moth would remain on its background; but if they were different, "contrast conflict" would result and the moth would move off again. That moths tend to be restless when the colors conflict is certainly borne out by recent field observations.

It is evident, then, that industrial melanism is much more than a simple change from light to dark. Such a change must profoundly upset the balance of hereditary traits in a species, and the species must be a long time in restoring that balance. Taking into account all the favorable and unfavorable factors at work in this process, let us examine the spread of a mutation similar to the dark form of the peppered moth. To do so we must consult Figure 2–5.

According to the mutation rate and the size of the population, the new mutation may not appear in a population for a period varying from one to 50 years. This is represented by AB on the diagram. Let us now assume the following: that the original successful mutation took place in 1900, that subsequent new mutations failed to survive, that the total local popu-

1. Several later experiments (still on a small scale) appear to confirm this conclusion, but the author says that still more experimentation is needed before the result can be accepted as truly significant. Ed.

lation was one million, and that the mutant had a 30-per-cent advantage over the light form. (By a 30-per-cent advantage for the dark form we mean that, if in one generation there were 100 light moths and 100 dark, in the next generation there would be 85 light moths and 115 dark.)

On the basis of these assumptions there would be one melanic moth in 1,000 only in 1929 (BC). Not until 1938 would there be one in 100 (BD). Once the melanics attain this level, their rate of increase greatly accelerates.

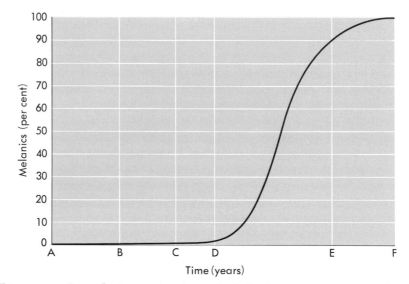

Figure 2–5. Spread of mutation from the light form to the dark (melanic) is expressed by this curve, discussed in detail in the text. The mutation occurs in the period AB, spreads slowly during BD and spreads rapidly during DE. During EF the light form is either gradually eliminated, as indicated by the curve, or remains at a level of about 5 per cent of the population.

In the period between 1900 and 1938 (BD) natural selection is complicated by other forces. Though the color of the dark form gives it an advantage over the light, the new trait is introduced into a system of other traits balanced for the light form; thus the dark form is at first at a considerable physiological disadvantage. In fact, when moths of the dark form were crossed with moths of the light form 50 years ago, the resulting broods were significantly deficient in the dark form. When the same cross is made today, the broods contain more of the dark form than one would expect. The system of hereditary traits has become adjusted to the new trait.

There is evidence that other changes take place during the period BD. Specimens of the peppered moth from old collections indicate that the earliest melanics were not so dark as the modern dark form: they retained some of the white spots of the light form. Today a large proportion of the moths around a city such as Manchester are jet black. Evidently when the early melanics inherited one gene for melanism, the gene was not entirely dominant with respect to the gene for light coloration. As the gene complex adjusted to the mutation, however, the new gene became almost entirely dominant.

When the dark form comprises about 10 per cent of the population, it may jump to 90 per cent in as little as 15 or 20 years. This is represented by period DE on the graph. Thereafter the proportion of the dark form increases at a greatly reduced rate.

Eventually one of two things must happen: either the light form will slowly be eliminated altogether, or a balance will be struck so that the light form continues to appear as a small but definite proportion of the population. This is due to the fact that the moths which inherit one gene for dark coloration and one for light (heterozygotes) have an advantage over the moths which inherit two genes for dark coloration (homozygotes). And when two heterozygotes mate, a quarter of their offspring will have two genes for light coloration, *i.e.*, they will be light. Only after a very long period of time, therefore, could the light forms (and with them the gene for light coloration) be entirely eliminated. This period of removal, represented by EF on the diagram, might be more than 1,000 years. Indications so far suggest, however, that complete removal is unlikely, and that a balance of the two forms would probably occur. In this balance the light form would represent about 5 per cent of the population.

The mechanisms I have described are without doubt the explanation of industrial melanism: normal mutation followed by natural selection resulting in an insect of different color, physiology and behavior. Industrial melanism involves no new laws of nature; it is governed by the same mechanisms which have brought about the evolution of new species in the past. . . .

The melanism of moths occurs in many parts of the world that are not industrialized, and in environments that are quite different. It is found in the mountain rain forest of New Zealand's South Island, which is wet and dark. It has been observed in arctic and subarctic regions where in summer moths must fly in daylight. It is known in very high mountains, where dark coloration may permit the absorption of heat and make possible increased activity. In each case recurrent mutation has provided the source of the change, and natural selection, as postulated by Darwin, has decided its destiny.

Melanism is not a recent phenomenon but a very old one. It enables us to appreciate the vast reserves of genetic variability which are contained within each species, and which can be summoned when the occasion arises. Had Darwin observed industrial melanism he would have seen evolution occurring not in thousands of years but in thousands of days—well within his lifetime. He would have witnessed the consummation and confirmation of his life's work.

Reasonable Doubts?

_____

How Right Was Darwin? [*]
by Joseph Wood Krutch (1956)

That a mere layman has no right to make authoritative pronouncements is granted. But since the authorities do sometimes seek to convince him he ought to be permitted to confess that of certain things he has not been convinced without being visited with the wrath of an _odium theologicum_. In science no less than in religion honest doubt is worthy of respect and it is for honest doubts that I plead.

This being the case, it is well to be as specific as possible. Here, then, are things which as a layman I have been able to believe or not believe: (1) That the higher plants and animals "evolved" from the lower seems to me as certain as circumstantial evidence can make anything. It is, to use the legal phrase, beyond reasonable doubt. (2) That natural selection, working mechanically, has contributed very importantly to the process seems almost equally certain, because the circumstantial evidence is again very strong. (3) On the other hand, the statement that no factor except natural selection has ever influenced the course of that evolution seems to me a statement based wholly on negative evidence and one making the whole story so nearly incredible that negative evidence alone is not sufficient to rule out completely certain alternate possibilities.

A doubt that orthodox Darwinism really does embody the whole truth was expressed by Alfred Russel Wallace from the beginning; [1] and similar doubts about the adequacy of the various neo-Darwinisms which have followed upon the original theory have been shared by respectable thinkers. . . .

That the case for natural selection as a factor—and probably a very important one—is almost unassailably strong I have already admitted. It does take place, and unaided it may possibly be capable of producing organisms better fitted for survival than their parents. But whether or not it alone is sufficient to account for _everything_ that has happened is quite a

[*] From Joseph Wood Krutch, _The Great Chain of Life_, Houghton Mifflin Company, Boston. Copyright © 1956 by Joseph Wood Krutch. Reprinted by permission.
   1. See Part 5, pp. 257–63; also Darwin's own statement, p. 82. Ed.

113

different question, and Darwinians themselves have never been entirely comfortable with their own interpretations.

No one today would claim that the theory as Darwin first set it forth is adequate. Further studies have modified the original conception and discovered, or imagined, new ways in which it might be supposed to work better. Such is, of course, the usual history of scientific theories. But the repeated modifications of the original theory do testify also to a certain gnawing doubt whether the theory as it has stood at any one time was fully adequate. Time and again it has been tinkered with, modified, made more elaborate. Yet early Darwinians had accepted it as a dogma not to be questioned before it had the support of later discoveries. They asserted its adequacy in a form no one today believes adequate. And if they believed it before it was credible, then can one be certain that it would not be generally believed today even if it were not, at last, unassailable?

Even now the tale as told by the most careful expositors is a very tall one and there are few who will not admit that it would be easier to swallow if only we could believe that it was not exclusively a story of the miracles which pure chance, complete blindness, and utter mindlessness are supposed to have accomplished. But it is usually added that facts are facts and that no intervention of will, purpose, or intention did occur—or at least did not until man or some near-man had emerged with the mind which had been mindlessly created.

But why and how justifiably can anyone be so sure? Because, so the orthodox will reply, mechanical selection can be discovered at work whereas nothing except mechanical selection ever has been so discovered. Yet this argument is purely negative and it overlooks an important fact, namely, that mechanical operations are relatively easy to observe, whereas those of the will or the spirit would be difficult if not impossible to detect by the methods commonly used.

Take for example one of the classical experiments which has been repeatedly described. You tether somewhere various individuals of the same species of insect or small mammal which vary markedly among themselves in color. Presently it is found that those which blend with the background are picked off less promptly by birds or other predators. Obviously, then, protective coloration has survival value. The longer an individual survives the better his chance to reproduce and to pass on to another generation his advantageous color. Hence, so it is concluded, natural selection has produced protective coloration as well as all the much more striking examples of disguise and mimicry nature provides.

So far so good. That the protectively colored insects and the sand-colored mice in desert regions were produced by this process is credible enough. The creation by the same process of a mind that devises such experiments and ponders the conclusions to be drawn from them is not quite so easy to believe in! But that is not the principal point. The principal point is that the methods employed in the classic experiments are such that they are necessarily incapable of detecting anything except what they do detect.

The proof that in these cases natural selection actually does select is purely statistical. But statistical proof never can eliminate the possibility that factors other than those which are the subject of the experiment exist. Suppose that some individual insects or mice were in some way aware of the advantage selective coloration gave them. Suppose they could give themselves a slight push in that direction. The fact would not be detected in the course of the experiment. It would increase somewhat the survival rate of those favorably colored. But since you do not know what that rate would be on the basis of the merely mechanical assumption, the contribution of the nonmechanical factor would not be detected.

Grant that in this particular case the known factors do seem as though they might be sufficient to explain the facts and that there is no reason for suspecting the existence of any others. Even so one should nevertheless admit in all honesty that their nonexistence is not conclusively demonstrated even though one might let it go at that. But when it comes to the whole story of evolution, the situation is very different. It is a much taller story and one is less ready to believe that natural selection alone is sufficient to account for it.

In this case, then, we have a far better reason for caution and some real excuse for the suspicion that some other factor has been at work. The mere negative argument against the assumption is no longer sufficient. There is no longer any excuse for denying positively, dogmatically, indignantly, and even irascibly that natural selection is the only possible explanation of everything which has happened to evolving life since the beginning of time. Doubt is here at least legitimate and respectable. . . .

Unfortunately, however, all tendency to question in any way the adequacy of the purely mechanistic theory of evolution is now labeled either Lamarckian or vitalistic; and no less unfortunately both these words have become epithets to be hurled rather than terms indicating a position to be analyzed. To raise such reasonable questions as we have been raising is to be labeled Lamarckian and vitalist and to be so labeled is to be accused of holding two absurd ideas: (1) that life is something distinguishable from the matter which a living creature animates, and (2) that a giraffe, to take one example, got a very long neck because for a long time he had been thinking how convenient it would be to have one. . . .

But there is a considerable difference between so simplified a fable and the suggestion that choice, consciousness, awareness, and will have sometimes been able to intervene in an evolutionary process which has, therefore, not been wholly the result of blind accident. Such a suggestion does not at all depend upon any assumption that acquired characteristics are inherited and does not, in fact, imply anything more than that the generally admitted ability of the human being to intervene may be assumed to have existed to some extent for a very long time—perhaps even for as long as any form of life has existed. And if to suggest that as a possibility is in itself sufficient to earn the label "Lamarckian," then perhaps a Lamarckian is not obviously an absurd thing to be.

Evolution *

by Ludwig von Bertalanffy (1952)

THE TIBETAN PRAYER-WHEEL

The overwhelming majority of modern biologists, and especially the investigators connected with genetics, accept the theory of selection. This attitude is based, empirically, on the success of genetics, as such and in its application to evolution; methodologically it is based on the principle that only factors which are known and exactly and experimentally demonstrable should be admitted. Some morphologists and palaeontologists, however, hold opposite opinions. Workers in these fields, more so than experimenters, are incessantly confronted with the marvellous architecture of living organization, its integration, and the correspondence of structure and function, and are reluctant to believe that these are a product of mere chance. Similarly, the physiologist looks at the amazing complexity of the system of catalysers in a cell wherein the loss of even one member may make it degenerate into a cancer-cell, or surveys the number of conditions necessary for the normal function of a gland, or of the nervous connections necessary even for a simple reflex; he feels somewhat uncomfortable to accept the explanation that this is all due to chance.

As already mentioned, the mechanisms assumed—mutation, selection, isolation—are experimentally verified. However, apart from some cases occurring in polyploid plants, no new species has ever arisen within the sphere of observation, let alone "macro-evolutionary" changes. Selection

* From Ludwig von Bertalanffy, *Problems of Life, An Evaluation of Modern Biological and Scientific Thought,* C. A. Watts and Co., London. © 1952 by C. A. Watts and Co., Ltd. Reprinted by permission.

Theory is an extrapolation, the boldness of which is made acceptable by the impressiveness of its basic conception. With a less picturesque theory, one would doubtless hesitate to extend cosmically and universally a principle which is controlled experimentally only to a rather limited extent. The pros and cons of the theory of selection have been discussed on countless occasions. Indeed, the controversy with more or less noteworthy "objections," constitutes a main part of every presentation of the theory— a procedure that would be looked for in vain in a treatise on, say, physics or physiology. It is a methodological maxim, which is unreservedly correct, to take into account only known and experimentally demonstrable factors, to extend according to the actuality principle, their application as far as possible, and to exclude, in the sense of Occam's razor, factors unknown and experimentally unproved. On the other hand, for not quite fifty years we have carried out genetical research on some dozens of animal and plant species, the mutations of which never transgressed the limits of the species. It is a bold extrapolation that nothing else has happened in a billion years or so of evolution "from amoeba to man." So the dispute concerns ways of thinking rather than factual evidence.

Innumerable characters of the organisms are seemingly useless. To a great extent they are precisely those that the taxonomist considers most important, because on account of their functional neutrality they show a high degree of constancy. The multiplicity of leaf-forms, the arrangement of leaves, the numbers three, four, or five in flowers, the number seven characteristic of the cervical vertebrae in mammals, which is found in the neckless whale as well as in the long-necked giraffe; all such morphological peculiarities which form the framework of taxonomy seem to represent "types" that in themselves have nothing to do with usefulness but can be fitted to different exigencies and habitats much in the same way as churches, town halls, or castles can be built equally well in gothic, baroque, or rococo styles. Goebel, the great botanist, has emphasized that the manifoldness of organic forms is much greater than the manifoldness of environmental conditions. In the same part of the sea, in a thoroughly uniform environment, hundreds of species of foraminifera or radiolaria can be found; "natural art-forms," the fantastic diversity of shapes of which has obviously nothing to do with usefulness. The selectionist, however, is not embarrassed by these arguments. Structures which are useless in themselves can be preserved in uniform surroundings where selection-pressure is low. They may arise by chance, according to the Sewall-Wright principle,[1] in a species subdivided into small isolated populations;

1. [This is the] isolation or the "drift" principle, studied especially by Sewall Wright. If a species is subdivided into small populations isolated from each other, then the mere accident of gene-combinations can lead to the establishment of

or useless characters were linked with others having a selective advantage, perhaps mere differences in viability, and thus were perpetuated.

In the living world the maxims "Why make it simple, when it can be complicated?" and "It can be done this way, but it can also be done the other way" seem to be much in vogue. Often astonishingly roundabout ways are taken to reach a goal that could be reached far more simply and with less risk. Consider the life-cycles of many parasites, the relatively simple one of the human Ascaris, for example. The young larvae enter the intestine with the food; they pass through the intestinal wall into the blood and into the circulatory system, through the portal vein into the liver, then into the lung and throat, where they must be again swallowed, to arrive back, at the stage of sexual maturity, in the intestine, where, in our humble opinion, they could quite well have remained. In the remarkable trap flowers, such as Lady's Slipper or the wild Arum, the insects that help to pollinate the flowers are captured and imprisoned by intricate devices so that they can be used for pollination in due time. The net outcome of these intricate mechanisms, however is that in our climates the Lady's Slipper and Arum are among the more rare plants almost on the verge of extinction. Even in their tropical centres, the families to which they belong do no better than plants utilizing simple wind-pollination. "It can be done this way, but it can also be done the other way." The ruminants have acquired an extraordinarily complicated multiple stomach, doubtless an organ of the greatest usefulness and highest selective value for animals digesting vegetable food; but horses have a simple stomach, and attain the same remarkable body size and geographical distribution. With an artist's industry, nature paints in mimicry on the wings of a butterfly the copy of a protected model in order to make the birds believe that the imitation also tastes bad. But the ordinary Black Archers, although readily taken by birds, are as numberless as the sands of the seashore, and strip the forests bare. The same survival in the struggle for existence, retorts the selectionist, can be attained by different means, and the very existence of all these forms shows their fitness.

The same holds for absurd and even apparently disadvantageous formations, such as the giant and baroque forms of the ammonites, the monstrous horns of the titanotheres, the antlers of the megaceros, which, being heavy and hindering a forest animal in its movements, presumably accounted for its extinction. Ludwig [2] has stated that selectionism offers some fourteen to twenty possible ways of explaining such disadvanta-
different mutations in these populations, irrespective of their selective value; and this may lead to the splitting-up of the originally uniform species into different subspecies and eventually species.

2. See W. Ludwig and J. Krywienczyk: "Koerpergroesse, Koerperzeiten und Energiebilanz," 111. Z. vgl. Physiol., 32, 1950.

geous characters. This shows that they are not a refutation of the theory of selection, but equally that a decisive judgment is impossible. For one hypothesis that can be neatly proved or disproved has much greater value than a host of possibilities. To mention some of these explanations: a character that is disadvantageous or neutral at present may have been advantageous in the past; a disadvantageous character may be correlated with one having selective value . . . ; the character in question owes its origin to sexual selection; within a species the existence of which is secured interspecifically, intraspecific selection may take place, which can lead eventually to developments harmful to the species itself; and so on.

A lover of paradox could say that the main objection to selection theory is that it cannot be disproved. With a good theory, it must be possible to indicate an *experimentum crucis*, the negative outcome of which would refute it. If the gravitational force exerted by the planets was proportional to $1/r^3$ instead of $1/r^2$, Newtonian mechanics would be wrong; if a cuneiform text of logarithmic tables was found somewhere, we would have to revise our notions on Babylonian mathematics. In the case of Selection Theory, however, it appears impossible to indicate any biological phenomenon that would plainly refute it.

Consider the construction of an accommodating eye; a soft lens, a ciliary body, ciliary muscles, nerves leading to and from corresponding centres must all be present and work together to allow its functioning. There is a big difference between the "characters" that are studied in experimental genetics, and which represent small variations of, for instance, the size of an organ, its colour, and the like, and the origin of "systems" which are useful and have survival value only as organized wholes. In this sense, single and accidental mutations cannot lead to a gradual development or improvement of the apparatus, but can only spoil it; the absence of one single part makes the whole system into a sort of useless or even deleterious tumour. Thus it is calculated how infinitely improbable it is that such co-adaptation, the origin of apparatus capable of functioning only as a whole, resulted from accidental mutations. The selectionist's answer: Remember that the lens-eye is the product of an enormously long evolution. Imagine sufficiently minute intermediate stages from a simple pigment-spot through the socket-eye up to the lens-eye, stages as they are actually shown by comparative anatomy, each one giving a small selective advantage. Then you will understand how such formations have been added up in the long time of phylogenetic history. And the selectionist also gives nice calculations that support his view.

With the same stubbornness the uniformity of micro-evolution and macro-evolution, of the origin of the multiplicity of forms within a "type" and the origin of these types themselves, is challenged and defended. The rise, say, of a winged insect from unwinged ancestors is on a different level from the known mutations in Drosophila, which concern only differences of formation in already existing wings. The macro-evolutionary origin of a "type" is not the product of a gradual accumulation of small changes, but of "macro-mutations" that give rise to far-reaching transformations in the early embryonic stages. This view is supported by palaeontology, which indicates two phases in evolution; a first one, the sudden rise of a new type, which, immediately after its appearance, splits up, explosively as it were, into the main classes or orders; a second one, the slowly progressing speciation and adaptation to different environmental conditions within those groups. The selectionist's answer is: a clear definition of what shall be called "type" is not possible, so a boundary cannot be drawn between "macro-evolution" and "micro-evolution." A number of profoundly transforming mutations are known in experiment, such as the four-winged mutation *tetraptera* of the dipter Drosophila or mutations of the snapdragon with radial instead of bilateral symmetry of the corolla. The fact that transition stages between different "types" are rare or often missing is easily explained, for new types go back to thin ancestral roots, and accordingly the probability of the preservation of these ancestors as fossils is small. Nevertheless, we have such transition stages in good measure, for example, in *Archaeopteryx* from reptile to bird, or in the almost unbroken series that leads from reptiles through the *Theriodontia* to mammals. Indeed, there is no reason to assume a basic distinction between micro-evolution and macro-evolution, or that the laws of heredity were in the past different from those of today.

It is often asserted that evolutionary progress cannot be understood in terms of "usefulness." If higher organization means selective advantage, the higher organisms should have supplanted the lower ones. Every cross-section of nature, however, shows the most diverse levels of organization from unicellulars up to vertebrates, all maintaining themselves perfectly, and indeed all necessary for the maintenance of the biocoenosis. The selectionists retort: When man invented the bow and arrow the simpler method of clubbing became obsolete; the introduction of firearms condemned the armoured knights to extinction; tanks make cavalry attacks of doubtful value. In the present struggle for existence among nations, survival is granted only by aircraft—until in the near future perfected atomic bombs will save mankind from bothering further about questions of selection, both with respect to theory and to their own survival. In this wonderful progress earlier stages may well be preserved. The backwoodsman may persist on the level of civilization of the Merovingians, and less efficient methods of killing each other can be preserved in the hinterlands of Central Africa or New Guinea. Similarly, the sluggish saurians were replaced by the more versatile, warm-blooded mammals, the marsupials by the *Placentalia,* but this does not alter the fact that lizards, snakes, tortoises, and crocodiles have survived to remind us of the former reptilian splendour, or that marsupials still exist in Australia, where no higher mammals had come in.

Discussions of this sort can be continued to mutual exhaustion, but do not, however, lead to the conviction of the respective opponents. The reason is understandable. We have one or two dozen experiments in which the "usefulness" of a character is experimentally demonstrated, as when, for example, those individuals of a species of insect which have a colour identical with the background are less eaten by birds than those with contrasting colour. But for the extrapolation that evolution was controlled by "usefulness" there is no way of experimental verification or falsification. If any species has survived and has undergone further evolution, then the change must have been either advantageous or related to advantageous changes or at least not deleterious, for otherwise the species would have simply been extinguished. But this ever remains a *vaticinatio post eventum.* Like a Tibetan prayer-wheel, Selection Theory murmurs untiringly: "Everything is useful." But as to what actually happened and which lines evolution has actually followed, selection theory says nothing, for the evolution is the product of "chance," and therein obeys no "law."

But is this correct?

CHANCE AND LAW

Is evolution a process accidental in itself and directed only through outside factors, namely, is it a product of random mutation and equally accidental environmental conditions resulting in the struggle for existence and selection, plus the accidental effect of isolation and subsequent speciation? Or is evolution determined or co-determined by laws lying in the organisms themselves?

This is a question that leads us out of the tug-of-war of opinions, different interpretations and hypotheses, and which can be judged on the basis of facts.

We must start from a fundamental statement. Mathematical analysis shows that selection-pressure is greatly superior to mutation-pressure: even a small selective advantage in a positive or negative way is much more effective than directed mutation without selection, even if the mutation in question should appear at a high rate and repeatedly. Therefore "directiveness" of evolution in the sense that it works against selection would be impossible; in the sense that it works without selection, it would be effective only over exceedingly long periods of time.

From these statements and the "undirectiveness" of mutations selectionism concludes that the direction of evolution is determined only by external factors. But this conclusion does not follow from the premises. If selection represents a *necessary* condition of evolution, it does not follow that it indicates a *sufficient* condition. . . .

If we examine the mutations in Drosophila, for example, we find that they give the impression of an uncontrolled multiplicity of variations. Furthermore, spontaneous mutations, as well as those induced by outside factors, are "accidental" with respect to external conditions, that is, they show no adaptive character. It is not the case that mutations which appear, say, at increased temperature represent adaptations to that higher temperature. Only the rate of mutations which also appear otherwise is increased. However, the multiplicity of mutations and their lack of adaptiveness and direction with respect to external influences do not necessarily mean that they are entirely fortuitous. Rather there are indications that mutations and evolutionary changes have many, but not an infinite number, of degrees of freedom.

Of course, mutation is limited, in the first place, by the nature of the genes present and the possibilities of their variation. Putting it crudely, a vertebrate will never produce a mutation that leads to the formation of a chitinous cover such as is found in insects, because this is not within the range of a vertebrate's anatomical and physiological ground-plan. The same holds for finer details. For instance, green colour is found rather seldom among butterflies, though it is common at the caterpillar stage and is, moreover, an excellent protective colour. Blue roses and black tulips could not be produced in spite of the florists' efforts over many centuries.

Investigation into the nature of the gene and of mutation leads to a similar conclusion. Since a gene is a physico-chemical structural unit of the nature of a large protein molecule, and since a mutation represents a transition to a new stable state by way of isomerization or a change in side chains or the like, a change will certainly be possible in a number of directions but not in all directions, in a similar way as only certain quantum states are permitted to an atom. In both cases, quantization is at the basis of the jump-like character of the change, as well as of the high stability and organization of the system. An atom cannot take up any small quantities of energy it is exposed to under the constant bombardment of the heat movements of the surrounding particles; only a complete quantum jump will induce a change in it. This ensures that it may remain unchanged over an indefinite time.[3] In the same way, the "quantized"

3. For an interesting discussion of the molecular basis for the stability of the gene see the Erwin Schrödinger selection in *The Mystery of Matter*, edited by Louise B. Young, Oxford University Press, 1965, pp. 433–60. Ed.

character of mutations is at the basis, first, of their discontinuity, secondly, of the great stability of the gene and the relative rarity of mutations, and thirdly, it follows that the number of possible mutations is not infinite, since only certain stable states are "allowed."

Here we come to an important problem. The theory of evolution, based upon an enormous amount of factual evidence, states that the animal and plant kingdoms have arisen, in the course of geological time, from simpler and more primitive forms to more complicated and more highly organized ones. Genetical experience leads us to accept as a fact that this has happened by way of step-like mutations. Actually, however, we find no evidence either in the living world of today or of past geological epochs for a continuous transition. What we actually find are separate and well-distinguished species. Even the existence of more or less numerous mutations, races, subspecies, etc., within the species does not alter the basic fact that intermediate stages from one species to another which should be found if there were a gradual transition, are not met with. The worlds of organisms, living and extinct, do not represent a continuum but a discontinuum.

The discontinuity of species is based presumably on the fact that certain conditions of stability exist not only for the individual genes but also for genomes.[4] In individual genes the transition from one state to another is discontinuous, and this is the reason for the jump-like character of mutation. As regards the conditions of stability for genomes, the following conception may be stated. A "species" represents a state in which a harmoniously stabilized "genic balance" has been established, that is, a state in which the genes are internally so adapted to each other that an undisturbed and harmonious course of development is guaranteed. If there are no external disturbances, stability is ensured for a theoretically unlimited number of generations. If a mutation occurs it means a disturbance of this pattern; therefore in the majority of cases, mutations will lead to unfavourable, even lethal, results, even before selection is taken into account. But every gene acts not only in its own right but more or less also as a modifying factor influencing the action of the rest of the genome. Species always differ from one another in a great number of genes. The more numerous the mutations that have occurred, however, the greater is the chance of a disturbance of the established genic balance, even if these mutations in themselves would be favourable. Hence a form which is on the way from one species to another will be in a state of instability, and it will be particularly exposed to selection. So this transition must be passed through quickly, or speaking statistically, such transitional stages will be comparatively rare. Finally, the form, provided it does not in the mean-

4. A *genome* is an interacting system of genes. Ed.

time become extinct, reaches a new state of genic balance in which it can remain again for a long time. This gives precisely the state of affairs we find in nature, namely, in general, well-defined species showing mutations but not continuous inter-gradation. . . .

Every law of nature implies a statement of recurrences. If, therefore, we wish to establish laws of evolution we must look for such recurrences. We find them in the phenomenon that parallel changes appear in many groups of organisms. . . . To cite the most famous example, lens-eyes constructed on the principle of the camera are found in very similar form in the scallop, the cuttle-fish, and vertebrates. . . . In invertebrates the eye is derived from the epidermis, in vertebrates it is a derivative of the brain. Anyway, once the way towards formation of a complicated eye is taken in phylogeny,[5] nature obviously has no other course than to pass through the successive stages of flat-eye, socket-eye, and lens-eye, and thus we find them in the most divergent classes of animals. . . . The same thing applies to physiological characters. For instance, there exists only a small number of respiratory pigments in the animal kingdom, mainly haemoglobin, chlorocruorin, haemerythrin, and haemocyanin; they are found independently in different animal groups, such as in vertebrates, ram's-horn snails (*Planorbis*), and bloodworms (larvae of the gnat *Chironomous*). The number of possible proteins has been estimated at 10^{2700}, contrasted with a total number of 10^{79} electrons in the universe. But although the formation of respiratory pigments has happened a number of times, it could take only one of these few directions.

Thus the changes undergone by organisms in the course of evolution do not appear to be completely fortuitous and accidental; rather they are restricted, first by the variations possible in the genes, secondly, by those possible in development, that is, in the action of the genic system, thirdly, by general laws of organization.

These factors seem to be responsible for the fact that evolution often conveys the impression of "orthogenesis," that is, of inherent trends in definite directions. As already stated, mathematical analysis shows that the effect of selection-pressure is much greater than that of mutation-pressure within the known rates of mutation. Therefore, orthogenesis, in the sense of trends to determine the evolutionary process in the teeth of selection, will be an exceptional phenomenon, or perhaps not exist at all. But there is orthogenesis in the sense that evolution is not determined merely by accidental factors of the environment and the resulting struggle for existence, but also by internal factors. . . .

While fully appreciating modern selection theory, nevertheless we ar-

5. *Phylogeny* is the development or evolution of a type of organism. *Ontogeny* is the development of an individual organism. Ed.

rive at an essentially different view of evolution. It appears to be not a series of accidents, the course of which is determined only by the change of environments during earth's history and the resulting struggle for existence, which leads to selection within a chaotic material of mutations. No more is it, to be sure, the work of mysterious factors, such as a striving for perfection or a tendency towards purposiveness or adaptation. But it appears as a process essentially co-determined by organic laws, which in suitable cases can be formulated in an exact way. . . . and we believe that the discovery of these laws constitutes one of the most important tasks of the future theory of evolution.

UNSCIENTIFIC INTERLUDE

The first Darwinist to explain organic "fitness" on the basis of events at random was the pre-Socratic Empedocles. In a famous fragment it is said that life sprang from the moist earth under the influence of fire; first single heads, eyes, limbs were formed, which united into monsters, such as creatures with many hands, cattle with men's heads, human bodies with bull's heads, until creatures capable of survival also appeared accidentally; these were the ancestors of the plants and animals of today. It is also known that Darwin's theory was an application of contemporary national economy to biological science. A most important starting point for Darwin was Malthus's statement that living beings multiply at a higher rate than the quantity of food available increases. Similarly, the consideration of biological phenomena in terms of "profit" and "competition" corresponded to the national economy of the Manchester school. In these general conceptions the emotional antipathy against "Darwinism" is rooted. . . .

The utilitarian conception in biology was in line with the general ideology. The machine theory of life is the perfect expression of the *Zeitgeist* of an epoch which, proud of its technological mastery of inanimate nature, also regarded living beings as machines.

Realization of the limits of the mechanistic conception led first to vitalism, which assumed the aggregate of parts and machine-structures to be controlled by purposive agents. In turn, the recognition of the inadequacy of both views led to organismic conceptions, which attempt to give a scientific meaning to the concept of wholeness. This is the common trend which we can find equally in biology, medicine, and psychology. . . .

Not only must the parts of the organism and the individual processes be considered, but also their mutual interactions and the laws governing the latter. These are manifest everywhere in the phenomena of regulation following disturbances, as well as in the normal functioning of the organism. Further, we have the concept of organization. The basic feature of the organic world is the tremendous hierarchy which extends from the molecules of organic compounds through self-multiplying biological units to cells and multicellular organisms, and finally to biological communities. New laws appear at each level of organization, and it is the task of biological research progressively to unveil them. Finally, there is the concept of dynamics. Living structures are not in being, but in becoming. They are the expression of a ceaseless stream of matter and energy, passing through the organism and forming it at the same time. . . .

For naïve and unbiased contemplation nature does not look like a calculating merchant; rather she looks like a whimsical artist, creative out of an exuberant fantasy and destroying her own work in romantic irony. The principles of "economy" and of "fitness" are true only in a Pickwickian sense. On the one hand, nature is a niggard when she insists on abolishing, say, an already minute rudimentary organ; this little economy having, as maintained by the theory of selection, enough advantage to be decisive in the struggle for existence. On the other hand, she produces a wealth of colour, form, and other creations, which, as far as we can see, is completely useless. Consider, for example, the exquisite artistry of butterflies' wings, which has nothing to do with function, and cannot even be appreciated by their bearers with their imperfect eyes.

This productivity and joy of creation seems to be expressed in the "horizontal" multiplicity of forms on the same level of organization as well as in the "vertical" progress of organization, which can, but need not necessarily, be considered as "useful." We have already said that such considerations are not a refutation of the theory of selection. There are, however, two aspects where we remain dissatisfied.

From the viewpoint of science, we are not satisfied with the meagre answer that all this is possible within the range of the established factors of evolution, that it has arisen somehow by a selection of the advantageous or been allowed to survive merely by the accidents of gene distribution of indifferent mutations. We rather want to know the "secret law" at which "the chorus hints."

On the other hand, being people of our time, we are inclined to consider the utilitarian theory as a sort of living fossil like the tuatara of New Zealand, as a relic of Victorian Upper-Middle class philosophy. It is the projection of the sociological situation of the nineteenth and early twentieth centuries into two billion years of the earth's history. "Progress" is "useful," therefore, "usefulness" is the "cause of progress." . . .

So, evolution appears to be more than the mere product of chance governed by profit. It seems a cornucopia of *évolution créatrice,* a drama full of suspense, of dynamics and tragic complications. Life spirals laboriously upwards to higher and ever higher levels, paying for every step. It develops from the unicellular to the multicellular, and puts death into the world at the same time. It passes into levels of higher differentiation and centralization, and pays for this by the loss of regulability after disturbances. It invents a highly developed nervous system and therewith pain. It adds to the primeval parts of this nervous system a brain which allows consciousness that by means of a world of symbols grants foresight and control of the future; at the same time it is compelled to add anxiety about the future unknown to brutes; finally, it will perhaps have to pay for this development with self-destruction. The meaning of this play is unknown, unless it is what the mystics have called God's attaining to awareness of Himself.

From the standpoint of science, however, the history of life does not appear to be the result of an accumulation of changes at random but subject to laws. This does not imply mysterious controlling factors that in an anthropomorphic way strive towards progressive adaptation, fitness, or perfection. Rather there are principles of which we already know something at present, and of which we can hope to learn more in the future. Nature is a creative artist; but art is not accident or arbitrariness but the fulfilment of great laws.

Editor's Note. The previous selection from Bertalanffy and the one following from Whyte are representative of a small but articulate minority of scientists who believe that biological processes can be understood best by considering each organism as a whole and by studying those properties which are present only when the parts interact to form an organized system.[1] They suggest that these coordinating factors may influence the direction of evolutionary change. This school of thought is called *organicism*. The reader should be aware, however, that these views do not represent the opinion of the majority of scientists today. Most biologists believe that Darwin's theory of natural selection modified by the later discoveries of the mechanism of mutation and heredity (sometimes called the "synthetic" theory) is sufficient to explain the evolutionary process.

Internal Factors in Evolution [*]
by Lancelot Law Whyte (1965)

It is now being suggested that beside the well-established external competitive selection of the "synthetic" theory of evolution, an internal selec-

---

1. For further readings on this subject see *The Mystery of Matter*, pp. 526–43.
[*] From Lancelot Law Whyte, *Internal Factors in Evolution*. Copyright © 1965 by Lancelot Law Whyte. Reprinted by permission of George Braziller, Inc.

tion process acts directly on mutations, mainly at the molecular, chromosomal, and cellular levels, in terms not of struggle and competition, but of the system's capacity for coordinated activity. The Darwinian criterion of fitness for external competition has to be supplemented by another: that of good internal coordination. Internal *co*adaptation is necessary as well as external adaptation. . . .

Viewed at any level, from organ and tissue down to cell and component molecules, the organism is a highly ordered system. This is true both of structures and processes. The macroscopic organs and their functions fit together. Similarly the basic structures, such as the chromosomes, form a coherent pattern and undergo collective motions and transformations displaying a high degree of spatial and temporal coordination. This coordination is pervasive at all levels. The ultra structure of the living cell is an intricate differentiated network undergoing global pulsations and transformations under some law of ordering which preserves the unity of the system. These are clumsy words beside the elegant harmony which they seek to describe. . . . Only those changes which result in a mutated system that satisfies certain stringent physical, chemical, and functional conditions will be able to survive the complex chromosomal, nuclear, and cellular activities involved in the processes of cell division, growth, and function.

"The struggle for survival of mutations begins at the moment mutation occurs," [1] but this is a "struggle" to conform, not to compete.

The greater we assume the degree of order in the cell, the more stringent these conditions will be, and they certainly must be severe, that is highly restrictive of the permissible changes. For example, at the chemical level only a limited number of stable macromolecular arrangements are possible—say in the pattern of atoms constituting a gene—and these in turn must conform to unknown principles of over-all ordering which must be satisfied if the various chromosomal activities are to be carried through without confusion. Only certain changes in the composition of the DNA and the arrangement of the chromosomes will be possible without loss of the faculty of replication or of other chromosomal functions. All the parts of the genetic system must be adapted to one another so that they work together. . . .

Internal selection has hitherto been neglected for two main reasons: little has been known about it, and it is probably mainly associated with those changes in general organization about which we are also ignorant.

These are the provisional conclusions to which one is inescapably led by taking seriously the high internal order of organisms and in particular of genotypes.[2]

1. From L. L. Whyte, *Science*, 132, p. 954 (1960).
2. A *genotype* is the hereditary structure of an organism. Ed.

The mutated genotype which satisfies the co-ordinative conditions (C.C.) develops successfully and proceeds to the second test of adaptive selection. But how are those genotypes which do not satisfy the C.C. eliminated? This may, it seems, occur in at least three ways:

1. An inappropriate molecular configuration may be forced back to its original state by a *return* mutation.

2. The mutated or disturbed configuration may, in the course of the earliest chromosomal activities, be molded into a *new* configuration which does satisfy the C.C. This hypothetical prefunctional adjustment of a non-conforming disturbed genotype will be called its *reformation*. Though there is as yet no direct evidence for this effect, it is a type of assimilation and is similar to the "adjustment" of mutations which has been discussed in the literature, and it appears to be a probable consequence of the operation of powerful C.C. A complex ordered system in the course of its functional activities is likely under known physical principles (e.g. the tendency towards configurations of minimal potential energy) to possess the power to adjust parts that do not properly conform. This is the self-repairing property of organisms viewed at the basic structural level, but with this important difference: a *new* functional configuration results. As a possible source of favorable mutations this mechanism is of great importance and requires more detailed consideration.

Medawar [3] has suggested that there is in systems of polymers a "repetitiousness" or tendency towards elaboration, so that evermore complex sets of genetical instructions are offered for trial, and that this may provide a basis for advance in complexity. . . .

This tendency towards the formation of more complex unified patterns does not imply any obscure vitalistic factor, since in appropriate circumstances it can be the direct result of the tendency towards arrangements of minimal potential energy. Thus the potential energy principle can, in complex low temperature quantum mechanical systems, produce a structuring or formative tendency which, under certain conditions, will shape the genetic system towards novel, stable, unified arrangements. Arbitrary changes in the genetic system may thus be reformed into favorable mutations satisfying the C.C.

3. Such return mutations and reformation will only be possible when certain critical thresholds have not been passed. In other cases the defective genotype will result in the death of the resulting cells, tissues, or organisms.

These lethal mutations have been known for several decades. But it has only recently become evident that the existence of such lethal mutations, if they cannot be compensated by a change in the environment, implies the operation of a second radically distinct kind of selective process, one

3. See P. B. Medawar, *The Future of Man* (London: Methuen, 1960, p. 38).

in which the criterion is not a matter of competition but of coordination. Deleterious or lethal mutations, expressing a failure of coordination and not merely the failure to produce some metabolite which the environment might supply, involve a new type of selection. . . .

Some who have considered the operation of internal factors have argued that what is here called internal selection is merely Darwinian selection extended to cover the internal environment. It is suggested that any attempt to distinguish the two fails because on a truly relational view in which interactions are recognized as pervasive there is no distinction between the internal and external environments. "Natural selection" occurs at all levels, everywhere.

Darwinian adaptive selection is a relative, statistical, external process effective between pairs of individuals or populations, a matter of degree determined only by comparative fitness and frequencies in some particular environment. Internal selection is an intrinsic, usually all-or-none process operating within single individuals, determined by observing the history of one organism in the most favorable environment. Adaptive selection depends on a high death rate and expresses competition; internal selection rests on coordination within individuals.

The evidence for internal selection is not a relative rate of increase in a given environment, but the direct structural observation of the successful (or unsuccessful) internal coordination of a single mutated organism.

Suggestions for Further Reading

* Asimov, Isaac: *The Genetic Code*, Signet Science Library, The New American Library of World Literature, 1963. A biochemist and popular science writer explains the biochemistry of the cell and the process of heredity with particular emphasis on the chemistry of DNA.

* ———: *The Wellsprings of Life*, Signet Science Library, The New American Library of World Literature, 1960. A good historical account of our growing understanding of the origin and development of life. The structure of the cell and the biochemistry of its food and energy processes are examined, together with the experimental findings concerning regeneration, evolution, and inheritance.

Beadle, George and Muriel: *The Language of Life*, Doubleday & Co., 1966. An excellent introduction to genetics and evolution.

* Huxley, Julian: *Evolution, The Modern Synthesis*, Science Editions, John Wiley & Sons, 1964. A reprint with a new Introduction of an older book (1942) which is still one of the most comprehensive and authoritative presentations of the "synthetic" theory of evolution.

* Paperback edition

————: *Heredity, East and West,* Henry Schuman, Inc., 1949. Contains a very interesting account of the peculiar brand of Lamarckism styled "Michurinism" and given official sanction in the U.S.S.R. under the influence of Lysenko.

Moore, Ruth: *The Coil of Life,* Alfred A. Knopf, 1960. Describes in historical and biographical form the development of modern concepts of the biochemical nature of life.

* Simpson, George Gaylord: *The Meaning of Evolution,* Yale Paperbound, Yale University Press, 1960. A very fine general survey of the principles of evolution and the meaning of evolution for human beings.

Whyte, Lancelot Law: *Internal Factors in Evolution,* George Braziller, 1965. Readers who were interested in the selection on p. 127 will do well to read Whyte's complete essay presenting this idea.

3

Does Evolution Imply Progress or Purpose?

Ancient man believed in a cosmic purpose but not in
the idea of progress. Modern man believes in progress
but not in a cosmic purpose. How has the theory of
evolution influenced this change in beliefs?

Does Evolution Imply Progress or Purpose?

Introduction

THE GREEKS believed that the world represented a static hierarchy of different levels of organic form, like steps leading from the
lowest and simplest to the highest and most complex. Man, of course, was
on the highest level. Each level was the result of a creative act which had
taken place at the beginning of the world, and each level fulfilled a specific purpose. The whole represented a great cosmic design whose purpose was the creation of man. In this world-view, progress existed only in
the sense of an increasing approximation to perfection at each level of
being. But this progress had occurred once and was crystallized into a design which was static and complete within itself.

This explanation of the relationship of living forms was a reasonable
one within the context of the scientific knowledge of ancient and even
medieval times. There was very little evidence then that species had developed by an historical process from one organic form to another. Christian doctrine which was formulated during the ascendancy of this view
quite naturally adopted the static explanation of the creation of living
things. The alternative hypothesis that creation had taken place by a
dynamic process in time is not really antithetical to the basic tenets of
Christian thought, but, by the time that evidences of evolution began to
be discovered, Christianity was already deeply committed to the former
explanation of the origin of life.

The publishing of Darwin's theory in 1859 and the subsequent uncovering of fossil evidence supporting the evolutionary hypothesis rapidly un-
dermined the accepted religious view of creation. Furthermore, the cause
of evolution which Darwin proposed seemed to rule out any possibility of
a divine plan or purpose. It was not so much that Darwin suggested a
natural cause for evolution (natural law can be interpreted as a manifestation of divine will), but the random nature of the force that was postulated was difficult to reconcile with any concept of purpose. Darwinism,
as interpreted early in the twentieth century, seemed to replace an intelligent design of creation with a "blind, deaf, dumb, heartless, senseless mob
of forces." [1] There were many who found it difficult to accept such a
mechanistic view of the origin of living things.

A number of rival theories were proposed and gained popular support
in the early part of this century when evolutionary theory was floundering
between the excessively mechanistic ideas of the neo-Darwinists and the
wishful thinking of the neo-Lamarckists. The science of genetics was not

1. See Bernard Shaw selection, p. 150.

yet well enough advanced to resolve these arguments. In this atmosphere of doubt a number of biologists turned to alternative explanations, suggesting non-materialistic causes for evolution, in addition to the materialistic causes which had already been identified. Finalism, vitalism, orthogenesis, holism, *élan vital,* goal-seeking, and many other evolutionary theories were proposed.

The finalists and, to a lesser extent, the believers in orthogenesis suggested that the end, although later in time, is the cause of the preceding events. An inner destiny of the species is held to be responsible for the direction of evolutionary change. These theories imply a cosmic plan or purpose and relegate progress to a minor role, the unwinding of a preconceived design.

The vitalists, on the other hand, believed that there is a special life substance or force which is different from the substances and forces found in inorganic matter. This vital essence is seen as an impulsion affecting the direction of evolution, but in these theories the end is not predetermined. Progress is achieved through a spontaneous and inventive process.

These hypotheses and the other similar ones all invoke forces which are not accessible to scientific verification. The only point that can be settled scientifically is whether such theories can be shown to be contradictory to known facts. For instance, the randomness of mutational change is thought to be in direct opposition to the idea of finalism, which postulates the working out of a preconceived design.

We do not have the space in these readings to give a fair presentation of each of these theories. Henri Bergson's *Creative Evolution* has been chosen as representative of this type of thought. His theory, although not generally accepted by biologists today, has had a wide influence on both science and philosophy.

As the modern synthetic theory was developed (by men like Haldane, Fisher, Wright, Dobzhansky, Waddington, and Huxley), the process of evolution was seen to be a much more subtle combination of forces than had been supposed by the neo-Darwinists. It was recognized that specific characteristics are not inherited, as such. What is inherited is a series of determiners for a developmental system. The characteristics that result from this inheritance depend on the interaction of the determiners, the environment during development, and the activities of the organism. Although mutation does appear to be non-adaptive in relation to the needs of the organism, natural selection is seen to be an anti-chance agency, making adaptive sense out of the relative chaos of countless combinations of mutant genes. The modern evolutionists also emphasize the creativity in evolution. Some of them see in this process, creation as a living reality of the present and man "not merely as a witness but a participant and partner, as well." [2]

Obviously, this explanation can be given a theistic interpretation. Christianity also believes that mankind is progressing toward a higher degree of perfection. St. Paul said: ". . . the creature itself also shall be delivered from the bondage of corruption into the glorious liberty of the children of God. For we know that the whole creation groaneth and travaileth in pain until now." [3]

Teilhard de Chardin was one of the few religious thinkers in recent years to attempt to reconcile religious and evolutionary thought. But his *Phenomenon of Man,* written in 1938, was refused publication by the Church until after his death in 1955. It is interesting that his thoughts were championed first by scientists like Julian Huxley.

As Bernard Shaw points out, the evolutionary view of creation also offers an explanation of the presence of evil, a problem which had embarrassed religion and philosophy since the beginning. If creation must "struggle with matter and circumstance by the method of trial and error, then the world must be full of unsuccessful experiments . . ." [4] It is worth noting that Shaw was appealing to Bergson's theory to support his views, but Dobzhansky's position as outlined here would have supported them just as well.

As the reader will see, there is some difference of opinion among contemporary scientists about the concept of progress in evolution. Julian Huxley speaks out clearly in support of biological progress. Haldane and others conclude that it is not so much progress as it is simply change. They all find, however, that along certain lines some progress has occurred: increasing awareness, increasing power to master and mold the environment. And according to some definitions man can be said to be the present summit of the process. However, there are also many cases where no progress has occurred, and there are cases where the direction of change has been toward simplification or parasitism.

The fact that progress is not a universal and inevitable feature of evolutionary change is considered by most biologists to disprove the existence of a predetermined cosmic plan. "Plan there may be," says Donald Culross Peattie, "but only a working plan, a vast experimentation still in course." [5]

2. Theodosius Dobzhansky, *The Biological Basis of Human Freedom,* Columbia University Press, New York, 1956, p. 5.

3. Rom. 8: 21–22.

4. See Bernard Shaw selection, p. 151.

5. Donald Culross Peattie, *An Almanac for Moderns,* G. P. Putnam's Sons, New York, p. 64.

"The purpose manifested in evolution," says Julian Huxley, "whether in adaptation, specialization, or biological progress, is only an apparent purpose. It is just as much a product of blind forces as is the falling of a stone to earth or the ebb and flow of the tides." [6]

George Gaylord Simpson is even more emphatic: "Man is the result of a purposeless and materialistic process that did not have him in mind. He is a state of matter, a form of life, a sort of animal . . ." [7] But, he goes on to say (and this is another point on which all the biologists agree) man does represent the highest form of organization of matter and energy that has yet appeared. With his coming, purpose, intelligence, and the knowledge of good and evil have entered the cosmic process. Thus, purpose which man searched for throughout nature has been found enshrined within his own mind.

Charles Sherrington caps this statement with a final question: Does the fact that the cosmic process has evolved mind and purpose tell us anything significant about the nature of the universe?

6. See Julian Huxley selection, p. 170.
7. See George G. Simpson selection, p. 172.

Chance or Design in Evolution

Belief in a Cosmic Purpose *

by Donald Culross Peattie (1935)

First to grasp biology as a science, Aristotle thought that he had also captured the secret of life itself. From the vast and original body of his observation, he deduced a cosmology like a pure Greek temple, symmetrical and satisfying. For two millenniums it housed the serene intelligence of the race.

Here was an absolute philosophy; nothing need be added to it; detraction was heretical. It traced the ascent of life from the tidal ooze up to man, the plants placed below the animals, the animals ranged in order of increasing intelligence. Beyond man nothing could be imagined but God, the supreme intelligence. God was all spirit; the lifeless rock was all matter. Living beings on this earth were spirit infusing matter.

Still this conception provides the favorite text of poet or pastor, praising the earth and the fullness thereof. It fits so well with the grandeur of the heavens, the beauty of the flowers at our feet, the rapture of the birds! The Nature lover of today would ask nothing better than that it should be true.

Aristotle was sure of it. He points to marble in a quarry. It is only matter; then the sculptor attacks it with his chisel, with a shape in his mind. With form, soul enters into the marble. So all living things are filled with soul, some with more, some with less. But even a jellyfish is infused with that which the rock possesses not. Thus existence has its origin in supreme intelligence, and everything has an intelligent cause and serves its useful purpose. That purpose is the development of higher planes of existence. Science, thought its Adam, had but to put the pieces of the puzzle together, to expose for praise the cosmic design, all beautiful.

✽   ✽   ✽

* From Donald Culross Peattie, *An Almanac for Moderns,* G. P. Putnam's Sons, New York. Copyright 1935 by Donald Culross Peattie. Reprinted by permission.

Archaic and obsolete sounds the wisdom of the great old Greek. Life, his pronouncement ran, is soul pervading matter. What, soul in a jellyfish, an oyster, a burdock? Then by soul he could not have meant that moral quality which Paul of Tarsus or Augustine of Hippo were to call soul. Aristotle is talking rather about that undefined but essential and precious something that just divides the lowliest microörganism from the dust; that makes the ugly thousand-legged creature flee from death; that makes the bird pour out its heart in morning rapture; that makes the love of man for woman a holy thing sacred to the carrying on of the race.

But what is this but life itself? In every instance Aristotle but affirms that living beings are matter pervaded by a noble, a palpitant and thrilling thing called life. This is the mystery, and his neat cosmology solves nothing of it. But it is not Aristotle's fault that he did not give us the true picture of things. It is Nature herself, as we grow in comprehension of her, who weans us from our early faith.

❋   ❋   ❋

There is no certainty vouchsafed us in the vast testimony of Nature that the universe was designed for man, nor yet for any purpose, even the bleak purpose of symmetry. The courageous thinker must look the inimical aspects of his environment in the face, and accept the stern fact that the universe is hostile and deathy to him save for a very narrow zone where it permits him, for a few eons, to exist.

❋   ❋   ❋

As it was Aristotle himself who taught us to observe, investigate, deduce what facts compel us to deduce, so we must concede that it is Nature herself, century after century—day after day, indeed, in the whirlwind progress of science—that propels us farther and farther away from Aristotelian beliefs. At every point she fails to confirm the grand old man's cherished picture of things. There are persons so endowed by temperament that they will assert that if Nature has no "soul," purpose, nor symmetry, we needs must put them in the picture, lest the resulting composition be scandalous, intolerable, and maddening. To such the scientist can only say, "Believe as you please."

Evidence of a Purpose in Nature [*]

by William Paley (1802)

In crossing a heath, suppose I pitched my foot against a *stone*, and were asked how the stone came to be there: I might possibly answer, that for any thing I knew to the contrary, it had lain there for ever: nor would it perhaps be very easy to shew the absurdity of this answer. But suppose I had found a *watch* upon the ground, and it should be inquired how the watch happened to be in that place; I should hardly think of the answer which I had before given, that, for any thing that I knew, the watch might have always been there. Yet why should not this answer serve for the watch as well as for the stone? Why is it not as admissible in the second case, as in the first? For this reason, and for no other, viz. that, when we come to inspect the watch, we perceive (what we could not discover in the stone) that its several parts are framed and put together for a purpose, e.g., that they are so formed and adjusted as to produce motion, and that motion so regulated as to point out the hour of the day; that, if the different parts had been differently shaped from what they are, of a different size from what they are, or placed after any other manner, or in any other order, than that in which they are placed, either no motion at all would have been carried on in the machine, or none which would have answered the use that is now served by it. To reckon up a few of the plainest of these parts, and of their offices, all tending to one result:—We see a cylindrical box containing a coiled elastic spring, which, by its endeavour to relax itself, turns round the box. We next observe a flexible chain (artificially wrought for the sake of flexure), communicating the action of the spring from the box to the fusee. We then find a series of wheels, the teeth of which catch in, and apply to each other, conducting the motion from the fusee to the balance, and from the balance to the pointer; and at the same time, by the size and shape of those wheels so regulating that motion, as to terminate in causing an index, by an equable and measured progression, to pass over a given space in a given time. We take notice that the wheels are made of brass in order to keep them from rust; the springs of steel, no other metal being so elastic; that over the face of the watch there is placed a glass, a material employed in no other part of the work, but in the room of which, if there had been any other than a transparent substance, the hour could not be seen without opening the case. This mechanism being observed (it requires in-

[*] From William Paley, *Natural Theology,* 1802.

deed an examination of the instrument, and perhaps some previous knowledge of the subject, to perceive and understand it; but being once, as we have said, observed and understood), the inference, we think, is inevitable, that the watch must have had a maker; that there must have existed, at some time, and at some place or other, an artificer or artificers, who formed it for the purpose which we find it actually to answer; who comprehended its construction, and designed its use.

There cannot be a design without a designer; contrivance without a contriver; order without choice; arrangement without anything capable of arranging; subserviency and relation to a purpose, without that which could intend a purpose; means suitable to an end, and executing their office in accomplishing that end, without the end ever having been contemplated, or the means accommodated to it. Arrangement, disposition of parts, subserviency of means to an end, relation of instruments to a use, imply the presence of intelligence and mind.

Sturmius held, that the examination of the eye was a cure for atheism.

The gist of Darwin's theory is this simple idea: that the Struggle for Existence in Nature evolves new Species *without* design just as the Will of Man produces new Varieties in Cultivation *with* design.

Ernst Haeckel
The Evolution of Man, New York, 1896

Concluding Remarks *

by Charles Darwin (1872)

I see no good reason why the views given in this volume should shock the religious feelings of any one. It is satisfactory, as showing how transient such impressions are, to remember that the greatest discovery ever made by man, namely, the law of the attraction of gravity, was also attacked by Leibnitz, "as subversive of natural, and inferentially of revealed, religion." A celebrated author and divine has written to me that "he has gradually learnt to see that it is just as noble a conception of the Deity to believe that He created a few original forms capable of self-development into other and needful forms, as to believe that He required a fresh act of creation to supply the voids caused by the action of His laws." . . .

* From Charles Darwin, *The Origin of Species*, 6th edition, 1872.

When I view all beings not as special creations, but as the lineal descendants of some few beings which lived long before the first bed of the Cambrian system was deposited, they seem to me to become ennobled. Judging from the past, we may safely infer that not one living species will transmit its unaltered likeness to a distant futurity. And of the species now living very few will transmit progeny of any kind to a far distant futurity; for the manner in which all organic beings are grouped, shows that the greater number of species in each genus, and all the species in many genera, have left no descendants, but have become utterly extinct. We can so far take a prophetic glance into futurity as to foretell that it will be the common and widely-spread species, belonging to the larger and dominant groups within each class, which will ultimately prevail and procreate new and dominant species. As all the living forms of life are the lineal descendants of those which lived long before the Cambrian epoch, we may feel certain that the ordinary succession by generation has never once been broken, and that no cataclysm has desolated the whole world. Hence we may look with some confidence to a secure future of great length. And as natural selection works solely by and for the good of each being, all corporeal and mental endowments will tend to progress towards perfection.

It is interesting to contemplate a tangled bank, clothed with many plants of many kinds, with birds singing on the bushes, with various insects flitting about, and with worms crawling through the damp earth, and to reflect that these elaborately constructed forms, so different from each other, and dependent upon each other in so complex a manner, have all been produced by laws acting around us. These laws, taken in the largest sense, being Growth with Reproduction; Inheritance which is almost implied by reproduction; Variability from the indirect and direct action of the conditions of life, and from use and disuse: a Ratio of Increase so high as to lead to a Struggle for Life, and as a consequence to Natural Selection, entailing Divergence of Character and the Extinction of less-improved forms. Thus, from the war of nature, from famine and death, the most exalted object which we are capable of conceiving, namely, the production of the higher animals, directly follows. There is grandeur in this view of life, with its several powers, having been originally breathed by the Creator into a few forms or into one; and that, whilst this planet has gone cycling on according to the fixed law of gravity, from so simple a beginning endless forms most beautiful and most wonderful have been, and are being evolved.

Creativity and Orientation in Evolution *
by Theodosius Dobzhansky (1967)

THE CREATIVITY OF NATURAL SELECTION

The occurrence of mutation and of natural selection is said to be due to "chance." Can one possibly believe that the *humanum* arose through summation of a series of accidents? De Cayeux,[1] a French paleontologist, emphatically disagrees: "The lottery does not suffice. Mutationism is a double failure. The evolution of species cannot be due to chance alone, even if corrected by the selection." Protestations to the same effect have been made with considerable eloquence by such writers as Joseph Wood Krutch, Barzun, and others. Now, a goodly portion of all this is sheer misunderstanding. What, indeed, is meant by "chance," or "randomness," in evolution? Reference has already been made above to the fact that mutations are adaptively ambiguous. Nature has not seen fit to make mutations arise where needed, when needed, and only the kind that is needed. A geneticist would be hard put to envisage a way to accomplish this. But not even mutations are random changes, because what mutations can take place in a given gene is evidently decreed by the structure of that gene. Mutations do not, however, constitute evolution; they are only raw materials for evolution. They are manipulated by natural selection.

Natural selection is a chance process (despite the misplaced superlative "fittest" in the "survival of the fittest") only in the sense that most genotypes have not absolute but only relative advantages or disadvantages compared to others. Natural selection may act if one genotype leaves 100 surviving offspring compared to 99 left by the carriers of another genotype. Otherwise natural selection is an antichance agency. It makes adaptive sense out of the relative chaos of the countless combinations of mutant genes. And it does so without having a will, intention, or foresight. . . . The best analogue of natural selection is a cybernetic mechanism; it transmits "information" about the state of the environment to the genotype.

The point so central that it must be pressed is that natural selection is

* From Theodosius Dobzhansky, *The Biology of Ultimate Concern,* Perspectives in Humanism Series, Edited by Ruth Nanda Anshen, The New American Library, New York. Copyright © 1967 by Theodosius Dobzhansky. Reprinted by permission of the World Publishing Company.
    1. A. de Cayeux, *Trente Millions de siècles de vie,* André Bonne, Paris, 1958.

in a very real sense creative. It brings into existence real novelties—genotypes which never existed before. Moreover, these genotypes, or at least some of them, are harmonious, internally balanced, and fit to live in some environments. Writers, poets, naturalists, have often declaimed about the wonderful, prodigal, breathtaking inventiveness of nature. They have seldom realized that they were praising natural selection. But a creative process runs the risk of failure and miscreation; this is where creation is less safe than mass production from a set pattern. Miscreation in biology is death, extinction. Paleontology abundantly shows that extinction is the most usual end of evolutionary lines. Some biologists who thought that natural selection is too soulless and mechanical believed instead in orthogenesis, a theory according to which evolution is merely an uncovering of the preformed and predetermined organic configurations. Yet, with orthogenesis evolution would create nothing really new, and extinction would be a mystery or plain nonsense.

There are, broadly speaking, two kinds of interpretations of evolution. One kind supposes that any and all evolutionary changes that ever occurred were predestined to occur. The other kind recognizes that there may be many different ways of solving the problems of adaptation to the same environment; which one, if any, of these ways is in fact adopted in evolution escapes predetermination. . . .

ORIENTATION IN EVOLUTION

Evolution is not a collection of independent and unrelated happenings; it is a system of interrelated events. Life could not have arisen until cosmic evolution had produced at least one planet capable of supporting life. A being such as man, with a capacity for symbolic thinking and for self-awareness, could not have appeared until biological evolution had generated organisms with highly developed brains. Since certain evolutionary events could have happened only on the foundation of a series of preceding events, the history of the universe may be said to have an "orientation." We may choose to call the evolutionary line that produced man the "privileged axis" of the evolutionary process.

"Orientation" may, however, be understood also in a different manner. The process of evolution may be oriented, guided, and propelled by some natural or supernatural agency. Evolution was then able to follow only a single path, so that its final outcome, as well as all the stages through which it had to pass, were predestined and have appeared in a certain fixed order in time. Some minority schools among biologists believe in such a foreordained orientation. For example, the finalists posit that all evolution occurred for the express purpose of producing man, and that evolutionary changes at all times were guided toward this goal by some supernatural power or powers. Another school maintains that evolution, at least biological evolution, is orthogenesis. The changes that occur are sequences of events determined by factors inside the organism, by the structure of its genetic endowment, and proceed straight toward a fixed objective, such as man. The evolutionary development follows, then, a predetermined path, and its final outcome is likewise predetermined.

In contradistinction to finalism, orthogenesis does not necessarily assume supernatural forces guiding evolution. The favorite "explanation," which is really nothing more than an attractive analogy, is that evolutionary development (phylogeny) is predetermined in the same way as is the development of an individual (ontogeny). A fertilized human egg cell does not contain a homunculus, a little human figure, and yet from this egg cell arises an embryo, which undergoes many complex transformations and growth and finally becomes an adult man. Did the bodies of our remotest ancestors, or even primordial life, contain all the rudiments needed to produce all evolutionary developments? If evolution is orthogenesis, then it is what the etymology of the word "evolution" implies, i.e., unfoldment of preexisting rudiments, like the development of a flower from a bud. Finalism and orthogenesis have this much in common: the evolutionary history of the living world was predestined at the beginning of life and even in primordial matter. If true, this would make evolution a rather dull affair. Evolution produced nothing really new, since all that it did produce was ordained to happen. It lacked all freedom and creativity. . . .

The harmfulness of most mutation is a dramatic demonstration of the absence of guidance in evolution. At the level of mutation, evolution is neither directional nor oriented nor progressive. It is the very antithesis of orthogenesis. Mutation alone would cause chaos, not evolution. Natural selection redresses the balance. Harmful genes are reduced in frequency, and useful ones perpetuated and multiplied. . . . Natural selection works not with genes but with whole genetic endowments; what survives or dies, begets progeny or remains childless, is not a gene but a living individual. A gene useful in combination with some genes may be harmful in combination with others. The changes which natural selection promotes at present depend upon the changes that occurred in the past. Natural selection is comparable not to a sieve but to a regulatory mechanism in a cybernetic system. The genetic endowment of a living species receives and accumulates information about the challenges of the environments in which the species lives. The evolutionary changes are creative responses to the challenges of the environment. . . .

The process of organic evolution can be epitomized in the terms used by Toynbee in his monumental *Study of History* to describe the origins of civilizations and their subsequent histories: challenge and response. Challenges come from the environment, the responses occur through the agency of natural selection. In the absence of challenge there is evolutionary stagnation; hence the "living fossils," species that show no perceptible change over long stretches of geological time. But a challenge does not guarantee that a response will be given, nor does it determine exactly what the response will be. A species which fails to respond because of the absence of suitable genetic materials will go into decline or become extinct. A response may be the transformation of the species into a new state (anagenesis), or splitting into several species, adaptively specialized for different environments (cladogenesis). Cladogenesis enhances organic diversity; it may be said to create numerous trial parties "exploring" the possibilities of different environments ("groping" in Teilhard de Chardin's more picturesque simile).

The evolution of life took a long time, two billion years at least, because it was not a simple matter of unfolding or uncovering something which was there from the beginning. The history of life is comparable to human history in that both involve creation of novelty. Both proceed by groping, trial and error, many false starts, being lost in blind alleys, failures ending in extinction. Both had, however, also their successes, master strokes, and both achieved an overall progress. There has been a general acceleration of evolution. Cosmic evolution took more time than the biological, and human evolution is of shortest duration. The concepts of creativity and freedom are not directly applicable below the human level. It may nevertheless be argued that the rigid determinism is becoming gradually relaxed as the evolution of life progresses. The elements of creativity are more perceptible in the evolution of higher than in that of lower organisms.

Adherents of finalism and orthogenesis contend that since it is quite incredible that evolution could all be due to "chance," one must assume that it has had a design which it has followed. The reality is, however, more complex and more interesting than the chance vs. design dichotomy suggests. The statement that life must have had from the very beginning the potentialities for all the evolutionary developments which did in fact occur is obviously true but just as obviously trivial. If it were otherwise evolution could not have done what it did. What is more important is that life had also innumerable other potentialities which remained unrealized. This follows from the fact that only a minuscule fraction of the possible gene combinations can ever be actualized. It is misleading to think that all organisms, including man, were preformed in the primordial life, and merely needed time to unfold, like flowers from buds. The real problem lies in a different dimension.

A child receives one-half of the genes of his father, and one-half of the

maternal ones; which particular maternal and paternal genes are transmitted to a given child is a matter of chance. Which mutations occur, and
when and where, is also a matter of chance. And yet, evolution is a matter
of chance no more than a mosaic picture is a fortuitous aggregation of
stones. (Some modern "painters" make "pictures" by pelting a canvas
with paint, but there seems to be no biological analogy to this stunt.) In
evolution, chance is bridled in by an antichance agency, which is natural
selection responding to environmental challenges. Let us be reminded
that natural selection does not act as a sieve, but as a much more sophisticated regulating and guiding device. What is most remarkable is that the
"guidance" does not amount to a rigid determinism. Especially in the evolution of higher organisms there are discernible elements of creativity and
freedom.

How is this possible? Many environmental challenges can be met successfully in more than one way. For example, all desert plants must cope
with dryness. Different plants do so, however, by different means. Some
have leaves reduced to spines, others have leaves protected by waxy or
resinous secretions, others shed their leaves when humidity becomes deficient, and still others germinate, grow, flower, and mature seeds all within
a short span of time when water is available. Animals above a certain
body size must have some organs of respiration; these may be gills, or
tracheae, or lungs of a variety of kinds. Several groups of animals have
evolved flight, and again did so in different ways. To live in cold countries, animals may have warm fur and a metabolism which maintains a
constant body temperature, or they may be dormant during the cold season, or they may migrate to warmer countries. Some animals avoid their
enemies by concealing shapes and colorations, which make them hard to
see; others have gaudy warning colorations, which advertise their presence and their real or pretended dangerous or obnoxious qualities to their
potential enemies. All these methods of adaptation are sometimes successful, and none seems intrinsically superior to the others.

The multiplicity of ways of becoming adapted to similar environments
is not in accord with hypotheses of design and orthogenesis in evolution;
these hypotheses would lead one rather to expect that a single, and presumably most perfect method, will be used everywhere. On the contrary,
natural selection is more permissive. Nineteenth-century evolutionists
called natural selection "the survival of the fittest"; we prefer to delete the
superlative: what is fit to survive survives.

If evolution follows a path which is predestined (orthogenesis), or if it
is propelled and guided toward some goal by divine interventions (finalism), then its meaning becomes a tantalizing, and even distressing, puzzle. If the universe was designed to advance toward some state of abso-

lute beauty and goodness, the design was incredibly faulty. Why, indeed, should many billions of years be needed to achieve the consummation? The universe could have been created in the state of perfection. Why so many false starts, extinctions, disasters, misery, anguish, and finally the greatest of evils—death? The God of love and mercy could not have planned all this. Any doctrine which regards evolution as predetermined or guided collides head-on with the ineluctable fact of the existence of evil.

Human Values *

by Julian Huxley (1953)

There is, of course, a problem of values lurking in the background. How, for instance, can we reconcile the existence of so much cruelty and suffering with the concept of progress?

The female ichneumon fly lays her eggs deep within the body of a caterpillar by means of her long sharp ovipositor—a beautiful adaptation. The young grubs on hatching out start to eat up the living tissues, rather as if a brood of rats were to eat their way through a living sheep. But the ichneumon grubs show a further adaptation. At first they spare the vital organs, subsisting only on the reserve stores of fat, the connective tissues, and the like. Only when the caterpillar has become full-grown and turned into a chrysalis do the ichneumons attack it radically, eating up all that is left inside the protective shell. Finally, burrowing their way out, they too pupate, turning into white cocoons on the outside of the empty shell of their victim.

In general, parasites have evolved at the expense of their hosts, often causing great suffering, as with the fly maggots that live in the noses of various animals.

Furthermore, the adaptations of parasites often involve large-scale degeneration in the parasites themselves, especially of locomotor and sensory organs. Sacculina is descended from a free-living crustacean: today it is little more than a bag of reproductive cells attached to a system of root-like processes draining nourishment from the body of its crab host. How can this be reconciled with the notion of advance?

There is no simple answer. The specializations of Sacculina or a liver fluke for a parasitic existence are improvements from the angle of the parasites' own evolution. Pain and suffering are part of the wastage involved in the workings of the selective process. We must not expect to find human values at work in nature's day-to-day operations.

The Humanitarians and the Problem of Evil *
by Bernard Shaw (1921)

Yet the humanitarians were as delighted as anybody with Darwinism at first. They had been perplexed by the Problem of Evil and the Cruelty of Nature. They were Shelleyists, but not atheists. Those who believed in God were at a terrible disadvantage with the atheist. They could not deny the existence of natural facts so cruel that to attribute them to the will of God is to make God a demon. Belief in God was impossible to any thoughtful person without belief in the Devil as well. The painted Devil, with his horns, his barbed tail, and his abode of burning brimstone, was an incredible bogey; but the evil attributed to him was real enough; and the atheists argued that the author of evil, if he exists, must be strong enough to overcome God, else God is morally responsible for everything he permits the Devil to do. Neither conclusion delivered us from the horror of attributing the cruelty of nature to the workings of an evil will, or could reconcile it with our impulses towards justice, mercy, and a higher life.

A complete deliverance was offered by the discovery of Circumstantial Selection: that is to say, of a method by which horrors having every appearance of being elaborately planned by some intelligent contriver are only accidents without any moral significance at all. Suppose a watcher from the stars saw a frightful accident produced by two crowded trains at full speed crashing into one another! How could he conceive that a catastrophe brought about by such elaborate machinery, such ingenious preparation, such skilled direction, such vigilant industry, was quite unintentional? Would he not conclude that the signal-men were devils?

Well, Circumstantial Selection is largely a theory of collisions: that is, a theory of the innocence of much apparently designed devilry. In this way Darwin brought intense relief as well as an enlarged knowledge of facts to the humanitarians. He destroyed the omnipotence of God for them;

but he also exonerated God from a hideous charge of cruelty. Granted that the comfort was shallow, and that deeper reflection was bound to shew that worse than all conceivable devil-deities is a blind, deaf, dumb, heartless, senseless mob of forces that strike as a tree does when it is blown down by the wind, or as the tree itself is struck by lightning. That did not occur to the humanitarians at the moment: people do not reflect deeply when they are in the first happiness of escape from an intolerably oppressive situation. Like Bunyan's pilgrim they could not see the wicket gate, nor the Slough of Despond, nor the castle of Giant Despair; but they saw the shining light at the end of the path, and so started gaily towards it as Evolutionists.

And they were right; for the problem of evil yields very easily to Creative Evolution. If the driving power behind Evolution is omnipotent only in the sense that there seems no limit to its final achievement; and if it must meanwhile struggle with matter and circumstance by the method of trial and error, then the world must be full of its unsuccessful experiments. Christ may meet a tiger, or a High Priest arm-in-arm with a Roman Governor, and be the unfittest to survive under the circumstances. Mozart may have a genius that prevails against Emperors and Archbishops, and a lung that succumbs to some obscure and noxious property of foul air. If all our calamities are either accidents or sincerely repented mistakes, there is no malice in the Cruelty of Nature and no Problem of Evil in the Victorian sense at all. The theology of the women who told us that they became atheists when they sat by the cradles of their children and saw them strangled by the hand of God is succeeded by the theology of Blanco Posnet, with his "It was early days when He made the croup, I guess. It was the best He could think of then; but when it turned out wrong on His hands He made you and me to fight the croup for Him."

Editor's Note. The theory of Creative Evolution to which Shaw was referring was the theory published in 1911 by Henri Bergson.

Creative Evolution [*]

by Henri Bergson (1911)

The neo-Darwinians are probably right, we believe, when they teach that the essential causes of variation are the differences inherent in the

[*] From Henri Bergson, *Creative Evolution,* translation by Arthur Mitchell. Copyright 1911, 1939, by Holt, Rinehart and Winston. Reprinted by permission.

germ borne by the individual, and not the experiences or behavior of the individual in the course of his career. Where we fail to follow these biologists, is in regarding the differences inherent in the germ as purely accidental and individual. We cannot help believing that these differences are the development of an impulsion which passes from germ to germ across the individuals, that they are therefore not pure accidents, and that they might well appear at the same time, in the same form, in all the representatives of the same species, or at least in a certain number of them. . . .

We thus arrive at a hypothesis . . . , according to which the variations of different characters continue from generation to generation in definite directions. . . . Of course, the evolution of the organic world cannot be predetermined as a whole. We claim, on the contrary, that the spontaneity of life is manifested by a continual creation of new forms succeeding others. But this indetermination cannot be complete; it must leave a certain part to determination. An organ like the eye, for example, must have been formed by just a continual changing in a definite direction. Indeed, we do not see how otherwise to explain the likeness of structure of the eye in species that have not the same history.[1] Evolution will thus prove to be something entirely different from a series of adaptations to circumstances, as mechanism claims; entirely different also from the realization of a plan of the whole, as maintained by the doctrine of finality.

That adaptation to environment is the necessary condition of evolution we do not question for a moment. It is quite evident that a species would disappear, should it fail to bend to the conditions of existence which are imposed on it. But it is one thing to recognize that outer circumstances are forces evolution must reckon with, another to claim that they are the directing causes of evolution. This latter theory is that of mechanism. It excludes absolutely the hypothesis of an original impetus, I mean an internal push that has carried life, by more and more complex forms, to higher and higher destinies. Yet this impetus is evident, and a mere glance at fossil species shows us that life need not have evolved at all, or might have evolved only in very restricted limits, if it had chosen the alternative, much more convenient to itself, of becoming anchylosed in its primitive forms. Certain Foraminifera have not varied since the Silurian epoch. Unmoved witnesses of the innumerable revolutions that have upheaved our planet, the Lingulae are to-day what they were at the remotest times of the paleozoic era.

The truth is that adaptation explains the sinuosities of the movement of evolution, but not its general directions, still less the movement itself.[2] The road that leads to the town is obliged to follow the ups and downs of

1. Refer back to Bertalanffy's remarks on this point, Part 2, pp. 119–20. Ed.
2. This view of adaptation has been noted by M. F. Marin in a remarkable article

the hills; it *adapts itself* to the accidents of the ground; but the accidents of the ground are not the cause of the road, nor have they given it its direction. At every moment they furnish it with what is indispensable, namely, the soil on which it lies; but if we consider the whole of the road, instead of each of its parts, the accidents of the ground appear only as impediments or causes of delay, for the road aims simply at the town and would fain be a straight line. Just so as regards the evolution of life and the circumstances through which it passes—with this difference, that evolution does not mark out a solitary route, that it takes directions without aiming at ends, and that it remains inventive even in its adaptations.

But, if the evolution of life is something other than a series of adaptations to accidental circumstances, so also it is not the realization of a plan. A plan is given in advance. It is represented, or at least representable, before its realization. The complete execution of it may be put off to a distant future, or even indefinitely; but the idea is none the less formulable at the present time, in terms actually given. If, on the contrary, evolution is a creation unceasingly renewed, it creates, as it goes on, not only the forms of life, but the ideas that will enable the intellect to understand it, the terms which will serve to express it. That is to say that its future overflows its present, and can not be sketched out therein in an idea.

There is the first error of finalism. It involves another, yet more serious.

If life realizes a plan, it ought to manifest a greater harmony the further it advances, just as the house shows better and better the idea of the architect as stone is set upon stone. If, on the contrary, the unity of life is to be found solely in the impetus that pushes it along the road of time, the harmony is not in front, but behind. The unity is derived from a *vis a tergo:* it is given at the start as an impulsion, not placed at the end as an attraction. In communicating itself, the impetus splits up more and more. Life, in proportion to its progress, is scattered in manifestations which undoubtedly owe to their common origin the fact that they are complementary to each other in certain aspects, but which are none the less mutually incompatible and antagonistic. So the discord between species will go on increasing. Indeed, we have as yet only indicated the essential cause of it. We have supposed, for the sake of simplicity, that each species received the impulsion in order to pass it on to others, and that, in every direction in which life evolves, the propagation is in a straight line. But, as a matter of fact, there are species which are arrested; there are some that retrogress. Evolution is not only a movement forward; in many cases we observe a marking-time, and still more often a deviation or turning back. It must be so . . . and the same causes that divide the evolution movement

on the origin of species, "L'Origine des espèces" (*Revue scientifique*, Nov. 1901, p. 580).

often cause life to be diverted from itself, hypnotized by the form it has just brought forth. Thence results an increasing disorder. No doubt there is progress, if progress mean a continual advance in the general direction determined by a first impulsion; but this progress is accomplished only on the two or three great lines of evolution on which forms ever more and more complex, ever more and more high, appear; between these lines run a crowd of minor paths in which, on the contrary, deviations, arrests, and set-backs, are multiplied. The philosopher, who begins by laying down as a principle that each detail is connected with some general plan of the whole, goes from one disappointment to another as soon as he comes to examine the facts; and, as he had put everything in the same rank, he finds that, as the result of not allowing for accident, he must regard everything as accidental. For accident, then, an allowance must first be made, and a very liberal allowance. We must recognize that all is not coherent in nature. By so doing, we shall be led to ascertain the centres around which the incoherence crystallizes. This crystallization itself will clarify the rest; the main directions will appear, in which life is moving whilst developing the original impulse. True, we shall not witness the detailed accomplishment of a plan. Nature is more and better than a plan in course of realization. A plan is a term assigned to a labor: it closes the future whose form it indicates. Before the evolution of life, on the contrary, the portals of the future remain wide open. It is a creation that goes on for ever in virtue of an initial movement. This movement constitutes the unity of the organized world—a prolific unity, of an infinite richness, superior to any that the intellect could dream of, for the intellect is only one of its aspects or products.

Progress in Evolution

The Nature of the Process *

by J. B. S. Haldane (1931)

Now the hypothesis that mind has played very little part in evolution horrifies some people. Shaw's preface to "Back to Methuselah" is a good example of a strong emotional reaction. He admits that Darwinism cannot be disproved, but goes on to state that no decent-minded person can believe in it. This is the attitude of mind of the persecutor rather than the discoverer.

Bergson attributed evolution to an *élan vital,* or vital impulse, which pushed organisms forward along the path of evolution. He laid special stress on convergence, *i.e.* the production of very similar structures by different means in different lines of descent. For example, he pointed out that vertebrates and molluscs have independently developed eyes with a lens and retina, and regarded this as disproving Darwinism. Now, as far as we can see, there are only four possible types of eye, if we define an eye as an organ in which light from one direction stimulates one nerve fibre. There is the insect type of eye, a bundle of tubes pointing in different directions, and three types analogous to three well-known instruments, the pinhole camera, the ordinary camera with a lens, and the reflecting telescope. A straightforward series of small steps leads through the pinhole type to that with a lens, and it is quite easy to understand how this should have been evolved several times. On the other hand, the type with a reflector would be little use in its early stages, and has never been evolved. However, if I were designing an animal as a construct with no historical background, like the ideal state, I should very probably give it an eye with a concave mirror rather than a lens.

But the main objection to *élan vital* is that it is so very erratically distributed. That sturdy little creature, the limpet, has watched the legions of evolution thunder by for some three hundred million years without changing its shell form to any serious extent. And the usual course taken by an evolving line has been one of degeneration. . . .

* From J. B. S. Haldane, *The Causes of Evolution,* Harper & Row Publishers, Inc., New York, 1932. Reprinted by permission.

Occasionally true progress was made, as when insects and birds developed wings, but for every form which has improved, dozens have degenerated. Probably all the birds are derived from one ancestral species which took to the air, but very many have independently lost the power of flight. The ostriches and their allies, the dodo, the kiwi, the flightless parrots and rails of New Zealand, have all lost their flying power and gained nothing in exchange. Only the penguins have transformed their wings into fairly effective fins.

Very numerous groups whose ancestors were motile have taken to sessile habits or internal parasitism. Degeneration is a far commoner phenomenon than progress. It is less striking because a progressive type, such as the first bird, has left many different species as progeny, while degeneration often leads to extinction, and rarely to a widespread production of new forms. Just the same is true with plants. Many primitive forms have not progressed. A few have done so, but relapses of various kinds are equally common. Certainly the study of evolution does not point to any general tendency of a species to progress. The animal and plant community as a whole does show such a tendency, but this is because every now and then an evolutionary advance is rewarded by a very large increase in numbers, rather than because such advances are common. But if we consider any given evolutionary level we generally find one or two lines leading up to it, and dozens leading down.

I have been using such words as "progress," "advance," and "degeneration," as I think one must in such a discussion, but I am well aware that such terminology represents rather a tendency of man to pat himself on the back than any clear scientific thinking. The change from monkey to man might well seem a change for the worse to a monkey. But it might also seem so to an angel. The monkey is quite a satisfactory animal. Man of to-day is probably an extremely primitive and imperfect type of rational being. He is a worse animal than the monkey. His erect posture leads to all sorts of mechanical troubles, such as hernia and a narrowing of the pelvis which makes childbirth painful and dangerous. The last stage in man's evolution certainly has its dark side. You will find a highly symbolic account of it in the second chapter of the Bible. Our first parents are represented as living in a state of ignorance, and then suddenly acquiring the knowledge of good and evil. This may conceivably be true. A decisive step from animal to human mentality may have occurred by mutation. . . . Perhaps it is more likely that it occurred in several steps. But this change is not chronicled as "Man's ascent to reason" or "Man's new nature" but "Man's shameful fall." The writer of Genesis very clearly felt "la honte de penser et l'horreur d'être homme," and we must remember that when we speak of progress in evolution we are already leaving

Figure 3–1. The intricate shapes of coiled ammonite shells are often lustrous with iridescent color. Some shells exceed six feet in diameter. (Photograph courtesy of the American Museum of Natural History. Figure added by editor.)

the relatively firm ground of scientific objectivity for the shifting morass of human values. . . .

If I were compelled to give my own appreciation of the evolutionary process as seen in a great group such as the Ammonites, where it is completed, I should say this: In the first place, it is very beautiful. In that beauty there is an element of tragedy. On the human time-scale the life of a plant or animal species appears as the endless repetition of an almost identical theme. On the time-scale of geology we recapture that element of uniqueness, of *Einemaligkeit*, which makes the transitoriness of human life into a tragedy. In an evolutionary line rising from simplicity to complexity, then often falling back to an apparently primitive condition before its end, we perceive an artistic unity similar to that of a fugue, or the life work of a painter of great and versatile genius like Picasso, who began with severe line drawing, passed through cubism, and is now, in the intervals between still more bizarre experiments, painting somewhat in the manner of Ingres. Possibly such artistic work gives us as good insight into the nature of the reality around us as any other human activity. To me at least the beauty of evolution is far more striking than its purpose.

Evolutionary Progress [*]
by Julian Huxley (1942)

IS EVOLUTIONARY PROGRESS A SCIENTIFIC CONCEPT?

There still exists a very great deal of confusion among biologists on the subject [of evolutionary progress]. Indeed the confusion appears to be greater among professional biologists than among laymen. This is probably due to the common human failing of not seeing the wood for the trees; there are so many more trees for the professional! [1]

The chief objections that have been made to employing *progress* at all as a biological term, and to the use of its correlates *higher* and *lower* as applied to groups of organisms, are as follows. First, it is objected that a bacillus, a jellyfish, or a tapeworm is as well adapted to its environment as a bird, an ant, or a man, and that therefore it is incorrect to speak of the latter as higher than the former, and illogical to speak of the processes

[*] From Julian Huxley, *Evolution: The Modern Synthesis.* Harper & Row Publishers, Inc., New York. Copyright 1942 by Julian Huxley. Reprinted by permission.
1. For a fuller discussion of certain aspects of the problem see . . . H. G. Wells, J. S. Huxley, and G. P. Wells, *The Science of Life* (1930), London, Book 5, chap. 6, #5.

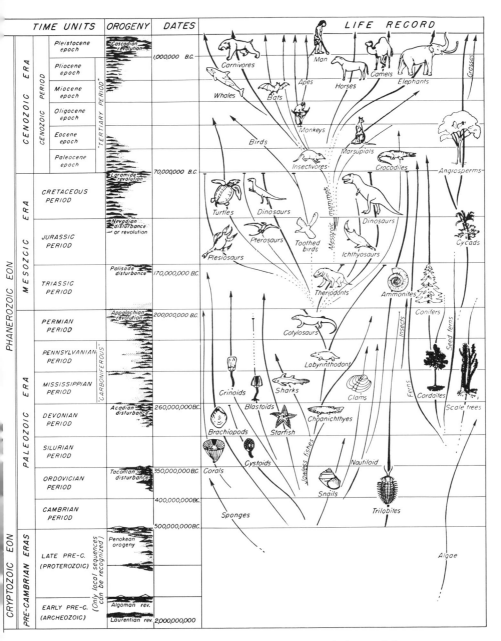

Figure 3–2. Diagram showing four geological eras and their subdivisions into periods and epochs. The time scale is not uniform; for instance, the Pre-Cambrian eras cover 1,500,000,000 years while the Cenozoic covers about 70,000,000. (Reprinted courtesy of the American Museum of Natural History. Chart added by editor.)

leading to their production as involving progress. An even simpler objection is to use mere survival as criterion of biological value, instead of adaptation. Man survives: but so does the tubercle bacillus. So why call man the higher organism of the two?

A somewhat similar argument points to the fact that evolution, both in the fossil record and indirectly, shows as numerous examples of specialization leading to increased efficiency of adaptation to this or that mode of life; but that many of such specialized lines become extinct, while most of the remainder reach an equilibrium and show no further change.

This type of objection, then, points to certain fundamental attributes of living things or their evolution, uses them as definitions of progress, and then denies that progress exists because they are found in all kinds of organisms, and not only in those that the believers in the existence of progress would call progressive.

A slightly less uncompromising attitude is taken up by those who admit that there has been an increase of complexity or an increase in degree of organization, but deny that this has any value, biological or otherwise, and accordingly refuse to dignify this trend by a term such as progress, with all its implications.

Some sociologists, faced with the problem of reconciling the objective criteria of the physical sciences with the value criteria with which the sociological data confronts them, take refuge in the ostrich-like attitude of refusing to recognize any scale of values. Thus Doob in a recent book writes:

"In this way, the anthropologist has attempted to remove the idea of progress from his discipline. For him, there is just change, or perhaps a tendency towards increasing complexity. Neither change nor complexity is good or bad; there are differences in degree, not in quality or virtue. . . . The sweep of historical progress reveals no progressive trend. . . ."[2]

By introducing certain objective criteria into our definition of progress, as we do in the succeeding section, this objection can be overcome, at least for pre-human evolution. In regard to human evolution, however . . . the nettle must be grasped, and human values given a place among the criteria of human progress.

The second main type of objection consists in showing that many processes of evolution are not progressive in any possible sense of the word, and then drawing the conclusion that progress does not exist. For instance, many forms of life, of which the brachiopod *Lingula* is the best-known example, have demonstrably remained unchanged for enormous periods of several hundreds of millions of years; if a Law of Progress

2. From L. W. Doob, *The Plans of Men*, New Haven, 1940. Ed.

exists, the objectors argue, how is it that such organisms are exempt from its operations?

A variant of this objection is to draw attention to the numerous cases where evolution has led to degeneration involving a degradation of form and function, as in tapeworms, *Sacculina* and other parasites, in sea-squirts and other sedentary forms: how, it is asked, can the evolutionary process be regarded as progressive if it produces degeneration?

This category of objections can be readily disposed of. Objectors of this type have been guilty of setting up an Aunt Sally of their own creation for the pleasure of knocking her down. They have assumed that progress must be universal and compulsory: when they find, quite correctly, that universal and compulsory progress does not exist, they state that they have proved that progress does not exist. This, however, is an elementary fallacy. The task before the biologist is not to define progress *a priori*, but to proceed inductively to see whether he can or cannot find evidence of a process which can legitimately be called progressive. It may just as well prove to be partial as universal. Indeed, human experience would encourage search along those lines; the fact that man's progress in mechanical arts, for instance, in one part of the world is accompanied by complete stagnation or even retrogression in other parts, is a familiar fact. Thus evolution may perfectly well include progress without being progressive as a whole.

The first category of objections, when considered closely, is seen to rest upon a similar fallacy. Here again an Aunt Sally has been set up. Progress is first defined in terms of certain properties: and then the distribution of those properties among organisms is shown not to be progressive.

These procedures would be laughable, if they were not lamentable in arguing a lack of training in logical thought and scientific procedure among biologists. Once more, the elementary fact must be stressed that the only correct method of approach to the problem is an inductive one. Even the hardened opponents of the idea of biological progress find it difficult to avoid speaking of higher and lower organisms, though they may salve their consciences by putting the words between inverted commas. The unprejudiced observer will accordingly begin by examining various types of "so-called higher" organisms and trying to discover what characters they possess in common by which they differ from "lower" organisms. He will then proceed to examine the course of evolution as recorded in fossils and deduced from indirect evidence, to see what the main types of evolutionary change have been; whether some of them have consistently led to the development of characters diagnostic of "higher" forms; which types of change have been most successful in producing new groups, dominant forms, and so forth. If evolutionary prog-

ress exists, he will by this means discover its factual basis, and this will
enable him to give an objective definition.

THE DEFINITION OF EVOLUTIONARY PROGRESS

Proceeding on these lines, we can immediately rule out certain characters of organisms and their evolution from any definition of biological
progress. Adaptation and survival, for instance, are universal, and are
found just as much in "lower" as in "higher" forms: indeed, many higher
types have become extinct while lower ones have survived. Complexity of
organization or of life-cycle cannot be ruled out so simply. High types *are*
on the whole more complex than low. But many obviously low organisms
exhibit remarkable complexities, and, what is more cogent, many very
complex types have become extinct or have speedily come to an evolutionary dead end.

Perhaps the most salient fact in the evolutionary history of life is the
succession of what the paleontologist calls dominant types.[3] These are
characterized not only by a high degree of complexity for the epoch in
which they lived, but by a capacity for branching out into a multiplicity
of forms. This radiation seems always to be accompanied by the partial or
even total extinction of competing main types, and doubtless the one fact
is in large part directly correlated with the other.

In the early Paleozoic the primitive relatives of the Crustacea known as
the trilobites [see Figure 3–3] were the dominant group. These were succeeded by the marine arachnoids called sea-scorpions or eurypterids, and
they in turn by the armoured but jawless vertebrates, the ostracoderms,
more closely related to lampreys than to true fish. The fish, however, were
not far behind, and soon became the dominant group. Meanwhile, groups
both from among the arthropods and the vertebrates became adapted to
land life, and towards the close of the Paleozoic, insects and amphibians
could both claim the title of dominant groups. The amphibia shortly gave
rise to the reptiles, much more fully adapted to land life, and the primitive early insects produced higher types, such as beetles, hymenoptera
and lepidoptera. Higher insects and reptiles were the dominant land
groups in the Mesozoic, while among aquatic forms the fish remained pre-
eminent, and evolved into more efficient types: from the end of the
Mesozoic onwards, however, they show little further change.

Birds and mammals began their career in the Mesozoic, but only became dominant in the Cenozoic. The mammals continued their evolution
through the whole of this epoch, while the insects reached a standstill
soon after its beginning. Finally man's ancestral stock diverged, probably

3. For fuller summary, see Wells, Huxley, and Wells, *The Science of Life,* Book 5.

Figure 3–3. During the Cambrian Period trilobites lived on the ocean floor. Their fossil remains comprise about half the known fauna of that era. Like their relatives the lobsters, they shed their outgrown shells. After reigning for 300 million years, they mysteriously became extinct. (Photograph courtesy of the American Museum of Natural History. Figure added by editor.)

towards the middle of the Cenozoic, but did not become dominant until the latter part of the Ice Age.

In these last two cases, the rise of the new type and the downfall of the old was without question accompanied and facilitated by world-wide climatic change, and this was probably true for other biological revolutions, such as the rise of the reptiles to dominance.

When the facts concerning dominant groups are surveyed in more detail, they yield various interesting conclusions. In the first place, biologists are in substantial agreement as to what were and what were not dominant groups. Secondly, some groups once dominant have become wholly extinguished, like the trilobites, eurypterids and ostracoderms, while others survive only in a much reduced form, many of their sub-groups having been extinguished, as with the reptiles or the monotremes, or their numbers enormously diminished, as with the larger non-human placentals. Those which do not show reduction of one or the other sort have remained to all intents and purposes unchanged for a longer or shorter period of geological time, as with the insects or the birds. Finally, later dominant groups normally arise from an unspecialized line of an earlier dominant group, as the birds and reptiles from among the early reptiles, man from the primates among the mammals. They represent, in fact, one among many lines of adaptive radiation; but they differ from the others in containing the potentiality of evolving so as to become dominant on a new level, with the aid of new properties. Usually the new dominance is marked by a fresh outburst of radiation: the only exception to this rule is Man, a dominant type which shows negligible radiation of the usual structurally-adapted sort, but makes up for its absence by the complexity of his social life and his division of labour.

If we then try to analyse the matter still further by examining the characters which distinguish dominant from non-dominant and earlier from later dominant groups, we shall find first of all, efficiency in such matters as speed and the application of force to overcome physical limitations. The eurypterids must have been better swimmers than the trilobites, the fish, with their muscular tails, much better than either; and the later fish are clearly more efficient aquatic mechanisms than the earlier. Similarly the earlier reptiles were heavy, and clumsy, and quite incapable of swift running. Sense-organs also are improved, and brains enlarged. In the latest stages the power of manipulation is evolved. Through a combination of these various factors man is able to deal with his environment in a greater variety of ways, and to apply greater forces to its control, than any other organism.

Another set of characteristics concerns the internal environment. Lower marine organisms have blood or body-fluids identical in saline concentrations with that of the seawater in which they live; and if the composition of their fluid environment is changed, that of their blood changes correspondingly. The higher fish, on the other hand, have the capacity of keeping their internal environment chemically almost constant. Birds and mammals have gone a step further: they can keep the temperature of their internal environment constant too, and so are independent of a wide range of external temperature change.

The early land animals were faced with the problem of becoming independent of changes in the moisture-content of the air. This was accomplished only very partially by amphibia, but fully by adult reptiles and insects through the development of a hard impermeable covering. The freeing of the young vertebrate from dependence on water was more difficult. The great majority of amphibians are still aquatic for the earlier part of their existence: the elaborate arrangements for rendering the reptilian egg cleidoic [4] were needed to permit of the whole life-cycle becoming truly terrestrial.

There is no need to multiply examples. The distinguishing characteristics of dominant groups all fall into one or other of two types—those making for greater control over the environment, and those making for greater independence of the environment. Thus advance in these respects may provisionally be taken as the criterion of biological progress.

THE NATURE AND MECHANISM OF EVOLUTIONARY PROGRESS

It is important to realize that progress, as thus defined, is not the same as specialization. Specialization, as we have previously noted, is an improvement in efficiency of adaptation for a particular mode of life: progress is an improvement in efficiency of living in general. The latter is an all-round, the former a one-sided advance. We must also remember that in evolutionary history we can and must judge by final results. And there is no certain case on record of a line showing a high degree of specialization giving rise to a new type. All new types which themselves are capable of adaptive radiation seem to have been produced by relatively unspecialized ancestral lines.[5]

Looked at from a slightly different angle, we may say that progress must in part at least be defined on the basis of final results. These results have consisted in the historical fact of a succession of dominant groups. And the chief characteristic which analysis reveals as having contributed to the rise of any one of these groups is an improvement that is not one-sided but all-round and basic. Temperature-regulation, for instance, is a property which affects almost every function as well as enabling its possessors to extend their activities in time and their range in space. Placental reproduction is not only a greater protection for the young—a placental mother, however hard-pressed, cannot abandon her unborn embryo—but this additional protection, together with the later period of maternal care, makes possible the extension of the plastic period of learning which then served as the basis for the further continuance of progress.

It might, however, be held that biological inventions such as the lung

4. See J. Needham, *Chemical Embryology*, Cambridge, 1931. *Cleidoic* means closed-up, having little more than gaseous exchange with the environment. Ed.

5. If Garstang's suggestion be true that "clandestine evolution" has enabled new large-scale radiations to start by utilizing a *larval* organization and driving the adult organization off the stage, we have here an apparent exception. It is only fair to say, however, that this view is still highly speculative, and that in any case we would presume that a relatively unspecialized larval type would have served as the new starting-point. See W. Garstang, "The Theory of Recapitulation . . . ," *J. Linn. Soc.* (Zoology), 35:81.

and cleidoic shelled egg, which opened the world of land to the vertebrates, are after all nothing but specializations. Are they not of the same
nature as the wing which unlocked the kingdom of the air to the birds, or
even to the degenerations and peculiar physiological changes which made
it possible for parasites to enter upon that hitherto inaccessible habitat
provided by the intestines of other animals? This is in one sense true; but
in another it is untrue. The bird and the tapeworm, although they did
conquer a new section of the environment, in so doing were as a matter of
actual fact cut off from further progress. Theirs was only a specialization,
though a large and notable one. The conquest of the land, however, not
only did not involve any such limitations, but made demands upon the
organism which could be and in some groups were met by further
changes of a definitely progressive nature.[6] Temperature-regulation, for
instance, could never have arisen through natural selection except in an
environment with rapidly-changing temperatures: in the less changeable
waters of the sea the premium upon it would not be high enough.[7]

Of course a progressive advance may eventually come to a dead end, as
has happened with the insects, when all the biological possibilities inherent in the type of organization have been exploited. From one point of
view it might be permissible to call such a trend a long-range specialization; but it would appear more reasonable to style it a form of progress,
albeit one which is destined eventually to be arrested. It is limited as opposed to unlimited progress.

A word is needed here on the restricted nature of biological progress.
We have seen that evolution may involve downward or lateral trends, in
the shape of degeneration or certain forms of specialization, and may also
leave certain types stable. Further, lower types may persist alongside
higher, even when the lower are representatives of a once-dominant
group that includes the higher types. From this, it will first be seen, as we
already mentioned, that progress is not compulsory and universal; and
secondly that it will not be so marked in regard to the average of biological efficiency as to its upper limit. Progress, in other words, can most readily be studied by examining the *upper levels* of biological efficiency (as
determined by our criteria of control and independence) attained by life
at successive periods of its evolution.

6. Morley Roberts gives numerous interesting examples in which new and in a
sense abnormal demands upon organisms result eventually in adjustments which are
more or less adaptive in relation to the new situation. Unfortunately he postulates a
Lamarckian transmission of modifications which vitiates or obscures much of his
evolutionary discussion. See Morley Roberts, *Warfare in the Human Body*, London,
1920, and *The Serpent's Fang*, London, 1930.
7. Once evolved on land, however, it proved its value in even the sea, as
evidenced by the success of the Cetacea and other secondary aquatics among
mammals.

For this, during the earlier part of life's history, we must rely upon the indirect evidence of phylogeny, drawn from comparative morphology, physiology, and embryology, while for the last thousand million years this is further illuminated by the light of paleontology, with its direct evidence of fossils.

We have thus arrived at a definition of evolutionary progress as consisting in a raising of the upper level of biological efficiency, this being defined as increased control over and independence of the environment.[8] As an alternative we might define it as a raising of the upper level of all-round functional efficiency and of harmony of internal adjustment.[9]

This brings us to a further objection which is often raised to the idea of progress, namely, that it is a mere anthropomorphism. This view asserts that we judge animals as higher or lower by their greater or lesser resemblance to ourselves and that we give the name of progress to the evolutionary trend which happens to have culminated in ourselves. If we were ants, the objectors continue, we should regard insects as the highest group and resemblance to ants as the essential basis of a "high" organism: while if we were eagles our criterion of progress would be an avian one.

Even Haldane has adopted this view. He writes, "I have been using such words as 'progress,' 'advance,' and 'degeneration', as I think one must in such a discussion, but I am well aware that such terminology represents rather a tendency of man to pat himself on the back than any clear scientific thinking. . . . Man of to-day is probably an extremely primitive and imperfect type of rational being. He is a worse animal than the monkey. . . . We must remember that when we speak of progress in Evolution we are already leaving the relatively firm ground of scientific objectivity for the shifting morass of human values."

This I would deny. Haldane has neglected to observe that man possesses greater power of control over nature, and lives in greater independence of his environment than any monkey. The use of an inductive method of approach removes all force from such objections. The definitions of progress that we were able to name as a result of a survey of evolutionary facts, though admittedly very general, are not subjective but objective in their character. That the idea of progress is not an anthropomorphism can immediately be seen if we consider what views would be taken by a philosophic tapeworm or jellyfish. Granted that such organisms could reason, they would have to admit that they were neither dominant

8. Herbert Spencer recognized the importance of increased independence as a criterion of evolutionary advance: see references in J. Needham, "Integrative Levels: A Revaluation of the Idea of Progress." The Herbert Spencer Lecture, Oxford, 1937.

9. See also R. W. Gerard (1940), "Organism, Society, and Science," *Sci. Mo.,* 50: 340.

types, nor endowed with any potentiality of further advance, but that one was a degenerate blind alley, the other a specialization of a primitive type long left behind by more successful forms of life. And the same would be equally true, though not so strikingly obvious, of ant or eagle. Man *is* the latest dominant type to be evolved, and this being so, we are justified in calling the trends which have led to his development *progressive*. We must, however, of course beware of subjectivism and of reading human values into earlier stages of evolutionary progress. Human values are doubtless essential criteria for the steps of any future progress: but only biological values can have been operative before man appeared.

The value of such a broad biological definition of progress may be illustrated by reference to a recent definition of human progress by Professor Gordon Childe.[10] Professor Childe, too, is seeking for an objective criterion for progress; but the criterion he adopts is increase of numbers. Quite apart from the logical difficulty that increase in population must, on a finite earth, eventually approach a limit, it is clear that this criterion is at once invalidated by the facts of general biology.

There are many more of various common plankton organisms than of men or of any bird or mammal. There are in all probability many more houseflies than human beings, more bacteria, even of a single species, than of any metazoan. If we apply our criterion of increased control and independence, we see that it would be theoretically quite possible (though difficult with our present type of economy) to obtain progressive changes in human civilization with an accompanying decline in population.

Here let me interject a further word concerning objective and subjective criteria for progress. As regards human progress, it is clear that subjective criteria cannot and should not be neglected; human values and feelings must be taken into account in deciding on the future aims for advance. But in comparing human with pre-human progress, we must clearly stick to objective standards. I would thus like to make a distinction between biological or evolutionary progress and human progress. The former is a biological term with an objective basis: it includes one aspect of human progress. Human progress, on the other hand, has connotations of value as well as of efficiency, subjective as well as objective criteria.[11]

10. See V. G. Childe, *Man Makes Himself,* London, 1936.

11. On the other hand, to confine the term *progress* entirely to human affairs, and to contrast it with *evolution* in pre-human history, as does Marett (1933, 1939), is to restrict the meaning of progress unduly, while distorting that of evolution. On three successive pages Marett describes or defines progress in three different ways: (1) the moral of human history and pre-history would seem to be that "progress in the direction of the spiritual is implicit in normal human endeavour". (2) Progress in spirituality in the future "may be conceived in terms of the greatest self-realization of the greatest number". (3) "Real progress is progress in

Returning to biology, we may sum up as follows. Progress is all-round biological improvement. Specialization is one-sided biological improvement: it always involves the sacrificing of certain organs or functions for the greater efficiency of others. It is the failure to distinguish between these two types of evolutionary process that vitiates the generalizations of many biologists.[12]

Degeneration is a form of specialization in which the majority of the somatic organs are sacrificed for greater efficiency in adaptation to a sedentary or a parasitic life. Locomotor organs disappear, sensory and nervous systems are much reduced, and in parasites the digestive system may be abolished. Reproductive mechanisms, however, may be inordinately specialized, as in certain parasites.

Besides these types of evolutionary process, we may have stability, as in the lamp-shell *Lingula,* or in ants during most of the Cenozoic epoch. Stable types are presumably either extremely well-adapted to a permanent biological niche or have reached the limit of specialization or of progress possible to them.

Finally, we may have the type of evolutionary trend best known among the Ammonites, of increasing complication followed by simplification. . . .

As revealed in the succession of steps that have led to new dominant types, progress has taken diverse forms. At one stage, the combination of cells to form a multicellular individual, at another the evolution of a head; later the development of lungs, still later of warm blood, and finally the enhancement of intelligence by speech. But all, though in curiously different ways, have enhanced the organism's capacities for control and for independence; and each has justified itself not only in immediate results but in the later steps which it made possible.

We have now dealt with the fact of evolutionary progress, and with the philosophical and biological difficulties inherent in the concept. What of its mechanism? It should be clear that if natural selection can account for adaptation and for long-range trends of specialization, it can account for progress too. Progressive changes have obviously given their owners advantages which have enabled them to become dominant. Sometimes it may have needed a climatic revolution to give the progressive change full play, as seems to have been the case at the end of the Cretaceous with the

charity." It should be clear how important it is to give greater universality and concreteness to the idea of progress by considering human progress as a special case of biological progress. Marett, R. R. (1933). "Progress as a Sociological Category." *Sociol. Rev.* 1933: Jan.–Apr.:3. (1939). "Charity and the Struggle for Existence." J. Roy. Anthr. Inst. 69: part 2.

12. For example, H. L. Hawkins, "Palaeontology and Humanity," *Rep. Brit. Ass.,* 1936:59.

mammal-reptile differential of advantage; but when it came, the advantage had very large results—wholesale extinctions on the one hand, wholesale radiation of new types on the other. It seems to be a general characteristic of evolution that in each epoch a minority of stocks give rise to the majority in the next phase, while conversely the majority of the rest become extinguished or are reduced in numbers.

There is no more need to postulate an *élan vital* or a guiding purpose to account for evolutionary progress than to account for adaptation, for degeneration or any other form of specialization.[13]

One point is of importance. Although we can quite correctly speak of evolutionary progress as a biological fact, this progress is of a particular and limited nature. It is, as we have seen, an empirical fact that evolutionary progress can only be measured by the upper level reached: for the lower levels are also retained. This has on numerous occasions been used as an argument against the existence of anything which can properly be called progress; but its employment in this connection is fallacious. It is on a par with saying that the invention of the automobile does not represent an advance, because horse-drawn vehicles remain more convenient for certain purposes, or pack animals for certain localities. A progressive step in evolution will normally and probably invariably bring about the extermination of some types at a lower level; but the variety of environments and of the available modes of filling them is such that it is extremely unlikely to exterminate them all. The fact that protozoa should be able to exist side by side with metazoa, or a considerable army of the "defeated" group of reptiles together with their mammalian "conquerors", is not in any way surprising on selectionist principles: it is to be expected. . . .

The ordinary man, or at least the ordinary poet, philosopher, and theologian, is always asking himself what is the purpose of human life, and is anxious to discover some extraneous purpose to which he and humanity may conform. Some find such a purpose exhibited directly in revealed religion; others think that they can uncover it from the facts of nature. One of the commonest methods of this form of natural religion is to point to evolution as manifesting such a purpose. The history of life, it is asserted, manifests guidance on the part of some external power; and the usual deduction is that we can safely trust that same power for further guidance in the future.

I believe this reasoning to be wholly false. The purpose manifested in

13. A small minority of biologists, such as R. Broom, still feel impelled to invoke "spiritual agencies" to account for progressive evolution, but their number is decreasing as the implications of modern selection theories are grasped. See R. Broom, "Evolution—Is There Intelligence Behind It?" *S. Afr. J. Sci.* 30:1.

evolution, whether in adaptation, specialization, or biological progress, is only an apparent purpose. It is just as much a product of blind forces as is the falling of a stone to earth or the ebb and flow of the tides. It is we who have read purpose into evolution, as earlier men projected will and emotion into inorganic phenomena like storm or earthquake. If we wish to work towards a purpose for the future of man, we must formulate that purpose ourselves. Purposes in life are made, not found.

But if we cannot discover a purpose in evolution, we can discern a direction—the line of evolutionary progress. And this past direction can serve as a guide in formulating our purpose for the future. Increase of control, increase of independence, increase of internal co-ordination; increase of knowledge, of means for co-ordinating knowledge, of elaborateness and intensity of feeling—those are trends of the most general order. If we do not continue them in the future, we cannot hope that we are in the main line of evolutionary progress any more than could a sea-urchin or a tapeworm. . . .

After the disillusionment of the early twentieth century it has become as fashionable to deny the existence of progress and to brand the idea of it as a human illusion, as it was fashionable in the optimism of the nineteenth century to proclaim not only its existence but its inevitability. The truth is between the two extremes. Progress is a major fact of past evolution; but it is limited to a few selected stocks. It may continue in the future, but it is not inevitable; man, by now become the trustee of evolution, must work and plan if he is to achieve further progress for himself and so for life.

This limited and contingent progress is very different from the *deus ex machina* of nineteenth-century thought, and our optimism may well be tempered by reflection on the difficulties to be overcome. None the less, the demonstration of the existence of a general trend which can legitimately be called progress, and the definition of its limitations, will remain as a fundamental contribution of evolutionary biology to human thought.

Mankind and Cosmic Purpose

Purpose Arises Only in Man [*]
by George Gaylord Simpson

Man is the result of a purposeless and materialistic process that did not have him in mind. He was not planned. He is a state of matter, a form of life, a sort of animal, and a species of the Order Primates, akin nearly or remotely to all of life and indeed to all that is material. It is, however, a gross misrepresentation to say that he is *just* an accident or *nothing but* an animal. Among all the myriad forms of matter and of life on the earth, or as far as we know in the universe, man is unique. He happens to represent the highest form of organization of matter and energy that has ever appeared. Recognition of this kinship with the rest of the universe is necessary for understanding him, but his essential nature is defined by qualities found nowhere else, not by those he has in common with apes, fishes, trees, fire, or anything other than himself.

It is part of this unique status that in man a new form of evolution begins, overlying and largely dominating the old, organic evolution which nevertheless also continues in him. This new form of evolution works in the social structure, as the old evolution does in the breeding population structure, and it depends on learning, the inheritance of knowledge, as the old does on physical inheritance. Its possibility arises from man's intelligence and associated flexibility of response. His reactions depend far less than other organisms' on physically inherited factors, far more on learning and on perception of immediate and of new situations.

This flexibility brings with it the power and the need for constant choice between different courses of action. Man plans and has purposes. Plan, purpose, goal, all absent in evolution to this point, enter with the coming of man and are inherent in the new evolution, which is confined to him. With them comes the need for criteria of choice. Good and evil, right and wrong, concepts largely irrelevant in nature except from the human viewpoint, become real and pressing features of the whole cosmos

[*] From George Gaylord Simpson, *The Meaning of Evolution*, Yale University Press, New Haven. Copyright © 1949 by Yale University Press. Reprinted by permission.

as viewed by man—the only possible way in which the cosmos can be viewed morally because morals arise only in man. . . .

I was at this point in the summary, when I coincidentally came across some highly pertinent remarks in a recent example of the legions of articles deploring the decline of religious faith.[1] The author, a distinguished philosopher, finds himself agreeing with certain ecclesiastical dignitaries that chaos and bewilderment in the world today result from loss of faith in God and religion. This has become almost banal by constant repetition (although I beg leave to note that repetition does not establish truth). From this point, however, the author takes a less beaten path and one not likely to comfort the godly. He finds that, indeed, the old religious faith was unjustified and that the truth is quite otherwise. He does not question or even particularly deplore the fact that the universe does not operate by divine plan, but he thinks it a great pity that we ever found this out. He is a little petulant with scientists for discovering that the world is purposeless and thus forcing abandonment of religions that require the postulate of purpose. He can only face the fact that childish dreams of a meaningful universe must be laid aside, and he exhorts mankind to become adult and to live as honorably as may be in a stark and bleak world.

The honesty of this philosophical acceptance of truth as we see it must be admired, but it does seem that the total surrender is premature and that the argument is still not entirely adult, in the author's sense of the word. It seems still a little childish to regret that vain dreams have ended. Perhaps it is even more juvenile to blame the loss rather on the scientists (unkind adults) who exposed the sham than on the falseness of the dreams or on the dreamers. Most important of all, the fully adult reaction to the loss of false ideals would seem to be not simply to regret them and to determine grimly to face the world without them, but to see whether true values may not exist and whether they may not indeed be easier to come by now that the false are cleared away. . . . such values do, indeed, exist.

The ethical need is within and peculiar to man, and its fulfillment also lies in man's nature, relative to him and to his evolution, not external or unchanging.

[1] W. T. Stace, "Man against Darkness," *Atlantic Monthly*, 182, No. 3 (1948), 53–58.

Nature's Apostrophe *

by Charles Sherrington (1950)

If we accept the story behind us, the planet, which being blind never before had purpose, now is lent a purpose and—anthropomorphism of anthropomorphisms!—by man. It pleads excuse because a broken light. Hume's Philo spoke in bitterness of Nature. He reproached it with being "a vivifying principle bringing forth its countless offspring into suffering." He spoke in bitterness of its blindness. But he himself was an outcome of that Nature; and he was at least a part of Nature beginning to understand itself.

Might not Nature, so bitterly apostrophized by him, make answer. "You and yours once thought me rational; you said I had foresight and a purpose; you declared I had plan while I created; you read into me a super-intelligence and design. Now that you have undeceived yourself about me, you seem to take it ill that where you used to say I showed intelligence, I show none; where you used to find design you find now none. But what would you? That I aped intelligence was the making of you. Aping intelligence as I did, it required intelligence to read my pseudo-logical doings. If I had been chaos what good would it have been for you to have intelligence? Intelligence amid chaos would have had no survival value. You would therefore not have had intelligence. Intelligence is yours because it helped you to understand what you misunderstood me to be. That justified your having it and so you have it. It was my pseudo-intelligence sharpened your wits. You thought me rational with 'forces' and 'causes' and 'purposes' and all the rest and so you got your endowment of reason, reason being a means for dealing with what seemed reason. And now your thus begotten intelligence seems to revile its only true begetter! At least you should thank me for being 'law.' That it was which made you and has given you mind."

And Nature might continue: "For you it will be to remember, having got your measure of reason, you are now in fact in competition against a measure of reason as well as of mere semblance of reason. You will have to deal with it as a new situation. For you have your own fellow-folk to deal with. Remember too that it is not going to stop where it is—the mod-

* From Charles Sherrington, *Man on His Nature*, Cambridge University Press, 1953. Reprinted by permission.

icum of reason which you and yours have—because an extra dose of it can have survival value. You have now as a species had a long inning by reason of your intelligence. It is perhaps a million years since first you were knocking stones into shape.

"You thought me moral, you now know me without moral. How can I be moral being, you say, blind necessity, being mechanism. Yet at length I brought you forth, who are moral. Yes, you are the only moral thing in all your world, and therefore the only immoral.

"You thought Nature intelligent, even wise. You now know her devoid of reason, most of her even of sense. How can she have reason or purpose being pure mechanism? Yet at length she made you, you with your reason. If you think a little you with your reason can know that; you, the only reasoning thing in all your world, and therefore the only mad one.

"You are my child. Do not expect me to love you. How can I love—I who am blind necessity? I cannot love, neither can I hate. But now that I have brought forth you and your kind, remember you are a new world unto yourselves, a world which contains in virtue of you, love and hate, and reason and madness, the moral and immoral, and good and evil. It is for you to love where love can be felt. That is, to love one another.

"Bethink you too that perhaps in knowing me you do but know the instrument of a Purpose, the tool of a Hand too large for your sight as now to compass. Try then to teach your sight to grow."

Suggestions for Further Reading

* Bertalanffy, Ludwig von: *Problems of Life, An Evaluation of Modern Biological and Scientific Thought*, Harper Torchbook, Harper & Brothers, 1960. The organismic conception of life is explained by one of its most distinguished supporters.

Dobzhansky, Theodosius: *The Biology of Ultimate Concern*, Perspectives in Humanism Series, planned and edited by Ruth Nanda Anshen, The New American Library of World Literature, 1967. An examination of the philosophical implications of modern evolutionary theory.

Eiseley, Loren: *The Firmament of Time*, Atheneum, 1962. A poetic description of the way the theory of evolution became accepted and how it has affected man's view of himself and his place in nature.

* ————: *The Immense Journey*, Vintage Books, Random House, Inc., 1957. A series of imaginative and beautifully written essays on the implications of the theory of evolution.

* Gillispie, Charles Coulston: *Genesis and Geology: The Decades Before Darwin*, Harper Torchbooks, Harper & Brothers, 1957. The impact of scientific discoveries upon religious beliefs in the years before Darwin.

* Paperback edition

* Simpson, George Gaylord: *The Meaning of Evolution,* A Yale Paper-bound, Yale University Press, 1960. Chapter XV discusses the concept of biological progress and arrives at a different conclusion than Julian Huxley does in these readings.

* Sinnott, Edmund W.: *The Biology of the Spirit,* The Viking Press, 1955. An "unorthodox" book in which a botanist presents his own personal theory of the goal-seeking qualities of living matter, implying the reality of purpose in living things.

* ————: *Matter, Mind and Man: The Biology of Human Nature,* Atheneum, 1962. The intellectual and spiritual qualities of mankind are considered as an essential part of the whole process of evolution.

* Teilhard de Chardin, Pierre: *The Phenomenon of Man,* Harper Torchbook, Harper & Brothers, 1961. The author, a palaeontologist and a Jesuit, finds in the evolutionary process a coherent vision of both biological and spiritual significance. Several short selections from his work appear throughout this book.

4

What Is the Origin of Man?

A summary of the present knowledge of the evolution of man is followed by a discussion of the ways in which this knowledge affects man's understanding of his own nature. Is man nothing but an animal or are there ways in which man is unique?

What Is the Origin of Man?

Introduction

WE HAVE ALREADY SEEN that the Darwinian theory met with great opposition because it contradicted the religious interpretation of the origin of the species and because it seemed to substitute a blind and meaningless mechanism for a divinely conceived cosmic plan. Furthermore, the most violent objections to the theory centered around the new evolutionary view of the ancestry of man.

In *The Origin of Species,* Darwin was careful to avoid discussion of the application of his theory to human evolution. While he was still working on the manuscript, he wrote to Wallace: "I think I shall avoid the whole subject, as so surrounded with prejudices, though I fully admit that is the highest and most interesting problem for the naturalist." At the conclusion of his book he mentioned the subject only in a very brief and noncommittal manner: "Light," he said, "will be thrown on the origin of man and his history." [1]

Other scientists, however, immediately recognized the human implications. Thomas Huxley's *Evidence as to Man's Place in Nature* appeared in 1863 and Charles Lyell's *The Antiquity of Man* in the same year. In 1874 Darwin finally came out into the open with the publication of *The Descent of Man.* In this book he spelled out his view that man had descended from sub-human forms, a view which he realized "will be highly distasteful to many."

Although most of the scientific world was won over to this theory of man's origin, opposition from religious and lay groups continued throughout the first half of the twentieth century and has not been entirely laid to rest today.[2]

By far the most famous confrontation between the literal Biblical story of the creation of man and the evolutionary theory took place in Tennessee in 1925 at the trial of John Scopes. Although Scopes and his eloquent lawyer, Clarence Darrow, technically lost the trial, the publicity and discussion it generated brought knowledge of evolution to a much wider public. And the reaction to bigotry brought many new converts to the evolutionary theory. This ironic result is another example of a principle which has been demonstrated several times in the history of science: attempts to repress knowledge serve only to bring it out into the open. Public curiosity is aroused by arguments in the press (and in radio and tele-

1. *Life and Letters of Charles Darwin,* 3 vols., edited by Francis Darwin, London, John Murray, 1888. Vol. 2, p. 109.
2. See Part 1, p. 60.

vision) which bring the subject into many a household that had not been conscious of its existence, thus achieving a rapid dissemination of the "forbidden" knowledge. Galileo's long struggle with the Roman Catholic Church, Margaret Sanger's battles with the government and the Church, as well as the Bradlaugh-Besant trial in England [3]—all these attempts at repression resulted in the more rapid spread of scientific knowledge.

At the time that Darwin extended his theory to the origin of man, only two sub-human fossils had been discovered and neither of these finds were studied by Darwin. Thus his conclusions were based entirely on theory. But the impetus his books gave to paleontology led to a remarkable series of discoveries during the succeeding century. These discoveries presented the world with an impressive body of evidence supporting the thesis that man evolved from more primitive members of the Primate order. The evidence was accumulated in spite of organized resistance of religious groups throughout the world and in spite of the much more subtle psychological resistance of people, scientists as well as laymen, to the uncovering of factual evidence of man's ape-like ancestors.

It is my genuine belief [says Loren Eiseley] that no greater act of the human intellect, no greater gesture of humility on the part of man has been or will be made in the long history of science. The marvel lies not in the fact that the bones from the caves and river gravels were recognized in trepidation and doubt as beings from the half-world of the past; the miracle, considering the nature of the human ego, occurs in the circumstance that we were able to recognize them at all, or to see in these remote half-fearsome creatures our long-forgotten fathers who had cherished our seed through the ages of ice and loneliness before a single lighted city flickered out of the darkness of the planet's nighttime face.[4]

Some scientists may have secretly hoped that their researches would turn out differently, that they would discover a more special and respectable ancestor for mankind. Instead, fossils were found which seemed to show that near-man had been carnivorous and cannabilistic, undoubtedly more predatory than the monkeys or apes of today. These discoveries caused some scientists to fall into a deeply pessimistic frame of mind, and to conclude that if man is descended from such ancestors, he must carry a genetic heritage that predisposes him to violence and brutality. Raymond Dart was one of the most prominent exponents of this view. His arguments, although not generally shared by his fellow anthropologists, have been brought to public attention recently by Robert Ardrey in *African*

3. The trial of Charles Bradlaugh and Mrs. Annie Besant in 1877 contested their right to publish a pamphlet on birth control.
4. From Loren Eiseley, *Darwin's Century*, Doubleday & Company, Garden City, N.Y., p. 257.

Genesis. The Naked Ape by Desmond Morris presents another version of this same interpretation of the nature of man. It is significant that both books have become best-sellers, perhaps because they seem to offer an explanation of the violence and irrationality abroad in the world today. Is it possible that we have been misled by too exalted an image of ourselves? Should we not take a more realistic view and recognize the primitive instincts that derive from our animal heritage?

Julian Huxley, however, and a large number of other biologists and anthropologists believe that there is a fundamental fallacy in this view.

We have been guilty [Huxley says] of mistaking origins for explanations— what we may call the "nothing but" fallacy; if sexual impulse is at the base of love, then love is to be regarded as nothing but sex; if it can be shown that man originated from an animal, then in all essentials he is nothing but an animal. This, I repeat, is a dangerous fallacy.

We have tended to misunderstand the nature of the difference between ourselves and animals. We have a way of thinking that if there is continuity in time there must be continuity in quality. A little reflection would show that this is not the case. When we boil water there is a continuity of substance between water as a liquid and water as steam; but there is a critical point at which the substance H_2O changes its properties. This emergence of new properties is even more obvious when the process involves change in organization . . .[5]

The article by Julian Huxley in these readings describes the properties which, he believes, make man unique. The discovery of tool-making and the invention of language are cited as two of the new abilities which gave man his dominant position. Tool-making became possible when hands were freed by upright posture. Tools, in turn, made possible the securing of a more concentrated food supply, the meat of large animals. These new skills freed some of early man's time and energy, enabling him to invent more efficient tools and a means of communicating with others. Through language he was able to build a cumulative tradition which could be passed along from generation to generation, in addition to the genetic heritage. At this point, a qualitative change took place in the evolutionary process. Thus Huxley, as well as Leakey and Eiseley, support the contention that tool-making was a crucial step in the transformation of near-man to man.

Ardrey, on the other hand, considers the term "tool" to be a euphemism. Tools were primarily weapons, he says, developed by a carnivorous animal for the purpose of killing. And the instincts that made him a killer are still powerful within him today. If man is a biological invention

5. From Julian Huxley, *Evolution in Action*, The New American Library, New York, p. 115.

evolved to suit the purposes of the weapon then human history must be read in those terms.[6]

Some scientists, however, believe that in man instinct has been largely replaced by rational thought. They point to the fact that most other organisms are born fully equipped with instinctive knowledge and ability to cope with the usual life situation of that particular species, while the human infant is completely helpless. For many years he must be cared for and taught the cumulative tradition and wisdom of the race. Is it not, then, fallacious to seek an explanation of modern man's problems and behavior in the instinctual makeup of his Primate ancestors?

What man believes about himself is a most potent factor in influencing what he will become. For many centuries man believed that he was a special creation and that he had an exalted destiny to fulfill. Perhaps this belief helped to develop some of the traits that as humanists we value most.

The crisis of our day [suggests Edmund Sinnott] comes from the fact that this traditional appraisal of man's nature is now gravely challenged, and from several directions. Whether in the light of modern knowledge he can maintain this high estimate of himself or must give it up for something different, something far less exalted and godlike, is the deepest question he has to face . . . It is not the change in our ideas about God that is so ominous, important as this is, but the change in what we think about ourselves.[7]

6. See Robert Ardrey selection, p. 230.
7. See Edmund Sinnott selection, p. 237.

The Discovery of Man's Origins

The Reception of Darwin's Theory [*]
by Ashley Montagu (1959)

On November 24, 1859, Charles Darwin's *On the Origin of Species* was published by John Murray. A pre-publication copy had been sent by Darwin to Huxley, who on November 23 wrote an enthusiastic letter to Darwin enjoining him not to allow himself to be disgusted by the considerable abuse and misrepresentation which he had no doubt was in store for him, and promising to stand by him. "I am," he wrote, "sharpening up my claws and beak in readiness."

On Saturday, June 30, 1860, at the annual meeting of The British Association for the Advancement of Science, which was held at Oxford, Huxley's great opportunity presented itself. This was in the form of Bishop Wilberforce, otherwise known as "Soapy Sam" because of his ingratiating manners. Bishop Wilberforce was out to crush the Darwinists. In order to do so the more effectively he had been coached by an archenemy of Darwinism, Richard Owen, the distinguished paleontologist of the British Museum. The meeting-place was crowded to the rafters, there being more than 700 people present. When Wilberforce's turn came to speak, he rose and in dulcet tones and well-turned periods spoke for half an hour with what one of his more sympathetic listeners describes as "inimitable spirit, emptiness and unfairness." Finally he turned to Huxley and with smiling insolence begged to know "Was it through his grandfather or his grandmother that he claimed descent from a monkey?"

Huxley, turning to Sir Benjamin Brodie who was seated near him, struck his hand on his knee and quietly remarked, "The Lord hath delivered him into mine hands." When the applause for the Bishop subsided Huxley was called upon by the chairman to speak. In a letter written to his friend, Dr. Frederick Dyster, Huxley has in his own words given us an account of what he said upon that famous occasion. He writes, "When I got up I spoke pretty much to the effect—that I had listened with great

[*] From Ashley Montagu, Introduction to *Man's Place in Nature* by Thomas H. Huxley, University of Michigan Press, Ann Arbor. Copyright © 1959 by the University of Michigan. Reprinted by permission.

attention to the Lord Bishop's speech but had been unable to discover either a new fact or a new argument in it—except indeed the question raised as to my personal predilections in the matter of ancestry—That it would not have occurred to me to bring forward such a topic as that for discussion myself, but that I was quite ready to meet the Right Rev. prelate even on that ground. If then, said I, the question is put to me would I rather have a miserable ape for a grandfather or a man highly endowed by nature and possessing great means and influence and yet who employs those faculties and that influence for the mere purpose of introducing ridicule into a grave scientific discussion—I unhesitatingly affirm my preference for the ape.

"Whereupon there was unextinguishable laughter among the people, and they listened to the rest of my argument with the greatest attention . . . and I happened to be in a very good condition and said my say with perfect good temper and politeness—and I assure you of this because all sorts of reports [have] been spread about, *e.g.,* that I had said I would rather be an ape than a bishop, etc."

Never before or since has a scientist more ringingly made a monkey out of a Bishop. The occasion also caused the wife of another Bishop to deliver herself of an immortal remark. It is said that when the Bishop of Worcester communicated the intelligence to his wife that the horrid Professor Huxley had announced that man was descended from the apes, she exclaimed: "Descended from the apes! My dear, let us hope that it is not true, but if it is, let us pray that it will not become generally known." Alas, for the dear lady, it *has* become generally known. As Huxley remarked, *"Veritas praevalebit."* Truth, indeed, has prevailed, and in no small part this has been due to the voice and labors of Thomas Henry Huxley.

The Search for Mankind's Ancestors *

by Ruth Moore (1962)

Was there somewhere in the earth a record of what man had been in the past, if indeed he had been any different in the past than he was in the eighteenth and early nineteenth centuries?

George Cuvier, who was astounding France with the bones of the ex-

* From Ruth Moore and The Editors of Life, *Evolution,* Time Incorporated, New York. Copyright © 1962 by Time Inc. First three paragraphs from *Race* to be published by American Heritage in co-operation with the Smithsonian Institution.

tinct and strange animals he was digging from the soil of the Paris area itself, denied that there was any such thing as fossil man. "L'homme fossile n'existe pas," he declared with all of his authority.

Others had reported the existence of very odd ancestors. . . . Tracking down early notions about missing links, Thomas Huxley found an account written in 1598 by a Portuguese sailor named Eduardo Lopez. "In the Songan country on the banks of the Zaire [Congo]," Lopez related, "are multitudes of apes which afford great delight to the nobles by imitating human gestures." By way of documenting this he had drawn two tailless, long-armed apes, cavorting like clowns—on the same page with a winged, two-legged, crocodile-headed dragon. Huxley was interested but not impressed.

The logical Carl Linnaeus correctly classified man with the mammals—and with what are now called the anthropoids—in his *Systema Naturae* published in 1735. But the illustrations by one of his pupils, in a later book, show that the master had some strange concepts of the anthropoidal order's other members. Several of them combined human heads that could have been copied from a medieval tapestry with bodies as shaggy as a bear's. Another one, hairless except for a circular brush around the face, was drawn with a foot-long tail. [See Figure 1–2]

Both extreme versions of man's origins—that he was without earthly predecessors or was endowed with fanciful ones—suffered a rude upset in 1856. In the little Neander Valley near Düsseldorf, Germany, a limestone cave yielded an extraordinary skull fragment and some associated long bones. The skullcap was thick, with massive, lowering ridges over the eyes, and immediate scientific reaction to its discovery was as confused as the case of the five blind men and the elephant. It was clearly manlike—but what kind of a man? Rudolf Virchow, Europe's leading pathologist, dismissed the Neanderthal "man" as a not very ancient pathological idiot. Another physician declared that the deceased had suffered from "hypertrophic deformation." A German authority theorized that the remains were those of "one of the Cossacks who came from Russia in 1814."

Darwin heard about these remarkable bones, yet never investigated them, but Huxley undertook a thorough study of the unprecedented skull. In the condition in which it was discovered, the cranium could hold 63 cubic inches of water; complete, it would have contained 75 cubic inches, or as much as the skulls of living primitive tribesmen. So the brain must have been of modern size too; and the limb bones, though on the bulky side, were "quite those of an European of middle stature. . . ." "Under whatever aspect we view this cranium," wrote Huxley in 1863, in his book *Zoological Evidences as to Man's Place in Nature*, "whether we regard its vertical depression, the enormous thickness of the supraciliary ridges, its sloping occiput, or its long and straight squamosal suture, we meet with apelike characteristics, stamping it as the most pithecoid [apelike] of human crania yet discovered." Neanderthal man, Huxley concluded, was more nearly allied to the higher apes than the latter are to the lower apes, but for all of that he was a man, a human or a species set aside from its predecessors by its upright posture and higher skills. (A most successful man, as later finds were to prove: he dominated Europe for some 35,000 years until *Homo sapiens,* modern man, took over about 40,000 to 50,000 years ago.) "In some older strata," Huxley wondered, "do the fossilized bones of an ape more anthropoid, or a man more pithecoid, than any yet known await the researches of some unborn paleontologist?"

The question obsessed a young Dutch doctor, Eugène Dubois, who determined that he would be the paleontologist such bones awaited. Dubois reasoned that any in-between form would in all probability have originated either in Africa where the gorilla and chimpanzee still exist, or in Malaya where the orangutan survives. He therefore eagerly accepted an appointment as a surgeon with the Royal Dutch Army in Sumatra. His hopes were high, for Sumatra and neighboring Java had escaped extensive glacial earth-scraping during the ice ages and there was a good prospect that fossils might have been preserved there. Thus Dubois sailed for Padang, Sumatra, in 1887, to seek man's ancestors.

The Sumatra caves where he began his explorations yielded nothing but some teeth of the orangutan, which was then extinct on that island. But word came of the discovery of a very ancient skull at Wadjak in Java. Dubois persuaded the government to send him there to search for "fossil vertebrate fauna." Java, like Sumatra, once had formed a part of the Asiatic mainland. Animals could have wandered down from the north freely and dry of foot. Later the seas had rolled in, inundating the lower land and turning the mountaintops into islands. [See Figure 4–13]

At Wadjak, Dubois managed to buy the skull of which he had heard, and then promptly found another. But they were not the skulls of the missing link of which he was dreaming; they were too recent. He said nothing about them, and pressed on. At Trinil, roughly in the center of Java, the Solo River was cutting its slow way through a plain covered deep with ashes and tufa from the surrounding volcanoes. In some places the volcanic debris had piled up to a depth of 350 feet. For years the natives had been prying huge ancient bones out of the riverbanks, bones reputed to be those of the giant rakshasas, the spirits which guarded all the temples of Java.

In a stratum about four feet thick and exposed just above the stream level, Dubois in his turn came upon a rich store of animal fossils: a stegodon, an extinct hippopotamus, a small axis deer, an antelope. It was a favorable omen. Soon he also uncovered a fragment of a lower jaw. He felt certain it was human rather than animal. Before he could pursue this highly interesting find, the rains set in and he had to abandon his excavations until the following autumn. In September 1891 he once more set to work. At first he found a right molar tooth that again looked human or near-human. A month later he brushed the earth away from the fossil he had crossed half the world to find, a thick, chocolate-brown cranium. From its resting place of untold millennia Dubois lifted out a piece of a skull unlike any ever seen before. Clearly it was too low and flat to be the cranium of a modern man, and yet in its conformation and other features, it could not be the cranium of an ape.

"The amazing thing had happened," wrote G. Elliot Smith, an English paleontologist. "Dubois had actually found the fossil his scientific imagination had visualized."

But Dubois himself was not so sure. Was this perhaps only the cranium of some unknown, extinct ape? Once more the rains halted his work. The next season he cut a new excavation about 33 feet from where the strange cranium had been buried, and there he found a thigh bone, a femur. As a skilled anatomist, Dubois could tell what it was: an essentially human thigh bone, belonging to a being that had walked upright.

The implications were staggering. Dubois had to make certain. He studied and measured the apelike skull and the humanlike leg bone, for a momentous decision was forming in his mind. Not until 1894 was he ready to make a public statement of it. Then he announced to the world that the low skull and the human leg bone had belonged to the same creature. Nothing could have been more incredible, an apelike head and the upright posture of a man. Deliberately and almost provocatively, Dubois named this creature he had materialized from the past *Pithecanthropus erectus: pithekos* from the Greek for ape, and *anthropos* for man. Some years earlier the German scientist Ernst Haeckel had hypothesized the existence of an in-between creature to which he gave this name. By appropriating it for his Java find, Dubois boldly filed his claim to have found the missing link.

To those unwilling to acknowledge any link with any form of anthropoid ancestor, *Pithecanthropus* was insult added to injury. Clergymen hastened to assure their flocks that Adam, and not the crude, half-ape, half-human brute unearthed in Java, was the true ancestor of man. Dubois was denounced from pulpit and platform. Scientists were almost as angry and skeptical. The combination of apelike head and upright posture ran directly contrary to the belief that the development of a

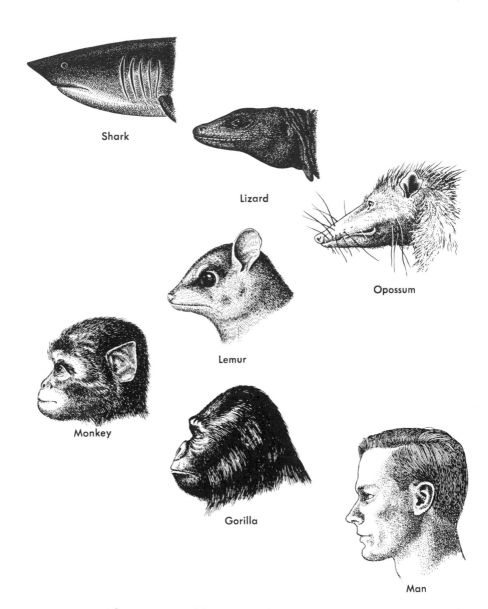

Figure 4–1. The face: from fish to man. The development of the face, from the fish to contemporary man, is shown in successive stages. In the lower orders, here represented by the shark, the jaw is simply an underslung mouth, the nose just a snout, the forehead flat, and the head long. Progressing through reptiles, marsupials and primates to man, the jaw pushes down and forward, the eyes work their way to the front of the head for true binocular vision. The head becomes increasingly spherical—the most efficient shape for maximum brain in minimum skull. (Reproduced by permission of Time-Life Books from Evolution. Copyright © 1964 by Time Incorporated.)

larger, better brain had come first in the separation of the human stock from earlier anthropoids. A being with a human head and an apelike body was expected, not the other way around.

Dubois patiently defended his *Pithecanthropus*. He exhibited the fossil's bones at scientific meetings throughout Europe, presenting detailed measurements and the full data about his discovery. The attacks, nevertheless, did not lessen. Disheartened and hurt, Dubois, by then a professor of geology at the University of Amsterdam, withdrew the remains of *Pithecanthropus* from the public realm. In 1895 he locked the fossils in a strongbox in the Teyler Museum of Haarlem, his home town, and permitted no one to see them for the next 28 years.

In 1920 the discovery of an ancient skull in Australia led scientists to urge that *Pithecanthropus* be let out of solitary confinement, but Dubois was obstinate. In fact, he added to the problem by announcing for the first time that he had the two Wadjak skulls; no one could see them either. At this point Dr. Henry Fairfield Osborn, head of the American Museum of Natural History, appealed to the president of the Dutch Academy of Sciences in the hope that this material, essential to science, would be made available. Soon afterward, in 1923, Dubois opened his strongboxes for Dr. Aleš Hrdlička of the Smithsonian Institution, and thereafter again exhibited *Pithecanthropus* at scientific meetings. Dubois also released a cast of the *Pithecanthropus* skull, indicating a brain of about 900 cubic centimeters, well above the 325- to 685-cubic-centimeter range of the apes, and below the average 1,200 to 1,500 of modern man.

While the bones of Java man lay locked away, a few other fossils of apparently early origin were turning up in other parts of the world. England was filled with excitement and controversy over a remarkable skull and some other fossil fragments unearthed on the Piltdown Common between 1908 and 1915. The Piltdown skull differed radically from that of *Pithecanthropus*. It rounded into almost as high a dome as a modern skull, although the jaw was that of an ape. Thus Piltdown came much closer than *Pithecanthropus* to the prevailing expectations of what early man ought to have looked like.

A young Canadian physician and biologist, Dr. Davidson Black, went to England at about this time to study and to assist with the restoration of the Piltdown skull—a find which many years later embarrassingly turned out to be a hoax.[1] The work deepened Black's already ardent interest in the history and background of man. He thought that man might well have originated in Asia, and when he was offered an appointment as pro-

1. See Part 5, p. 266 for further discussion of this hoax. Ed.

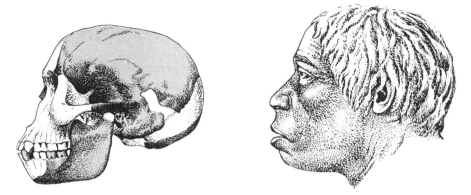

Figure 4–2. The Piltdown man puzzled anthropologists for decades because he combined a chimpanzee-like jaw and teeth with fragments of a human skull. Not until 1953 did scientists prove conclusively that the fossil was a hoax perpetrated by Charles Dawson, an amateur biologist. Dawson had artfully put together unrelated fragments found in a gravel bed near the Piltdown Common in England, and claimed the discovery of a missing link. (Reproduced by permission of Time-Life Books from Evolution. *Copyright © 1964 by Time Incorporated.)*

fessor of anatomy at Peking Union Medical College, he eagerly accepted. It would place him on the scene.

At first, though, Black's careful investigations yielded nothing. Not a trace of early man could be found on the whole great continent until Dr. J. G. Andersson, a Swedish geologist, walked into Black's office one day with two teeth an associate had found protruding from a rock face at Choukoutien. For centuries the Chinese had been mining "dragon bones" in the clefts of the hill. When pulverized they made a prized medicine. As Black examined the teeth from this reputed repository of ancient monsters, his hopes soared. He was certain that they came from a human of great antiquity. The Rockefeller Foundation also was impressed. At Black's urging the foundation agreed to finance a full-scale scientific exploration of Choukoutien.

The hill reared its low, rounded crest about 40 miles southwest of Peking. Just below it lay the little village of Choukoutien, and beyond that the wide expanse of the Hopei plain, crossed by both a modern railroad and the old sunken road along which camels plodded their slow way to the city. At some remote time in the past, water had honeycombed the limestone of the hill with caves and fissures. The caves in turn had filled with the deposits of running water and with the debris of collapsing roofs.

By the 20th century, when modern quarrying cut away one face of the hill, the former caves appeared only as clefts, or in some cases as hard dikes, or solidified fills.

It was here in 1927 that the first scientific expedition in the search for man's origins began its spadework. Digging in the hard, compacted stone proved difficult; blasting was often necessary. Just as much of a problem was the troubled political condition of China. Antiforeign riots were flaring and Chiang Kai-shek's armies, moving to the relief of Shanghai, still were far from the city. Bandits controlled the countryside around Peking. For weeks at a time they isolated the dig from the city.

Three days before the first season's work was to end, Dr. Birgir Bohlin, field supervisor, found another early human tooth. As he hurried to Peking to take it to Black, soldiers stopped him several times without suspecting that he carried a scientific treasure in his pocket. Black pored over the new tooth night after night. It differed so markedly from all the others in his large collection of casts that he decided upon a bold step. He set up a new genus and species for the man from whom it came—*Sinanthropus pekinensis* (Chinese man of Peking). It was a large classification to be based on one small tooth, and the scientific world reserved judgment.

The work in 1928 and through most of 1929 yielded more than 1,000 boxes of fossil animal bones, a few additional human teeth and several small fragments of human bone. By December 2 the weather was bitingly cold and the work was about to be closed for the year, when Dr. W. C. Pei opened up two caves at the extreme end of the fissure. On the floor of one was a large accumulation of debris. Pei brushed some of it away, and suddenly there lay revealed the object of all their work and searching, a nearly complete skullcap. It was partly surrounded by loose sand and partly embedded in travertine, a water-formed rock. Even at first glance, Pei felt certain that it was a skullcap of *Sinanthropus*. He rushed the momentous news to Black.

At a dinner party on the night of December 7, 1929, the night after the skull reached Peking, Black whispered to Roy Chapman Andrews, the American scientist: "Roy, we've got a skull! Pei found it December second." As soon as the two men could get away, they hurried to Black's laboratory. "There it was, the skull of an individual who had lived half a million years ago," Andrews wrote. "It was one of the most important discoveries in the whole history of human evolution. He could not have been very impressive when he was alive, but dead and fossilized, he was awe-inspiring."

The news made headlines all around the world. The work at the hill then was reorganized on a broader basis, and in the 1930s the pieces of a second *Sinanthropus* skull came to light. Black continued to work day and night, organizing the work, keeping detailed records of all the finds, classifying, making casts, drawings and photographs of the heavy volume of material pouring into Peking. He suffered a heart attack one day as he was climbing around the hill at Choukoutien. Without saying anything about the seriousness of his condition, he made plans for the next season and put his affairs in order. Death came to the dedicated scientist in 1934.

The Rockefeller Foundation sought carefully for a successor. It found him in Dr. Franz Weidenreich, then a visiting professor of anatomy at the University of Chicago. Before the Nazis drove him from his native Germany, Weidenreich had completed world-famous studies of the evolutionary changes in the pelvis and foot that made possible man's upright posture. His studies underwrote the contention of Darwin and Huxley that man is a descendant of some ancient anthropoid stock, but not of any recent genus of one.

In China, Weidenreich began a classic series of studies of Peking man —*The Dentition of Sinanthropus, The Extremity Bones of Sinanthropus, The Skull of Sinanthropus*. All three scientifically supported the conclusion of Black: *Sinanthropus* was indeed a human, though a very primitive one. He was not a link between apes and men. What placed him solidly in the human race was his undoubted ability to walk upright on two legs. "Apes, like man, have two hands and two feet, but man alone has acquired an upright position and the faculty of using his feet exclusively as locomotor instruments," said Weidenreich. "Unless all signs are deceiving, the claim may even be ventured that the change in locomotion and the corresponding alteration of the organization of the body are the essential specialization in the transformation of the prehuman form into the human form."

The teeth and dental arch of *Sinanthropus* testified further to his status. The canines were not the projecting fangs of the ape; they did not come together like the blades of a pair of scissors. And the dental arch was curved, not oblong. Still another proof lay in the skull. Weidenreich arranged the skulls of a gorilla, of Peking man and of modern man in an enlightening, haunting row. Even a glance revealed their striking differences: the extremely flat skull of the gorilla, the somewhat higher skull of Peking man, the rounded skull of modern man. In the low pate of the gorilla a brain of about 450 cubic centimeters was housed; in the higher dome of Peking man, one of about 1,000 cubic centimeters; and in the high cranium of modern man, a brain of about 1,350 cubic centimeters.

With a brain so small, some scientists questioned the human status of Peking man. Weidenreich cautioned that brain size alone is no absolute determinant, pointing out that one species of whale has a brain of about 10,000 cubic centimeters. But this amounts to one gram of brain for each 8,500 grams of body weight, compared to man's one gram for each 44 grams of weight. "Neither the absolute nor the relative size of the brain can be used to measure the degree of mental ability in animals or man," he added. "Cultural objects are the only guide as far as spiritual life is concerned. They may be fallacious guides too, but we are completely lost if these objects are missing."

At Choukoutien cultural objects were not missing. The continuing excavations produced thousands of chipped-stone tools. They were simple, with only a few chips removed, but they were made to a pattern. Some of them lay with charred bits of wood and bone. And the charring did not result from some accidental fire, for hard-baked red and yellow clay— hearths, that is—often underlay the carbon. Peking man had mastered the use of fire.

The bones of thousands of animals were strewn about in the former caves. More than three quarters of them belonged to deer, which must have been the favorite meat of Peking man. There also were bones of the bighorn sheep, the boar, the bison, the ostrich, and even of such river dwellers as the otter. All of the mammalian bones came from species long since gone from the earth. About 20 feet below the lowest outer threshold of the big cave, the expedition also found Peking man's garbage dump, by now a stony amalgam of thousands of scraps of bone, stone chips and hackberry seeds. All in all, by his fires and his handiwork as well as by his bodily structure, Peking man indubitably established his right to a place in the human family, if not in that of modern man.

Soon after the big excavations began in China, the Geological Survey of the Netherlands East Indies invited a young German paleontologist, G. H. R. von Koenigswald, to resume the search for early man in Java. On an upper terrace of the same Solo River whose banks had harbored the bones of Dubois' *Pithecanthropus*, Von Koenigswald in 1937 found fragments of 11 somewhat more recent "Solo skulls," and in a region to the west, called Sangiran, the skull of another *Pithecanthropus*. At first only a few pieces of the latter were found. Von Koenigswald offered his native helpers 10 cents for each additional piece they could discover. When it was all assembled, the second skull could scarcely have been more like the first. "It was a little eerie," said Von Koenigswald, "to come upon two skulls . . . which resembled each other as much as two eggs."

In 1939 Von Koenigswald went to Peking to compare the Java finds with those from Choukoutien. Again two widely separated discoveries proved conspicuously alike. "In its general form and size [the Peking skull] agrees with the Java skull to such an extent that it identifies *Pithe-*

canthropus too as true man and a creature far above the stage of an ape," said Weidenreich, upsetting the judgment of Dubois that *Pithecanthropus* came before man.

The first judgment of the men in the field was later corroborated by Wilfrid E. Le Gros Clark, professor of anatomy at the University of Oxford. *Pithecanthropus* appeared slightly more primitive, with his brain of about 900 cubic centimeters, and a shade heavier jaw. The animals he killed and ate also were a little older than those at Peking and no tools were found with *Pithecanthropus*. Despite these differences the two were strikingly alike.

Von Koenigswald and Weidenreich agreed that Java man and Peking man differed little more than "two different races of present mankind," and Clark came to the same conclusion. The Oxford authority proposed dropping the *Sinanthropus* classification which said the Peking man constituted another genus. It was doubtful that the two even formed separate species. Clark therefore suggested that both should be identified as *Pithecanthropus,* and distinguished only by their place names, *Pithecanthropus erectus* of Java, and *Pithecanthropus pekinensis,* or more simply as Java man and Peking man.

The modern world was about to prove too hazardous for these early men emerging from the long-hidden reaches of the past. By the autumn of 1941 the scientists working at Choukoutien could not misread the signs of war. There were daily reports of Japanese troop movements. Weidenreich was in the United States on a visit, but the director of the Geological Survey appealed to Dr. Henry S. Houghton, president of Peking Union Medical College, to have the irreplaceable remains of Peking man taken to safety. Dr. Houghton was doubtful that the United States should assume responsibility, but nevertheless arranged to have the collection taken to the United States with a Marine detachment that was leaving by special train in a few days.

At 5 a.m. on December 5 the train pulled out of Peking. It was to meet the American liner *President Harrison* at the small coastal town of Chingwangtao. This rendezvous was never kept, for on December 7 Japanese bombs crashed down on Pearl Harbor and total war came violently to the Pacific. In the maelstrom, all that existed of Peking man—fossil pieces representing about 40 individuals—disappeared, never to be seen again.

In Java, Von Koenigswald knew that it was only a matter of time until the island would be seized. So he quietly gave some of his most valuable fossils to friends and substituted casts for them in his museum. Soon after the occupation of Java, Von Koenigswald, his wife and daughter were sent to separate concentration camps. The Japanese became suspicious

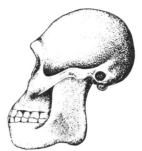

a. Original fragment   b. Reconstructed skull

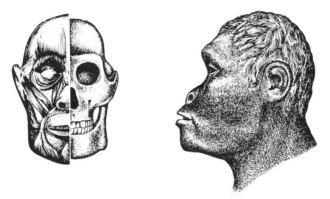

c. Deep-muscle addition   d. Finished head

Figure 4–3. Steps in reconstructing a fossil head.

(*a*) *Original fragment. The reconstruction of a fossil head is a three-dimensional jigsaw puzzle in which bone fragments (such as those above) are pieced together, and layers of clay to simulate muscles, tissues and skin added until a facsimile of the original is obtained.*

(*b*) *Reconstructed skull. In the first stage of fossil reconstruction, the molder extends the contours indicated by the original fragments by adding clay to the skull. The more extensive the original fragment, the less work there is and the more accurate the representation.*

(*c*) *Deep-muscle addition. Next the molding artist—who must have a solid understanding of anatomy—uses more clay to form the deep muscles, which are attached to the original bone and to the newly reconstructed parts. Old muscle markings on the original bone guide him.*

(*d*) *Finished head. After clay for the fatty tissue has been added to the tip of the nose, the chin and to the ears, the skin covering is applied. However, details like the distribution and amount of hair and minute surface configurations can only be assumed by the artist. (Reprinted by permission of Time-Life Books from* Evolution. *Copyright © 1964 by Time Incorporated.)*

about the fossil collection, but never discovered the truth. On the whole they took good care of the museum, for they expected to keep it after winning the war, and they understood its value. Only one skull was removed, one of the Solo skulls, which was sent as a special birthday gift to the Emperor of Japan.

At the end of the war, Von Koenigswald learned that his wife and daughter were safe. "My happiness was complete," he said, "when I learned that my precious specimens had been saved. Large parts of my collections, many of my books, and all of my clothes had been stolen, but Early Man had survived the disaster." Von Koenigswald later exhibited Java man in New York and then took the precious bones to The Netherlands, where he became professor of paleontology and historical geology at the State University of Utrecht.

With the end of war, the second search for Peking man began. Three nations joined it, but the story of what happened to the fossils in December 1941 has never been clarified. It is certain only that the *President Harrison* was beached to prevent its seizure and that the Japanese captured the special train. One account holds that the Japanese loaded the small remains of Peking man on a lighter for transfer to a Japanese ship, and that the lighter sank.

It is unlikely that any Japanese who recognized the uniqueness of the collection came into possession of it. After the occupation of Peking the Japanese ransacked the city searching for Peking man. "They knew what they were looking for, and knew that the relics had been in Peking," said Miss Agnes Pearce, secretary of the China Medical Board. "The controller of the college was taken into custody and questioned for five days."

On the chance that Peking man might have been taken secretly to Japan, the Looted Properties Division of the Far East Command combed Japan for his bones. The Communist government of China in its turn took up the search. No trace was found, but Dr. Pei, the finder of the first skull, broadcast the charge that the United States was secreting the remains of Peking man at the American Museum of Natural History in New York. Dr. Harry L. Shapiro, chairman of the museum's Department of Anthropology, pointed out in answer that the museum has only casts of the Peking fossils. Weidenreich had made them and sent them to a number of museums before the start of World War II.

The fate of Peking man remains one of the great international mysteries, but his standing is secure. However fleeting and violent his reappearance, his bones and those of Java man offered incontrovertible proof of mankind's lengthy existence. These ancestors emerging from the past were not what their descendants expected them to be, for they were crude, primitive and low of brow. But they walked like men, they were intelligent, and they lived successfully in their environments of 360,000 years ago. They were men, not forms transitional between animals and men. Huxley's haunting question still had to be answered: was there in some older strata an ape more anthropoid or a man more pithecoid than any yet known?

DAWN MAN AND HIS BROTHERS

Dr. Raymond A. Dart, professor of anatomy at South Africa's University of Witwatersrand, impatiently pried the lid off a big wooden box of rocks that had just been delivered at his Johannesburg home. He hoped the rough fragments inside the box might include a fossil baboon skull. And luck was with him, for on the very top of the heap lay the cast, or mold, of the interior of a skull. Such a fossilized "brain cast" of any species of ape would have been a notable discovery, but this one was no ordinary anthropoidal brain. It was three times as big as a baboon's and larger than that of an adult chimpanzee. When he came to a chunk of rock into which the cast fitted perfectly, the tremendous thought struck Dart that he might be holding the missing link in his hands.

The box had come from a limestone quarry near Taung, a railroad station not far from Johannesburg. A rare fossil baboon skull had been found there recently, and by way of cooperating with science, the quarry operator had agreed to ship any bone-bearing rock to the doctor. Dart's first problem was to free the rest of the strange, larger skull from the stone matrix. Working with a hammer, chisels and a sharpened knitting needle, he "pecked, scraped, and levered" bits of stone from the front of the skull and the eye sockets.

After days of this painstaking dissection an incredible face began to emerge. It was not surmounted by the beetling brow of an ape but by a true forehead, and the upper jaw, instead of jutting forward as in all true apes, was shortened and retracted under the skull. Was this ape, or man, or both? From then on, Dart lay awake nights "in a fever of thoughts" about what kind of ape might have lived long ago in this semidesert plateau. For more than 70 million years, while ice had advanced and retreated over much of the earth and while mountains rose along the continental coasts, South Africa had stood as a dry, relatively undisturbed veld, much as it is today. It had never been clothed with the kind of jungle in which an ape could live, and the nearest natural habitat of apes was more than 2,000 miles away from Taung. Could some different kind of ape have found a way to adapt itself to life in an arid, open land?

Dart was up daily at dawn to continue his exacting labor with the baffling skull. On his 73rd day of work the stone parted and he saw before him the skull of a six-year-old child with a full set of milk teeth. The permanent molars were just beginning to erupt and the canines, as in humans, were quite small. The set of the skull indicated that this child had walked upright. No ape can take more than a few steps without going down on all fours.

All previous discoveries of human predecessors had proved in the end to be men—authentic, if early, men. This was true of the thick-skulled but large-brained Neanderthal man, of Java man, of Peking man. All were men with certain apelike features. But the child's face before Dart seemed the reverse, an ape with human features. He set up a new genus for it, with the formidable name *Australopithecus africanus* (*Austral*, for south, and *pithekos*, for ape). Dart wrote a full scientific report for the February 3, 1925, issue of the British magazine *Nature*. It included the provocative statement, "The specimen is of importance because it exhibits an extinct race of apes intermediate between living anthropoids and man." He agreed to release the news to the Johannesburg *Star* on the same date. At the last moment *Nature* decided it could not publish so startling an article without referring it to experts for comment. The *Star* would wait no longer, so on that day, in South Africa and around the world, the headlines proclaimed that the missing link was found. Actually it was the first link in a long chain of discoveries in which Africa, and not the Middle East or Asia, would be established as the presumed birthplace of the human race.

The attacks on Dart were not long in coming. Such English authorities as Sir Arthur Keith, Elliot Smith and Sir Arthur Smith Woodward expressed skepticism and disagreement. Dart's "baby," several critics suggested, was only "the distorted skull of a chimpanzee." Taung became something of a byword; it figured in songs and was heard on every music-hall stage. But Dart was encouraged by a warm letter of congratulation. It came from Dr. Robert Broom, a Scottish physician who had hunted fossils all over the world before being drawn to Africa by reports of the finding of mammallike reptiles. In the Karroo, Broom had unearthed fossils that almost completely filled the gap between the egg-laying reptiles and the small early mammals. His researches were called "as masterly as any in the century" by the scholarly prime minister of South Africa, Field Marshal Jan C. Smuts. Two weeks after the arrival of his letter, Broom showed up at Dart's laboratory. He spent a weekend in intensive study of the child's skull, and was convinced that as "a connecting link between the higher apes and one of the lowest human types," it was the most important fossil find made up to that time. He firmly said so in an article in *Nature*.

However, after the first flare-up of attention, Dart's "baby" was either forgotten or dismissed by most scientists. Still Dart and Broom continued to study the skull. In 1929 Dart succeeded in separating the lower jaw from the upper, and for the first time was able to see the entire pattern of the teeth. Most of them could have belonged to a child of today, though the molars were larger. In the minds of the two physicians any lingering doubts were removed.

Broom was eager to take up the search for another *Australopithecus*, an adult one. It would be the only way, he was sure, to dissolve the disbelief. Smuts opened the way by offering Broom a post as curator of vertebrate paleontology and physical anthropology at the Transvaal Museum in Pretoria. For the next 18 months Broom was occupied in digging out and identifying half a dozen extinct species of rats and moles. He also unearthed a baboon jaw which at first appeared to be australopithecine. It was not, but the publicity led two students of Dart's to tell Broom about some small skulls they had found in a quarry at Sterkfontein, a town not far from Pretoria. Broom arranged to go there with them.

Ever since the first mining camps were opened during the Witwatersrand gold rush of 1886, the people of the Sterkfontein area had been picking up the fossilized remains of baboons, monkeys and perhaps, unknowingly, ape men. The limeworks had even issued a little guidebook, "Come to Sterkfontein and Find the Missing Link." Dr. Broom came on August 9, 1936, and promptly did so. G. W. Barlow, the quarry manager, had formerly worked at Taung and knew about the australopithecine skull. When Broom asked if he had ever seen anything like it at Sterkfontein, Barlow said he "rather thought" he had. He explained that he usually sold "any nice bones or skulls" to Sunday visitors to the quarry. He promised Broom that he would keep a sharp lookout for anything resembling an ape-man skull.

Nine days later when Broom returned, Barlow asked, "Is this what you're after?" and handed him two thirds of a "beautiful fossil brain cast." It had been blasted out only that morning. Broom anxiously dug into the debris to try to find the skull that had served as the mold. Though he worked until dark, he found nothing. The next day, as he sorted the piles of stone, he not only recovered the base of the skull and both sides of the upper jaw, but also fragments of the brain case. When the fragments were pieced together, Broom had most of the skull of an adult australopithecine.

For three years the doctor continued to visit his fossil gold mine. One June day in 1938, Barlow met him with, "I have something nice for you this morning." He handed Broom an ape-man upper jaw with one molar in place. He had obtained it from a schoolboy, Gert Terblanche, who lived on a farm at Kromdraai, less than a mile away. The doctor drove over to Kromdraai. Gert was at school, but his mother and sister showed Broom the hillside where he had found the fossil piece. The doctor hurried on to the school and found Gert. At his first questions, the boy pulled out of his pocket "four of the most beautiful teeth ever found in the world's history." Back at the site, Gert opened his private cache and gave the doctor an excellent piece of a lower jaw. During the next two days, Broom and the boy sifted earth and found a number of scraps of bone and teeth. When the pieces were put together, Broom had most of another ape-man skull, the third. The face was flatter than the Sterkfontein *Australopithecus'*, the jaw heavier and the teeth larger, though more human in conformation.

When the newest findings were published, some of the old skepticism began to dissipate. Kromdraai man differed so markedly from both the Taung child and the Sterkfontein adult that Broom set up a new genus for him, *Paranthropus robustus* (robust near-man). But the claim again upset his fellow scientists. Broom was thought to be going beyond all bounds in setting up another genus for one small African area.

"Of course the critics did not know the whole of the facts," said Broom. "When one has jealous opponents one does not let them know everything." What he had not disclosed was that each of the skulls was associated with its own distinctive set of animals. Ancient horses abounded at Kromdraai. None occurred at Sterkfontein, only a mile away. At Kromdraai, Broom found the bones of jackals, baboons and saber-toothed tigers. The Sterkfontein quarry yielded an extinct wart hog. None was to be found at Kromdraai. If the sites were occupied at different times, Taung some two million years ago, Sterkfontein about 1,200,000 and Kromdraai 800,000, the doctor pointed out, each might well have a different group of extinct animals—and a different genus of ape man.

At this point, World War II halted work at the South African sites. Broom and his assistant, G. W. H. Schepers, made use of the war years to study, measure and describe the astounding volume of material they had collected. After more than 20 years of ignoring or deriding the South African creatures with their astonishing mixture of ape and human characteristics, science was brought to with a shock by a book published by Broom and Schepers. Their data left almost no question that the fossils were near men, representing an important stage in the evolution of humanity. Exactly what their position might be—ancestors or cousins—still remained moot.

Soon after the end of the war, Broom resumed digging at Sterkfontein. One day a blast in some seemingly unpromising cave debris revealed the first of a series of important discoveries. When the smoke cleared away, the upper half of a perfect skull sparkled brilliantly in the sunlight striking the pinkish-gray stone of the quarry. Lime crystals incrusting its inner surface caught and reflected the light like diamonds. The lower half lay embedded in a block of stone that had broken away. The glittering skull was that of an adult female. Her jaw was heavy, her forehead low, but there was an unmistakable quality of humanness about her. Her discovery was a worldwide sensation. It was followed by other important finds: a male jaw with an intact canine tooth worn down in line with the other teeth as the canines are in men, and then in 1948 a nearly perfect pelvis. If the pelvis belonged to an ape man, he had walked upright, like man. Other finds confirmed this posture many times over.

By 1949 the remains of more than 30 individuals had been recovered from the South African caves, and Le Gros Clark at Oxford undertook an impartial, definitive study. He meticulously studied the South African fossils themselves, and compared them to a series of 90 skulls of modern apes. His verdict was unqualified: "It is evident that in some respects they [the Australopithecinae] were definitely ape-like creatures, with small brains and large jaws.

"But in the details of the construction of the skull, in their dental morphology, and in their limb bones, the simian features are combined with a number of characters in which they differ from recent or fossil apes and at the same time approximate quite markedly to the *Hominidae* [the family of man]. All those who have had the opportunity of examining the original material are agreed on these hominid characters: the real issue to be decided is the question of their evolutionary and taxonomic significance."

Were the ape men among the direct ancestors of *Homo sapiens?* The evolutionary question to which Le Gros Clark referred was complicated by the absence of cultural objects—tools—which Franz Weidenreich had pointed out are essential to determining human status. Not one chipped-stone tool had been found with the numerous remains of the ape men. Then in 1953 on a terrace in the Vaal valley laid down in the same dry period in which the ape men lived, some simple pebble tools were discovered—fist-sized pieces of stone from which a few chips had been removed. It seemed impossible that the australopithecines, with their brains of only 450 to 650 cubic centimeters, no larger than those of the apes, could have made them.

In the spring of 1957, pebble tools were found in the upper part of australopithecine debris of Sterkfontein, and a little later J. T. Robinson, who had succeeded Dr. Broom, and Revil Mason dug into a layer of red-brown rock which contained several ape-men teeth and more than 200 pebble tools. To the untrained eye they would have looked like naturally fractured stone. But close examination showed that chips had been flaked off two sides; the head of the stone was left round. A hammer stone held in a hand and guided by understanding had shaped them to perform certain tasks of cutting, scraping and, probably, killing. And the tools were not made of the same stone as the caves where they lay buried. The pebbles had been carried there, purposefully, from some other place.

It appeared that before they learned to *make* tools, the ape men may have been tool-users. In 1947 Dr. Dart, after 18 years as dean of the Faculty of Medicine at the university, returned to the search for "dawn man." Analyzing thousands of fossilized animal bones found in the cave deposits, the discoverer of *Australopithecus* concluded that the ape men had employed tusks and teeth for cutting tools, jaws for saws and scrapers, and leg bones for bludgeons.

"The Australopithecines must have originated as apes that by walking upright became adapted to life in the open country," said Kenneth Page Oakley in his study *Tools Makyth Man.* "There are many reasons to suppose, as Dart, George A. Bartholemew, and others have shown, that the earliest hominids must have been tool *users.* Bipedalism is initially disadvantageous biologically unless there is some compensating factor—in the case of the hominids this was the ability to use tools and wield weapons while moving. . . . The earliest tools and weapons would have been improvizations with whatever lay ready to hand. Although the hominids must have begun as occasional tool users, ultimately they were only able to survive in the face of rigorous natural selection by developing a system of communication among themselves which enabled cultural tradition to take the place of heredity. At this point systematic tool making replaced casual tool using, and it may be that this changeover took place in the Australopithecine stage."

Far to the north and east of the desert land of the australopithecines, in Tanganyika's Oldoway Gorge, Louis S. B. Leakey also was finding pebble tools. He first came upon them in 1931, long before they were discovered in South Africa, and wondered if they could have been made by a creature similar to the australopithecines. If so, Leakey was unable to find him. In vain he searched the clearly stacked strata of the gorge for the maker of the tools.

Oldoway Gorge is an abrupt rent in the earth, some 25 miles long and 300 feet deep, a "little Grand Canyon." A German entomologist named Wilhelm Kattwinkel found it in 1911 when he almost fell into its depths as he broke through some bush on the edge. A hasty exploration showed it to be a storehouse of fossils. Some which Kattwinkel took back to Berlin

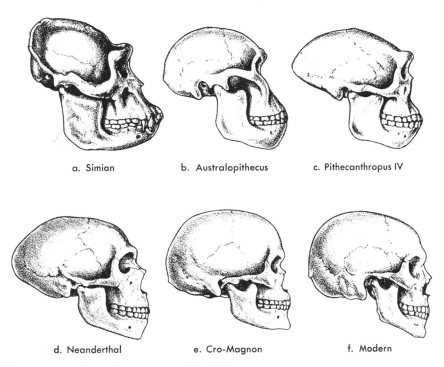

| a. Simian | b. Australopithecus | c. Pithecanthropus IV |
| d. Neanderthal | e. Cro-Magnon | f. Modern |

Figure 4–4. The evolution of man's skull.

(a) *Simian.* The skull of a gorilla bears a superficial resemblance to that of man, but its brain capacity is only about 450 c.c. The thick bony crest on top serves as a support for the muscles that are needed to open and close the heavy jaw laden with large teeth.

(b) *Australopithecus.* Though basically simian in appearance, the skull of Australopithecus, an extinct African ape man that walked upright and may have used tools, lacks the large, sharp canine teeth of the gorilla. The brain capacity ranged from about 450 to 650 c.c.

(c) *Pithecanthropus IV.* This primitive man had a brain capacity of 900 c.c. As man's jaw became lighter and his teeth smaller, there was more space for the tongue. And because lighter muscles operated the lighter jaw, the skull could become thinner and the brain larger.

(d) *Neanderthal.* The profile of a Neanderthal skull shows a retreating forehead, heavy eyebrow ridges and an elongated brain case, which varied in size from 1,400 c.c. to 1,600 c.c. The chin, though sloping, is less muzzlelike than that of more primitive forms of man.

(e) *Cro-Magnon.* Cro-Magnon's skull approaches modern man's in appearance, but has an even larger brain capacity—approximately 1,590 c.c. as compared to 1,500 c.c. for the average European. Its high forehead contrasts with the depressed one of Neanderthal man.

(f) *Modern.* Modern man's skull houses evolution's finest product—a richly convoluted, intelligent brain. It is believed that the brain, in continuing to evolve, will grow larger and that the man of the future will have a brain capacity of at least 2,000 c.c. (Reproduced by permission of Time-Life Books from Evolution. Copyright © 1964 by Time Incorporated.)

were so unusual that a German expedition headed by Hans Reck was sent out in 1913 to explore further. Its investigations were ended by the First World War, and after the war Reck was unable to raise funds to resume operations. Eventually he wrote to Leakey, the young curator of the Coryndon Memorial Museum at Nairobi, urging him to take over.

The search for early man was one to which Leakey was already dedicated. He was born, in 1903, in a wattle hut near Kabete. His parents were the first missionaries to the Kikuyu tribe, and Leakey was the first white baby most of the tribesmen had ever seen. As Leakey grew up with Kikuyu playmates, he thought of himself as one of them. Sent to England to school and college, Leakey wrote his Cambridge thesis on the Stone Age in Kenya. He returned to Africa in 1924 on an archeological expedition, and later was appointed curator of the Coryndon museum. Not until 1931, however, could he raise the funds for an expedition to Oldoway. It took seven days to cover the 500 miles from Nairobi to the gorge, a wild place frequented by inquisitive lions.

One season was enough to convince Leakey that Oldoway was a site "such as no other in the world." [See Figure 4–6.] Pebble tools were so abundant at the bottom of the gorge and in its lowest exposed strata that Leakey named their type Oldowan, after the place of their discovery. In these old strata too lay the bones of many extinct animals; ultimately Leakey identified more than 100.

Leakey returned to the gorge year after year, and on later expeditions took along his wife and their two sons. They could seldom spend more than seven weeks a year there. The trips were costly, the heat intense, and water had to be hauled from a spring 35 miles away. There was only one way to search systematically for fossils and that was to crawl along on the hands and knees, inspecting every inch of ground. Through such work, the Leakey collection of animal fossils and stone tools grew and grew. But 28 years went by with no sign of the men, or ape men, who manufactured the tools. It began to seem that the most primitive of hominids, the ape men, had lived at one place, South Africa, and that the most primitive stone tools had been made at another, East Africa. At least this was the baffling situation until July 17, 1959.

That morning Leakey awakened with a fever and headache. His wife insisted that he remain in camp. But their season was drawing to an end and the day could not be lost. Mary Leakey, accompanied by two of their Dalmatians, drove to the point where the party was working. As she crept along the hillside, a bit of bone lodged in a rockslide caught her eye. She recognized it as a piece of skull. She searched higher along the slope, and suddenly saw two big teeth, brown-black and almost iridescent, just erod-

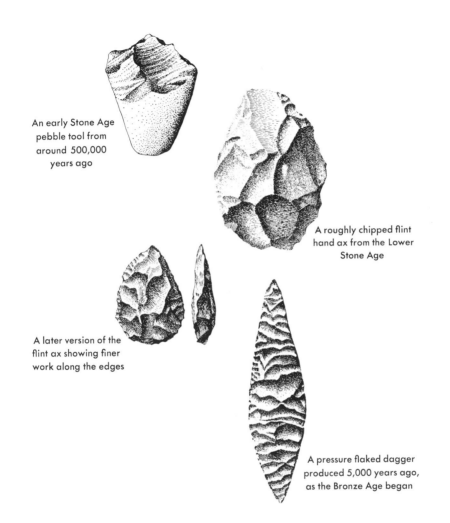

An early Stone Age pebble tool from around 500,000 years ago

A roughly chipped flint hand ax from the Lower Stone Age

A later version of the flint ax showing finer work along the edges

A pressure flaked dagger produced 5,000 years ago, as the Bronze Age began

Figure 4–5. The evolution of stone age tools.
The making of tools began some 500,000 years ago when men first gathered stones from river beds and beaches and fashioned them into crude, chipped pebble tools for cutting and scraping. By the end of the period, just prior to the metal ages, primitive men had learned to make delicately wrought hand axes and other tools, flaked from a variety of materials, including obsidian, flint—which was sometimes mined from pits and seams—bones and wood. (Reproduced by permission of Time-Life Books from Evolution. *Copyright © 1964 by Time Incorporated.)*

Figure 4–6. South Africa's 25-mile-long Oldoway Gorge, a part of the Great Rift Valley, holds primate remains spanning 25 million years. (Photograph courtesy of the Trustees of The British Museum, Natural History.)

ing from the hill. She marked the spot with a small cairn of stone, ran to their Land Rover and sped back to camp.

Leakey heard the car racing up the road and sprang up in alarm, his first thought that his wife had been bitten by a snake that had slipped by the guard of the dogs. But as the car stopped he heard his wife shout, "I've got him! I've got him!" The "him," she felt sure, was man, the early man they had been seeking so many years. Leakey's fever and headache forgotten, they jumped in the car and sped down the trail as far as they could drive. The last half mile they covered at a run.

Mary Leakey's first impression had been right: the teeth were human or near-human. The two dropped to their hands and knees to examine them with the most minute care. Leakey could see that the dark molars glinting in the afternoon sun were twice as wide as the molars of modern man, but human in shape.

"I turned to look at Mary, and we almost cried with sheer joy, each seized by that terrific emotion that comes rarely in life," said Leakey. "After all our hoping and hardship and sacrifice at last we had reached

206

our goal . . . we had discovered the world's earliest known human." The Leakeys went to work with camel's-hair brushes and dental picks. The palate to which the teeth were affixed came into view and then fragments of a skull. In order not to lose a single precious scrap they removed and sieved tons of scree, a fine rock debris, from the slope below the find. At the end of 19 days they had about 400 fragments.

While the delicate task of assembling the bits and pieces went on, the Leakeys continued to excavate the site. They had not only discovered the oldest near-man skull to be known in the whole of eastern and central Africa, but also a campsite of this ancient creature's. Scattered on what had been the shores of an ancient lake were nine pebble tools, not the most primitive Leakey had found at Oldoway, but very early ones. Lying about too were the fossil bones of animals the residents of the campsite had killed and eaten—rats, mice, frogs, lizards, birds, a snake, a tortoise, some young pigs, a juvenile giant ostrich. Nearly all these bones were broken, while the near-human skull and a tibia, a leg bone that also appeared in the stone of "Bed I" of the site were not. So ape man had killed the animals. But here were no giant beasts. Even with their pebble tools the beach dwellers evidently could not cope with the big animals of their day. Their prey were the young and the small.

The skull taking form from the fragments was that of an 18-year-old male. The unworn wisdom teeth and the fact that the suture joining the two halves of the skull had not yet closed indicated a young adult. In many ways the young male resembled the larger ape men of the south. In brain size and in general appearance he was at least a member of the australopithecine family. Yet as Leakey studied the skull more closely he saw significant differences. The face was not as apelike as that of the South Africans; the cheek had almost the curve of the human cheek. The palate was deeper and more modernly arched. The molars Mrs. Leakey had seen protruding from the hill were unusually large and heavy; they could have cracked nuts. But detailed study confirmed that they were like human teeth. Leakey made a bold decision to set up a new genus for this early Stone Age tool-user, and named him *Zinjanthropus boisei. Zinj* meant eastern Africa in Arabic; the *boisei* honored Charles Boise who had helped to finance the search for early man.

"Zinjanthropus, in spite of being classified in the sub-family *Australopithecinae* already exhibits specializations which foreshadow *Homo* . . ." wrote Leakey.

But how old was he? Luckily for science, not long after *Zinjanthropus* died a stream of lava poured in over the silt that covered him. Two scientists from the University of California, Jack Evernden and Garniss H.

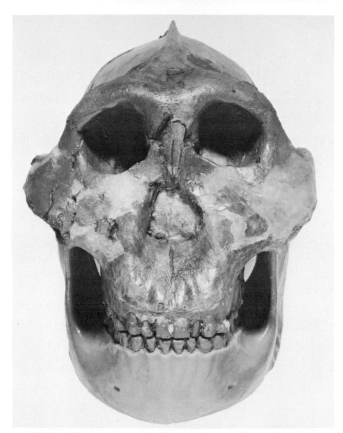

Figure 4–7. Zinjanthropus, one of the earliest tool makers, preyed on other mammals' young in Africa more than a million years ago. (Photograph courtesy of the Trustees of The British Museum, Natural History.)

Curtis, extracted radioactive potassium from this volcanic cover and also from an older volcanic bed that underlay the *Zinj* site. By measuring the potassium's slow decay into argon they first fixed the startling age of about 1,750,000 years for *Zinj*. Later tests indicated that this might be at least half a million years too great; but in any event *Zinjanthropus* was of record-breaking age.

In the years while the Leakeys were probing Oldoway without finding man, they saw that the gorge nevertheless was virtually a book of life. In the five main deposits that cropped out in the 300-foot wall, there was a graduation of tools—from the chipped pebbles of Bed I at the bottom to the tools of contemporary men at the surface. There was a similar progression from the extinct animals at the bottom level to the animals of today roaming the Serengeti plains at the top. Among the pebble tools in the next-to-bottom bed Leakey found, early in his digging, some stone hand axes flaked on both sides.

They were known as Chellean hand axes, and there was nothing unique about their discovery. A century earlier similar axes had been found at Chelles-sur-Marne, a little east of Paris. Since then the same kind of axes had turned up in "digs" all over the world. They were found among the bones of giant animals—elephants, rhinoceroses and the like. But for all the thousands of Chellean axes uncovered at all the different sites, nobody had ever found Chellean man, the maker of the tools and the slayer of the big animals. Who he was and what he looked like remained one of the mysteries of archeology.

On December 1, 1960, Louis Leakey happened to take a backward glance along a slope where he was working in Oldoway's Bed II. From his vantage point he saw a small patch of deposits that had been overlooked. The afternoon light was failing, but first thing the next morning he went to investigate. "Imagine my joy when I walked into the tiny exposure and saw, sticking out of the bank, parts of a human skull," Leakey related. He soon freed the skull from its matrix—and there, to his even greater joy, was Chellean man! The vault of the skull was higher than in *Zinjanthropus* and the eyebrow ridges thrust out into an overhanging shelf.

"The new skull," Leakey reported, "is more or less the contemporary of Java Man and Peking Man and has certain very definite resemblances to these. . . . But it is much larger and differs from them in . . . significant characters."

All about the site lay Chellean tools. Like the earlier Oldowan chipped pebbles, they were made of oval or pear-shaped pieces of stone. But no longer did they have an untouched rounded end to fit into the palm of the hand. They were chipped all around the edges, first in one direction and then in the reverse, so that they had two faces and a zigzag or sinuous edge. With this potent implement, Chellean man could fashion spears to bring down and skin the largest of the animals. No longer was he limited to preying upon the young and small. Now he was a truly formidable hunter. The bones of the giant beasts he slew in many parts of the world testify to his new prowess.

Proconsul (see Figure 4–8), *Australopithecus* (including *Zinjanthropus*), Chellean man—like the tools and animals, they succeeded one another in the deep African strata. Nowhere else in the world was there any other such progression or any other discovered evidence of man's earliest primate ancestors.

In 1952, even before the discovery of *Zinjanthropus* and Chellean man, the French paleontologist, Père Pierre Teilhard de Chardin, speaking at the Wenner Gren Foundation's International Symposium on Anthropology at New York, had pointed to the full meaning of the African finds: "It

*Figure 4–8. Proconsul—or a creature much like him—was the common an-
cestor of men and apes. He prowled near Oldoway 20 to 25 million years ago.
Hairy and slouching, he had less brain than a chimpanzee, but he was built
to lead an agile life on or off the ground. He could have evolved either toward
tree-living, like apes, or toward erect life on the grasslands, like early men.
(Paintings by Maurice Wilson. Reprinted courtesy of the Trustees of the
British Museum, Natural History.)*

Figure 4–9. Australopithecus, an ape man of about one million years ago, walked erect, probably used his arms and hands to make simple pebble tools, and possibly mastered a few speech sounds. (Painted by Maurice Wilson.)

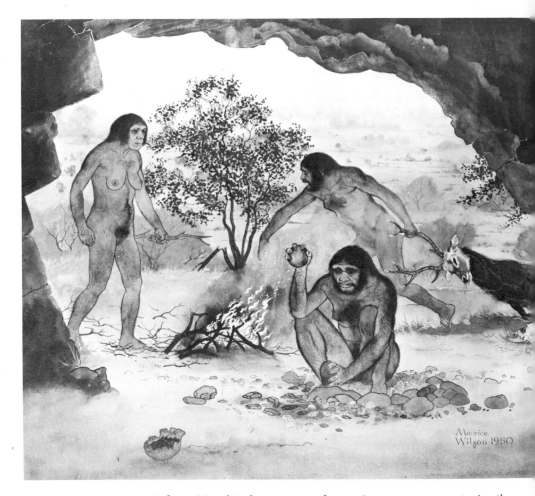

Figure 4–10. Peking Man left his remains about 360,000 years ago in fossil-famous Choukoutien cave in Northern China. There he lit fires, killed deer, flaked out stone tools and probably talked. His own bones—not gnawed but split open—indicate that he also practiced cannibalism. (Painted by Maurice Wilson.)

becomes both difficult and unscientific not to accept the idea that the Dark Continent . . . acted as the main laboratory for the zoological development and the earliest establishment of man on this planet. It is apparently in the depths of Africa (and not on the shores of the Mediterranean Sea or on the Asiatic plateau), therefore, that the primeval center of human expansion and dispersion must have been located. . . ." By all indications man at last had found his birthplace—and his earliest ancestors. He was no longer a backgroundless, unrelated, unproved creature-from-nowhere on this planet.

THE EMERGENCE OF HOMO SAPIENS

A million years or so ago, some of those near-men of South Africa—who may or may not have been among our direct ancestors—took to globe-trotting. The world beckoned them; perhaps the game always looked to be a little more plentiful a little farther on, and in their time "farther on" was uniquely reachable. The wide continental ice sheets had crept down from the north, and as vast quantities of water were locked up in the miles-deep ice, the sea level receded. All through the world the shores widened and land bridges were revealed, wide and crossable, between some of the earth's major land masses.

So early man pushed on, dry-shod, out of Africa into southern and western Europe, where some of his primitive stone tools have been found; into India south of the Himalayas, where he left more of his enduring flints to mark his passing; and along a warm, dry corridor into a southern extension of the Asiatic continent that is now the island of Java (see Figure 4–13). There the skull and leg bones of *Pithecanthropus* testified to his arrival and survival. Probably a little later the migrators turned north into China.

By the time these pioneering primates reached Java, their brain capacity had increased from the 450 to 650 cubic centimeters of their forebears to about 900; and by the time of their establishment at Choukoutien in China, brain capacity was up to about 1,100 cubic centimeters. It is likely that those with the larger brains had the best chance of surviving to reproduce themselves among the perils of new lands, and of finding ways to adapt to new modes of living. The selective pressures for better brains must have been very intense. Their locomotor skeleton, while it remained essentially that of the earlier australopithecines, was already human.

While Java man and Peking man flourished in the Far East, other and similar early men from central Africa were getting established in Europe and northwest Africa, though there is only scant evidence of their presence. In all of Europe only two skulls, one of them half-preserved, show that they were there and tell of what they were. In North Africa so far, only fragmentary jaws and teeth go back to the Middle Pleistocene epoch, a time now roughly and uncertainly estimated at some 500,000 to 150,000 years ago. The anthropoids' family tree branched out richly in the Pleistocene, and until much more evidence is dug up, the question of exactly who begat whom among all the possible predecessors of modern man is highly debatable.

Though the provable record of man and near-man is still obscure through the long middle period of the last million years, the geologic record is generally explicit and clear. Three more times the ice thrust down from the north, covering much of Europe, Asia and North America with a solid icecap much like that which mantles Greenland today. It is certain that each time the cold and ice advanced, man was driven back south. Between the glaciations the climate waxed warmer and more hospitable and man moved north again, finding rich hunting in the wake of the receding ice. Late in the Pleistocene a group of men distinctly different from all the more primitive types whose traces have so far been discovered gradually became predominant in Europe and western Asia. The bones of more than 100 of them have been found in places ranging from southwest France and Gibraltar to Italy, Germany, Yugoslavia, southern Russia, Iraq, Iran and Palestine.

These were the Neanderthals, so called for the first of their kind, found in the Neander valley in Germany. They averaged five foot five in height and were ruggedly built. As we have seen, their brains, lodged in massive heads, were in the upper size range of modern human brains. They made skillfully chipped stone tools and with them hunted a wide range of game, even bringing down such massive animals as the mammoth and woolly rhinoceros. Without doubt they were men of dawning religious feeling—men who could *reflect*—for the arrangement of some of their bones bespeaks a ceremonial burial. In one Swiss cave they left the skulls of bears in the elevated dignity of objects of worship, in what appears to have been a shrine.

In Neanderthal man's case, the relative abundance of fossil material has produced confusion instead of clarity; there are almost as many theories of his history, dispersal and mysterious fate as there are skulls or scholars. After extended study of the whole group, F. Clark Howell of the University of Chicago was convinced, however, that "a single, variable, racial group seems to have been everywhere present in Europe." The coming of the last ice age, it is believed, disrupted this continuity. As the cold increased and the Scandinavian ice sheet again edged southward, the glaciers that filled the mountain valleys grew far out across the plains. The ice drove a wide and impassable barrier between the Neanderthals of western and southern Europe and those of eastern Europe and the Middle East. Gene interchange was barred as effectively as if the colonies had been isolated on widely separated islands.

Thus imprisoned by the ice and the sea, the western Neanderthals developed in their own distinctive way. They could not get out, and no new racial stocks could get in to interbreed with them. The concentration of one genetic pattern, many anthropologists believe, caused the westerners to become more Neanderthalian, with heavier eyebrow ridges and more massive jaws. Conditions were different for the Neanderthals in eastern Europe and the Middle East. There the glaciers lay far to the north or were localized in the Balkan and Carpathian Mountains. The wide, flat steppes and high plateaus supported large herds of animals, and roots and berries grew in the wooded valleys. Bands of men could move about freely, and there was nothing to hinder a free genetic interchange over this whole area—the regions bordering the Black Sea and southwestern Asia. Neanderthal-like people wandered as far east as Samarkand and as far south as Haifa.

Some of the Neanderthals lived in hillside caves looking out on the Mediterranean, and others in the caves bordering the "fertile crescent" of the Tigris and Euphrates Rivers. One such cave, in a rocky slope of the Zagros Mountains, about 250 miles north of Baghdad, turned its arched mouth to the south. In winter the sun shone warm and bright on part of its floor, and the chilly winds whistled by. In 1951 when Ralph S. Solecki, an anthropologist at Columbia University, first saw Shanidar Cave, Kurdish goatherds and their animals were living in small brush huts and corrals on its capacious floor. Undoubtedly their forefathers had lived there in much the same way. Solecki thought that the cave would be a promising place to search for early man.

In three seasons of work, an expedition sent a shaft through 45 feet of floor deposits before bedrock was reached. The thick load of fallen rocks and earth had accumulated over a period of about 100,000 years.

In Layer D, 16 feet below the surface and extending down to bedrock, lay the skeletons of seven Neanderthals, one of them an infant. Several had been killed by rocks falling from the cave ceiling. In many ways these people of Shanidar resembled the early Neanderthal men of Europe— there was the massive jaw, the sloping forehead and the bulging brow ridge. Solecki reported that there were "few suggestions of progress toward the features of *Homo sapiens*," modern man. In one case, the brow ridge was not a continuous one running across the forehead. It had a depression between the eyes, as does the forehead of contemporary man. In other caves in the Middle East were other remains with a curious mixture

Figure 4–11. Neanderthals, a rugged early human breed, roamed Europe and the Middle East about 75,000 years ago. They may have interbred with early Homo sapiens, who was their contemporary, competitor and possibly their exterminator. This family is enjoying a moment of comparative peace and safety in an open space before its cave home at the foot of the Rock of Gibraltar. At that time, much of the earth's water was tied up in icecaps and the lowered oceans left many sea caves high and dry. The Neanderthal father is trying on one deerskin cape while the mother dresses another—and the children quarrel. The dead birds at left are a great auk—an extinct, flightless, penguin-sized diving bird—and an alpine chough, a relative of the crow. (Painted by Maurice Wilson.)

Figure 4–12. Cro-Magnons, the Neanderthals' survivors and successors, hunted in Europe during the late ice ages and immortalized their prowess in many cave paintings. They were big, essentially modern Caucasoids who wore sewn clothes of hide and fur, wielded spear throwers and antler-tipped spears, and differed little from barbarian tribesmen of Julius Caesar's day. On this warm spring day three warriors march into camp, hot from the hunt, half-stripped and laden with game. The veteran at left carries an alpine antelope, or chamois, the younger man at right an arctic fox. Beside the fire, fish are being smoked from a pole. Below at left lies one horn of a saiga antelope. The rocky overhang at the right serves as the hunters' temporary shelter. (Painted by Maurice Wilson.)

of Neanderthal and modern features. Those from Palestine most nearly resembled modern human skeletons.

Had a different, more modern people moved into this eastern area, or was the change a result of the evolution of the people there? As yet there is no answer, but Howell suggests: "It is difficult to escape the conclusion, although there is still inadequate evidence to fix exactly the region and time, that the southwest Asian area, and including Southern Russia, was a primary one in the evolutionary transformation of protosapiens people, who had some general early Neanderthal affinities, into anatomically modern man. . . ." Only in this area, Howell argues, was there an an appropriate base for the rise of the modern men who were soon to take over the earth.

Though the full story of the Neanderthals is hazy and uncertain, even deeper mystery surrounds their disappearance. After thriving some 75,000 to 50,000 years ago, while the northern part of the Eurasian continent still was ice-covered and bleak, the Neanderthals vanished. In the caves of East and West their tools, the bones of the animals they killed and their own sturdy bones come to an abrupt end. At Shanidar Cave the end comes in deposits 29 feet above bedrock. Higher levels contain only the tools and remains of another kind of man. In some caves a "sterile" layer, one without bones or tools and indicating an accumulation during years of severe cold or flooding, comes briefly between.

What brought about the demise, if it was a demise, of the men who had earlier spread with such hardihood over a large section of the earth? Carleton S. Coon, professor of anthropology at the University of Pennsylvania, and himself the excavator of seven of the caves in which early man lived, suggests that wet, bitter cold may largely have wiped out Neanderthal man. Earlier peaks of cold had been dry and easier to survive. There is some belief that Neanderthal lost out to newer, abler hunters in the competition for game, also that he was killed off by newcomers, much as were the aborigines of Australia and the American Indians.

In the caves, the remains of *Homo sapiens* nearly always lie immediately above those of Neanderthal man; there is almost no intermingling. A few small fragments of bone have cast some doubt, however, upon this abrupt succession and suggest that the first of the modern men may have appeared in Europe 250,000 or more years ago, even before the Neanderthals held sway. A nearly complete skull was found at Steinheim, near Stuttgart, in 1933. Two years later some skull fragments were found deep in very old gravels on the banks of the Thames River at Swanscombe, not far from London. The three bones from Swanscombe have a *sapiens* conformation which impresses many authorities. If Swanscombe man lived at the time gravels were laid down by the river, he would be about 250,000 years old; and if he is indeed a modern man, then his kind, *Homo sapiens,* would be far older than the cave deposits have indicated. Other fragments that indicate an earlier appearance for him are an incomplete skullcap and a *sapiens*-appearing piece of frontal bone discovered in the grotto of Fontéchevade, in southwest France. Associated with them are the bones of animals which lived before the last ice age.

But the question is not settled. Until more complete material is available, at least one scientific journal has urged anthropologists to withhold judgment and avoid basing broad and far-reaching theories of evolution on such fragmentary materials as those found at Swanscombe and Fontéchevade. There is no doubt, however, that a modern, European form of *Homo sapiens* and his different way of life had replaced Neanderthal man by 40,000 years ago, and perhaps by 50,000. At this point the record becomes full and easy of interpretation.

The green valleys in the beautiful Dordogne region of France are walled in by gentle limestone cliffs, which at many points are honeycombed with caves, many of them the longtime homes of Neanderthal man. Sometime after he disappeared from these caves, modern *Homo sapiens* moved in. From the cave ledges he could safely survey the whole valley for game or enemies. To this day, people still build their homes against the sheltering cliffs and use the rock face for a rear wall, or the caves themselves for rear rooms.

The men who took over in the Dordogne, and who quickly occupied many other parts of Europe and the Near East, were somewhat taller than those Neanderthals whose stature is known. Their skulls were thinner and higher, their foreheads nearly vertical, their features finer, their mouths retracted, and their posture and carriage as upright as anyone's. They were Cro-Magnon men, named for the cave near the little riverside village of Les Eyzies where their traces were first discovered.

Though their brains were in the same size range as those of Neanderthal men, these new men put their brains to new uses. They made a wide variety of greatly improved tools. They knew how to select a sizable piece of flint, shape it into cylindrical form and with one skillful blow knock off a long flake blade, in effect a long knife with two razor edges coming together in a point. By applying pressure to the cylinder with another piece of stone or bone, they removed additional chips exactly as they wanted. The cylinder was portable and a new tool could be made on the hunt if an old one broke. With this new arsenal—quite advanced over any ever made by Neanderthal man—the largest of animals could be brought down with a new efficiency, and their skins turned into warm clothing. Hun-

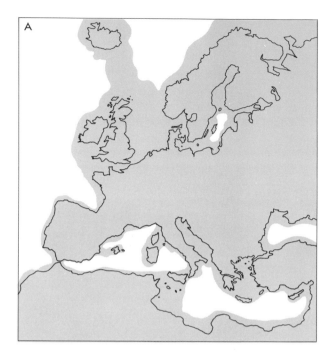

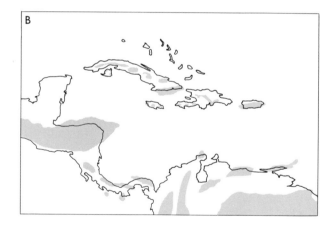

Figure 4–13. Land bridges of the past and the present.

(a) Greater Europe. If the water around Europe were lowered 300 feet, the gray areas on the map would emerge as land, permitting passage of animals. Such a shift in sea level during the ice ages is believed to have linked Britain to Europe, Spain and Italy to Africa.

(b) The Isthmus of Panama. A bridge now exists between North and South America, but for long periods during the past, Central America was nothing but a series of islands (gray areas). The present bridge has broken and reappeared often during the past million years.

220

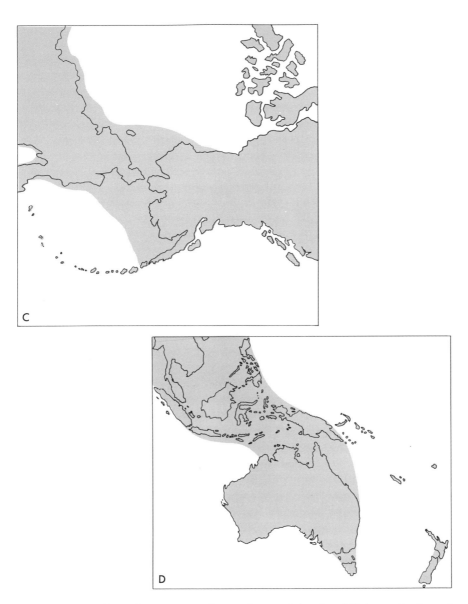

(c) *Asia and North America. The similarity of fossil and living animals and plants in Asia and the Americas suggests the presence of an ancient bridge across the Bering Strait. It is believed that the first men to inhabit the North American continent came over it from Siberia.*

(d) *Southeast Asia and Australia. The oldest bridge of all, and probably the first to disappear, was one that connected Australia with Southeast Asia. It vanished even before the advent of placental mammals, leaving Australia with more primitive types which still persist today. (Reproduced by permission of Time-Life Books from* Evolution. *Copyright © 1964 by Time Incorporated.)*

dreds of awls and needles used in the making of clothing were found in
the caves.

In some of the cave passages running back into the cliffs, these mighty
hunters began to scratch the outlines of the animals they pursued: mammoths, horses, bison, wild oxen and rhinoceroses as well as their favorite,
reindeer. Often they selected a rounded place in the rock to give a sculp-
tured effect to their art, for it was art. Soon they were outlining the animals in black on the stone walls of the caves, and in time they began to
paint in earth colors. At Lascaux in the main cave and several adjoining
galleries, nearly 500 bulls, horses, antelopes and even a herd of reindeer
swimming a river are painted with an artistry, power and accuracy of
movement that have not been surpassed. The paintings there must be
considered one of the world's great assemblages of art. By their art these
men of the caves, the Cro-Magnons, proved their possession of the highest
capacities yet achieved in the long course of evolution. Their cultural objects proved their status without question and abundantly. And as reflective beings they had come a long, long way: many of their caves were
used only as places of ritual worship, and they supported some of their
fellows as full-time artists. [See Figure 4–14.]

These new men—*Homo sapiens* like ourselves—may have arisen in one
locality and spread over the globe. Or, by another theory, several subspecies in different parts of the world may independently have progressed to
Homo sapiens status, and gone on to develop as the present races of man.
Whatever their birthplace or places, the new men absorbed or extinguished all others who had come before them. Modern men of the Mongoloid race, for example, moved along the eastern edge of Asia and up to
the Bering Strait, where they crossed the land bridge into Alaska. In the
New World they worked their way down the west coast and to the very
tip of the southern continent. In Europe as the ice began its latest retreat,
some 12,000 years ago, men pushed on into the thawing north. The only
lands these new men failed to occupy were the bleak tundra of Canada,
the forbidding shores of Greenland, the oceanic islands and Antarctica.

Even at the time of their earliest known appearance, these men who
were inheriting the earth were biologically as advanced as any men who
have been born since. In brain size, in posture and in physical organization, the three billion human beings who are their descendants of today
have not basically changed the patterns that evolution had already built
into their bodies.

Today as in the past all are members of one species. For many years
scientists debated the point. As men had spread around the world and
had been isolated by oceans and other barriers, they had changed in ap-

Figure 4–14. Lascaux Cave paintings. (Photographs courtesy of French Embassy Press & Information Division.)

223

pearance. Skins and hair, particularly, had taken on different colors in different areas. Some scientists insisted that men who varied so obviously should be classified as subspecies, if not as separate species. Darwin largely ended the argument with the unequivocal proof that all could, and often did interbreed—the crucial test of a species. Darwin also showed that all, even the barbaric Fuegians living at the stormy tip of South America, quickly could learn the languages and manners of civilized man. There was no intellectual difference, except in degree.

Modern genetics, with its discovery of the hereditary basis of difference, soon proved, in turn, that a very few genes or sections of the basic material of heredity, DNA, could account for all the differences between races.

The racial differences were obvious, but minor in the total of man.

Australoid

Negroid

Capoid

Caucasoid

Mongoloid

Figure 4–15. The many faces of man. Man is easily the most various of the animals. Though each human holds most of his tens of thousands of genes in common with all other men, about 2,000 genes account for the wide spectrum of physical differences between the races. Above, the five basic types.

Two Views of Man

Editor's Note. The emotional response to the discovery of mankind's ancestors produced a controversy which still rages among biologists and humanists today. The controversy involves different interpretations of the significance of this scientific fact. One view emphasizes man's unity with the Primate order from which he sprang, and seeks in his biological and instinctual inheritance the explanation of man's nature. The other view emphasizes the ways in which man is unique, the ways in which he has transcended his animal origins.

An argument which has been widely used in attempting to reconcile the general public to the fact of man's origin has been to say that men were not descended from monkeys but that they both were descended from a common ancestor. Although this statement is true in one sense, it offers very shallow comfort as George Gaylord Simpson points out in the following selection.

The World into Which Darwin Led Us [*]
by George Gaylord Simpson (1960)

No one doubts that man is a member of the order Primates along with the lemurs, tarsiers, monkeys, and apes. Few doubt that his closest living relatives are the apes. On this subject, by the way, there has been too much pussyfooting. Apologists emphasize that man cannot be a descendant of any living ape—a statement that is obvious to the verge of imbecility— and go on to state or imply that man is not really descended from an ape or monkey at all, but from an earlier common ancestor. In fact, that common ancestor would certainly be called an ape or monkey in popular speech by anyone who saw it. Since the terms *ape* and *monkey* are defined by popular usage, man's ancestors *were* apes or monkeys (or successively both). It is pusillanimous if not dishonest for an informed investigator to say otherwise.

---

[*] From George Gaylord Simpson, "The World into Which Darwin Led Us," *This View of Life*, Harcourt, Brace & World, Inc. Copyright © 1960 by George Gaylord Simpson. Reprinted by permission. Originally published in *Science*, 131, 1960.

Man's Place in Nature *
by Thomas H. Huxley (1863)

We are indeed told by those who assume authority in these matters, . . . that the belief in the unity of origin of man and brutes involves the brutalization and degradation of the former. But is this really so? Could not a sensible child confute, by obvious arguments, the shallow rhetoricians who would force this conclusion upon us? Is it, indeed, true, that the Poet, or the Philosopher, or the Artist whose genius is the glory of his age, is degraded from his high estate by the undoubted historical probability, not to say certainty, that he is the direct descendant of some naked and bestial savage, whose intelligence was just sufficient to make him a little more cunning than the Fox, and by so much more dangerous than the Tiger? Or is he bound to howl and grovel on all fours because of the wholly unquestionable fact, that he was once an egg, which no ordinary power of discrimination could distinguish from that of a Dog? Or is the philanthropist or the saint to give up his endeavours to lead a noble life, because the simplest study of man's nature reveals, at its foundations, all the selfish passions and fierce appetites of the merest quadruped? Is mother-love vile because a hen shows it, or fidelity base because dogs possess it?

The common sense of the mass of mankind will answer these questions without a moment's hesitation. Healthy humanity, finding itself hard pressed to escape from real sin and degradation, will leave the brooding over speculative pollution to the cynics and the righteous "overmuch" who, disagreeing in everything else, unite in blind insensibility to the nobleness of the visible world, and in inability to appreciate the grandeur of the place Man occupies therein.

Nay more, thoughtful men, once escaped from the blinding influences of traditional prejudice, will find in the lowly stock whence man has sprung, the best evidence of the splendour of his capacities; and will discern in his long progress through the Past, a reasonable ground of faith in his attainment of a nobler Future.

* From Thomas H. Huxley, *Evidence as to Man's Place in Nature*, London, 1863.

The Descent of Man: Conclusion *

by Charles Darwin (1886)

The main conclusion arrived at in this work, namely, that man is descended from some lowly organized form, will, I regret to think, be highly distasteful to many. But there can hardly be a doubt that we are descended from barbarians. The astonishment which I felt on first seeing a party of Fuegians on a wild and broken shore will never be forgotten by me, for the reflection at once rushed into my mind—such were our ancestors. These men were absolutely naked and bedaubed with paint, their long hair was tangled, their mouths frothed with excitement, and their expression was wild, startled, and distrustful. They possessed hardly any arts, and like wild animals lived on what they could catch; they had no government, and were merciless to every one not of their own small tribe. He who has seen a savage in his native land will not feel much shame, if forced to acknowledge that the blood of some more humble creature flows in his veins. For my own part I would as soon be descended from that heroic little monkey, who braved his dreaded enemy in order to save the life of his keeper, or from that old baboon, who, descending from the mountains, carried away in triumph his young comrade from a crowd of astonished dogs—as from a savage who delights to torture his enemies, offers up bloody sacrifices, practices infanticide without remorse, treats his wives like slaves, knows no decency, and is haunted by the grossest superstitions.

Man may be excused for feeling some pride at having risen, though not through his own exertions, to the very summit of the organic scale; and the fact of his having thus risen, instead of having been aboriginally placed there, may give him hope for a still higher destiny in the distant future. But we are not here concerned with hopes or fears, only with the truth as far as our reason permits us to discover it; and I have given the evidence to the best of my ability. We must, however, acknowledge, as it seems to me, that man with all his noble qualities, with sympathy which feels for the most debased, with benevolence which extends not only to other men but to the humblest living creature, with his god-like intellect which has penetrated into the movements and constitution of the solar

* From Charles Darwin, *The Principal Works of Charles Darwin.* John B. Alden, Publisher, New York, 1886.

system—with all these exalted powers—Man still bears in his bodily frame the indelible stamp of his lowly origin.

The Naked Ape *
by Desmond Morris (1968)

There are 193 living species of monkeys and apes. One hundred and ninety-two of them are covered with hair. The exception is a naked ape self-named Homo sapiens. This unusual and highly successful species spends a great deal of time examining his higher motives and an equal amount of time studiously ignoring his fundamental ones. He is proud that he has the biggest brain of all the primates but attempts to conceal the fact that he also has the biggest penis, preferring to accord this honor falsely to the mighty gorilla. He is an intensely vocal, acutely exploratory, overcrowded ape, and it is high time we examined his basic behavior.

I am a zoologist and the naked ape is an animal. He is therefore fair game for my pen, and I refuse to avoid him any longer simply because some of his behavior patterns are rather complex and impressive. My excuse is that in becoming so erudite Homo sapiens has remained a naked ape nevertheless; in acquiring lofty new motives, he has lost none of the earthy old ones. This is frequently a cause of some embarrassment to him, but his old impulses have been with him for millions of years, and his new ones only a few thousand at the most—and there is no hope of quickly shrugging off the accumulated genetic legacy of the whole evolutionary past. He would be a far less worried and more fulfilled animal if only he would face up to this fact. . . . We naked apes are, despite all our great technological advances, still very much a simple biological phenomenon. Despite our grandiose ideas and our lofty self-conceits, we are still humble animals, subject to all the basic laws of animal behavior. We tend to suffer from a strange complacency that there is something special about us, that we can never collapse as a dominant species, that we are somehow above biological control. But we are not. Many exciting species have become extinct in the past, and we are no exception. Sooner or later we shall go, making way for something else. If it is to be later rather than sooner, then we must take a long, hard look at ourselves as biological specimens and gain some understanding of our limitations. . . .

Optimism is expressed by some who feel that since we have evolved a high level of intelligence and a strong inventive urge we shall be able to twist any situation to our advantage; that we are so flexible that we can remold our way of life to fit any of the new demands made by our rapidly rising species status . . . that our intelligence can dominate all our basic biological urges.

I submit that this is rubbish. Our raw animal nature will never permit it. Of course we are flexible. Of course we are behavioral opportunists. But there are severe limits to the form our opportunism can take. By stressing our biological features here, I have tried to show the nature of these restrictions. By recognizing them clearly and submitting to them, we shall stand a much better chance of survival. This does not imply a naive "return to nature." It simply means that we should tailor our intelligent opportunist advances to our basic behavioral requirements. We must somehow improve in quality rather than in sheer quantity. If we do this, we can continue to progress technologically in a dramatic and exciting way without denying our evolutionary inheritance. If we do not, then our suppressed biological urges will build up and up until the dam bursts and the whole of our elaborate existence is swept away in the flood.

Cain's Children *

by Robert Ardrey (1961)

When Raymond Dart in 1953 prepared his paper, *The Predatory Transition from Ape to Man,* he stated the thesis that man's animal ancestry was carnivorous and predatory. The hunting life in a creature ill-armed by nature made necessary the use of weapons. The use of weapons and the predatory way perfected the specialized human anatomy, and demanded complex nervous co-ordination never before experienced in the animal world. And so, as a final answer to evolutionary necessities of unprecedented complexity, came the big brain and man. The human being in the most fundamental aspects of his soul and body is nature's last if temporary word on the subject of the armed predator. And human history must be read in these terms.

So profoundly did Dart violate our fundamental human assumptions that the most eminent anthropologist in Africa could not obtain publication of his paper by any reputable scientific journal. It was published by a

journal of small prestige and less circulation, and remains today all but
unread. . . . Science and all branches of modern thought must at last
reckon with Dart's terrifying thesis.

Weapons preceded man. Whether man is in fact a biological invention
evolved to suit the purposes of the weapon must be a matter of future
debate. . . . Many factors contributed to the Pleistocene's supreme evolutionary invention. And certainly our animal inheritance cannot be
summed up in terms as simple as those expressed in the *Predatory Transition*. Other forces of enormous power, all similarly derived from the animal world, play their instinctual roles in the drama of human conduct.
We have investigated a few of them: the drive to acquire private property; social groupings based on the defence of a territory held in common;
the commandment to gain and hold individual dominance within such a
society; the contest between males for superior territory or superior
status; sexual choice exercised by the female in terms of the male's acquisition of property or status; the hostility of territorial neighbours whether
individual or group; and the dual code of behaviour, prevailing in the
members of a group, demanding amity for the social partner and enmity
for individuals outside the territorial bond. All these are human instincts
derived from ancient animal patterns. But to them must now be added
those particular attributes of the hominid line: the way of the predator,
and the dependence upon weapons.

Man is unique, says modern thought; and all babies are born innocent.
Science has done its mighty best to uphold the tenets of the romantic fallacy. If anthropology now finds itself pressed to the brink of failure, then
the scientist cannot be blamed. He has given his all. With a thousand
thumbs and consummate heroism he has blocked a thousand leaks in the
scientific dike. No better example of the genius which he has brought to
this desperate chore can be found than in his euphemistic use of the word
"tool." I have suggested that we glance through the illustrations in the
British Museum's edition of Kenneth Oakley's authoritative *Man the Tool-
Maker*. With few exceptions, they are pictures of weapons. . . .

The time has long passed when modern thought can afford the luxury
of heart-warming but obsolete assumptions. Those ancient fish dragged
up from some Madagascar deep may go their appointed, watery rounds
holding firmly to the assumption that Cretaceous times have not passed.
Ponderous tortoises may waddle about the margins of the Galápagos Islands, musing on past, distant glories and convinced that the reptile still
reigns. The Loch Ness monster, for all we know, may be fathering baby
Loch Ness monsters beneath some ledge in his Scottish lake, happily dedicated to the proposition that for Loch Ness monsters day is about to
dawn. But for a sapient being, the mightiest predator that the world has

ever known, to attempt to resolve his difficulties through an assumption that his species is both innocent and unique, is to court a fate more severe. Evolution may abandon the experiment with the enlarged brain, serve him with the extinction that he so grossly deserves, and turn natural selection to the more hopeful merits of the ancient citizens of the Madagascar deeps.

We are Cain's children. The union of the enlarging brain and the carnivorous way produced man as a genetic possibility. The tightly packed weapons of the predator form the highest, final, and most immediate foundation on which we stand. How deep does it extend? A few million, five million, ten million years? We do not know. But it is the material of our immediate foundation as it is the basic material of our city. And we have so far been unable to build without it.

Man is a predator whose natural instinct is to kill with a weapon. The sudden addition of the enlarged brain to the equipment of an armed already-successful predatory animal created not only the human being but also the human predicament.

The Significance of Tool-Making *

by L. S. B. Leakey (1961)

The stone-age culture which is found in association with *Zinjanthropus* in deposits of the Lower Pleistocene Age was first discovered at Olduvai Gorge (at the very same site which has now yielded the fossil *Zinjanthropus* material) and it was found as long ago as 1931. It was named the "Oldowan culture" and it has long been recognized as being the oldest well-authenticated stone-age culture so far known in the world. This fact, of itself, is important to our theme, for it suggests that the African continent was the scene of the most significant step that was ever made in the whole animal kingdom—the step of beginning to "make tools to a set and regular pattern." It was this step which lifted "near-man" from the purely animal level to that of human status.

Let us consider for a moment why this step was of such great significance.

Near-man, as an advanced higher primate, seems to have been mainly a

* From L. S. B. Leakey, *The Progress and Evolution of Man in Africa*, Oxford University Press, London. Copyright © 1961 by Oxford University Press. Reprinted by permission.

herbivorous mammal, feeding upon nuts, berries, edible roots, leaves, and shoots. Like chimpanzees, baboons and some of the monkeys, near-man was probably not averse to an occasional carnivorous addition to his diet on those rare occasions when he could catch young animals and birds, or when he could succeed in scavenging some part of a small animal that had been killed and partly eaten by the true carnivores. But without cutting instruments near-man was quite unable to avail himself of an immense potential addition to his source of food, namely, the meat of the larger animals whose skins were too tough to be penetrated or torn by near-man's somewhat blunt teeth and very inadequate fingernails.

As soon as near-man began to make tools—chopping and cutting tools of stone—he opened up for himself an immense new source of food. The legs of the larger herbivores, such as a pig or an antelope which had been killed and partly eaten by a lion or a sabre-tooth tiger, could now be detached and carried home; for the Oldowan chopper was fully adequate for cutting through thick skin as well as the tendons which hold the limbs to a carcass. Moreover, we may, I think, assume that the smaller antelope could, almost certainly, be stalked and killed with the bare hands, either by a single skilled individual, or else by a "hunting pack." The skin of such an animal could now be flayed with an Oldowan chopper and the flesh cut into joints. These were feats which could not be achieved by the tool-less near-man, although even they probably hunted and ate many smaller creatures such as rats and mice, as well as beasts with soft skins that could be torn apart without actual cutting instruments.

This major step of "making tools to a set and regular pattern" not only increased man's food potential as we have seen but, if I interpret correctly the available evidence, it also had another very much more important effect upon humanity.

Science has every reason to believe that genetic mutation provides the heritable variation essential to evolutionary change. Moreover, the rate of evolution under natural conditions is seldom, if ever, regular, and it is usually relatively slow. Often it takes place by fits and starts, so that we can recognize what seem to us to be "explosive phases" as well as "quiescent ones" for any given group of organisms.

Under purely natural conditions, even at those times when the rate of mutation appears to be at its highest during a so-called explosive phase, the speed at which mammalian evolution actually takes place is controlled and considerably slowed down by the fact that, usually, there is less likelihood that a mutant will survive until capable of reproduction than that a more normal non-mutant member of the community will do so. It is only those mutations representing an improvement in the genotype in which they happen to appear that are of evolutionary importance, and then only if they live long enough to be able to reproduce themselves, and then with the aid of natural selection.

This position seems, however, to be in large part reversed when natural conditions are replaced by those which we call domestication, and there is abundant evidence that the mere fact of domesticating various wild animals has caused them to accelerate their rate of evolution. This seems to be because domestication, of itself, results in the setting up of conditions under which those mutants that do appear from time to time have a much higher chance of surviving to an age when they can propagate themselves than they ever had before. Dogs and cattle provide excellent examples of this. Later, of course, the effect of domestication upon the speed of evolution was still further enhanced by deliberate selective breeding of most domestic animals.

All too often we overlook the fact that once near-man became man, through that initial step of beginning to make tools he thereby not only provided himself with a much wider field of food supply but he did in fact create for himself the earliest known conditions of domestication. Man had indeed domesticated himself and, by so doing, I suggest that immediately he greatly increased the speed of his own evolution.

I believe, therefore, that when near-man became a tool-making man, he not only took the most significant step in all the animal kingdom but he also started for himself a new phase of evolution in which the results of the natural mechanism of change were greatly accelerated.

How Man Became Natural *

by Loren Eiseley (1960)

The nature of the original animal-man is still a matter of some debate. The likelihood of man's origin on some idyllic Bornean island has been abandoned long since. However man managed his transformation, it was achieved amidst the great mammals of the Old World land mass, and most probably in Africa. Here the discoveries of the last few decades have revealed small-bodied, upright-walking man-apes whose brains appear little, if at all, larger than those of the existing great apes. Whether they could speak is not known, but it now appears that they were capable of at least some crude tool-making capacities.

As might be expected, these bipedal apes are quite different from our arboreal cousins, the great apes of today. Their teeth are heavy-molared, but they lack completely the huge canine teeth with which the nineteenth century, using gorilla models, endowed our ancestors. In spite of a powerful jaw musculature the creatures are light-bodied and apparently pygmoid, compared with modern man.

In spite, also, of exaggerated guesses when the first massive jaws were recovered, the several forms of these creatures in no case indicate man's descent from a giant primate—quite the reverse, in fact. Instead, we seem to be confronted with a short-faced, big-jawed ape of quite moderate body dimensions and a brain qualitatively, if not quantitatively, superior to that of any existing anthropoid.

It is unlikely that all of the several species known became men. Some, indeed, are late enough in time that primitive men were probably already in existence. Surveyed as a whole, however, they suggest an early half-human or near-human level, revealing clearly that the creature who became man was a ground-dwelling, well-adapted ape long before his skull and brain underwent their final transformation.

Interestingly, the hesitation exhibited by Darwin [1] over the psychical nature of the human forerunner continues into these remote time depths of the lower ice age. Raymond Dart, one of the pioneer discoverers of these ape-men, regards them as successful carnivores and killers of big game. He sees them as brutal and aggressive primates, capable of killing their own offspring, and carrying in their genetic structure much of the sadism and cruelty still manifested by modern man. They are club-swingers par excellence, already ably balanced on their two feet, and the terror of everything around them.

One cannot help but feel, however, that Dart tends to minimize the social nature which even early man must have had to survive and care for young who needed ever more time for adaptation to group life. He paints a picture too starkly overshadowed by struggle to be quite believable. It would seem that into his paleontological studies has crept a touch of disillusionment and distaste which has been projected backward upon that wild era in which the human predicament began. This judgment is, of course, a subjective one. It reveals that the hesitations of Darwin's day, in spite of increased knowledge, follow man back into the past. Perhaps it is a matter of temperament on the part of the observer. Perhaps man has always been both saint and sinner—even in his raw beginnings.

---

1. "We do not know," Darwin said, "whether man is descended from some small species, like the chimpanzee, or from one as powerful as the gorilla. Therefore we cannot say whether man has become larger and stronger or smaller and weaker than his ancestors." Loren Eiseley, *The Firmament of Time*, p. 102. Ed.

As we press farther back in time, however, back until the long, desolate years of the fourth ice age lie somewhere far in the remote future, we come upon something almost unbelievable. We come upon man, near-man, "the bridge to man," estimated as over seven hundred thousand years removed from us, in the Olduvai Gorge in Tanganyika. He is a step beyond Dart's man-apes, according to Dr. L. S. B. Leakey, his discoverer. Like them, he is small, huge-jawed, with a sagittal crest like that of a gorilla, but a true shaper of tools. Unlike Dart, Leakey describes his specimen as a semivegetarian eater of nuts, small rodents and lizards. Perhaps massive jaws and molar teeth, just as in the case of the living gorilla, were developed for other food than flesh.

The thing which appears the strangest of any news to come down from that far epoch, however, is the report from London that the youthful ape-man's body had apparently been carefully protected from scavenging hyenas until rising lake waters buried him among his tools.

Three years ago, in a symposium on the one hundredth anniversary of the discovery of Neanderthal man, I made these remarks: "When we consider this creature of 'brute benightedness' and 'gorilloid ferocity,' as most of those who peered into that dark skull vault chose to interpret what they saw there, let us remember what was finally revealed at the little French cave near La Chapelle-aux-Saints in 1908. Here, across millennia of time, we can observe a very moving spectacle. For these men whose brains were locked in a skull foreshadowing the ape, these men whom scientists had contended to possess no thoughts beyond those of the brute, had laid down their dead in grief.

"Massive flint-hardened hands had shaped a sepulcher and placed flat stones to guard the dead man's head. A haunch of meat had been left to aid the dead man's journey. Worked flints, a little treasure of the human dawn, had been poured lovingly into the grave. And down the untold centuries the message had come without words: 'We too were human, we too suffered, we too believed that the grave is not the end. We too, whose faces affright you now, knew human agony and human love.'

"It is important to consider," I said then, "that across fifty thousand years nothing has changed or altered in that act. It is the human gesture by which we know a man, though he looks out upon us under a brow reminiscent of the ape."

If the London story is correct, an aspect of that act has now been made distant from us by almost a million years. The creature who made it could only be identified by specialists as human. He is far more distant from Neanderthal man than the latter is from us.

Man, bone by bone, flint by flint, has been traced backward into the night of time more successfully than even Darwin dreamed. He has been traced to a creature with an almost gorilloid head on the light, fast body of a still completely upright, plains-dwelling creature. In the end he partakes both of Darwinian toughness, resilience, and something else, a humanity—if this story is true—that runs well nigh as deep as time itself.

Man has, in scientific terms, become natural, but the nature of his "naturalness" escapes him. Perhaps his human freedom has left him the difficult choice of determining what it is in his nature to be. Perhaps the two sides of the dark question Darwin speculated upon were only an evolutionary version of man's ancient warfare with himself—a drama as great in its hidden fashion as the story of the Garden and the Fall.

The Paragon of Animals [*]
by Edmund W. Sinnott (1957)

For generations . . . man has thought of himself as that for which the long laborious ages have prepared; a being set apart and significant in the universe; lifted by his reason and his spirit above all created things; a little lower than the angels, to be sure, but alight with a touch of the divine flame and clothed with glory and honor. Created from the dust though he might be, he still believed himself a child of God and precious in the divine scheme of things.

This exalted opinion of his position has not been simply an expression of his vanity and egotism. It was vitally necessary. From it he drew courage and comfort in times of tribulation. Without this conviction of his worth and this belief in his relationship to the Divine, it is doubtful whether man would have been able to survive the ordeals of his eventful history. A sense of his own significance has constantly reassured him.

The crisis of our day comes from the fact that this traditional appraisal of man's nature is now gravely challenged, and from several directions. Whether in the light of modern knowledge he can maintain this high estimate of himself or must give it up for something very different, something far less exalted and godlike, is the deepest question that he has to face. Here the great traditions of the past come most sharply into conflict with man's new scientific insights about himself and the universe. How this conflict will be decided and what man's true nature will finally prove to be are questions that shake the world today. It is not the change in our ideas about God that is so ominous, important as this is, but the change in what we think about ourselves, for this will finally be reflected in our philosophy and our religion.

Is man simply the paragon of animals or is he something more? Everywhere he is trying, as never in the past, to find what manner of creature he really is. In this search all other questions have a part. Here finally converge the disciplines of science. Here come to focus the problems of philosophy. Here is an essential element in every religion. Until man comes to know *himself,* all other knowledge that he gains is incomplete.

The Uniqueness of Man *

by Julian Huxley (1944)

Man's opinion of his own position in relation to the rest of the animals has swung pendulum-wise between too great or too little a conceit of himself, fixing now too large a gap between himself and the animals, now too small. The gap, of course, can be diminished or increased at either the animal or the human end. One can, like Descartes, make animals too mechanical, or, like most unsophisticated people, humanize them too much. Or one can work at the human end of the gap, and then either dehumanize one's own kind into an animal species like any other, or superhumanize it into beings a little lower than the angels.

Primitive and savage man, the world over, not only accepts his obvious kinship with animals but also projects into them many of his own attributes. So far as we can judge, he has very little pride in his own humanity. With the advent of settled civilization, economic stratification, and the development of an elaborate religion as the ideological mortar of a now class-ridden society, the pendulum began slowly to swing into the other direction. Animal divinities and various physiological functions such as fertility gradually lost their sacred importance. Gods became anthropomorphic and human psychological qualities pre-eminent. Man saw himself as a being set apart, with the rest of the animal kingdom created to serve his needs and pleasure, with no share in salvation, no position in eternity. In Western civilization this swing of the pendulum reached its

* From Julian Huxley, *Man in the Modern World,* Harper & Row Publishers, Inc., New York. Copyright 1939 by Julian S. Huxley. Reprinted by permission.

limit in developed Christian theology and in the philosophy of Descartes:
both alike inserted a qualitative and unbridgeable barrier between all
men and any animals.

With Darwin, the reverse swing was started. Man was once again regarded as an animal, but now in the light of science rather than of unsophisticated sensibility. At the outset, the consequences of the changed
outlook were not fully explored. The unconscious prejudices and attitudes
of an earlier age survived, disguising many of the moral and philosophical
implications of the new outlook. But gradually the pendulum reached the
furthest point of its swing. What seemed the logical consequences of the
Darwinian postulates were faced: man is an animal like any other; accordingly, his views as to the special meaning of human life and human
ideals need merit no more consideration in the light of eternity (or of evolution) than those of a bacillus or a tapeworm. Survival is the only criterion of evolutionary success: therefore, all existing organisms are of equal
value. The idea of progress is a mere anthropomorphism. Man happens to
be the dominant type at the moment, but he might be replaced by the ant
or the rat. And so on.

The gap between man and animal was here reduced not by exaggerat-
ing the human qualities of animals, but by minimizing the human qualities of men. Of late years, however, a new tendency has become apparent.
It may be that this is due mainly to the mere increase of knowledge and
the extension of scientific analysis. It may be that it has been determined
by social and psychological causes. Disillusionment with *laisser-faire* in
the human economic sphere may well have spread to the planetary system of *laisser-faire* that we call natural selection. With the crash of old
religious, ethical, and political systems, man's desperate need for some
scheme of values and ideals may have prompted a more critical re-
examination of his biological position. Whether this be so is a point that I
must leave to the social historians. The fact remains that the pendulum is
again on the swing, the man-animal gap again broadening. After Darwin,
man could no longer avoid considering himself as an animal; but he is beginning to see himself as a very peculiar and in many ways a unique animal. The analysis of man's biological uniqueness is as yet incomplete. This
essay is an attempt to review its present position.

The first and most obviously unique characteristic of man is his capacity for conceptual thought . . .

This basic human property has had many consequences. The most important was the development of a cumulative tradition. . . . The existence of a cumulative tradition has as its chief consequence—or if you
prefer, its chief objective manifestation—the progressive improvement of

human tools and machinery. Many animals employ tools; but they are always crude tools employed in a crude way. Elaborate tools and skilled
technique can develop only with the aid of speech and tradition.

In the perspective of evolution, tradition and tools are the characters
which have given man his dominant position among organisms. This biological dominance is, at present, another of man's unique properties. In
each geological epoch of which we have knowledge, there have been
types which must be styled biologically dominant: they multiply, they ex-
tinguish or reduce competing types, they extend their range, they radiate
into new modes of life. Usually at any one time there is one such type—
the placental mammals, for instance, in the Cenozoic Epoch; but sometimes there is more than one. The Mesozoic is usually called the Age of
Reptiles, but in reality the reptiles were then competing for dominance
with the insects: in earlier periods we should be hard put to it to decide
whether trilobites, nautiloids, or early fish were *the* dominant type. Today, however, there is general agreement that man is the sole type merit-
ing the title. Since the early Pleistocene, widespread extinction has diminished the previously dominant group of placental mammals, and man has
not merely multiplied, but has evolved, extended his range, and increased
the variety of his modes of life.

Biology thus reinstates man in a position analogous to that conferred on
him as Lord of Creation by theology. There are, however, differences, and
differences of some importance for our general outlook. In the biological
view, the other animals have not been created to serve man's needs, but
man has evolved in such a way that he has been able to eliminate some
competing types, to enslave others by domestication, and to modify physical and biological conditions over the larger part of the earth's land area.
The theological view was not true in detail or in many of its implications;
but it had a solid biological basis.

Speech, tradition, and tools have led to many other unique properties
of man. These are, for the most part, obvious and well known, and I propose to leave them aside until I have dealt with some less familiar human
characteristics.[1] For the human species, considered as a species, is unique
in certain purely biological attributes; and these have not received the attention they deserve, either from the zoological or the sociological standpoint.

In the first place, man is by far the most variable wild species known.
Domesticated species like dogs, horse, or fowl may rival or exceed him in
this particular, but their variability has obvious reasons, and is irrelevant
to our inquiry.

In correlation with his wide variability, man has a far wider range than

1. These properties will be discussed in Parts 5 and 6. Ed.

any other animal species, with the possible exception of some of his parasites. Man is also unique as a dominant type. All other dominant types have evolved into many hundreds or thousands of separate species, grouped in numerous genera, families, and larger classificatory groups. The human type has maintained its dominance without splitting: man's variety has been achieved within the limits of a single species.

Finally, man is unique among higher animals in the method of his evolution. Whereas, in general, animal evolution is divergent, human evolution is reticulate. By this is meant that in animals, evolution occurs by the isolation of groups which then become progressively more different in their genetic characteristics, so that the course of evolution can be represented as a divergent radiation of separate lines, some of which become extinct, others continue unbranched, and still others divergently branch again. Whereas in man, after incipient divergence, the branches have come together again, and have generated new diversity from their Mendelian recombinations, this process being repeated until the course of human descent is like a network.

All these biological peculiarities are interconnected. They depend on man's migratory propensities, which themselves arise from his fundamental peculiarities, of speech, social life, and relative independence of environment. They depend again on his capacity, when choosing mates, for neglecting large differences of colour and appearance which would almost certainly be more than enough to deter more instinctive and less plastic animals. Thus divergence, though it appears to have gone quite a long way in early human evolution, generating the very distinct white, black, and yellow subspecies and perhaps others, was never permitted to attain its normal culmination. Mutually infertile groups were never produced; man remained a single species. Furthermore, crossing between distinct types, which is a rare and extraordinary phenomenon in other animals, in him became normal and of major importance. According to Mendelian laws, such crosses generate much excess variability by producing new recombinations. Man is thus more variable than other species for two reasons. First, because migration has recaptured for the single interbreeding group divergences of a magnitude that in animals would escape into the isolation of separate species; and secondly, because the resultant crossing has generated recombinations which both quantitatively and qualitatively are on a far bigger scale than is supplied by the internal variability of even the numerically most abundant animal species. . . .

Another biological peculiarity of man is the uniqueness of his evolutionary history. Writers have indulged their speculative fancy by imagining other organisms endowed with speech and conceptual thought— talking rats, rational ants, philosophic dogs, and the like. But closer analysis shows that these fantasies are impossible. A brain capable of conceptual thought could not have been developed elsewhere than in a human body.

The course followed by evolution . . . is like a maze in which almost all turnings are wrong turnings. The goal of the evolutionary maze, however, is not a central chamber, but a road which will lead definitely onwards.

If now we look back upon the past history of life, we shall see that the avenues of progress have been steadily reduced in number, until by the Pleistocene period, or even earlier, only one was left. Let us remember that we can and must judge early progress in the light of its latest steps. The most recent step has been the acquisition of conceptual thought, which has enabled man to dethrone the non-human mammals from their previous position of dominance. It is biologically obvious that conceptual thought could never have arisen save in an animal, so that all plants, both green and otherwise, are at once eliminated. As regards animals, I need not detail all the early steps in their progressive evolution. Since some degree of bulk helps to confer independence of the forces of nature, it is obvious that the combination of many cells to form a large individual was one necessary step, thus eliminating all single-celled forms from such progress. Similarly, progress is barred to specialized animals with no blood-system, like planarian worms; to internal parasites, like tapeworms; to animals with radial symmetry and consequently no head, like echinoderms.

Of the three highest animal groups—the molluscs, the arthropods, and the vertebrates—the molluscs advanced least far. One condition for the later steps in biological progress was land life. The demands made upon the organism by exposure to air and gravity called forth biological mechanisms, such as limbs, sense-organs, protective skin, and sheltered development, which were necessary foundations for later advance. And the molluscs have never been able to produce efficient terrestrial forms: their culmination is in marine types like squid and octopus.

The arthropods, on the other hand, have scored their greatest successes on land, with the spiders and especially the insects. Yet the fossil record reveals a lack of all advance, even in the most successful types such as ants, for a long time back—certainly during the last thirty million years, probably during the whole of the Tertiary epoch. Even during the shorter of these periods, the mammals were still evolving rapidly, and man's rise is contained in a fraction of this time.

What was it that cut the insects off from progress? The answer appears to lie in their breathing mechanism. The land arthropods have adopted the method of air-tubes or tracheae, branching to microscopic size and conveying gases directly to and from the tissues, instead of using the dual mechanism of lungs and bloodstream. The laws of gaseous diffusion are such that respiration by tracheae is extremely efficient for very small animals, but becomes rapidly less efficient with increase of size, until it ceases to be of use at a bulk below that of a house mouse. It is for this reason that no insect has ever become, by vertebrate standards, even moderately large.

It is for the same reason that no insect has ever become even moderately intelligent. The fixed pathways of instinct, however elaborate, require far fewer nerve-cells than the multiple switchboards that underlie intelligence. It appears to be impossible to build a brain mechanism for flexible behaviour with less than a quite large minimum of neurones; and no insect has reached a size to provide this minimum.

Thus only the land vertebrates are left. The reptiles shared biological dominance with the insects in the Mesozoic. But while the insects had reached the end of their blind alley, the reptiles showed themselves capable of further advance. Temperature regulation is a necessary basis for final progress, since without it the rate of bodily function could never be stabilized, and without such stabilization, higher mental processes could never become accurate and dependable.

Two reptilian lines achieved this next step, in the guise of the birds and the mammals. The birds soon, however, came to a dead end, chiefly because their forelimbs were entirely taken up in the specialization for flight. The subhuman mammals made another fundamental advance, in the shape of internal development, permitting the young animal to arrive at a much more advanced stage before it was called upon to face the world. They also (like the birds) developed true family life.

Most mammalian lines, however, cut themselves off from indefinite progress by one-sided evolution, turning their limbs and jaws into specialized and therefore limited instruments. And, for the most part, they relied mainly on the crude sense of smell, which cannot present as differentiated a pattern of detailed knowledge as can sight. Finally, the majority continued to produce their young several at a time, in litters. As J. B. S. Haldane has pointed out, this gives rise to an acute struggle for existence in the prenatal period, a considerable percentage of embryos being aborted or resorbed. Such intra-uterine selection will put a premium upon rapidity of growth and differentiation, since the devil takes the hindmost; and this rapidity of development will tend automatically to be carried on into postnatal growth.

As everyone knows, man is characterized by a rate of development which is abnormally slow as compared with that of any other mammal. The period from birth to the first onset of sexual maturity comprises nearly a quarter of the normal span of his life, instead of an eighth, a tenth or twelfth, as in some other animals. This again is in one sense a unique characteristic of man, although from the evolutionary point of view it represents merely the exaggeration of a tendency which is operative in other Primates. In any case, it is a necessary condition for the evolution and proper utilization of rational thought. If men and women were, like mice, confronted with the problems of adult life and parenthood after a few weeks, or even, like whales, after a couple of years, they could never acquire the skills of body and mind that they now absorb from and contribute to the social heritage of the species.

This slowing (or "foetalization," as Bolk has called it, since it prolongs the foetal characteristics of earlier ancestral forms into postnatal development and even into adult life) has had other important by-products for man. Here I will mention but one—his nakedness. The distribution of hair on man is extremely similar to that on a late foetus of a chimpanzee, and there can be little doubt that it represents an extension of this temporary anthropoid phase into permanence. Hairlessness of body is not a unique biological characteristic of man; but it is unique among terrestrial mammals, save for a few desert creatures, and some others which have compensated for loss of hair by developing a pachydermatous skin. In any case, it has important biological consequences, since it must have encouraged the comparatively defenseless human creatures in their efforts to protect themselves against animal enemies and the elements, and so has been a spur to the improvement of intelligence.

Now, foetalization could never have occurred in a mammal producing many young at a time, since intra-uterine competition would have encouraged the opposing tendency. Thus we may conclude that conceptual thought could develop only in a mammalian stock which normally brings forth but one young at a birth. Such a stock is provided in the Primates—lemurs, monkeys, and apes.

The Primates also have another characteristic which was necessary for the ancestor of a rational animal—they are arboreal. It may seem curious that living in trees is a prerequisite of conceptual thought. But Elliot Smith's analysis has abundantly shown that only in an arboreal mammal could the forelimb become a true hand, and sight become dominant over smell. Hands obtain an elaborate tactile pattern of what they handle, eyes an elaborate visual pattern of what they see. The combination of the two kinds of pattern, with the aid of binocular vision, in the higher centres of the brain allowed the Primate to acquire a wholly new richness of knowledge about objects, a wholly new possibility of manipulating them. Tree

life laid the foundation both for the fuller definition of objects by conceptual thought and for the fuller control of them by tools and machines.

Higher Primates have yet another pre-requisite of human intelligence—
they are all gregarious. Speech, it is obvious, could never have been
evolved in a solitary type. And speech is as much the physical basis of
conceptual thought as is protoplasm the physical basis of life.

For the passage, however, of the critical point between subhuman and
human, between the biological subordination and the biological primacy
of intelligence, between a limited and a potentially unlimited tradition—
for this it was necessary for the arboreal animal to descend to the ground
again. Only in a terrestrial creature could fully erect posture be acquired;
and this was essential for the final conversion of the arms from locomotor
limbs into manipulative hands. Furthermore, just as land life, ages previously, had demanded and developed a greater variety of response than
had been required in the water, so now it did the same in relation to what
had been required in the trees. An arboreal animal could never have
evolved the skill of the hunting savage, nor ever have proceeded to the
domestication of other animals or to agriculture.

We are now in a position to define the uniqueness of human evolution.
The essential character of man as a dominant organism is conceptual
thought. And conceptual thought could have arisen only in a multicellular
animal, an animal with bilateral symmetry, head and blood-system, a
vertebrate as against a mollusc or an arthropod, a land vertebrate among
vertebrates, a mammal among land vertebrates. Finally, it could have
arisen only in a mammalian line which was gregarious, which produced
one young at a birth instead of several, and which had recently become
terrestrial after a long period of arboreal life.

There is only one group of animals which fulfills these conditions—a
terrestrial offshoot of the higher Primates. Thus not merely has conceptual
thought been evolved only in man: it could not have been evolved except
in man. There is but one path of unlimited progress through the evolutionary maze. The course of human evolution is as unique as its result. It
is unique not in the trivial sense of being a different course from that of
any other organism, but in the profounder sense of being the only path
that could have achieved the essential characters of man. Conceptual
thought on this planet is inevitably associated with a particular type of
Primate body and Primate brain.

A further property of man in which he is unique among higher animals
concerns his sexual life. Man is prepared to mate at any time: animals are
not. To start with, most animals have a definite breeding season; only during this period are their reproductive organs fully developed and functional. In addition to this, higher animals have one or more sexual cycles

within their breeding season, and only at one phase of the cycle are they prepared to mate. In general, either a sexual season or a sexual cycle, or both, operates to restrict mating.

In man, however, neither of these factors is at work. There appear to be indications of a breeding season in some primitive people like the Eskimo, but even there they are but relics. Similarly, while there still exist physiological differences in sexual desire at different phases of the female sexual cycle, these are purely quantitative, and may readily be over-ridden by psychological factors. Man, to put it briefly, is continuously sexed: animals are discontinuously sexed. If we try to imagine what a human society would be like in which the sexes were interested in each other only during the summer, as in songbirds, or, as in female dogs, experienced sexual desire only once every few months, or even only once in a lifetime, as in ants, we can realize what this peculiarity has meant. In this, as in his slow growth and prolonged period of dependence, man is not abruptly marked off from all other animals, but represents the culmination of a process that can be clearly traced among other Primates. What the biological meaning of this evolutionary trend may be is difficult to understand. One suggestion is that it may be associated with the rise of mind to dominance. The bodily functions, in lower mammals rigidly determined by physiological mechanisms, come gradually under the more plastic control of the brain. But this, for what it is worth, is a mere speculation. . . .

Still another point in which man is biologically unique is the length and relative importance of his period of what we may call "post-maturity." If we consider the female sex, in which the transition from reproductive maturity to non-reproductive post-maturity is more sharply defined than in the male, we find, in the first place, that in animals a comparatively small percentage of the population survives beyond the period of reproduction; in the second place, that such individuals rarely survive long, and so far as known never for a period equal to or greater than the period during which reproduction was possible; and thirdly, that such individuals are rarely of importance in the life of the species. The same is true of the male sex, provided we do not take the incapacity to produce fertile gametes as the criterion of post-maturity, but rather the appearance of signs of age, such as the beginnings of loss of vigour and weight, decreased sexual activity, or greying hair.

It is true that in some social mammals, notably among ruminants and Primates, an old male or old female is frequently found as leader of the herd. Such cases, however, provide the only examples of the special biological utility of post-mature individuals among animals; they are confined to a very small proportion of the population, and it is uncertain to what extent such individuals are post-mature in the sense we have defined. In any event, it is improbable that the period of post-maturity is anywhere near so long as that of maturity. But in civilized man the average expectation of life now includes over ten post-mature years, and about a sixth of the population enjoys a longer post-maturity than maturity. What is more, in all advanced human societies a large proportion of the leaders of the community are always post-mature. All the members of the British War Cabinet are post-mature.

This is truly a remarkable phenomenon. Through the new social mechanisms made possible by speech and tradition, man has been able to utilize for the benefit of the species a period of life which in almost all other creatures is a mere superfluity. We know that the dominance of the old can be over-emphasized; but it is equally obvious that society cannot do without the post-mature. To act on the slogan "Too old at forty"—or even at forty-five—would be to rob man of one of his unique characteristics, whereby he utilizes tradition to the best advantage. . . .

So far we have been considering the fact of human uniqueness. It remains to consider man's attitude to these unique qualities of his. Professor Everett, of the University of California, in an interesting paper bearing the same title as this essay, but dealing with the topic from the standpoint of the philosopher and the humanist rather than that of the biologist, has stressed man's fear of his own uniqueness. Man has often not been able to tolerate the feeling that he inhabits an alien world, whose laws do not make sense in the light of his intelligence, and in which the writ of his human values does not run. Faced with the prospect of such intellectual and moral loneliness, he has projected personality into the cosmic scheme. Here he has found a will, there a purpose; here a creative intelligence, and there a divine compassion. At one time, he has deified animals, or personified natural forces. At others, he has created a superhuman pantheon, a single tyrannical world ruler, a subtle and satisfying Trinity in Unity. Philosophers have postulated an Absolute of the same nature as mind.

It is only exceptionally that men have dared to uphold their uniqueness and to be proud of their human superiority to the impersonality and irrationality of the rest of the universe. It is time now, in the light of our knowledge, to be brave and face the fact and the consequences of our uniqueness. That is Dr. Everett's view, as it was also that of T. H. Huxley in his famous Romanes lecture. I agree with them; but I would suggest that the antinomy between man and the universe is not quite so sharp as they have made out. Man represents the culmination of that process of organic evolution which has been proceeding on this planet for over a thousand million years. That process, however wasteful and cruel it may be, and into however many blind alleys it may have been diverted, is also in one aspect progressive. Man has now become the sole representative of life in that progressive aspect and its sole trustee for any progress in the future.

Meanwhile it is true that the appearance of the human type of mind, the latest step in evolutionary progress, has introduced both new methods and new standards. By means of his conscious reason and its chief off-spring, science, man has the power of substituting less dilatory, less wasteful, and less cruel methods of effective progressive change than those of natural selection, which alone are available to lower organisms. And by means of his conscious purpose and his set of values, he has the power of substituting new and higher standards for change than those of mere survival and adaptation to immediate circumstances, which alone are inherent in pre-human evolution. To put the matter in another way, progress has hitherto been a rare and fitful by-product of evolution. Man has the possibility of making it the main feature of his own future evolution, and of guiding its course in relation to a deliberate aim.

But he must not be afraid of his uniqueness. There may be other beings in this vast universe endowed with reason, purpose, and aspiration: but we know nothing of them. So far as our knowledge goes, human mind and personality are unique and constitute the highest product yet achieved by the cosmos. Let us not put off our responsibilities onto the shoulders of mythical gods or philosophical absolutes, but shoulder them in the hopefulness of tempered pride. In the perspective of biology, our business in the world is seen to be the imposition of the best and most enduring of our human standards upon ourselves and our planet. The enjoyment of beauty and interest, the achievement of goodness and efficiency, the enhancement of life and its variety—these are the harvest which our human uniqueness should be called upon to yield.

Suggestions for Further Reading

* Barnett, Antony: *The Human Species*, A Pelican Book, Penguin Books Inc., revised edition 1961. An able presentation of our biological knowledge of the nature of man and how biology may be expected to help solve some of man's social problems in the future.

* Bates, Marston: *Man In Nature*, Prentice-Hall Foundations of Modern Biology Series, Prentice-Hall, Inc., 1961. A concise, absorbing account of man as a biological phenomenon. A good introduction for the student and layman alike.

Dobzhansky, Theodosius: *Mankind Evolving: The Evolution of the Human*

* Paperback edition

Species, Yale University Press, 1962. An excellent general survey of the evolution of man.

* Dubos, René: *The Torch of Life, Continuity in Living Experience,* The Credo Series, edited by Ruth Nanda Anshen, Pocket Books, Inc. 1963. A philosophical discussion of man, his place in nature and the evolutionary process.

* ————: *Man Adapting,* A Yale Paperback, Yale University Press, 1965. These Silliman Foundation Lectures discuss the biological nature of man with particular emphasis on the subject of disease and man's relationship to his environment.

Leakey, L. S. B.: *The Progress and Evolution of Man in Africa,* Oxford University Press, 1961. Two lectures delivered by Dr. Leakey in Oxford and Birmingham supporting his thesis that man originated in Africa.

Lorenz, Konrad: *On Aggression,* Harcourt, Brace & World, 1963. A naturalist examines the instinct of aggression in birds, fish and animals, and suggests the relevance of these findings to an understanding of the aggressiveness in human nature.

Wendt, Herbert: *In Search of Adam,* translated by James Cleugh, Houghton Mifflin Company, 1956. An historical account of the development of man's knowledge of his own origins.

5

Did Mind Evolve by Natural Selection?

Darwin believed that mind, like physical evolution, occurred by natural selection. Wallace believed that natural selection alone was not sufficient to account for it.
These readings combine the historical argument with
recent insights into the nature and origin of mind.

Did Mind Evolve by Natural Selection?

Introduction

COULD MAN'S MIND have evolved by natural selection?" Many years ago Alfred Russel Wallace raised this question, answering it himself in the negative. The problem has haunted evolutionists ever since. "When Darwin received a copy of an article that Wallace had written on this subject, he was obviously shaken. It is recorded that he wrote in anguish across the paper, 'No!' and underlined the 'No' three times heavily in a rising fervor of objection." [1]

Since natural selection can operate only on those properties that confer a survival or reproductive advantage, Wallace wondered how selection could have caused the development of mental abilities that have nothing to do with biological fitness—such as the ability to compose great music, design artistic creations, and devise new mathematical systems. Wallace believed that a spiritual force must be at work in developing the human mind to these levels. But Darwin sought for natural explanations—the competition of tribe with tribe, the practical advantages that accrued to those tribes or individuals who were the most innovative.

Although much more is understood today about the mechanism of the human brain and more is known about the rate at which it evolved, there is still no complete agreement among anthropologists about the process that led to its present state of development. Research is proceeding along a number of separate lines, and the facts have not yet coalesced into a coherent theory.

The mind appears to have evolved with remarkable speed, as Wallace postulated. Loren Eiseley calls it an "explosive" development. But he points out that this fact does not necessarily imply that a divinely directed force was at work, as Wallace believed. Among the modern explanations of the speed of mental development, one postulates a change in the growth pattern which appears to be genetically controlled. This change made possible a longer post-natal period of development and training of the human brain. The human baby is born with a brain approximately the same size as the brain of the baby gorilla, but in the first year it trebles in size and continues to grow for more than twenty years. How much of this post-natal growth is genetically controlled, resulting from mutations that occurred way back at the dawn of human history, and how much is induced by cultural conditioning is still an open question. David Krech de-

1. From Loren Eiseley, *The Immense Journey*, Random House, New York, 1957, p. 79.

253

scribes experiments performed recently on rats which indicate that a stimulating and innovative environment produces growth in the cortex of the rat's brain. Quantitative chemical changes also take place. If the inferences drawn from these experiments can be applied to man (and this remains to be proven), it appears that the more we learn, the more we are capable of learning. If this is true, the human infant who is exposed to the cumulative wisdom of mankind may have a significant advantage over other organisms.

Once mankind developed a system of communication and was thereby able to build up a cultural tradition, he initiated a new kind of evolution involving a process in which the experience and knowledge of individuals could be directly passed on to the next generation without involving any genetic transfer. Man's ideas and inventions changed his environment, creating new challenges that were primarily intellectual challenges, and these in turn called forth more ideas and more inventions. This process also was cumulative.

Wallace had seen and understood that man, through his various innovations, was able to solve physical problems that would have taken eons to solve by the slow biological method of evolving new bodily characteristics. For instance, the discovery of fire and the invention of clothing made it unnecessary for men living in cold climates to evolve a heavy natural coat of fur. The art of sewing animal skins is believed to have been discovered about 35,000 years ago, and the first known use of fire is associated with Peking man about 300,000 years ago. These innovations and many similar ones were very important for survival and conferred a definite evolutionary advantage. We may wonder whether Wallace was right in thinking that the brain of primitive man was developed far beyond the needs of its possessor. How highly developed must a brain be to invent an important new way of doing things?

One of the most crucial inventions for the human phase of evolution was the invention of language, both words as symbols for things, and numbers as symbols of quantity. These symbols are tools for thought. The better the tool, the more effectively it can be used. The clumsy Roman numeral system, for instance, does not lend itself to the computations of higher mathematics. The Arabic system had to be invented before the calculus could be discovered. The new tool of language opened up a whole new field for exploitation just as surely as the invention of the first stone arrowhead and spear.

The discovery of tools is not unique to man. There are birds that use tools. (See Figure 1–5.) There are insects and animals that use a primitive sort of language. The difference is more a matter of degree than of kind.

Interesting modern experiments show that primitive organisms like the planarian can be taught a simple task such as following a maze in search of food. These experiments have thrown new light on the nature of the memory process which seems to be chemically and electrically controlled. Surprisingly, it appears that it may be possible to transfer by injection a chemical code for knowledge acquired by one organism into another organism that was ignorant of this knowledge. Greater understanding of the memory process may eventually show why man is blessed with such a prodigious memory compared with other organisms. This trait has made human education and training possible. Here again the difference is in degree rather than in kind. Is the difference based on a genetic shift that could have occurred by mutation long ago and thus have given man the advantage which started him on his own evolutionary path?

The ability for abstract thought, the realization that a symbol can stand for a thing or an idea—these abilities do seem to be uniquely human. "Admittedly," says Teilhard de Chardin, "the animal knows. *But it cannot know that it knows.*" [2] This distinction separates us by a chasm or threshold which Teilhard calls the threshold of reflection. Reflection occurs, he says, when a consciousness turns in upon itself. Once more we are faced with a fundamental question: What is consciousness?

Several of the authors in these readings maintain that consciousness may be simply an electrical phenomenon, "a function of a specific voltage in a given nucleus of the brain," [3] and that it could in fact be duplicated by mechanical devices. Others, scientists as well as philosophers, question whether physics and chemistry in their present form can account for consciousness or free will.

Are consciousness and self-awareness attributes only of the human species? If so, have they appeared suddenly at a particular moment in man's development? We have seen that evolution progresses in the direction of increasing awareness. Could consciousness be a certain stage of awareness? In this case, we might project consciousness way back to the most primitive forms of living matter, back even (some philosophers believe) to the smallest units of inorganic matter. On the other hand, it has been shown that evolution can produce real novelties, such as the ability to reproduce, such as life itself. As Julian Huxley said, the emergence of new properties does not require a break in the continuity of substance, but rather a change of organization resulting in a change of state. [4]

2. See Teilhard de Chardin selection, p. 316.
3. See John Rowan Wilson selection, p. 303.
4. See Julian Huxley quote, p. 181, Part 4.

The Historical Argument

Darwinism Applied to Man *
by Alfred Russel Wallace (1896)

THE ORIGIN OF THE MORAL AND INTELLECTUAL NATURE OF MAN

. . . I fully accept Mr. Darwin's conclusion as to the essential identity of man's bodily structure with that of the higher mammalia, and his descent from some ancestral form common to man and the anthropoid apes. The evidence of such descent appears to me to be overwhelming and conclusive. Again, as to the cause and method of such descent and modification, we may admit, at all events provisionally, that the laws of variation and natural selection, acting through the struggle for existence and the continual need of more perfect adaptation to the physical and biological environments, may have brought about, first that perfection of bodily structure in which he is so far above all other animals, and in co-ordination with it the larger and more developed brain, by means of which he has been able to utilise that structure in the more and more complete subjection of the whole animal and vegetable kingdoms to his service.

But this is only the beginning of Mr. Darwin's work, since he goes on to discuss the moral nature and mental faculties of man, and derives these too by gradual modification and development from the lower animals. Although, perhaps, nowhere distinctly formulated, his whole argument tends to the conclusion that man's entire nature and all his faculties, whether moral, intellectual, or spiritual, have been derived from their rudiments in the lower animals, in the same manner and by the action of the same general laws as his physical structure has been derived. As this conclusion appears to me not to be supported by adequate evidence, and to be directly opposed to many well-ascertained facts, I propose to devote a brief space to its discussion.

THE ARGUMENT FROM CONTINUITY

Mr. Darwin's mode of argument consists in showing that the rudiments of most, if not of all, the mental and moral faculties of man can be de-

* From Alfred Russel Wallace, *Darwinism*, London, 1896.

tected in some animals. The manifestations of intelligence, amounting in some cases to distinct acts of reasoning, in many animals, are adduced as exhibiting in a much less degree the intelligence and reason of man. Instances of curiosity, imitation, attention, wonder, and memory are given; while examples are also adduced which may be interpreted as proving that animals exhibit kindness to their fellows, or manifest pride, contempt, and shame. Some are said to have the rudiments of language, because they utter several different sounds, each of which has a definite meaning to their fellows or to their young; others the rudiments of arithmetic, because they seem to count and remember up to three, four, or even five. A sense of beauty is imputed to them on account of their own bright colours or the use of coloured objects in their nests; while dogs, cats, and horses are said to have imagination, because they appear to be disturbed by dreams. Even some distant approach to the rudiments of religion is said to be found in the deep love and complete submission of a dog to his master.[1]

Turning from animals to man, it is shown that in the lowest savages many of these faculties are very little advanced from the condition in which they appear in the higher animals; while others, although fairly well exhibited, are yet greatly inferior to the point of development they have reached in civilised races. In particular, the moral sense is said to have been developed from the social instincts of savages, and to depend mainly on the enduring discomfort produced by any action which excites the general disapproval of the tribe. Thus, every act of an individual which is believed to be contrary to the interests of the tribe, excites its unvarying disapprobation and is held to be immoral: while every act, on the other hand, which is, as a rule, beneficial to the tribe, is warmly and constantly approved, and is thus considered to be right or moral. From the mental struggle, when an act that would benefit self is injurious to the tribe, there arises conscience; and thus the social instincts are the foundation of the moral sense and of the fundamental principles of morality.[2]

The question of the origin and nature of the moral sense and of conscience is far too vast and complex to be discussed here, and a reference to it has been introduced only to complete the sketch of Mr. Darwin's view of the continuity and gradual development of all human faculties from the lower animals up to savages, and from savage up to civilised man. The point to which I wish specially to call attention is, that to prove continuity and the progressive development of the intellectual and moral faculties from animals to man, is not the same as proving that these faculties have been developed by natural selection; and this last is what Mr.

1. For a full discussion of all these points, see *Descent of Man,* chap. iii.
2. *Descent of Man,* chap. iv.

Darwin has hardly attempted, although to support his theory it was abso-lutely essential to prove it. Because man's physical structure has been developed from an animal form by natural selection, it does not necessar-ily follow that his mental nature, even though developed *pari passu* with it, has been developed by the same causes only.

To illustrate by a physical analogy. Upheaval and depression of land, combined with sub-aerial denudation by wind and frost, rain and rivers, and marine denudation on coastlines, were long thought to account for all the modelling of the earth's surface not directly due to volcanic action; and in the early editions of Lyell's *Principles of Geology* these are the sole causes appealed to. But when the action of glaciers was studied and the recent occurrence of a glacial epoch demonstrated as a fact, many phenomena—such as moraines and other gravel deposits, boulder clay, erratic boulders, grooved and rounded rocks, and Alpine lake basins— were seen to be due to this altogether distinct cause. There was no breach of continuity, no sudden catastrophe; the cold period came on and passed away in the most gradual manner, and its effects often passed insensibly into those produced by denudation or upheaval; yet none the less a new agency appeared at a definite time, and new effects were produced which, though continuous with preceding effects, were not due to the same causes. It is not, therefore, to be assumed, without proof or against inde-pendent evidence, that the later stages of an apparently continuous devel-opment are necessarily due to the same causes only as the earlier stages. Applying this argument to the case of man's intellectual and moral nature, I propose to show that certain definite portions of it could not have been developed by variation and natural selection alone, and that, therefore, some other influence, law, or agency is required to account for them. If this can be clearly shown for any one or more of the special faculties of intellectual man, we shall be justified in assuming that the same unknown cause or power may have had a much wider influence, and may have pro-foundly influenced the whole course of his development.

THE ORIGIN OF THE MATHEMATICAL FACULTY

We have ample evidence that, in all the lower races of man, what may be termed the mathematical faculty is, either absent, or, if present, quite unexercised. The Bushmen and the Brazilian Wood-Indians are said not to count beyond two. Many Australian tribes only have words for one and two, which are combined to make three, four, five, or six, beyond which they do not count. The Damaras of South Africa only count to three; and Mr. Galton gives a curious description of how one of them was hopelessly puzzled when he had sold two sheep for two sticks of tobacco each, and

received four sticks in payment. He could only find out that he was correctly paid by taking two sticks and then giving one sheep, then receiving
two sticks more and giving the other sheep. Even the comparatively intellectual Zulus can only count up to ten by using the hands and fingers. The
Ahts of North-West America count in nearly the same manner, and most
of the tribes of South America are no further advanced.[3] The Kaffirs have
great herds of cattle, and if one is lost they miss it immediately, but this is
not by counting, but by noticing the absence of one they know; just as in
a large family or a school a boy is missed without going through the process of counting. Somewhat higher races, as the Esquimaux, can count up
to twenty by using the hands and the feet; and other races get even further than this by saying "one man" for twenty, "two men" for forty, and so
on, equivalent to our rural mode of reckoning by scores. From the fact
that so many of the existing savage races can only count to four or five, Sir
John Lubbock thinks it improbable that our earliest ancestors could have
counted as high as ten.[4]

When we turn to the more civilised races, we find the use of numbers
and the art of counting greatly extended. Even the Tongas of the South
Sea islands are said to have been able to count as high as 100,000. But
mere counting does not imply either the possession or the use of anything
that can be really called the mathematical faculty, the exercise of which
in any broad sense has only been possible since the introduction of the
decimal notation. The Greeks, the Romans, the Egyptians, the Jews, and
the Chinese had all such cumbrous systems, that anything like a science of
arithmetic, beyond very simple operations, was impossible; and the
Roman system, by which the year 1888 would be written MDCCCLXXXVIII,
was that in common use in Europe down to the fourteenth or fifteenth
centuries, and even much later in some places. Algebra, which was invented by the Hindoos, from whom also came the decimal notation, was
not introduced into Europe till the thirteenth century, although the
Greeks had some acquaintance with it; and it reached Western Europe
from Italy only in the sixteenth century.[5] It was, no doubt, owing to the
absence of a sound system of numeration that the mathematical talent of

3. Lubbock's *Origin of Civilisation*, fourth edition, pp. 43–440; Tylor's *Primitive
Culture*, chap. vii.
4. It has been recently stated that some of these facts are erroneous, and that
some Australians can keep accurate reckoning up to 100, or more, when required.
But this does not alter the general fact that many low races, including the Australians, have no words for high numbers and never require to use them. If they are
now, with a little practice, able to count much higher, this indicates the possession
of a faculty which could not have been developed under the law of utility only,
since the absence of words for such high numbers shows that they were neither used
nor required.
5. Article Arithmetic in *Eng. Cyc. of Arts and Sciences*.

the Greeks was directed chiefly to geometry, in which science Euclid, Archimedes, and others made such brilliant discoveries. It is, however, during the last three centuries only that the civilised world appears to have become conscious of the possession of a marvellous faculty which, when supplied with the necessary tools in the decimal notation, the elements of algebra and geometry, and the power of rapidly communicating discoveries and ideas by the art of printing, has developed to an extent, the full grandeur of which can be appreciated only by those who have devoted some time (even if unsuccessfully) to the study.

The facts now set forth as to the almost total absence of mathematical faculty in savages and its wonderful development in quite recent times, are exceedingly suggestive, and in regard to them we are limited to two possible theories. Either prehistoric and savage man did not possess this faculty at all (or only in its merest rudiments); or they did possess it, but had neither the means nor the incitements for its exercise. In the former case we have to ask by what means has this faculty been so rapidly developed in all civilised races, many of which a few centuries back were, in this respect, almost savages themselves; while in the latter case the difficulty is still greater, for we have to assume the existence of a faculty which had never been used either by the supposed possessors of it or by their ancestors.

Let us take, then, the least difficult supposition—that savages possessed only the mere rudiments of the faculty, such as their ability to count, sometimes up to ten, but with an utter inability to perform the very simplest processes of arithmetic or of geometry—and inquire how this rudimentary faculty became rapidly developed into that of a Newton, a La Place, a Gauss, or a Cayley. We will admit that there is every possible gradation between these extremes, and that there has been perfect continuity in the development of the faculty; but we ask, What motive power caused its development?

It must be remembered we are here dealing solely with the capability of the Darwinian theory to account for the origin of the *mind*, as well as it accounts for the origin of the *body* of man, and we must, therefore, recall the essential features of that theory. These are, the preservation of useful variations in the struggle for life; that no creature can be improved beyond its necessities for the time being; that the law acts by life and death, and by the survival of the fittest. We have to ask, therefore, what relation the successive stages of improvement of the mathematical faculty had to the life or death of its possessors; to the struggles of tribe with tribe, or nation with nation; or to the ultimate survival of one race and the extinction of another. If it cannot possibly have had any such effects, then it cannot have been produced by natural selection.

It is evident that in the struggles of savage man with the elements and with wild beasts, or of tribe with tribe, this faculty can have had no influence. It had nothing to do with the early migrations of man, or with the conquest and extermination of weaker by more powerful peoples. The Greeks did not successfully resist the Persian invaders by any aid from their few mathematicians, but by military training, patriotism, and self-sacrifice. The barbarous conquerors of the East, Timurlane and Gengkhis Khan, did not owe their success to any superiority of intellect or of mathematical faculty in themselves or their followers. Even if the great conquests of the Romans were, in part, due to their systematic military organisation, and to their skill in making roads and encampments, which may, perhaps, be imputed to some exercise of the mathematical faculty, that did not prevent them from being conquered in turn by barbarians, in whom it was almost entirely absent. And if we take the most civilised peoples of the ancient world—the Hindoos, the Arabs, the Greeks, and the Romans, all of whom had some amount of mathematical talent—we find that it is not these, but the descendants of the barbarians of those days—the Celts, the Teutons, and the Slavs—who have proved themselves the fittest to survive in the great struggle of races, although we cannot trace their steadily growing success during past centuries either to the possession of any exceptional mathematical faculty or to its exercise. They have indeed proved themselves, to-day, to be possessed of a marvellous endowment of the mathematical faculty; but their success at home and abroad, as colonists or as conquerors, as individuals or as nations, can in no way be traced to this faculty, since they were almost the last who devoted themselves to its exercise. We conclude, then, that the present gigantic development of the mathematical faculty is wholly unexplained by the theory of natural selection, and must be due to some altogether distinct cause.

THE ORIGIN OF THE MUSICAL AND ARTISTIC FACULTIES

. . . As with the mathematical, so with the musical faculty, it is impossible to trace any connection between its possession and survival in the struggle for existence. It seems to have arisen as a *result* of social and intellectual advancement, not as a *cause;* and there is some evidence that it is latent in the lower races, since under European training native military bands have been formed in many parts of the world, which have been able to perform creditably the best modern music.

The artistic faculty has run a somewhat different course, though analogous to that of the faculties already discussed. Most savages exhibit some rudiments of it, either in drawing or carving human or animal figures; but, almost without exception, these figures are rude and such as would be executed by the ordinary inartistic child. In fact, modern savages are, in this respect hardly equal to those prehistoric men who represented the mammoth and the reindeer on pieces of horn or bone.[6] With any advance in the arts of social life, we have a corresponding advance in artistic skill and taste, rising very high in the art of Japan and India, but culminating in the marvellous sculpture of the best period of Grecian history. In the Middle Ages art was chiefly manifested in ecclesiastical architecture and the illumination of manuscripts, but from the thirteenth to the fifteenth centuries pictorial art revived in Italy and attained to a degree of perfection which has never been surpassed. This revival was followed closely by the schools of Germany, the Netherlands, Spain, France, and England, showing that the true artistic faculty belonged to no one nation, but was fairly distributed among the various European races.

These several developments of the artistic faculty, whether manifested in sculpture, painting, or architecture, are evidently outgrowths of the human intellect which have no immediate influence on the survival of individuals or of tribes, or on the success of nations in their struggles for supremacy or for existence. The glorious art of Greece did not prevent the nation from falling under the sway of the less advanced Roman; while we ourselves, among whom art was the latest to arise, have taken the lead in the colonisation of the world, thus proving our mixed race to be the fittest to survive.

INDEPENDENT PROOF THAT THE MATHEMATICAL, MUSICAL, AND ARTISTIC FACULTIES HAVE NOT BEEN DEVELOPED UNDER THE LAW OF NATURAL SELECTION

The law of Natural Selection or the survival of the fittest is, as its name implies, a rigid law, which acts by the life or death of the individuals submitted to its action. From its very nature it can act only on useful or hurtful characteristics, eliminating the latter and keeping up the former to a fairly general level of efficiency. Hence it necessarily follows that the characters developed by its means will be present in all the individuals of a species, and, though varying, will not vary very widely from a common standard. . . . all those characters in man which were certainly essential to him during his early stages of development, exist in all savages with some approach to equality. In the speed of running, in bodily strength, in skill with weapons, in acuteness of vision, or in power of following a trail, all are fairly proficient, and the differences of endowment do not probably exceed the limits of variation in animals. . . . So, in animal instinct or intelligence, we find the same general level of development. Every wren

6. The cave paintings of Lascaux (see Figure 4–14) were not discovered until about forty years after Wallace wrote the book from which this selection is taken. Ed.

makes a fairly good nest like its fellows; every fox has an average amount of the sagacity of its race; while all the higher birds and mammals have the necessary affections and instincts needful for the protection and bringing-up of their offspring.

But in those specially developed faculties of civilised man which we have been considering, the case is very different. They exist only in a small proportion of individuals, while the difference of capacity between these favoured individuals and the average of mankind is enormous. . . .

It appears then, that, both on account of the limited number of persons gifted with the mathematical, the artistic, or the musical faculty, as well as from the enormous variations in its development, these mental powers differ widely from those which are essential to man, and are, for the most part, common to him and the lower animals; and that they could not, therefore, possibly have been developed in him by means of the law of natural selection.

*   *   *

The special faculties we have been discussing clearly point to the existence in man of something which he has not derived from his animal progenitors—something which we may best refer to as being of a spiritual essence or nature, capable of progressive development under favourable conditions. On the hypothesis of this spiritual nature, superadded to the animal nature of man, we are able to understand much that is otherwise mysterious or unintelligible in regard to him, especially the enormous influence of ideas, principles, and beliefs over his whole life and actions. Thus alone we can understand the constancy of the martyr, the unselfishness of the philanthropist, the devotion of the patriot, the enthusiasm of the artist, and the resolute and persevering search of the scientific worker after nature's secrets. Thus we may perceive that the love of truth, the delight in beauty, the passion for justice, and the thrill of exultation with which we hear of any act of courageous self-sacrifice, are the workings within us of a higher nature which has not been developed by means of the struggle for material existence.

The Descent of Man *

by Charles Darwin (1871)

Man in the rudest state in which he now exists is the most dominant animal that has ever appeared on this earth. He has spread more widely than any other highly organised form: and all others have yielded before him. He manifestly owes this immense superiority to his intellectual faculties, to his social habits, which lead him to aid and defend his fellows, and to his corporeal structure. The supreme importance of these characters has been proved by the final arbitrament of the battle for life. Through his powers of intellect, articulate language has been evolved; and on this his wonderful advancement has mainly depended. As Mr. Chauncey Wright remarks: "A psychological analysis of the faculty of language shews, that even the smallest proficiency in it might require more brain power than the greatest proficiency in any other direction." [1] He has invented and is able to use various weapons, tools, traps, &c., with which he defends himself, kills or catches prey, and otherwise obtains food. He has made rafts or canoes for fishing or crossing over to neighbouring fertile islands. He has discovered the art of making fire, by which hard and stringy roots can be rendered digestible, and poisonous roots or herbs innocuous. This discovery of fire, probably the greatest ever made by man, excepting language, dates from before the dawn of history. These several inventions, by which man in the rudest state has become so pre-eminent, are the direct results of the development of his powers of observation, memory, curiosity, imagination, and reason. I cannot, therefore, understand how it is that Mr. Wallace [2] maintains, that "natural selection could only have endowed the savage with a brain a little superior to that of an ape." . . .

* From Charles Darwin, *The Descent of Man*, London, 1871.

1. "Limits of Natural Selection," *North American Review*, Oct., 1870, p. 295.

2. *Quarterly Review*, April, 1869, p. 392. This subject is more fully discussed in Mr. Wallace's *Contributions to the Theory of Natural Selection*, 1870, in which all the essays referred to in this work are republished. The "Essay on Man" has been ably criticised by Prof. Claparède, one of the most distinguished zoologists in Europe, in an article published in the *Bibliothèque Universelle*, June 1870. The remark quoted in my text will surprise every one who has read Mr. Wallace's celebrated paper on "The Origin of Human Races deduced from the Theory of Natural Selection," originally published in the *Anthropological Review*, May, 1864, p. clviii. I cannot here resist quoting a most just remark by Sir J. Lubbock (*Prehistoric Times,*

We can see, that in the rudest state of society, the individuals who were the most sagacious, who invented and used the best weapons or traps, and who were best able to defend themselves, would rear the greatest number of offspring. The tribes, which included the largest number of men thus endowed, would increase in number and supplant other tribes. Numbers depend primarily on the means of subsistence, and this depends partly on the physical nature of the country, but in a much higher degree on the arts which are there practised. As a tribe increases and is victorious, it is often still further increased by the absorption of other tribes.[3] The stature and strength of the men of a tribe are likewise of some importance for its success, and these depend in part on the nature and amount of the food which can be obtained. In Europe the men of the Bronze period were supplanted by a race more powerful, and, judging from their sword-handles, with larger hands; [4] but their success was probably still more due to their superiority in the arts.

All that we know about savages, or may infer from their traditions and from old monuments, the history of which is quite forgotten by the present inhabitants, shew that from the remotest times successful tribes have supplanted other tribes. Relics of extinct or forgotten tribes have been discovered throughout the civilised regions of the earth, on the wild plains of America, and on the isolated islands in the Pacific Ocean. At the present day civilised nations are everywhere supplanting barbarous nations, excepting where the climate opposes a deadly barrier; and they succeed mainly, though not exclusively, through their arts, which are the products of the intellect. It is, therefore, highly probable that with mankind the intellectual faculties have been mainly and gradually perfected through natural selection; and this conclusion is sufficient for our purpose.

1865, p. 479) in reference to this paper, namely, that Mr. Wallace, "with characteristic unselfishness, ascribes it (i.e. the idea of natural selection) unreservedly to Mr. Darwin, although, as is well known, he struck out the idea independently, and published it, though not with the same elaboration, at the same time."

3. After a time the members of tribes which are absorbed into another tribe assume, as Sir Henry Main remarks (*Ancient Law*, 1861, p. 131), that they are the co-descendants of the same ancestors.

4. Morlot, *Soc. Vaud. Sc. Nat.*, 1860, p. 294.

Modern Views of Mental Activity

The Immense Journey *

by Loren Eiseley (1957)

THE REAL SECRET OF PILTDOWN

Today the question asked by Wallace and never satisfactorily answered by Darwin has returned to haunt us. A skull, a supposedly very ancient skull, long used as one of the most powerful pieces of evidence documenting the Darwinian position upon human evolution, has been proven to be a forgery, a hoax perpetrated by an unscrupulous but learned amateur. In the fall of 1953 the famous Piltdown cranium, known in scientific circles all over the world since its discovery in a gravel pit on the Sussex Downs in 1911, was jocularly dismissed by the world's press as the skull that had "made monkeys out of the anthropologists." Nobody remembered in 1953 that Wallace, the great evolutionist, had protested to a friend in 1913, "The Piltdown skull does not prove much, if anything!"

Why had Wallace made that remark? Why, almost alone among the English scientists of his time, had he chosen to regard with a dubious eye a fossil specimen that seemed to substantiate the theory to which he and Darwin had devoted their lives? He did so for one reason: he did not believe what the Piltdown skull appeared to reveal as to the nature of the process by which the human brain had been evolved. He did not believe in a skull which had a modern brain box attached to an apparently primitive face and given, in the original estimates, an antiquity of something over a million years.

Today we know that the elimination of the Piltdown skull from the growing list of valid human fossils in no way affects the scientific acceptance of the theory of evolution. In fact, only the circumstance that Piltdown had been discovered early, before we had a clear knowledge of the nature of human fossils and the techniques of dating them, made the long survival of this extraordinary hoax possible. Yet in the end it has been the press, absorbed in a piece of clever scientific detection, which has missed the real secret of Piltdown. Darwin saw in the rise of man, with his unique, time-spanning brain, only the undirected play of such natural forces as had created the rest of the living world of plants and animals. Wallace, by contrast, in the case of man, totally abandoned this point of view and turned instead toward a theory of a divinely directed control of the evolutionary process. . . .

To explain the rise of man through the slow, incremental gains of natural selection, Darwin had to assume a long struggle of man with man and tribe with tribe.

He had to make this assumption because man had far outpaced his animal associates. Since Darwin's theory of the evolutionary process is based upon the practical value of all physical and mental characters in the life struggle, to ignore the human struggle of man with man would have left no explanation as to how humanity by natural selection alone managed to attain an intellectual status so far beyond that of any of the animals with which it had begun its competition for survival.

To most of the thinkers of Darwin's day this seemed a reasonable explanation. It was a time of colonial expansion and ruthless business competition. Peoples of primitive cultures, small societies lost on the world's margins, seemed destined to be destroyed. It was thought that Victorian civilization was the apex of human achievement and that other races with different customs and ways of life must be biologically inferior to Western man. Some of them were even described as only slightly superior to apes. The Darwinians, in a time when there were no satisfactory fossils by which to demonstrate human evolution, were unconsciously minimizing the abyss which yawned between man and ape. In their anxiety to demonstrate our lowly origins they were throwing modern natives into the gap as representing living "missing links" in the chain of human ascent.

It was just at this time that Wallace lifted a voice of lonely protest. The episode is a strange one in the history of science, for Wallace had, independently of Darwin, originally arrived at the same general conclusion as to the nature of the evolutionary process. Nevertheless, only a few years after the publication of Darwin's work, *The Origin of Species,* Wallace had come to entertain a point of view which astounded and troubled Darwin. Wallace, who had had years of experience with natives of the tropical archipelagoes, abandoned the idea that they were of mentally inferior cast. He did more. He committed the Darwinian heresy of maintaining that their mental powers were far in excess of what they really needed to carry on the simple food-gathering techniques by which they survived.

"How, then," Wallace insisted, "was an organ developed so far beyond the needs of its possessor? Natural selection could only have endowed the savage with a brain a little superior to that of an ape, whereas he actually possesses one but little inferior to that of the average member of our learned societies."

At a time when many primitive peoples were erroneously assumed to speak only in grunts or to chatter like monkeys, Wallace maintained his view of the high intellectual powers of natives by insisting that "the capacity of uttering a variety of distinct articulate sounds and of applying to them an almost infinite amount of modulation . . . is not in any way inferior to that of the higher races. An instrument has been developed in advance of the needs of its possessor." . . .

"If you had not told me you had made these remarks," Darwin said, "I should have thought they had been added by someone else. I differ grievously from you and am very sorry for it." He did not, however, supply a valid answer to Wallace's queries. Outside of murmuring about the inherited effects of habit—a contention without scientific validity today—Darwin clung to his original position. Slowly Wallace's challenge was forgotten and a great complacency settled down upon the scientific world.

For seventy years after the publication of *The Origin of Species* in 1859, there were only two finds of fossil human skulls which seemed to throw any light upon the Darwin-Wallace controversy. One was the discovery of the small-brained Java Ape Man, the other was the famous Piltdown or "dawn man." Both were originally dated as lying at the very beginning of the Ice Age, and, though these dates were later to be modified, the skulls, for a very long time, were regarded as roughly contemporaneous and very old.

Two more unlike "missing links" could hardly be imagined. Though they were supposed to share a million-year antiquity, the one was indeed quite primitive and small-brained; the other, Piltdown, in spite of what seemed a primitive lower face, was surprisingly modern in brain. Which of these forms told the true story of human development? Was a large brain old? Had ages upon ages of slow, incremental, Darwinian increase produced it? The Piltdown skull seemed to suggest such a development.

Many were flattered to find their anthropoid ancestry seemingly removed to an increasingly remote past. If one looked at the Java Ape Man, one was forced to contemplate an ancestor, not terribly remote in time, who still had a face and a brain which hinted strongly of the ape. Yet, when by geological evidence this "erect walking ape-man" was finally assigned to a middle Ice Age antiquity, there arose the immediate possibility that Wallace could be right in his suspicion that the human brain might have had a surprisingly rapid development. By contrast, the Piltdown remains seemed to suggest a far more ancient and slow-paced evolution of man. The Piltdown hoaxer, in attaching an ape jaw to a human skull fragment, had, perhaps unwittingly, created a creature which supported the Darwinian idea of man, not too unlike the man of today, extending far back into pre-Ice Age times.

Which story was the right one? Until the exposé of Piltdown in 1953, both theories had to be considered possible and the two hopelessly unlike fossils had to be solemnly weighed in the same balance. Today Piltdown is gone. In its place we are confronted with the blunt statement of two modern scientists, M. R. A. Chance and A. P. Mead.

"No adequate explanation," they confess over eighty years after Darwin scrawled his vigorous "No!" upon Wallace's paper, "has been put forward to account for so large a cerebrum as that found in man." [1]

We have been so busy tracing the tangible aspects of evolution in the *forms of animals* that our heads, the little globes which hold the midnight sky and the shining, invisible universes of thought, have been taken about as much for granted as the growth of a yellow pumpkin in the fall.

Now a part of this mystery as it is seen by the anthropologists of today lies in the relation of the brain to time. "If," Wallace had said, "researchers in all parts of Europe and Asia fail to bring to light any proofs of man's presence far back in the Age of Mammals, it *will be at least a presumption that he came into existence at a much later date and by a more rapid process of development.*" If human evolution should prove to be comparatively rapid, "explosive" in other words, Wallace felt that his position would be vindicated, because such a rapid development of the brain would, he thought, imply a divinely directed force at work in man. In the 1870's when he wrote, however, human prehistory was largely an unknown blank. Today we can make a partial answer to Wallace's question. Since the exposure of the Piltdown hoax all of the evidence at our command—and it is considerable—points to man, in his present form, as being one of the youngest and newest of all earth's swarming inhabitants.[2] . . .

Today, with the solution of the Piltdown enigma, we must settle the question of the time involved in human evolution in favor of Wallace, not Darwin; we need not, however, pursue the mystical aspects of Wallace's thought—since other factors yet to be examined may well account for the rise of man. The rapid fading out of archaeological evidence of tools in lower Ice Age times—along with the discovery of man-apes of human as-

1. *Symposia of the Society for Experimental Biology*, VII, Evolution (New York: Academic Press, 1953), p. 395.
2. Although new dating techniques and more recent discoveries at the Oldoway Gorge indicate that a very small-brained type of tool-making man was present in Africa as early as 1,750,000 years ago, this is still comparatively late in the Age of Mammals which extends back 80,000,000 or 60,000,000 years (see Fig. 3–2). Ed.

pect but with ape-sized brains, yet possessing a diverse array of bodily characters—suggests that the evolution of the human brain was far more rapid than that conceived of in early Darwinian circles. At that time it was possible to hear the Eskimos spoken of as possible survivals of Miocene men of several million years ago. By contrast to this point of view, man and his rise now appear short in time—explosively short. There is every reason to believe that whatever the nature of the forces involved in the production of the human brain, a long slow competition of human group with human group or race with race would not have resulted in such similar mental potentialities among all peoples everywhere. Something—some other factor—has escaped our scientific attention.

There are certain strange bodily characters which mark man as being more than the product of a dog-eat-dog competition with his fellows. He possesses a peculiar larval nakedness, difficult to explain on survival principles; his periods of helpless infancy and childhood are prolonged; he has aesthetic impulses which, though they vary in intensity from individual to individual, appear in varying manifestations among all peoples. He is totally dependent, in the achievement of human status, upon the careful training he receives in human society.

Unlike a solitary species of animal, he cannot develop alone. He has suffered a major loss of precise instinctive controls of behavior. To make up for this biological lack, society and parents condition the infant, supply his motivations, and promote his long-drawn training at the difficult task of becoming a normal human being. Even today some individuals fail to make this adjustment and have to be excluded from society.

We are now in a position to see the wonder and terror of the human predicament: man is totally dependent on society. Creature of dream, he has created an invisible world of ideas, beliefs, habits, and customs which buttress him about and replace for him the precise instincts of the lower creatures. In this invisible universe he takes refuge, but just as instinct may fail an animal under some shift of environmental conditions, so man's cultural beliefs may prove inadequate to meet a new situation, or, on an individual level, the confused mind may substitute, by some terrible alchemy, cruelty for love.

The profound shock of the leap from animal to human status is echoing still in the depths of our subconscious minds. It is a transition which would seem to have demanded considerable rapidity of adjustment in order for human beings to have survived, and it also involved the growth of prolonged bonds of affection in the sub-human family, because otherwise its naked, helpless offspring would have perished.

It is not beyond the range of possibility that this strange reduction of instincts in man in some manner forced a precipitous brain growth as a compensation—something that had to be hurried for survival purposes. Man's competition, it would thus appear, may have been much less with his own kind than with the dire necessity of building about him a world of ideas to replace his lost animal environment. . . .

Modern science would go on to add that many of the characters of man, such as his lack of fur, thin skull, and globular head, suggest mysterious changes in growth rates which preserve, far into human maturity, foetal or infantile characters which hint that the forces creating man drew him fantastically out of the very childhood of his brutal forerunners. Once more the words of Wallace come back to haunt us: "We may safely infer that the savage possesses a brain capable, if cultivated and developed, of performing work of a kind and degree far beyond what he ever requires it to do."

As a modern man, I have sat in concert halls and watched huge audiences floating dazed on the voice of a great singer. Alone in the dark box I have heard far off as if ascending out of some black stairwell the guttural whisperings and bestial coughings out of which that voice arose. Again, I have sat under the slit dome of a mountain observatory and marveled, as the great wheel of the galaxy turned in all its midnight splendor, that the mind in the course of three centuries has been capable of drawing into its strange, nonspatial interior that world of infinite distance and multitudinous dimensions.

Ironically enough, science, which can show us the flints and the broken skulls of our dead fathers, has yet to explain how we have come so far so fast, nor has it any completely satisfactory answer to the question asked by Wallace long ago. Those who would revile us by pointing to an ape at the foot of our family tree grasp little of the awe with which the modern scientist now puzzles over man's lonely and supreme ascent. As one great student of paleoneurology, Dr. Tilly Edinger, recently remarked, "If man has passed through a Pithecanthropus phase, the evolution of his brain has been unique, not only in its result but also in its tempo. . . . Enlargement of the cerebral hemispheres by 50 per cent seems to have taken place, speaking geologically, within an instant, and without having been accompanied by any major increase in body size."

The true secret of Piltdown, though thought by the public to be merely the revelation of an unscrupulous forgery, lies in the fact that it has forced science to reëxamine carefully the history of the most remarkable creation in the world—the human brain.

         ❋   ❋   ❋

THE DREAM ANIMAL

If this brain, a brain more than twice as large as that of a much bigger animal—the gorilla—is to be acquired in infancy, its major growth must take place with far greater rapidity than in the case of man's nearest living relatives, the great apes. It must literally spring up like an overnight mushroom, and this greatly accelerated growth must take place during the first months after birth. If it took place in the embryo, man would long since have disappeared from the planet—it would have been literally impossible for him to have been born. As it is, the head of the infant is one of the factors making human birth comparatively difficult. When we are born, however, our brain size, about 330 cubic centimeters, is only slightly larger than that of a gorilla baby. This is why human and anthropoid young look so appealingly similar in their earliest infancy.

A little later, an amazing development takes place in the human offspring. In the first year of life. its brain trebles in size. It is this peculiar leap, unlike anything else we know in the animal world, which gives to man his uniquely human qualities. When the leap fails, as in those rare instances where the brain does not grow, microcephaly, "pinheadedness," is the result, and the child is then an idiot. Somewhere among the inner secrets of the body is one which keeps the time for human brain growth. If we compare our brains with those of other primate relatives (recognizing, as we do, many similarities of structure) we are yet unable to perceive at what point in time or under what evolutionary conditions the actual human forerunner began to manifest this strange postnatal brain expansion. It has carried him far beyond the mental span of his surviving relatives. As our previously quoted authority, Dr. Tilly Edinger of Harvard, has declared, "the brain of *Homo sapiens* has not evolved from the brains it is compared with by comparative anatomy; it developed within the Hominidae, at a late stage of the evolution of this family whose other species are all extinct."

We can, in other words, weigh, measure and dissect the brains of any number of existing monkeys. We may learn much in the process, but the key to our human brain clock is not among them. It arose in the germ plasm of the human group alone and we are the last living representatives of that family. As we contemplate, however, the old biological law that, to a certain degree, the history of the development of the individual tends to reproduce the evolutionary history of the group to which it belongs, we cannot help but wonder if this remarkable spurt in brain development may not represent something roughly akin to what happened in the geological past of man—a sudden or explosive increase which was achieved in a relatively short period, geologically speaking. . . .

* * *

Such rapidity suggests other modes of selection and evolution than those implied in the nineteenth-century literature with its emphasis on intergroup "struggle" which, in turn, would have demanded large populations. We must make plain here, however, that to reject the older Darwinian arguments is not necessarily to reject the principle of natural selection. We may be simply dealing with a situation in which both Darwin and Wallace failed, in different ways, to see what selective forces might be at work in man. . . .

Although there is still much that we do not understand, it is likely that the selective forces working upon the humanization of man lay essentially in the nature of the socio-cultural world itself. Man, in other words, once he had "crossed over" into this new invisible environment, was being as rigorously selected for survival within it as the first fish that waddled up the shore on its fins. I have said that this new world was "invisible." I do so advisedly. It lay, not so much in his surroundings as in man's brain, in his way of looking at the world around him and at the social environment he was beginning to create in his tiny human groupings.

He was becoming something the world had never seen before—a dream animal—living at least partially within a secret universe of his own creation and sharing that secret universe in his head with other, similar heads. Symbolic communication had begun. Man had escaped out of the eternal present of the animal world into a knowledge of past and future. The unseen gods, the powers behind the world of phenomenal appearance, began to stalk through his dreams. . . .

If one attempts to read the complexities of the story, one is not surprised that man is alone on the planet. Rather, one is amazed and humbled that man was achieved at all. For four things had to happen, and if they had not happened simultaneously, or at least kept pace with each other, the bones of man would lie abortive and forgotten in the sandstones of the past:

1. His brain had almost to treble in size.

2. This had to be effected, not in the womb, but rapidly, after birth.

3. Childhood had to be lengthened to allow this brain, divested of most of its precise instinctive responses, to receive, store, and learn to utilize what it received from others.

4. The family bonds had to survive seasonal mating and become permanent, if this odd new creature was to be prepared for his adult role.

Each one of these major points demanded a multitude of minor biological adjustments, yet all of this—change of growth rate, lengthened age, increased blood supply to the head, moved apparently with rapidity. It is a dizzying spectacle with which we have nothing to compare. The event is complex, it is many-sided, and what touched it off is hidden under the leaf mold of forgotten centuries.

Somewhere in the glacial mists that shroud the past, Nature found a way of speeding the proliferation of brain cells and did it by the ruthless elimination of everything not needed to that end. We lost our hairy covering, our jaws and teeth were reduced in size, our sex life was postponed, our infancy became among the most helpless of any of the animals because everything had to wait upon the development of that fast-growing mushroom which had sprung up in our heads.

Now in man, above all creatures, brain is the really important specialization. As Gavin de Beer, Director of the British Museum of Natural History, has suggested, it appears that if infancy is lengthened, there is a correspondingly lengthier retention of embryonic tissues capable of undergoing change.[3] Here, apparently, is a possible means of stepping up brain growth. The anthropoid ape, because of its shorter life cycle and slow brain growth, does not make use of nearly the amount of primitive neuroblasts—the embryonic and migrating nerve cells—possible in the lengthier, and at the same time paradoxically accelerated development of the human child. The clock in the body, in other words, has placed a limit upon the pace at which the ape brain grows—a limit which, as we have seen, the human ancestors in some manner escaped. This is a simplification of a complicated problem, but it hints at the answer to Wallace's question of long ago as to why man shows such a strange, rich mental life, many of whose artistic aspects can have had little direct value measured in the old utilitarian terms of the selection of all qualities in the struggle for existence.

When these released potentialities for brain growth began, they carried man into a new world where the old laws no longer totally held. With every advance in language, in symbolic thought, the brain paths multiplied. Significantly enough, those which are most heavily involved in the life processes, and are most ancient, mature first. The most recently acquired and less specialized regions of the brain, the "silent areas," mature last. Some neurologists, not without reason, suspect that here may lie other potentialities which only the future of the race may reveal.

Even now, however, the brain of man, with all its individual never-to-be-abandoned richness, is becoming merely a unit in the vast social brain which is potentially immortal, and whose memory is the heaped wisdom

3. *Embryos and Ancestors,* rev. ed. (New York, Oxford, 1951), p. 93.

of the world's great thinkers. The scientist Haldane, brooding upon the future, has speculated that we will even further prolong our childhood and retard maturity if brain advance continues.[4]

It is unlikely, however, in our present comfortable circumstances, that the pace of human change will ever again speed at the accelerated rate it knew when man strove against extinction. The story of Eden is a greater allegory than man has ever guessed. For it was truly man who, walking memoryless through bars of sunlight and shade in the morning of the world, sat down and passed a wondering hand across his heavy forehead. Time and darkness, knowledge of good and evil, have walked with him ever since. It is the destiny struck by the clock in the body in that brief space between the beginning of the first ice and that of the second. In just that interval a new world of terror and loneliness appears to have been created in the soul of man.

For the first time in four billion years a living creature had contemplated himself and heard with a sudden, unaccountable loneliness, the whisper of the wind in the night reeds. Perhaps he knew, there in the grass by the chill waters, that he had before him an immense journey. Perhaps that same foreboding still troubles the hearts of those who walk out of a crowded room and stare with relief into the abyss of space so long as there is a star to be seen twinkling across those miles of emptiness.

4. Ironically, the trend in the past hundred years has been towards earlier maturity in man. There is evidence that in the United States, Britain, and many other European countries puberty is attained 2½ to 3½ years earlier than it was a century ago. See J. M. Turner, "Earlier Maturation in Man," *Scientific American*, January 1968. Ed.

The Development of Mental Activity *

by Julian Huxley (1953)

   Life has two aspects, a material and a mental. Its mental aspect increases in importance during evolutionary time. Later animal deployments have reached a higher level of mental organization than earlier
ones: the higher animals have a larger mental component in their make-
up. This fact leads to an important conclusion—that mind is not a pale
epiphenomenon, not a mere "ghost in the machine," to use Professor
Ryle's phrase, but an *operative* part of life's mechanism. For no evolutionary trend can be maintained except by natural selection, and natural
selection can only work on what is biologically useful to its possessors.
   Mental activity is intensified and mental organization improved during
evolution; like bodily organization, it is improved in different ways in relation to different needs. . . .
   For a biologist, much the easiest way is to think of mind and matter as
two aspects of a single, underlying reality—shall we call it world substance, the stuff out of which the world is made. At any rate, this fits more
of the facts and leads to fewer contradictions than any other view. In this
view, mental activities are among the inevitable properties of world substance when this is organized in the form of the particular kind of biological machinery we find in a brain. The electrical properties of living
substance provide us with a useful analogy. We now know that all
activities in the body are accompanied by electrical changes—but changes
so minute that they were only detected when special instruments were
invented during the nineteenth century. All living substance, indeed all
substance, inorganic as well as organic, has electrical properties, and all its
properties have an electrical aspect. But these minute electrical changes
can be intensified and utilized for biological ends. In nervous tissue
they are utilized to transmit messages within the body, and a few fish,
like the torpedo and the electric eel, possess organs for intensifying
them to such a degree that they can be used to give dangerous electric shocks. I find myself driven to assume that the analogy with mind
holds good—in other words that all living substance has mental, or we
had better say mindlike, properties; but that these are, for the most part,
far below the level of detection. They could only be utilized for biological
ends when organs were evolved capable of intensifying them. These or-

   * From Julian Huxley, *Evolution in Action*, Harper & Brothers, New York. Copyright 1953 by Julian S. Huxley. Reprinted by permission.

gans are certain specially developed parts of brains. It is by means of them that mind emerges as an operative factor in evolution. . . .

Knowledge of the material processes going on in the brain may also help in understanding the evolution of mind. There is increasing agreement that one very special kind of nervous organization is of great importance—the organization in the cerebral cortex of large groups of nerve cells and their connecting outgrowths into self-reinforcing circuits of excitation. Circuits of this sort are arrangements for maintaining an organized flow of excitation through the cortex, on its way to become translated into organized behavior. If they are interfered with, it seems that various disturbances arise, like pain and fear; while pleasure and what we may call integrative emotions, like love, are linked up with the maintenance or increase of their organized flow. Pain is thus in its origin a by-product of nervous structure: but, as I shall later point out, it can subsequently be utilized as part of the machinery of learning. . . .

Before the nervous system could be improved, it had to be invented: the lowest animals are without one. Even the largest sponge has no nervous system whatever. Its first manifestation, which we find in creatures like polyps, is a nerve net—an irregular network of nerve cells and interlacing fibres extending all over the body. Its latest improvement is our own nervous system, with all its incoming and outgoing wires of nerves, and its central exchange and office and control room, in the form of an enormous brain filling up most of our head.

During evolution, the speed at which messages are transmitted along nerve fibres has increased over six hundredfold, from below six inches a second in some nerve nets to over a hundred yards a second in parts of our own nervous system. The brain's complexity of organization is almost infinitely greater than that of any other piece of biological machinery in ourselves or in any other animal. And the step in improvement from ape brain to human brain was the basis for the latest and most remarkable deployment of living substance.

❀   ❀   ❀

The fact of mind's emergence may be simply demonstrated from the actual behavior of a few animals. Consider first *Paramecium*, the microscopic slipper animalcule—that compulsory study for every elementary student in biology. Its normal existence merely consists in swimming onward in a spiral path. It does not actively pursue its food, but simply sweeps bacteria into its gullet as it swims. Now and again, however, it checks its advance, backs, turns a little, and then continues in a new direction. This so-called *avoiding reaction* is executed whenever the animal meets with unfavorable conditions, such as water which is not of the right

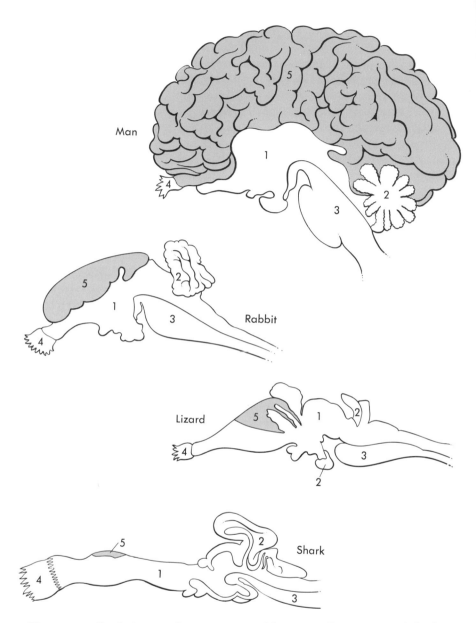

Figure 5–1. Evolutionary changes in animal brains. Different parts of the brain control different functions. Among orders like sharks, the inner brain (1), cerebellum (2), medulla (3) and frontal lobe (4), which control coordination plus automatic activities of the body and personality, dominate the brain area. Only a small portion is given to the cerebrum (5). In man the cerebrum is much larger and more convoluted, and areas governing instinct become proportionately smaller. (Reproduced by permission of Time-Life Books from Evolution. *Copyright* © *1964 by Time Incorporated. Figure added by editor.)*

degree of acidity for bacteria to grow in, and it repeats it until it finds itself in favorable conditions. This behavior is mere trial and error, and neither learning nor purposeful direction is involved. *Paramecium* must be in some way aware of the difference between more acid and less acid water, but this is almost the only sort of awareness we have any right to ascribe to it.

Euglena is another single-celled organism which swims in a spiral advance, but it has a little rudiment of an eye, a spot of pigment over a patch of specially light-sensitive substance. And so it is capable of what is broadly called a tropism; it directs its swimming in relation to the direction of the light that falls upon it. It has a primitive awareness of light, and it finds its way into the most favorable parts of its environment by utilizing this awareness, and not only by trial and error. The behavior of an earthworm is much more complicated. An earthworm is capable of a number of reflex actions, each resulting from some different awareness— awareness of light, of chemical stimuli, of touch and pressure. And finally it has some capacity for learning. If confronted with two alternative pathways, it can learn to choose the one which leads to a more favorable result. This is a very limited kind of learning; but still it *is* learning. Finally, let me jump from worms to mammals. A rat can learn to run through a maze with a dozen turnings. And we all know how elaborate the behavior of a dog can be. In dog life, mind has certainly emerged as a major factor.

But the evolutionary biologist is concerned more with the description of processes than with the demonstration of facts. He wants to understand something of the way in which the dual-aspect system of mind and behavior evolves; of how mental organization is specialized and improved during evolution. However, this is an enormous subject, and all I can do is to illustrate it by way of examples from the organization of animals' awareness, in a broad and perhaps rather loose sense of the word. In the first place, the range of awareness may be increased. This is achieved by the improvement of different receptor organs. They are windows, so to speak, letting in different kinds of awareness of different parts of reality. Vision lets in movement and shape and distance; smell and taste let in chemical properties; touch lets in a knowledge of what is in immediate contact with the body; sound is used as an indicator of the distant presence of something significant—enemy, or prey, or mate. An electrical sense is exceedingly rare; it is confined to a few fish which find their way about by means of electrical signals that they send out. Pit vipers have developed a unique type of sense organ, in the shape of a directional receptor, for heat radiations; this enables them to locate the warm-blooded small mammals on which they feed. Sometimes new ranges of sensations are evolved. Thus bats find their way about by means of ultrasonic vibrations, of a frequency far above that which can be detected by the human ear. These are emitted all the time while the bat is flying, and are reflected back from any solid object in the neighborhood—a kind of glorified echo sounding.

Bees do not use the same range of photic radiations as we do: they are blind to red, but can see a certain distance into the ultra-violet. Further, they are aware of something we can only detect by means of special apparatus—the plane of polarized light. This is because each of the units of their compound eyes is constructed in such a way that it can act roughly like a Nicol prism. This prerequisite has been turned to account, and the bees steer themselves in relation to the sun's position even on overcast days, by taking advantage of this capacity.

Many phenomena of nature have never entered into the effective world of animals because there were no sense organs to receive them. There are no animals with an awareness of X-rays or radio waves or magnetic fields; that had to wait for the construction of artificial sense organs by scientific man. In any particular animal, the performance of receptor organs—the kind of phenomena they admit—is a restricted one. It is partly restricted by the nature of living substance: thus the long electromagnetic waves of radio simply pass through living substance and its secretions without affecting them, so that it would be impossible to construct a radio receptor organ out of such materials. Then awareness may be organized by relating the *quality* of a sensation to the needs of the animal. Thus, to put the matter rather crudely, sweet things taste nice because sugars are an abundant and valuable source of food. Lead acetate, which does not exist as such in the natural environment, also tastes sweet, but it is a poison. We can be pretty sure that if it had been as common as sugar, and sugar as rare as lead acetate, sweet things would have tasted nasty.

Then the *range* of a sensation may be restricted, so that only a few significant events come to the notice of the animal. Thus, the smelling organs in the elaborately branched antennae in certain kinds of male moths, while incredibly acute in detecting the smell of a female of the same species, seem to be unresponsive to every other kind of smell.

The sense which is most in need of restriction is that of vision: it would merely be confusing if an animal were to pay attention to all the visible changes going on in its environment. One of the ways in which this restriction is accomplished is by building into the brain special channels of flow which let through certain patterns much more readily than others.

Such patterns of sensory awareness are called *releasers*, because they conduct the flow of excitation through the brain, to release a specific pattern of behavior; they are keys to unlock certain doors of action. They are found in relation to all the senses, but vision provides the best examples. Here I have space for only one or two. Let me first take the crouching reaction which young game birds practice, even at their first sight of a hawk overhead. It seemed difficult to account for this without appealing to some sort of Lamarckian inheritance of racial experience—until it was shown that the reaction could be produced by a crude four-armed model with one very blunt and one rather long arm. When this is towed overhead with the blunt arm forward, it makes a rough representation of a hawk in flight, with its long tail and short neck; whereas when the model is towed with the long arm forward, the resemblance is rather to a flying duck than a hawk, and elicits no crouching reaction from the young game birds. Sounds too have no effect. The releaser key, in this case, thus includes direction of movement as one of its necessary wards; but no detailed resemblance to a hawk is required. . . .

Releaser mechanisms are built into the animal by heredity; and they can only relate it to its environment in rather a crude way, and one which can easily become misleading. It is no accident that the only definite releaser known in man is the pattern made by a mother's smile to her infant. For more accurate adjustment, the animal must build up its patterns of awareness out of its own individual experience. I will give one example —our own perceptions. These are not, as is often thought, snapshot pictures of reality projected into our minds, but quite elaborate mental constructions.

People who recover their sight after being blind from infancy have to learn, by a long and tedious process, to build distinguishable objects and forms out of the kaleidoscopic patchwork which is all that they at first discern. For a long time, they cannot even distinguish triangles from circles except by tracing their outlines with their fingers. A combination of data from the senses of sight and touch is needed for putting the three dimensions of space into any perceptions. A baby spends a large part of its existence in constructing, out of the crude sensory experiences of handling, touching, looking, the three-dimensional world of objects in which it will later live. The same thing holds with animals like chimpanzees, as is demonstrated by the fact that they can be taken in by the same illusions as we. However, their spatial world is not exactly like ours—for instance, they cannot learn to see whether a mechanical construction is stable or not. They will pile boxes on each other to get at a banana hung from the ceiling; but they never seem to acquire any insight as to whether the pile of boxes will stand. To get it to stand, they have to work by trial and error.

The same sort of building-up process goes on in the perceptual processes of all higher vertebrates, although the elaborateness of the construction varies a great deal. Thus, a horse is unable to bring its sensations of touch and vision together in the same way as a monkey or a man; and so the objects which it constructs in its perceptual world are not nearly so well defined. This seems to be the reason why horses shy at a heap of stones by the roadside. They do not perceive it as a heap of stones as we do, but merely as something unfamiliar. In passing, it is worth noting that certain insects, like ants, must be able to construct perceptions of a quite different nature from anything of which we have experience. They feel and smell objects at one and the same time with the aid of one and the same organ —their antennae. Objects for them must be smells with shapes.

Our perceptions are thus based on a mass of assumptions derived from what we have learnt by experience. This is why it is easy to construct illusions. They introduce false assumptions, which then make us alter our total perception. . . .

There are, of course, many other kinds of learning mechanisms. One of the most obvious is based on a combination of pain with a conditioned reflex. Such a combination is a mechanism for avoiding harmful stimuli and ensuring the efficacy of useful ones. Professor Young's octopus learnt very quickly not to try to eat crabs after an electric shock had been associated with one or two attempts. The two parts of the arrangement are brought together in a special piece of nervous machinery, for when a particular part of the brain is cut out, the octopus will attack a crab over and over again, even if it gets a painful electric shock each time. Here we see how pain can be transformed from a by-product of physiology into an effective agency of behavior. On the other hand, sometimes the transformation is not readily possible, and then pain is not utilized in this way. Even severe damage will not cause pain in an organ which is not normally exposed to that kind of damage. Thus the tissue of the brain can be cut without any trace of pain.

Here I must mention one recent discovery, made by the German biologist, Professor Rensch—the fact that increased body size may be correlated with increased learning capacity; of two closely similar animals, like a raven and a jackdaw, the bigger will learn better, though the learning process takes longer. The chain of causation here is roughly as follows. Increased final size of the cerebral cortex is brought about by its growing at a faster rate than the rest of the brain. So an *absolutely* larger brain will have a *relatively* as well as an absolutely larger number of cells in its cortex. A larger number of cortical cells makes more elaborate learning possible; but more elaborate learning takes a longer time. It is interesting to note that this holds not only for birds and mammals but for beetles as well. This fact is of great interest, for it may help to explain the biological value of mere bulk—why so many lines of so many different deployments of higher animals have tended to increase in size during their evolution.

It seems to be a general rule that the greater the complexity of what can be learnt, the longer is the time taken to learn it. The human infant can learn to recognize extremely complicated shapes, but takes several years to do so. A rat, it seems from the latest experiments, cannot, or does not normally, learn to recognize even such a simple shape as a triangle; or at least it does not discriminate it as a unitary whole from the rest of the visual field. Professor Hebb, in his interesting but difficult book, *The Organization of Behaviour*, discusses this and other aspects of this complicated subject. But what the rat does learn about the spatial relations of its environment, it learns very quickly.

An immense amount of work has been done on the learning reaction of animals. Some learning is a matter of simple conditioned reflexes. As Lloyd Morgan showed many years ago, domestic chicks will begin by pecking at all sorts of small conspicuous objects, but have to learn by experience whether they are good to eat or not. In one set of experiments, Lloyd Morgan steeped one kind of grain in a bitter solution. The chicks pecked at it, rejected it, and on later occasions refused similar grains without pecking at them, rubbing their bills as if they already had the nasty taste in their mouths. Then he repeated this with other kinds of grain, until he had to give up the experiment for fear the chicks would starve. In nature, this negative conditioning leads young insect-eating birds to refuse nauseous and dangerous creatures such as wasps, after one experience. The wasps' distinctive color pattern makes it easier for the bird to recognize them in good time on later occasions.

Recently I saw a young magpie in Dr. Thorpe's laboratory which had been hand fed by its mistress long after it had fledged. It had not learnt to look for its food, and it was only with the greatest difficulty that Thorpe was teaching it to feed itself when hungry.

Broadly speaking, memory and the capacity to learn complex situations and make complex discriminations increase as we move up the animal scale. As I mentioned, an earthworm can be trained, though with difficulty, to choose one of two alternative paths in a Y-shaped tube. A rat can achieve success in an elaborate maze. An ape can learn to react to what it sees in a film; and some gifted apes can solve problems, such as fitting two sticks together by a fishing-rod joint, by pure insight instead of clumsy trial and error.

But the capacity for learning may be adaptively restricted or specialized. Animals often learn some things more easily than others. Thus herring gulls learn to distinguish their own chicks individually after a few days; but they seem unable to discriminate their own eggs, even from quite differently marked eggs belonging to another species. Similarly, male robins will attack a headless dummy so long as it has the releaser provided by the red breast; but they learn to distinguish their mates individually.

Conversely, the possibilities of learning may never be turned to biological account. The most obvious example is that of talking birds like parrots or mynahs. They can learn quite elaborate phrases, and certainly sometimes show "association of ideas" by saying the right phrase at the right time. (One parrot cited by Lorenz, after not seeing a hoopoe for nine years, at once remembered and spoke the appropriate words.) But they never learn to associate words with purposes—for instance to say "food" when hungry—in spite of intensive training. The potentialities and limitations of mind jostle each other in a strange and interesting way.

Sometimes animals in nature learn something both new and useful. A striking example is the way in which tits in Britain have learnt to open milk bottles, first with cardboard and then with metal tops, in the last few decades. Careful inquiry has shown that the discovery was made independently in several areas, presumably by a few exceptional birds, and then spread slowly round each center by imitation.

Some kinds of learning are quite different from anything in ourselves. Thus young geese which have been hatched in an incubator will attach themselves to birds of other species or even to human beings and follow them about as if they were real parents. This so-called "imprinting" has to take place during a critical period soon after hatching, only takes a minute or so, and is then irreversible; once it has happened, young geese will not switch over, even to their own parents.

In evolution, the greatest divergence as regards the organization of behavior is between insects and vertebrates. The insects rely much more on patterns built in by heredity, the vertebrates much more on patterns built up out of learning and individual experience, which means that their behavior can become much more flexible. Higher insects emerge from the pupa stage with their instincts fully formed and ready to come into action; all that experience can do is to adjust the performance of their instinctive actions to the immediate situation. Vertebrates, on the other hand, largely build their own behavior, and the more complicated its organization, the longer is the period of learning required. It is no accident that higher mammals pass through a longer period of dependence than any other animals, during which they learn the skills needed for adult life by experience and practice and play. Monkeys have a longer learning period than other mammals, apes a longer period than monkeys, and man the longest learning period of any organism. In other words, during the vertebrates' evolution, mental organization acquires a time dimension. Mental structure in insects hardly grows or develops at all; in higher mammals it grows rapidly and transforms itself radically until the adult phase is reached; and even after that slow growth may continue. And in human beings, mental structure may continue to develop throughout life, even up to old age—we need only think of Verdi or Titian. This increased capacity for incorporating experience in mental structure is naturally correlated with a trend to longer life. Once more, it is no accident that insects rarely live more than a year. Even when they live longer it is usually the grub stage which is prolonged: the adults' mental structure is still short-lived. This contrast comes to a head in the May fly, which lives as larva for years, but as adult only for a day. It is equally no accident that higher mammals on the whole live longer than lower types, and that their life span and their continuous mental development may extend over decades. This links up with another important subject—the role of communication and language in evolution.

The Austrian biologist, Von Frisch, has unearthed the secrets of the language of bees, and most extraordinary they are, enabling the animals to signify to their followers the type, abundance, distance, and direction of a source of nectar or pollen that they have discovered. This they do by variations in the elaborate dances they execute on returning to the hive. But bee language differs from human language or any communication system found in higher mammals in that it does not have to be learnt; and from any communication system found in higher vertebrates because it is for adults only. It does not have to be learnt because bees have no learning period: their language, like their instincts, has to be ready-made; it has to consist of genetically determined releasers, releasing genetically determined reactions. And it is for adults only because the young bees are mere limbless eyeless grubs, with which no real communication is possible. Human language, on the other hand, has its roots in the need for communication between individual parent and individual offspring in higher mammals, where there is a long period of dependence. When the father as well as the mother is drawn into the business of caring for the young, the unit of communication becomes the family; and where, as in apes, the family becomes the extended family group, communication becomes a comprehensive social function, at one and the same time an instrument for education, a vehicle of love, and an organ of group solidarity. This, at any rate, seems to have been the method by which the distinctive human system of communication actually arose.

The curious fact about apes is that they can learn a great variety of tricks, but not how to speak. With great difficulty, chimpanzees have been trained to say a few simple words, and to associate them broadly with certain situations; but they can neither acquire a large vocabulary by imitation, like parrots, nor learn to attach precise meanings to words. They have a rich and elaborate system of vocal sounds, which they use to express and communicate their feelings; but they have no words for things, no way of describing objects or situations. Their incapacity for true speech seems to be due in part to an inadequacy of the motor mechanisms of vocalization, in part to an inadequate faculty of forming concepts. . . .

The attainment of a high degree of complexity in behavior often has unexpected by-products and consequences. Higher vertebrates are capable of doing all sorts of things which they never actually do in their normal lives. Here is one surprising example. Some birds at least have a number sense and can count up to six or more, though they seem never to exercise that faculty in nature. Professor Otto Koehler set jackdaws the problem of taking a definite number of peas out of a series of boxes. Usually they mastered this problem fairly easily, but sometimes they made mistakes: and one jackdaw realized his mistake. He ought to have taken six peas—two out of the first box, then none, one, two and one. He went back to his cage after taking only five. But then he suddenly came back and counted out his task by bowing his head the right number of times in front of each box. When he got to five, he went on to the next box and picked up and ate the pea he had forgotten. The main reason why men can count better than jackdaws is that they have invented symbols as tools to count with, in place of merely repeating physical gestures.

An even more relevant example is that of the higher apes. Chimpanzees are capable of behaving in many human ways if they are placed in human situations. They enjoy the learning of all kinds of tricks, like driving a miniature motorcar, or riding a one-wheeled cycle. Indeed, the more difficult the trick the more they seem to enjoy performing it. They enjoy tobacco and alcohol, though they never could do this in nature. They can develop irrational fears and phobias. They can become a prey to mental illness, not only to neurosis but also psychosis. In this and many other ways they foreshadow human possibilities—both good and bad. But these possibilities were never realized in nature, because their mental organization had not undergone the final step of improvement. Apes have constructed for themselves a spatial world very nearly like ours; they sometimes show real insight; they can even organize some of their experience into concepts. But they have not yet built a symbolic world. For that, there was needed the enlargement of the association areas of the cerebral cortex.

In man's mental organization the two crucial novelties are speech and the creation of a common pool of organized experience for a group. . . . Man's language is unique in consisting of words—words for things and ideas instead of sounds or actions signifying a situation. Words, in fact, are symbols instead of signs. They are artificial constructions, tools for dealing more efficiently with the business of existence; so that language is properly speaking a branch of technology. Words are tools for thinking. Chimpanzees can construct some sort of concepts; but conceptual thought only became efficient and productive with the aid of proper tools, in the shape of verbal symbols. Like all tools, words need skill for their use. Human language is thus not merely a collection of words, but an elaborate technique. It is in fact the most complicated kind of skill in existence. The language of bees is a wonderful product of evolution; but in comparison with any human language it is as elementary as a mousetrap compared with a power station, or a primitive abacus as compared with an electronic calculating machine.

Words may be good or bad tools; some outgrow their usefulness, others have to be invented to fill new needs. But the detailed imperfections of words must not bind us to the unique value of words in general. Readers of Helen Keller's autobiography will remember the moving passage when the little creature, blind, deaf and dumb since the age of a year and half, suddenly realized that, as she put it, "everything has a name." She had already been taught various associations, had learnt to use simple signs, to recognize individual people and places, and to respond to some of the finger language of her teacher. But this day, when those fingers spelt out w-a-t-e-r on one of her hands, while her other hand was held under a spout of water in the well house, she realized that this particular combination of finger signs *"meant* the wonderful cool something" that she was feeling. As Professor Suzanne Langer writes in *Philosophy in a New Key*, it was no longer just a sign of wanting or of expecting water; it was a *name*, by which this substance could be mentioned, conceived, remembered, and thought about—a conceptual symbol. This was her first revelation of the meaning of things; it freed her from the prison of her frustrated and under-developed selfhood, and rapidly admitted her to a share in the possibilities of human existence.

Verbal language was perhaps the greatest technical invention of living substance. It enables human beings to communicate and share with each other, and in so doing it automatically gives rise to the second major uniqueness of man—a common pool of experience for a group. This is not a pool in the sense of a static water tank. It is something which can grow and develop. The pooled experience is organized, and its organization changes and evolves with time.

Nothing of the sort exists in any other organism. It provides a new kind of environment for life to inhabit. It needs a name of its own: following Père Teilhard de Chardin, the French paleontologist and philosopher, I shall call it the *nöosphere*, the world of mind. As fish swim in the sea and birds fly through the air, so we think and feel our way through this collective mental world. Our life is a voyage of exploration through its vast and varied landscape; as with all other kinds of exploration, hard work and passion and discipline are needed for success. Each one of us can only explore a limited area in any detail, but we can arrive at an idea of the whole, just as we can have an idea of the earth as a globe without physically journeying over all its surface. Only by exploring it and utilizing its resources can a man achieve the dual task of building a self and transcending the self that he has built. It is a world of possibilities, not merely of actualities. Though jackdaws do not usually practice the elementary mathematical art of counting, their mental organization makes it possible for them to do so when the opportunity is provided. In the same sort of way, primitive man did not practice higher mathematics. He did not even dream of its possibility: but it was an inherent potentiality of his mental organization. The difference between man and bird is that abundance of time was needed for its realization, as well as opportunity.

The great complexity of human mental organization gives it an enormous range and depth of new consequential possibilities. And evolution in the human phase is essentially the adventurous and stormy story of the emergence of more of these possibilities into actuality.

Editor's Note. Recent research in the biochemical and electrical properties of the brain is bringing new knowledge of the physiological nature of mental processes, knowledge which may also add to our understanding of the evolution of mind.

The Modern Search for the Engram *
by James V. McConnell (1968)

Man can learn more, and remember better, than any other animal—indeed, his prodigious memory is one of the chief attributes that sets man apart from the rest of the animal kingdom. That a ninety-year-old man may recall the halcyon days of his youth with surprising clarity is a com-

* From James V. McConnell, "The Modern Search for the Engram," *The Mind: Biological Approaches to Its Functions,* William C. Corning and Martin Balaban, editors, John Wiley & Sons, New York. Copyright © 1968 by John Wiley & Sons. Reprinted by permission. This is an updated version of an article that originally appeared in "n + m" (Science and Medicine) published by the Pharmaceutical Company, Boehringer-Mannhein, Germany.

monplace occurrence; yet, when one stops to think of it, how can this miracle be accomplished? How can a fragment of conversation heard 40, 60, or even 80 years ago still be remembered word for word after such a length of time? What kind of recording does the brain make that allows it to play back those dead voices years afterwards? Of what are our memories made, and where and how are they stored inside our bodies? Indeed, does it make sense to speak of "memory storage" at all, as if a memory were a physical quantity of some kind that can be stashed away in a cupboard or a refrigerator? The search for answers to these questions is one of the most challenging and fascinating undertakings in the biosciences today. But to understand the viewpoint that modern psychologists and biochemists take towards memory research, we must first place the subject in its proper historical perspective.

The early Greek philosophers believed that man was born with a mind like a blank tablet and that the fingers of experience "wrote" on the tablet much as a secretary copies down dictation in her notebook. When we "remembered," then, we merely searched through the mind's library of old notebooks until we found what it was we were hunting for. And indeed, when we are attempting to recall an old friend's name, or a lost fragment of poetry learned years ago, it does seem as if our minds flip through a catalog of past events. But such a viewpoint towards remembering is a limited one at best. To begin with, some memories are automatic—we don't have to "search" for them at all, for the external environment triggers off the proper bit of recall almost mechanically. In the second place, much of our remembering takes place at a totally unconscious level—we don't have to tell our hands and feet to press on the clutch and shift the gears in our automobiles; our feet and hands make the proper motions without our consciously directing them, almost as if "they had a mind of their own." The mentalistic "notebook" idea of memory, a viewpoint that science espoused for almost 20 centuries, was finally abandoned because it taught us precious little about memorial processes that common sense hadn't told us anyhow.

The real revolution in the study of memory began in Germany in the middle of the nineteenth century with the work of Hermann Ebbinghaus, who performed the first really scientific experiments on how men learned. Ebbinghaus' monumental handbook, *Ueber das Gedaechnis*, is still considered a classic by modern psychologists. But Ebbinghaus was interested in the external conditions that made remembering easy or difficult; his theoretical explanations of *why* men remembered as they did were tinged with mentalism. He did not try to look inside the human body to see what sorts of changes took place inside whenever someone learned.

The behaviorist revolution, which began in America with the work of

John B. Watson, was the next step forward. Watson saw very clearly that man often tends to attribute to himself almost magical mentalistic properties that he denies exist in animals. Desiring to get rid of magic and substitute science in its place, Watson espoused the study of animal behavior, since one could be much more objective about the way a white rat learned a maze than, say, about the way a pretty girl learns a popular song. But Watson's way of tossing out the magic of mentalism was to deny that anything of importance took place inside the body when a person learned. In Watson's theory, all animals became "empty organisms." If one but knew all of an animal's past behavior, and if one knew the present stimulus conditions the animal faced, one should be able to predict the animal's behavior completely. Watson then looked for correlations between stimuli and responses without ever taking into account the organism that sensed the stimuli and made the responses. The purely behavioristic viewpoint was a powerful one, for under the guidance of such brilliant theorists as B. F. Skinner at Harvard, we have learned a great deal about how to *control* the behavior of organisms in many rather complicated situations. But the behaviorists' stimulus–response correlations have only limited generality; they predict very poorly what animals will do when faced with new stimuli, and the correlations tell us nothing at all about *why* memory takes place, why some animals are better at remembering than others, why we forget when we're badly shocked, or a dozen other things.

Learning and remembering are such natural things to most of us that perhaps it is difficult for us to realize that these processes must have a physical representation inside our bodies. We see ourselves, like behaviorists, from the outside in. We do not sense the flow of blood through our veins, the movement of fluid in our spinal cord, nor the pulsation of electrical energy coursing through our nervous systems. Yet these processes exist—if ever they stopped, we would die almost at once. Life is a physiological process, and memory is a part of life. When we learn, there must be some actual, identifiable alteration in some part of our body; otherwise, we could not, at some later date, reconstruct with mere nervous energy that environmental condition we had experienced in the past.

At about the same time that the behaviorist revolution was beginning in America, when John B. Watson was telling his students that "mind doesn't exist," the noted Russian physiologist, Ivan Pavlov, was telling *his* students that "mind" must be translated into neurological terms to make sense. For Pavlov, the cerebral cortex, that massive envelope of nerve cells covering the entire brain, was the seat of learning. Pavlov thought that learning was a process, a rechanneling of the flow of nervous excitation from an old pathway into a new. What the structural changes were in the nervous system that caused the change of pathways was something neither Pavlov nor the neurophysiologists who followed him were able to determine. Physiological psychologists have, for half a century now, attempted to determine just exactly what the neurological changes are that take place within an organism when it learns; but for all their delicate and sensitive electronic gear, their surgical insults and their probing microelectrodes, they have failed almost completely. It is this failure of the physiologists to specify precisely where or how memory storage takes place, coupled with our recently gained knowledge of the biochemistry of the cell, that has led many of today's scientists to search for the "locus of memory" at a chemical rather than a purely physiological level.

It was in 1904 that a scientist named Richard Semon made a radical assumption. Suppose, he said, that each stimulus an organism experiences leaves a discrete material trace of some kind within the organism's nervous system. Semon went further—this material trace, which he called an "engram," he saw as being perhaps a chemical rather than a physiological entity. Each unique stimulus encountered by an organism, each thing that it learned, was to leave its own particular trace, its own unique engram. Many years later, the great American psychologist, K. S. Lashley, devoted an enormous amount of research time to what he called "the search for the engram." If the engram was unique, special, particular, then one ought to be able to localize the engram for a particular bit of learning in a specific place within the nervous system. Lashley taught rats and other animals to solve various problems, then cut out huge portions of the animals' brains, hoping to remove the engram along with the ablated tissue. Lashley never located the engram, for his animals showed only "relative" forgetting. If he removed 50% of their cortex, they forgot about half the problem; if he removed but 25% of the cortex, they remembered 75% or more of what they had learned. These results led Lashley to conclude that a single memory was represented in the brain by a diffuse *process* of some kind rather than stored in one discrete locus. What Lashley never realized, of course, was that an animal that learns even a "simple" problem actually learns thousands and thousands of things about the environment in which the experimenter puts it. A rat in a simple maze learns not only that it must turn right if it is to get to food, but also that the maze is of a certain color, certain texture, certain height, smell, width, and length; that the food is of a certain type and certain quantity; and a thousand other things. The rat must learn to "trust" the experimenter (that is, to become habituated to the experimental situation) or it will not eat, will not learn. All of these thousands of individual micro-memories go to make up a single "learning experience." Some of the memories are visual, some auditory, some kinesthetic, some olfactory; obviously the engrams would be widely scattered in various parts of the brain, and if one removed portions of the cortex at random, one would expect to obtain the results that Lashley did. And Lashley was searching for a neurophysiological rather than a biochemical locus of learning; apparently it never occurred to him that the engram might possibly be stored molecularly, as a complete entity, *within* a single cell.

Let us assume for the moment that each time an organism learns, a complicated biochemical change of some kind takes place within at least one nerve cell in the organism's brain. What chemicals might be involved, what chemicals might act as storage mechanisms? In the late 1940's Ward Halstead, a psychologist at the University of Chicago, theorized that the engram might exist as a molecular change in the nucleic acids found within a single cell. It was not until almost a decade later, however, that the Swedish biologist Holger Hydén [1] undertook a series of experiments that first implicated ribonucleic acid (RNA) as being the "memory molecule." The genes found in every cell are composed of a complicated molecule known as DNA, which acts as the commander-in-chief of all physiological processes in the cell. DNA directs the internal life of the cell by manufacturing RNA, a closely related nucleic acid molecule, which passes from the nucleus into the cytoplasm of the cell, where it directs the synthesis of proteins (among other things). DNA seems to contain the "genetic code." When the sperm and egg from a pair of white rats unite, the resulting embryo develops into a baby rat rather than into a baby elephant because the DNA in the rat sperm and egg contains a different "code" from that of the DNA found in elephant sperm and eggs.

From one point of view, then, we may say that DNA stores an organism's "ancestral memories" in coded form; that is, DNA "remembers" what an organism's progenitors were like. Professor Hydén theorized that RNA might encode or "remember" an organism's own personal memories, that RNA might well be the "tablet" on which the fingers of experience wrote (by changing the chemical "code" carried by the RNA molecule). To test his theory, Hydén performed experiments with rats and rabbits. First, Hydén developed a beautiful technique for taking large single cells from the nervous system, cutting them open individually by hand with an incredibly tiny scalpel, and scraping out the protoplasm inside each cell. With microanalytic techniques, he was then able to measure rather subtle changes in the RNA found in these cells. Then he took two groups of rats to work with. The first group was trained to balance on a taut wire in order to reach food; the second group was given passive ex-

1. See Hydén, H. and Egyházi, E. (1962). *Proc. N.A.S.*, 48(8): 1366. Hydén, H. and Egyházi, E. (1963). *Proc. N.A.S.*, 49: 618. Hydén, H. and Egyházi, E. (1964). *Proc. N.A.S.*, 52(4): 1030.

ercise but did not learn the balancing trick. Hydén found that the gross amount of RNA increased markedly in cells taken from both groups of animals, but qualitative changes in the RNA (that is, apparent changes in the "code" itself) took place only in the nuclear RNA taken from the trained animals. Hydén later found that the gross amount of RNA in the human brain increases from birth until age 40, remains at a relatively constant, high level until age 60, then rapidly begins to decline. These findings led Hydén to speculate that memories were stored by means of physical changes in the RNA molecule itself. Hydén's theory, which has actually never been proved or disproved, has sparked much of the recent but growing interest in the biochemistry of memory and has prompted dozens of other scientists to undertake related experiments on a wide variety of organisms ranging from the flatworm to the human.

PLANARIAN AS AN EXPERIMENTAL ANIMAL

Anyone at all familiar with the planarian, or common freshwater flatworm, might well wonder why a psychologist would pick this animal to experiment with. Seldom growing to more than an inch in length, it is a fragile creature that inhabits streams, ponds, lakes, and rivers throughout the world. The planarian is often difficult to maintain for long periods of time in a laboratory situation, and its nervous system is exceedingly primitive—so primitive indeed that its ability to learn even the simplest of tasks might be questioned on purely theoretical grounds. Yet the flatworm does have a rudimentary brain; its behavior is much more complicated than one might imagine at first glance, and it possesses certain other rather special characteristics that make it a most useful tool in the search for the biochemical correlates of memory.

The first time that I personally encountered a flatworm was in 1953, when I was a graduate student working towards my doctorate in experimental psychology at the University of Texas. At that time, psychologists in general were just becoming aware of the neurophysiological learning theories of D. O. Hebb, the Canadian psychologist, and of Sir John Eccles, the Nobel Prize winning physiologist from Australia. These two scientists had speculated that learning was a matter of physiological change at the synapse—that is, they believed that whenever an organism learns, some structural alteration must take place in its nervous system, probably at the junction point between two nerve cells which is called a synapse. Obviously, if Hebb and Eccles were correct, an organism would have to possess synapses in order to learn. Robert Thompson . . . suggested to me that since the flatworm is the lowest animal on the phyletic tree that possesses the type of synapse Hebb and Eccles had in mind, we might test their physiological learning theories by attempting to train a planarian. And so, working together, we set out to do just that. . . .

Our first training apparatus . . . consisted of a foot-long, half-inch-wide plastic tube that could be filled with water. Electrodes mounted at each end of the tube allowed us to pass an electric current through the water. A pair of lamps suspended above the trough provided the light which served as the conditioning stimulus. When a planarian was crawling smoothly along the bottom of the trough and we turned on the electric current, the animal would always give a vigorous contraction of its entire body. We trained the animals by pairing the light with the shock. First we turned on the light for two seconds; then we turned on the shock for an additional second; then we turned off both the light and the shock. At the beginning of training, the worms responded to the light (before the shock came on) no more than 25% of the time. After we had paired the light and the shock for 150 trials (all given in a period of two hours or so), the worms responded to the light 50% of the time, just as if they had now learned that the light was a signal that shock was coming. A group of animals given 150 trials of light without shock showed the expected phenomenon of habituation—that is, they responded less and less often to the light during training. Another group of worms, given 150 "shock-only" trials, also showed a decreased sensitivity to occasional test trials of light (never paired with the shock). Since the increased responsivity in our experimental animals was quite statistically significant, and since our various control groups showed us that this increase was presumably not due to sensitization, we were convinced that we had shown clear-cut classical conditioning in planarians. . . . In 1956, when I became an instructor at the University of Michigan, I set up my own laboratory and continued the planarian research.

To most people, I suppose, the flatworm is famous not because it is the lowest animal with a rudimentary brain and a true synaptic type of nervous system, but rather because it possesses enormous powers of regeneration. The species of planarian we most often use in our experiments is *Dugesia dorotocephala.* If one cuts this animal in half transversely, the animal, does not die; rather, the head grows a new tail and the tail grows a new head, and each regenerated piece will eventually become as large as the original organism. Indeed, under the best of conditions, one may cut this animal into as many as 50 pieces, and each section will regenerate into an intact, fully functioning organism.

It had occurred to Thompson and me that it might be interesting to train a planarian, then cut it in half and test the regenerated sections to see which portion (if any) would retain the memory of the training; Thompson and I never got to this study at Texas, however. But once my laboratory was set up at Michigan, two of my students (Allan Jacobson and Daniel Kimble) and I undertook that very study.

Giving planarians 150 trials all in a period of two hours is not a very effective way of training them; flatworms, like humans and rats, will usually learn much better if the trials are stretched out over a period of several days. By spacing the trials, we were able to get each of our worms to respond to the onset of the light at least 92% of the time. When each animal reached this criterion, we cut it in half and allowed it to regenerate for a period of a month. Then we retrained each regenerated half to the same 92% response level to determine whether it learned faster the second time than the first (this so-called "method of savings" was one of the pioneering methodological contributions of Hermann Ebbinghaus). We had expected the heads to remember at least some part of what they had been taught—after all, the head section retained the brain and most of the nervous system of the original animal. Our expectation was confirmed—the head regenerates remembered just as well as did planarians that were trained but not cut in half. But what of the tails? They had to regrow an entire new brain and replace most of their vital organs during the month's regeneration. How could one expect a regenerated tail to remember anything at all? The answer is, of course, that one could *not* expect the tails to remember at all—but they did. In fact, many of the tails showed almost perfect retention of the original training! It seemed as if their new brains were created with the old "learning" already "wired in," much the same way that an innate behavior pattern ("instinct") is "wired in" from birth. Furthermore, we soon found that if we cut the trained worm in thirds, or even fourths, each regenerated piece would show significant retention of the memory.

Our "retention following regeneration" results [2] were at first considered somewhat unbelievable. Since then, however, the study has been repeated in dozens of different laboratories all around the world. The results are apparently valid—but the nagging question remains, how could the tail remember anything at all?

As I thought about the question back in 1958, it occurred to me that the "engram" must be stored throughout the planarian's body, not just in the animal's brain, and that the storage mechanism probably was chemical in nature. Indeed, it seemed obvious that the same engram was probably duplicated hundreds of times throughout the animal's body, being stored within hundreds of individual cells. But what was the chemical (or chemicals) involved in memory formation? It was at this time that I became aware of Hydén's RNA theory of memory. Now, the cell that I thought

2. McConnell, J. V., Jacobson, A. L., and Kimble, D. P. (1959). *J. comp. physiol. Psychol,* 52: 1.

was probably involved in memory storage in the planarian was the neo-blast—a primitive, undifferentiated cell that can migrate anywhere in the animal's body and that has the property of being able to develop into any other type of cell whatsoever. According to such noted scientists as Prof. Etienne Wolfe of the College du France and Prof. H. V. Brondsted of the University of Copenhagen, the neoblast is the cell which is primarily responsible for the planarian's powers of regeneration. When the animal is cut in half, the neoblasts seem to migrate to the site of the wound and form a bud or blastema from which the missing parts of the animal are regenerated. The neoblast is also very rich in RNA, a fact that seemed to support my hypothesis that this primitive cell might act as the site of the engram in planarians.

The first experimental test of the RNA theory of memory with planarians was, however, performed by Prof. E. Roy John and his student, William Corning, at the University of Rochester. John and Corning[3] reasoned that if they would somehow destroy or alter the RNA in a trained planarian's cells, they would also destroy or alter the memory itself. Therefore, in 1960, they trained flatworms using the light-shock conditioning technique we had developed and, once the animals had reached criterion, cut them in half. Some of the animals were allowed to regenerate in the usual pond water; others were forced to regenerate in a weak solution of ribonuclease (an enzyme that hydrolyzes or breaks up RNA). They reported that both heads and tails that regenerated in pond water showed the usual retention, as did the heads that regenerated in ribonuclease. The tails that regenerated in the ribonuclease, however, showed almost complete forgetting of the original training, much as if the enzyme had somehow "erased the blackboard" on which the memory was written. A study by Fried and Horowitz[4] at UCLA indicates that much the same result obtains when one injects ribonuclease directly into the body cavity of trained planarians. Ribonuclease is a very powerful destructive agent, however, and there is still some question whether these studies show that the ribonuclease affected specifically the memory storage or whether the "forgetting" was due to a more general, debilitating effect on the animals' health.

While Corning and John were working with ribonuclease, we attacked the RNA hypothesis from rather a different direction. It had been known for a great many years that planarians lacked any "immune" reaction, and that one could often graft a small portion of one animal's body onto the body of another planarian. This non-specificity of planarian tissue sug-

3. Corning, W. C. and John, E. R. (1961). *Sci.* 134: 1363.
4. Fried, C. and Horowitz, S. (1964). *Worm Runner's Digest*, 6(2): 3, Planarian Research Group, Mental Health Research Institute, University of Michigan.

gested that we might be able to transfer engrams from one animal to another if we could somehow get chemicals from a trained animal into the body of a untrained planarian. Our first attempts at grafting large portions of one animal onto another were not particularly successful, however, and so we turned to what was a simpler but perhaps a more spectacular type of tissue transfer, that of "cannibalistic ingestion."

Many species of planarians, when hungry, cannibalize quite readily. The species we have used most often, *Dugesia dorotocephala*, is a particularly voracious cannibal. Beginning in 1960, my students and I conducted a series of experiments involving what we now call "the cannibalistic transfer of training" in planarians. First we trained some "victim" worms to criterion (using the now-standard light-shock conditioning technique). We then fed the trained victims to starved, untrained cannibals. At the same time, a set of untrained "victim" worms was fed to a different group of cannibals. Both sets of cannibals were then given light-shock training. The results were clear-cut: the cannibals that had ingested "trained" victims were, on the very first day of training, significantly superior to the cannibals that had eaten untrained animals.

We repeated the cannibalism study four times over, employing each time a "blind" testing technique (in which the person actually doing the training does not know the prior histories of any of the animals he trains). . . . By early 1965, some type of cannibalistic transfer in planarians had been reported by at least 10 other university laboratories (and by hundreds of American high school students working on what we call "Science Fair" projects).

Although we now jokingly refer to these studies as confirming what we call our "Mau Mau" hypothesis, and although they have been widely replicated, one obviously should not generalize these results to the human level too readily. To begin with, the planarian has a very simple digestive system; it lacks a true stomach and does not secrete the massive array of acids and enzymes necessary for digestion in higher organisms. RNA would surely stand little chance of being absorbed in an unaltered form in higher organisms. And in the second place, we still do not know just what it is that actually gets transferred via ingestion—whether the cannibal acquires a specific memory, or perhaps just a general response tendency, or perhaps both. However, there is a growing body of evidence suggesting that, at least under certain conditions, it is a particular and specific engram that gets passed from one worm to another.

No matter what it is that gets transferred by ingestion, however, we were still left with the problem of finding what the transfer mechanism itself is. If RNA were really involved, we obviously should be able to achieve the transfer if we trained some planarians, then extracted the

RNA from these animals and injected it into untrained planarians. We began on this problem in 1962 and, with the support of grants from the U.S. Atomic Energy Commission and the National Institute of Mental Health, are still continuing our studies today. . . .

In our first experiment, we had some 500 experimental planarians given the usual light-shock training (in a group), plus 500 animals in each of several control groups. After training was complete, we extracted the RNA from each group and injected it into untrained planarians and then tested the injectees in our usual "blind" fashion. The animals injected with RNA from the trained planarians were significantly better than any of the animals injected with "control" RNA. . . . Since 1964, there have been reported more than a dozen successful attempts to achieve the same sort of transfer in rats, mice, and other higher organisms. . . .

At the moment, it seems safe to conclude that RNA is intimately involved in the storage of memories, but it is obviously far too soon to conclude that RNA is indeed the "memory molecule" itself. As Dingman and Sporn [5] have pointed out, it is likely that RNA is but one small link in a complicated chain of cellular processes that, taken all together, functions to store our "remembrance of things past." It is my own opinion that RNA may play two distinct and somewhat different roles in memory formation. First, since Hydén's experiments indicate that RNA is quite sensitive to changes in the neuron's external environment, it may be that incoming sensory signals (that is, an organism's experiences) are "coded" first of all by physico-chemical changes in RNA molecules. The altered RNA would then produce different types of proteins than before the message was encoded and the neuron would function differently than before. But second, the "transfer" studies on planarians suggest that RNA may occasionally serve to transport coded information from one part of the body to another. . . .

Let me conclude by pushing my speculations to perhaps an incredible limit. If "coded" RNA is indeed Nature's "transfer agent," and if we can discover what that code is, we may eventually be able to implant some types of knowledge directly into a human brain by injection of suitable "memory molecules." Whether or not the day of "instant learning" ever dawns remains to be seen. But it is certain that the vistas opened by research on the biochemistry of memory will be fascinating ones for all of us to explore.

5. Dingman, W. and Sporn, M. B. (1961). *J. psychiat. Res.*, 1: 1.

The Chemistry of Learning [*]
by David Krech (1968)

Most adults who are not senile can repeat a series of seven numbers—8, 4, 8, 8, 3, 9, 9—immediately after the series is read. If, however, they are asked to repeat these numbers thirty minutes later, most will fail. In the first instance, we are dealing with the immediate memory span; in the second, with long-term memory. These basic behavioral observations lie behind what is called the two-stage memory storage process theory.

According to a common variant of these notions, immediately after every learning trial—indeed, after every experience—a short-lived electrochemical process is established in the brain. This process, so goes the assumption, is the physiological mechanism which carries the short-term memory. Within a few seconds or minutes, however, this process decays and disappears; but before doing so, if all systems are go, the short-term electrochemical process triggers a second series of events in the brain. This second process is chemical in nature and involves, primarily, the production of new proteins and the induction of higher enzymatic [1] activity levels in the brain cells. This process is more enduring and serves as the physiological substrate of our long-term memory. . . .

Dr. James L. McGaugh of the University of California at Riverside has argued that injections of central nervous system stimulants such a strychnine, picrotoxin, or metrazol should enhance, fortify, or extend the activity of the short-term electrochemical memory processes and thus increase the probability that they will be successful in initiating long-term memory processes. From this it follows that the injection of CNS stimulants immediately before or after training should improve learning performance. That is precisely what McGaugh found—together with several additional results which have important implications for our concerns today.

In one of McGaugh's most revealing experiments, eight groups of mice

[*] From David Krech, *Saturday Review*, January 20, 1968. Copyright Saturday Review, Inc., 1967. Reprinted by permission of Saturday Review, Inc. and the author. (This article is adapted from a speech to Three National Seminars on Innovation, sponsored by the U.S. Office of Education and the Charles F. Kettering Foundation, in Honolulu, July 1967.)

1. An enzyme is a protein which controls the rate of a chemical reaction in the living cell. The manufacture of each enzyme is believed to be controlled by a gene. Thus genes direct the chemical processes in the body through the enzymes. Ed.

from two different hereditary backgrounds were given the problem of learning a simple maze. Immediately after completing their learning trials, four groups from each strain were injected with a different dosage of metrazol—from none to five, 10, and 20 milligrams per kilogram of body weight. First, it was apparent that there are hereditary differences in learning ability—a relatively bright strain and a relatively dull one. Secondly, by properly dosing the animals with metrazol, the learning performance increased appreciably. Under the optimal dosage, the metrazol animals showed about a 40 per cent improvement in learning ability over their untreated brothers. The improvement under metrazol was so great, in fact, that the dull animals, when treated with 10 milligrams, did slightly better than their untreated but hereditarily superior colleagues.

In metrazol we not only have a chemical facilitator of learning, but one which acts as the "Great Equalizer" among hereditarily different groups. As the dosage was increased for the dull mice from none to five to 10 milligrams their performance improved. Beyond the 10-milligram point for the dull mice, however, and beyond the five-milligram point for the bright mice, increased strength of the metrazol solution resulted in a deterioration in learning. We can draw two morals from this last finding. First, the optimal dosage of chemical learning facilitators will vary greatly with the individual taking the drug (there is, in other words, an interaction between heredity and drugs); second, there is a limit to the intellectual power of even a hopped-up Southern Californian Super Mouse!

We already have available a fairly extensive class of drugs which can facilitate learning and memory in animals. A closer examination of McGaugh's results and the work of others, however, also suggests that these drugs do not work in a monolithic manner on something called "learning" or "memory." In some instances, the drugs seem to act on "attentiveness"; in some, on the ability to vary one's attacks on a problem; in some, on persistence; in some, on immediate memory; in some, on long-term memory. Different drugs work differentially for different strains, different individuals, different intellectual tasks, and different learning components.

Do all of these results mean that we will soon be able to substitute a pharmacopoeia of drugs for our various school-enrichment and innovative educational programs, and that most educators will soon be technologically unemployed—or will have to retool and turn in their schoolmaster's gown for a pharmacist's jacket? The answer is no—as our Berkeley experiments on the influence of education and training on brain anatomy and chemistry suggest. This research is the work of four—Dr. E. L. Bennett, biochemist; Dr. Marian Diamond, anatomist; Dr. M. R. Rosenzweig, psychologist; and myself—together, of course, with the help of graduate students, technicians, and, above all, government money.

Our work, started some fifteen years ago, was guided by the same general theory which has guided more recent work, but our research strategy and tactics were quite different. Instead of interfering physiologically or chemically with the animal to determine the effects of such intervention upon memory storage (as did Jarvik, Agranoff, and McGaugh), we had taken the obverse question and, working with only normal animals, sought to determine the *effects* of memory storage on the chemistry and anatomy of the brain.

Our argument was this: If the establishment of long-term memory processes involves increased activity of brain enzymes, then animals which have been required to do a great deal of learning and remembering should end up with brains enzymatically different from those of animals which have not been so challenged by environment. This should be especially true for the enzymes involved in trans-synaptic neural activity. Further, since such neural activity would make demands on brain-cell action and metabolism, one might also expect to find various morphological differences between the brains of rats brought up in psychologically stimulating and psychologically pallid environments.

I describe briefly one of our standard experiments. At weaning age, one rat from each of a dozen pairs of male twins is chosen by lot to be placed in an educationally active and innovative environment, while its twin brother is placed in as unstimulating an environment as we can contrive. All twelve educationally enriched rats live together in one large, wire-mesh cage in a well lighted, noisy, and busy laboratory. The cage is equipped with ladders, running wheels, and other "creative" rat toys. For thirty minutes each day, the rats are taken out of their cage and allowed to explore new territory. As the rats grow older they are given various learning tasks to master, for which they are rewarded with bits of sugar. This stimulating educational and training program is continued for eighty days.

While these animals are enjoying their rich intellectual environment, each impoverished animal lives out his life in solitary confinement, in a small cage situated in a dimly lit and quiet room. He is rarely handled by his keeper and never invited to explore new environments, to solve problems, or join in games with other rats. Both groups of rats, however, have unlimited access to the same standard food throughout the experiment. At the age of 105 days, the rats are sacrificed, their brains dissected out and analyzed morphologically and chemically.

This standard experiment, repeated dozens of times, indicates that as the fortunate rat lives out his life in the educationally enriched condition, the bulk of his cortex expands and grows deeper and heavier than that of his culturally deprived brother. Part of this increase in cortical mass is accounted for by an increase in the number of glia cells (specialized brain

Figure 5–2. As the fortunate rat lives out his life in the educationally enriched condition, the bulk of his cortex expands and grows deeper and heavier than that of his culturally deprived brother. . . . We have created superior prob-lem-solving animals. (Photograph by David Krech, University of California. Reprinted by permission.)

cells which play vital functions in the nutrition of the neurons and, per-haps, also in laying down permanent memory traces); part of it by an in-crease in the size of the neuronal cell bodies and their nuclei; and part by an increase in the diameters of the blood vessels supplying the cortex. Our postulated chemical changes also occur. The enriched brain shows more acetylcholinesterase (the enzyme involved in the trans-synaptic conduc-tion of neural impulses) and cholinesterase (the enzyme found primarily in the glia cells).

Finally, in another series of experiments we have demonstrated that these structural and chemical changes are the signs of a "good" brain. That is, we have shown that either through early rat-type Head Start pro-grams or through selective breeding programs, we can increase the weight and density of the rat's cortex and its acetylcholinesterase and cholinesterase activity levels. And when we do—by either method—we have created superior problem-solving animals.

What does all of this mean? It means that the effects of the psychological and educational environment are not restricted to something called the "mental" realm. Permitting the young rat to grow up in an educationally and experientially inadequate and unstimulating environment creates an animal with a relatively deteriorated brain—a brain with a thin and light cortex, lowered blood supply, diminished enzymatic activities, smaller neuronal cell bodies, and fewer glia cells. A lack of adequate educational fare for the young animal—no matter how large the food supply or how good the family—and a lack of adequate psychological enrichment results in palpable, measurable, deteriorative changes in the brain's chemistry and anatomy.

Returning to McGaugh's results, we find that whether, and to what extent, this or that drug will improve the animal's learning ability will depend, of course, on what the drug does to the rat's brain chemistry. And what it does to the rat's brain chemistry will depend upon the status of the chemistry in the brain to begin with. And what the status of the brain's chemistry is to begin with reflects the rat's early psychological and educational environment. Whether, and to what extent, this or that drug will improve the animal's attention, or memory, or learning ability, therefore, will depend upon the animal's past experiences. I am not talking about interaction between "mental" factors on the one hand and "chemical" compounds on the other. I am talking, rather, about interactions between chemical factors introduced into the brain by the biochemist's injection or pills, and chemical factors induced in the brain by the educator's stimulating or impoverishing environment. The biochemist's work can be only half effective without the educator's help. . . .

To be sure, all our data thus far come from the brains of goldfish and rodents. But is anyone so certain that the chemistry of the brain of a rat (which, after all, is a fairly complex mammal) is so different from that of the brain of a human being that he dare neglect this challenge—or even gamble—when the stakes are so high?

Mind and the Future of Man *
by John Rowan Wilson (1964)

The future of the mind is a subject to be approached expectantly but warily. Psychologists who have thought much about it seem disposed to expect some breathtaking changes in our notions about the brain and the mind, though they are not at all agreed on the nature of the changes. New information may lead to new insights into the nature of consciousness, perhaps the most baffling of all psychological phenomena. It may also reveal a great deal more about the nature of pleasure and pain, about the possibility of raising intelligence to previously undreamed-of levels, about the possibility, straight out of science fiction, of keeping the brain alive after its possessor dies—and still more.

Our new understanding of the brain is coming through two quite different avenues. One is the recent development of extraordinarily sophisticated electronic and chemical techniques that enable us to explore the brain in microscopic detail, to pinpoint the physical sources of specific emotions and thought processes, and to arouse or depress them artificially. The second is the new science of cybernetics, which has simulated the brain in an astonishing variety of functions, creating machines that, in some sense at least, are able to think and learn and respond "humanly" to a wide variety of situations. The implications of both kinds of research are so dramatic that it is easy to understand why the men working on them are wary, as well as expectant.

The electrochemical nature of activity in the human nervous system, like any electrical activity, can in principle be measured. However, the activity in the brain is so weak that recording it has become a fairly specialized job. Altogether, the brain has about 10 billion interconnected nerve cells, and while not all of them are discharging at any one time, one electrode placed on the scalp will record anywhere from five millionths to 50 millionths of a volt. Perhaps 60,000 scalps together might supply enough voltage to light a flashlight.

The existence of electrical currents in the brain was known as long ago as 1875, when they were discovered by an English physician, Dr. Richard Caton. The existence of brain waves—i.e., *rhythmic* currents with a variety of different cycles—was known as long ago as the 1920s, principally because of the pioneering work of Hans Berger, a German neurologist who recorded the waves on paper. Later it became clear that, in measuring mental activity in a new and objective way, Berger's electroencephalographs (from the Greek *enkephalos,* or brain) might open up a large new area of exploration for students of the mind. But Berger was not taken seriously for many years. He seems to have been a rather unscientific operator, often vague and sloppy in his reports, and quite ignorant of mechanics or electricity, with the result that he misinterpreted some of his own experimental data. Eventually, however, Berger's brainwave recordings were duplicated by English scientists, and the use of electroencephalographs (EEGs) came into common use.

An electrode placed on the scalp does not pick up one simple brain wave, but many waves of differing amplitudes and frequencies. The strongest, of course, will be generated by cells in the immediate vicinity of the electrode, but mixed in with these waves will be others from nearby areas of the brain. Meanwhile, there will be a persistent "background noise," reflecting random and barely discernible discharges that may be still farther away; alternatively, they may be nearby but weak. An EEG recorder may have as many as 32 different channels, but most of those now in use have eight; i.e., eight electrodes are used, and the result of a typical recording session will be eight rather wobbly lines. Each line must be broken down into its component waves, and there may be 30 or more waves represented in each wobbly "compound curve."

A FANTASTIC DELUGE

Thus the EEG researcher is confronted from the outset with a deluge of information that is fantastically difficult to sort out. The English neurologist Dr. W. Grey Walter, has observed that "the redundancy is . . . enormous. Information at the rate of about 3,600 amplitudes per minute may be coming through each of the eight channels during the average recording period of twenty minutes; so the total information in a routine record may be represented by more than half a million numbers; yet the usual description of a record consists only of a few sentences. Only rarely does an observer use more than one-hundredth of one percent of the available information."

Confronted with "the problem of redundancy" and unable to synthesize all the data that overwhelmed them, EEG researchers for many years had some difficulty finding practical applications of their work. To be sure, they could tell when gross and unambiguous damage had been inflicted on the brain, and they did in fact contribute a great deal to our understanding of epilepsy. Their readings of normal brains, however, were not able to do much more than characterize the subject's degree of alertness.

But now computer analysis has made the "background noise" largely meaningful. The usual procedure is to take a number of different readings of a subject's EEG responses to a given stimulus. In any one reading, the "noise" is apt to drown out meaningful signals. The computer can rapidly analyze the different readings, and get an average response; i.e., the brain wave related to the response can be sorted out from all the others. Waves are now readily found even for such trivial mental operations as responding to a gentle tap on the hand or to a flicker of light.

When the potential of computers was grasped by EEG researchers, a specially designed Average Response Computer (ARC) was built for

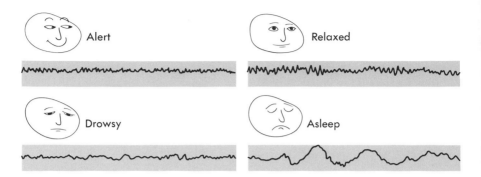

Figure 5–3. *Rhythms of mental life. Scientists have been able to correlate the electrical activity of the brain, as recorded by electroencephalograph* (EEG), *with various states of alertness in the normal human being. The brain waves illustrated above are from actual EEG records. When a subject is alert, a rapid spiky pattern appears. As a subject becomes drowsy, the EEG transcribes slower and slower electrical rhythms, which become long, rolling waves in sleep. (Reproduced by permission of Time-Life Books from* The Mind. *Copyright* © 1964 *by Time Incorporated.*)

them at the Lincoln Laboratory of the Massachusetts Institute of Technology. One of ARC's most fascinating and suggestive accomplishments involved the interpretation of data from deep within an animal's brain. A light was flashed at a constant rate into the animal's eyes. The report on the ARC oscilloscope showed that the wave involved in its response could be broken down into three distinct components. One of these three, however, gradually faded out.

The report on the experiment suggested this hypothesis: one of the relatively constant components may pass on to the visual cortex, thereby signifying to the animal that the stimulus is visual and not, say, olfactory or auditory. Perhaps the second component indicates that the stimulus is a recurrent one. The third and waning component may be signaling "unexpectedness" and, by dropping out, may carry the message that the stimulus is simply repeating over and over without change. It may be saying, in effect, that the stimulus is devoid of novelty (or information) and can be safely ignored.

CLOSER AND CLOSER TO COGNITIVE THOUGHT

EEG researchers have not yet been able to capture on paper the waves denoting any sort of involved human thought; for example, they have not been able to record the electrical discharges that might be given off by a student solving a problem in arithmetic. But if they believe they are already tracking down such concepts as "unexpectedness," they may be very close to cognitive thought.

Another extraordinary research development has made it clear that the brain's division of labor is rather more complex than anyone had previously realized. This new development, electrical stimulation of the brain (ESB), has proved, for example, that there are many different centers in the brain for both pleasure and pain. In general, the structure of the brain provides for a great deal of functional duplication. In other words, there is no individual nerve cell, or any small cluster of cells, that seems to be absolutely necessary to the performance of any given mental task, and if cells performing a given task are somehow destroyed, other cells are ordinarily able to take over.

The basic procedure in ESB research employs wires that are insulated except for an electrode at their tip. These are inserted into the brain, sometimes deeply. The wires are hooked up to sources of power, so that the electrode can pass a current into the brain at the point of contact. The effect of this stimulation is the same as if all the nerve cells reached had fired electrical discharges at the same time. If the experiment is to be a lengthy one, a terminal socket may be cemented onto the skull; to this are attached wires that stretch considerable distances, giving the subject of the experiment plenty of room to move around in. The procedure sounds frightening. It is in fact harmless and painless.

One famous series of ESB experiments on cats and monkeys showed that they could, in effect, be reduced to puppets by stimulating certain areas of the cerebral cortex. In general, the cortex, which forms the prominent convoluted outer layer of the brain, is concerned with the more intellectual of the mental processes. However, one section of the cortex governs motor responses, and by discharging currents at appropriate points within this section, the experimenters could make the animals move different parts of their bodies, or could make them "freeze" in whatever position they were already in—"like electrical toys," as Dr. José M. R. Delgado of Yale, one of the leaders in ESB research, puts it.

"TURNING ON" EMOTIONS

When Dr. Delgado and his associates at Yale pushed deeper into the brain, into the areas governing the emotions, they turned up even more startling data. The emotions are decisively affected by activity in the thalamus, in the very center of the brain; in the hypothalamus, just below it; in the limbic system, a series of structures rooted around them; and in the reticular system, a cluster of nerve cells in the brainstem, extending far toward the spinal cord. In stimulating points within these structures, it was dramatically shown, the researcher can frequently achieve quite opposite effects by moving his electrodes just a little. A current in one area of the limbic system will set off a frightful display of rage, with the stimulated animal turning ferociously on friends. Moving the electrode just a bit brings on demonstrations of affection. In the hypothalamus, Dr. Delgado found that stimulation at one point will "turn on" the animal's appetite. Stimulation at another point nearby will leave the animal totally uninterested in food, so much so that it would literally starve to death if the current were applied steadily. (Similar effects can be produced by stimulating areas of the limbic system.)

Another famous and fascinating series of experiments with ESB was carried on during the 1950s by James Olds at McGill University in Montreal. Working with rats, Olds accidentally discovered that a portion of the brain seemed to be a "pleasure center." It came about because Olds was somewhat off the mark in placing an electrode intended to shock the animal. The rat was placed in a large box in which it was free to roam about, but any time it got into one corner of the box the apparatus gave it a shock. To Olds's surprise, the rat appeared to enjoy the experience. Instead of avoiding that corner it kept returning to it repeatedly.

Researchers have also discovered the existence of what might be called "pain centers" in the brain, places where discharges are associated with agonizing pain, or with fearful or panicky reactions. Sometimes these centers may be as close as 0.02 inch from a pleasure center. The existence of both kinds of centers adjacent to one another has been found in human, as well as animal, brains.

ESB experiments on human subjects have in general been limited to those cases in which a serious ailment exists, but it is now clear that a wide range of healthy human emotions and sensations can be stimulated artificially. Anxiety, restlessness, panic and intense pain can be induced by electrical stimulation of some areas of the brain. More agreeable feelings, ranging from satisfaction to euphoria, can be induced in other nearby areas. Some of the pleasure centers seem to have the additional function of sharply reducing, or even suspending, any existing sensation of pain, at least temporarily; thus ESB promises to be extremely useful in relieving the rigors of extremely painful diseases like cancer.

Perhaps the most breathtaking of all the results achieved by ESB research are those concerned with the artificial stimulation of memory. Here the technique is to send electrical currents into parts of the cortex. This area of investigation is especially associated with Wilder Penfield. The famed neurosurgeon came upon it quite by accident, when he was experimenting with ESB treatment of some epileptic patients. Penfield was told by the patients that under treatment they suddenly recalled, with intense vividness, long-forgotten episodes out of their distant pasts. Moreover, they would continue recalling more and more about the episodes so long as the electrode providing the stimulation remained in the same place. There would seem to be a real possibility that brain researchers will some day be able to stimulate memories of events about which there is some special curiosity.

For all the remarkable discoveries it has led to, ESB research has some obvious limitations. Electrical stimulation of the brain is maddeningly imprecise; any nerve cells touched by an electrode will discharge, and the scientist trying to limit himself to the study of the appetite, let us say, is in danger of finding that he has triggered pain instead of (or along with) hunger. . . .

Our expanded understanding of the brain comes not only from chemical and electronic research but also from successful efforts to simulate the brain's operations with computers. More and more of the brain's functions have now been duplicated by computers. Like the brain, computers can of course be used to solve mathematical and other logical problems. Both brains and computers are, in effect, able to process the raw material of information they receive and to sort out what is relevant in a particular situation from what is not. Both work on electrical circuits; indeed, the brain's nerve cells are often compared to vacuum tubes, which can be turned on and off to govern the flow of electricity through a circuit. Both brains and computers have fairly elaborate procedures for making sure that errors do not occur in their calculations. The brain does this by recording the evidence of several different human senses on millions of different nerve cells, and noting whether there seem to be inconsistencies, while some computers have built-in "memory" devices to ensure that the data from past operations are consistent with more recent data. And computers are becoming more and more successful at making predictions, something the brain has been doing for a long time.

There is, to be sure, an extraordinary difference in the capacities of the brain and the computers. While all computers can solve some kinds of problems better than the brain, none of them has the brain's fantastic capacity to store information. Dr. Walter, perhaps the leading authority on duplicating brain functions mechanically, has succeeded in constructing electric models of brain cells. He has declared, however, that the cost of wiring them so that they are interconnected as cells are in the brain would be wildly uneconomical.

But Walter and others have pushed ahead with projects to produce a model of the brain that simulates many of its capabilities. Walter began by hypothesizing that cerebral functions might be duplicated with far fewer units than the total number of cells in the brain, provided only that the units used were elaborately and intricately interconnected. The final product of his and others' labors is a device christened *Machina speculatrix*, also known as *M. speculatrix*. . . . Dr. Walter speaks of this and similar machines as "creatures," and has said that devices of this kind can be constructed that will be capable of a "mimicry of life [that] will be valid and illuminating."

ARE THEY CREATURES?

It can be argued that, while such machines may indeed "mimic" life, they are not really "creatures" because they lack the essential ingredient of living creatures, i.e., a sense of consciousness. This is a matter about which students of the brain are in disagreement. Some of them feel that consciousness itself presents no particular problem to a "mechanistic" view of the brain. Dean Wooldridge, the distinguished American scientist, and one of the founders of Thompson Ramo Wooldridge, the giant electronics company, has argued in *The Machinery of the Brain* that we may soon come "to accept the sense of consciousness itself as a natural phenomenon suited to being described by and dealt with by the body of laws and methods of the physical sciences." We may find, for example, that consciousness may be a function of a specific voltage in a given nucleus of the brainstem.

It is true, of course, that by its nature consciousness is something the subject has to tell us about; consciousness is not something another person can see. This point presents a considerable difficulty for the researcher trying to gauge it without depending on another's subjective reactions. On the other hand, the point raises a startling question. Is it possible that consciousness might be duplicated in computers of the future, even if the machines could not tell us about it? Wooldridge raises this question even about existing computers: "Is it possible that, somewhere among their wires and transistors, there already stirs the dim glimmering of the same kind of sense of awareness that has become, for man, his most personal and precious possession?"

What Wooldridge asks is an electronic-age echo of a statement by Sir Charles Sherrington, who was a poet as well as a dedicated researcher of the nervous system in the early 1900s. The brain, Sir Charles decided at the age of 83, is "an enchanted loom where millions of flashing shuttles weave a dissolving pattern, always a meaningful pattern though never an abiding one."

Consciousness and Free Will

Consciousness *

by René Dubos (1962)

During recent years many physiologists have come to believe that other traits which are characteristic of man, such as self-consciousness and free will, can also be brought within the fold of natural sciences and of the theory of evolution. The understanding of feedbacks and the knowledge of servomechanisms have provided some basis for a tentative scientific interpretation of the operations of the mind. Professor Homer Smith, for example, likens consciousness "to a television tube with a so-called long decay-time such that it continues to glow for a substantial period after it has been excited, and thus affords a continuous rather than a flickering image. Within the parameters of this image the conscious animal can relate past experiences to anticipate the future and react accordingly, and with respect to one or another of various possibilities which may be presented simultaneously. Insofar as selectivity enters into this reaction we may speak of 'choice' without giving this much abused word any metaphysical implications; and insofar as any of the alternative modes of action promote the organism's welfare, we may designate the resulting activity as 'self-serving.'

"From the perspective of evolution, then, we may venture a tentative definition of consciousness as awareness of environment and of self-revealed objectivity, by self-serving neuromuscular activity which exhibits choice between alternative actions and relates past experience to anticipated future. Whether the time-binding activity extends over a period of seconds or of years is immaterial to the cogency of the definition."

It must be emphasized, however, that many are those who question that physicochemical sciences, in their present form, can account for free will and consciousness. The skeptics are found not only among philosophers but also among scientists. And it is entertaining in this regard to

* From René Dubos, *The Torch of Life*, the Credo series, edited by Ruth Nanda Anshen, Simon & Schuster, Inc. Copyright © 1962 by Pocket Books, Inc. Reprinted by permission.

contrast with the opinions of Professor Homer Smith, quoted above, those of another eminent American physiologist who believes that consciousness is an individual experience not yet explainable by physical theory. In Dr. Seymour Kety's view, not even the perception of color can be explained, even though we know so much about wave lengths and about the structure and properties of the eye and brain. "Where is the sensation of blueness?" asks Dr. Kety. "It is neither wave length, nor nerve impulse, nor spatial arrangement of impulses—it is richer and far more personal. A machine can be built to perform any function that a man can perform in terms of behavior, computation, or discrimination. Shall we ever know, however, what components to add or what complexity of circuitry to introduce in order to *make it feel?*" (italics mine).

Mind, Brain, and Humanist Values *
by Roger W. Sperry (1965)

HUMANIST IMPACTS OF BEHAVIORAL SCIENCE

At first glance the record achieved by the brain-behavior sciences during the past half-century must seem to the humanist to read less like a list of contributions and advancements than like a list of major criminal offenses. The accusations that antiscience can raise in this area are not exactly trivial. For example, before science, man had reason to believe that he possessed a mind that was potent and full of something called "consciousness." Our modern experimental objective psychology and the neurosciences in general would divest the human brain of this fantasy and, in doing so, would dispense not only with the conscious mind but with most of the other spiritual components of human nature, including the immortal soul. Before science, man used to think that he was a spiritually free agent, possessing free will. Science tells us free will is just an illusion and gives us, instead, causal determinism. Where there used to be purpose and meaning in human behavior, science now shows us a complex biophysical machine with positive and negative feedback, composed entirely of material elements, all obeying the inexorable and universal laws of physics and chemistry. Thanks to Freud, with some assistance from astrophysics, science can be accused further of having deprived the

* From Roger W. Sperry, "Mind, Brain, and Humanist Values," *New Views of the Nature of Man,* edited by John R. Platt, University of Chicago Press, Chicago & London. Copyright © 1965 by the University of Chicago. Reprinted by permission.

thinking man of his Father in heaven, along with heaven itself. Freud's devastating indictment is said by many to have reduced much of man's formalized religion to little more than manifestations of neurosis.

Man's inner self and his heritage have not fared much better. Thanks to Darwin, and to Freud again, man now enters this life, not trailing clouds of glory, as the poet once had it, but trailing instead clouds of jungle-ism and bestiality, full of carnal impulses and a predisposition to Oedipal and other complexes. The confining veneer of our civilization is seen to be superficial, and when it rubs thin or cracks, the animal within quickly shows through. These and related lesser onslaughts on the worth and the meaning of human nature tend to add up, one item reinforcing another, to yield a pretty dim over-all picture that is certainly not heartening to think about—and in science we generally don't think about it. Doubt and rejection of science by humanist thinkers in favor of other roads to truth is not hard to understand; and even for the scientist himself, the picture drawn by science imposes a severe test of his credo that it is better to know the truth, however ugly, and to live in accordance, than to live and die by false premises and illusory values.

But for myself, speaking as a brain researcher—and one not too familiar with matters ethical and philosophical and hence in a position to speak with some conviction—I find myself and my hypothetical working model for the brain to be in marked disagreement with many, if not the majority, of the foregoing implications especially and with that whole general picture of human nature that seems to emerge from the currently prevailing objective, materialistic approach of the brain-behavior sciences. When the humanist is led to favor the implications of modern materialism over the older idealistic values in these and related matters, I suspect that he has been taken, that science has sold society and itself a somewhat questionable bill of goods.

THE NATURE OF CONSCIOUSNESS: THE CENTRAL ISSUE

Most of the disagreements that I have referred to revolve around, or hinge either directly or indirectly upon, a central point of controversy that emerges from the following question: Is it possible, in theory or in principle, to construct a complete, objective explanatory model of brain function without including consciousness in the causal sequence?

If the prevailing view in neuroscience is correct, that consciousness and mental forces in general can be ignored in our objective explanatory model, then we come out with materialism and all its implications. On the contrary, if it turns out that conscious mental forces do in fact govern and direct the nerve-impulse traffic and other biochemical and biophysical events in the brain and, hence, do have to be included as important features in the objective chain of control, then we come out at the opposite pole, or at mentalism, and with quite a different and more idealistic set of values all down the line. We deal here, of course, with the old mind-body dichotomy, the age-old problem of mind versus matter, the issue of the spiritual versus the material, on which books and books have been written and philosophies have foundered ever since man started to think about his inner world and to question its relation to the outer "real" world.

Let us begin by stating the case against consciousness and mind as raised by today's objective experimental psychology, psychobiology, neurophysiology, and the related disciplines. The best way to deal with consciousness or introspective, subjective experience in any form, they tell us, is to ignore it. Inner feelings and thoughts cannot be measured or weighed; they cannot be centrifuged or photographed, chromatographed, spectrographed, or otherwise recorded or dealt with objectively by any scientific methodology. As some kind of introspective, private, inner something, accessible only to the one experiencing individual, they simply must be excluded by policy from any scientific model or scientific explanation.

Furthermore, the neuroscientist of today feels he has a pretty fair idea about the kinds of things that excite and fire the nerve cells of the brain. Cell membrane changes, ion flow, chemical transmitters, pre- and post-synaptic potentials, sodium pump effects and the like, may be on his list of acceptable causal influences—but not consciousness. Consciousness, in the objective approach, is clearly made a second-rate citizen in the causal picture. It is relegated to the inferior status of (*a*) an inconsequential by-product, (*b*) an epiphenomenon (a sort of outsider on the inside), or most commonly, (*c*) just an inner aspect of the one material brain process. Scientists can see the brain as a complex, electrochemical communications network, full of nerve impulse traffic and other causally directed chemical and physical phenomena, with all elements moved by respectable scientific laws of physics, chemistry, physiology, and the like; but few are ready to tolerate an interjection into this causal machinery of any mental or conscious forces.

This is the general stance of modern behavioral science out of which comes today's prevailing objective, mechanistic, materialistic, behavioristic, fatalistic, reductionistic view of the nature of mind and psyche. This kind of thinking is not confined to our laboratories and the classrooms, of course. It leaks and spreads, and though never officially imposed on the societies of the Western world, we nevertheless see the pervasive influence of creeping materialism everywhere we turn.

Once we have materialism squared off against mentalism in this way, I think we must all agree that neither is going to win the match on the basis of direct, factual evidence. The facts simply do not go far enough to provide the answer, or even to come close. Those centermost processes of the brain with which consciousness is presumably associated are simply not understood. They are so far beyond our comprehension at present that no one I know of has been able even to imagine their nature. We are speaking here of the brain code, the physiological language of the cerebral hemispheres. There is good reason to believe that this language is built of nerve impulses and related excitatory effects in nerve cells and fibers and perhaps also in those glia cells that are said to outnumber the nerve cells in the brain by about ten to one. And we would probably be safe in the further noncommittal statement that the brain code is built of spatiotemporal patterns of excitation. But when it comes to even imagining the critical variables in these patterns that correlate with the variables that we know in inner, conscious experience, we are still hopelessly lost. . . .

To conclude that conscious, mental or psychic, forces have no place in filling this gap in our explanatory picture is at least to go well beyond the facts into the realm of intuition and speculation. The objective, materialist doctrine of behavioral science, which tends to be identified with a rigorous scientific approach, is thus seen to rest, in fact, on an insupportable mental inference that goes far beyond the objective evidence . . . But the conviction [is] held by most brain researchers—up to some 99.9 per cent of us, I suppose—that conscious mental forces can be safely ignored, insofar as the objective, scientific study of the brain is concerned.

AN ALTERNATIVE MENTALIST POSITION

I am going to line myself up with the 0.1 per cent or so mentalist minority in a stand that admittedly also goes well beyond the facts. It is a position, however, that seems to me equally strong and somewhat more appealing than those we have just outlined. In my own hypothetical brain model, conscious awareness does get representation as a very real causal agent and rates an important place in the causal sequence and chain of control in brain events, in which it appears as an active, operational force. Any model or description that leaves out conscious forces, according to this view, is bound to be sadly incomplete and unsatisfactory. The conscious mind in this scheme, far from being put aside as a by-product, epiphenomenon, or inner aspect, is located front and center, directly in the midst of the causal interplay of cerebral mechanisms. Mind and consciousness are put in the driver's seat, as it were: They give the orders, and they push and haul around the physiology and the physical and chemical processes as much as or more than the latter processes direct them. . . . It is a brain model in which conscious mental psychic forces are recognized to be the crowning achievement of some five hundred million years or more of evolution.

The Threshold of Reflection *
by Pierre Teilhard de Chardin (1955)

From our experimental point of view, reflection is, as the word indicates, the power acquired by a consciousness to turn in upon itself, to take possession of itself *as of an object* endowed with its own particular consistence and value: no longer merely to know, but to know oneself; no longer merely to know, but to know that one knows.[1] By this individualisation of himself in the depths of himself, the living element, which heretofore had been spread out and divided over a diffuse circle of perceptions and activities, was constituted for the first time as a *centre* in the form of a point at which all the impressions and experiences knit themselves together and fuse into a unity that is conscious of its own organisation.

Now the consequences of such a transformation are immense, visible as clearly in nature as any of the facts recorded by physics or astronomy. The being who is the object of his own reflection, in consequence of that very doubling back upon himself, becomes in a flash able to raise himself into a new sphere. In reality, another world is born. Abstraction, logic, reasoned choice and inventions, mathematics, art, calculation of space and time, anxieties and dreams of love—all these activities of *inner life* are nothing else than the effervescence of the newly-formed centre as it explodes onto itself.

This said, I have a question to ask. If, as follows from the foregoing, it is the fact of being "reflective" which constitutes the strictly "intelligent" being, can we seriously doubt that intelligence is the evolutionary lot proper to man and to man *only?* If not, can we, under the influence of some false modesty, hesitate to admit that man's possession of it consti-

1. [*Non plus seulement connaître, mais se connaître; non plus seulement savoir, mais savoir que l'on sait.*]

tutes a radical advance on all forms of life that have gone before him? Admittedly the animal knows. *But it cannot know that it knows:* that is quite certain. If it could, it would long ago have multiplied its inventions and developed a system of internal constructions that could not have escaped our observation. In consequence it is denied access to a whole domain of reality in which we can move freely. We are separated by a chasm—or a threshold—which it cannot cross. Because we are reflective we are not only different but quite other. It is not a matter of change of degree, but of a change of nature, resulting from a change of state.

Suggestions for Further Reading

Asimov, Isaac: *The New Intelligent Man's Guide to Science*, Basic Books, 1965. The chapter "The Mind" provides a good introduction to our knowledge of the nervous system and the electrical nature of the mind.

* Eiseley, Loren: *Darwin's Century*, Anchor Book, Doubleday & Co., Inc., 1960. Chapters XI and XII contain a more detailed discussion of the conflicting views of Wallace and Darwin on the evolution of human characteristics.

* Huxley, Julian: *Evolution in Action*, Mentor Books, The New American Library, Inc., 1957. A very fine general survey of evolution. The last three chapters on the development of mental activity and the human phase (which have been excerpted here) are recommended in their entirety in connection with these readings.

Mind, Life Science Library, Time Inc., New York. Interesting illustrations and text (in addition to the portion included here) provide many fascinating sidelights on the nature of mind.

* Sherrington, Sir Charles: *Man on His Nature*, Anchor Books, Doubleday & Co., 1953. A discussion of the physical basis of life with particular emphasis on human consciousness, mind, and its origins in living substance.

* Sinnott, Edmund: *Matter, Mind and Man: The Biology of Human Nature*, Atheneum, 1962. The chapters "Mind" and "Self" are especially pertinent to the discussion of the development of awareness and consciousness in evolution.

Stettner, Lawrence Jay, and Kenneth A. Matynial: "The Brain of Birds," *Scientific American*, June 1968. Recent research suggests that birds may use parts of the brain other than the cerebral cortex to effect intelligent behavior.

* Paperback edition

6

Is Civilization a New Aspect of Evolution?

The readings in this part consider the ways in which civilization represents a new phase in the evolutionary process. Is psychosocial evolution leading to the emergence of a superorganism and, if so, how does the individual human being relate to this converging process?

Is Civilization a New Aspect of Evolution?

Introduction

M AN'S EVOLUTION shaped a social animal. Man's social characteristics have made possible the uniquely human, cultural phase of
evolution. As we have seen, the long period of helplessness of the human
infant made family ties necessary for the survival of the species. Thus,
family groups and tribes probably originated for mutual protection and
the rearing of the young. Many other species also developed societies but,
as far as we know, the step from a society to a culture has happened only
once in evolutionary history.

The differences, as well as the similarities, between human and insect
societies are described in several articles in these readings. The similarities
are indeed intriguing; the differences profound.

Species such as termites and bees have evolved a highly integrated and
stable social system in which the members carry out specialized functions
and subjugate their needs to the needs of the society. Behavior is not
learned but is transmitted by genetic inheritance or by some form of
chemical communication which appears to regulate the growth and organization of the total colony.

The advantages of this type of system are obvious. Each member of the
colony is born fully prepared by his genotype and is very well adapted to
the environment which the species has regularly encountered. The disadvantage is that there is little capacity for adaptation when confronted
with unusual environmental factors. Adaptation must take place by the
slow biological method of random mutation and selection.

On the other hand, the human phase of evolution which depends upon
the communication of cumulative experience is highly adaptive to changing circumstances. This evolution is Lamarckian in character. The environment acts directly upon the individual who is capable of changing
his behavior to suit the changing conditions. In this new style of evolution
the individual becomes an active participant in evolutionary change,
while in the older style, the individual was largely by-passed. Mutational
change acted upon the germ plasm and was passed directly to the germ
plasm of the succeeding generations. In the human phase of evolution,
each individual learns the cumulative tradition, changes it as he sees fit,
and passes it on with this individual cast to many others, not just his own
biological descendants.

To learn the cultural tradition is onerous and time-consuming, but it
confers a freedom which the member of an insect society does not

possess. Controlled by innate behavior patterns, the bee has no choice but to defend the community when it is threatened. Man has the choice and, therefore, morality enters the picture. Should the individual "always sacrifice himself to the interest of the group, and does the group always have the right to expect its members to do so? This is, of course, one of the greatest problems facing mankind." [1]

Altruism is considered by many biologists to be the product of biological evolution. Natural selection can work upon a society as a whole, favoring that society in which the individual members act for the good of the group. In a small, well-integrated colony altruism confers positive survival value. But in less homogeneous groups, such as a human society, biological evolution may not adequately account for self-sacrifice and other related qualities such as love and friendliness.

J. B. S. Haldane makes the point that altruism cannot become genetically fixed in a large society in which the majority of the members reproduce themselves. The most self-sacrificing individuals tend to die early and do not (on the average) leave as many offspring as the more selfish members of the community. It can be shown mathematically that under these conditions selection would not favor altruism. However, in cases like insect societies where the business of reproduction is carried on entirely by a few individuals and the rest of the colony works selflessly in the service of these queens, natural selection would favor co-operation and self-sacrifice.

"This goes a long way," says Haldane, "to account for the much completer subordination of the individual to society." [2] Would such a system of reproductive specialization bring out a more co-operative and altruistic behavior in mankind? Haldane reminds us that this improvement would be achieved only with some loss of freedom and individuality. Another disadvantage of reproductive specialization is that the whole colony is closely related, brothers and sisters, or at least, half-brothers and half-sisters. Genetic diversity is one of the great factors leading to adaptability and evolutionary progress.

The study of the remarkably well-integrated insect societies suggests an analogy to the cells of a multicellular organism and also suggests that such societies may represent stages in the evolution of larger, more complex organisms. If this analogy is meaningful in primitive species (there is some disagreement on this issue), can it also be applied to human societies? Will the state eventually become the organism and will the individual cease to be a person in his own right? G. G. Simpson believes that this analogy is "thoroughly erroneous," that the trend in human evolution has been toward greater individualization.

1. See Theodosius Dobzhansky selection, p. 339.
2. See J. B. S. Haldane selection, p. 345.

We know that there are basic differences between the older biological evolution and the psychosocial evolution that is unique to man. The individual is an active participant in human evolution. Since mind is the special vehicle of this cultural evolution, it may be reasonable to expect that mind rather than instinct would be the dominant characteristic of the next evolutionary phase. Evolution is tending, Teilhard de Chardin suggests, toward megasynthesis, and "the still un-named Thing which the gradual combination of individuals, peoples and races will bring into existence must needs be *supra-physical* . . . The idea is that of the earth not only becoming covered by myriads of grains of thought, but becoming enclosed in a single thinking envelope so as to form, functionally, no more than a single vast grain of thought of the sidereal scale, the plurality of individual reflections grouping themselves together and reinforcing one another in the act of a single unanimous reflection." [3]

In such a synthesis would each human being cease to be a conscious individual and become merged in a super-consciousness? It is interesting that this concept which seems fantastic even to the modern mind was also suggested by Francis Galton in 1883: "Our part in the universe," he said, "may possibly in some distant way be analogous to that of the cells in an organized body, and our personalities may be the transient but essential elements of an immortal and cosmic mind." [4]

3. See Teilhard de Chardin selection, p. 367.
4. See Francis Galton selection, p. 368.

Culture and Social Values

Heredity, Culture, and Society *
by Theodosius Dobzhansky (1956)

Evolutionists since Darwin have concentrated their efforts on proving that man is indeed a biological species, and, as such, is a product of organic evolution. This concentration was a natural, and perhaps a necessary, response to the antievolutionists who stubbornly challenged the validity of what, at least to competent biologists, became almost a commonplace. But the focusing of attention on the commonplace has obscured the biological singularity of man. Man is a most extraordinary product of evolution; he is so much unlike any other biological species that his evolution cannot be adequately understood in terms of only those causative factors which are operating in the biological world outside the human kind. The singularity of human evolution lies in the fact that the human species has evolved culture. Culture is the exclusive property of man. Once man became, to use the pithy words of Aristotle, "a political animal," his biological evolution was profoundly modified; he cannot any longer be understood except as a uniquely human phenomenon.

✿   ✿   ✿

BIOLOGICAL AND CULTURAL EVOLUTIONS

The appearance of culture signified the beginning of a hitherto nonexistent type of evolutionary development—the evolution of culture or human evolution proper. The date of this event is uncertain; it may have taken place from half a million to more than a million years ago. In any case the humanness of man arose gradually rather than overnight. The significance of this event is comparable with that of a much earlier event, the origin of life from inanimate matter, which happened perhaps two billion years ago. The appearance of life, of self-reproduction, was the be-

---

* From Theodosius Dobzhansky, *The Biological Basis of Human Freedom.* Columbia University Press, New York. Copyright © 1956 by Columbia University Press. Reprinted by permission.

324

ginning of a new type of evolution, biological evolution. Cultural evolution was added to biological evolution, just as biological evolution was superimposed upon the evolution of matter, cosmic evolution. The beginning of cosmic evolution is now placed by physicists at about five billion years ago. In neither case has the new evolution replaced or abolished the old; cosmic evolution continued after the appearance of life, and both cosmic and biological evolutions went on following the emergence of man. Nevertheless, the beginnings of the cosmic, biological, and human evolutions are crucial events in the history of Creation. In producing life, cosmic evolution overcame its own bounds; in giving rise to man, biological evolution transcended itself. Human evolution may yet ascend to a superhuman level.

Kluckhohn has defined culture as "the total life-way of a people, the social legacy the individual acquires from his group." [1] An alternative definition is "Culture is the learned portion of human behavior." The key word in this last definition is "learned," as it draws a clear distinction between biological and cultural heredity. Indeed, the mechanisms of transmission of biological heredity do not differ appreciably in man from those in other organisms which reproduce by sexual unions. Biological heredity is transmitted by genes; consequently it is handed down exclusively from parents to their children and other direct descendants. Culture is transmitted by teaching and learning. At least in principle, "the social legacy" can be transmitted by anybody to anyone, regardless of biological descent.

Man may be said to have two heredities, a biological one and a cultural one; all other organisms have only the biological one. But this statement rather oversimplifies the actual situation and may prove misleading. It is important to realize that biological heredity and culture, the inborn and the learned behavior, are not independent or isolated entities. They are interacting processes. The acquisition and maintenance of culture by learning is possible only owing to the biological organization with which the human species is genetically endowed. Culture is said to be "superorganic." But the superorganic stands in a relation of interdependence with the organic. Human evolution is a singular product of interaction between biology and culture.

CULTURAL CONDITIONING

A quaint medieval belief had it that infants who do not hear spoken any human language come to talk in ancient Hebrew, which is the lan-

1. C. Kluckhohn and O. M. Mowrer, "Culture and Personality," *American Anthropologist,* 46 (1944), 1–29. Ed.

guage of God. The basis of this legend is evidently the naive idea that there is such a thing as an innate or intrinsic language of mankind, which one may come to know without learning. Of course there is no such language. The language or languages which a person speaks are not contained in his genes; they must be acquired by learning in childhood or in later years. No less naive are the attempts to explain various cultural characteristics, and even all human behavior, as biologically determined ways to satisfy the bodily functions and drives of the human animal. Modern cultural anthropology has shown very clearly that there is no "culture" planted by nature in the genes of any members of the human species. A human individual acquires only gradually, by means of the processes of socialization and acculturation, the habits, skills, and beliefs which integrate him into his society. Socialization and acculturation consist of conditioning and learning; they confer upon a person a competence to deal with his human and his physical environment.

An infant is born with a genotype which makes him receptive to conditioning by other human beings with whom he is associated—parents, relatives, playmates, teachers, and, directly or indirectly, most or all members of the society or the group to which he belongs. Children isolated from contacts with other humans grow up having only a physical resemblance to human beings; human mind and human behavior are products mainly of the accommodation the individual makes to the habits, standards, tastes, opinions, beliefs, feelings, and values imposed by the community. In the words of Herskovits: "This is because every human being is born into a group whose customs and beliefs are established before he arrives on the scene. Through the learning process he acquires these customs and beliefs; and he learns his cultural lessons so well that much of his behavior in later years takes the form of automatic responses to the cultural stimuli with which he is presented." [2]

Culturally induced traits make human beings behave predictably in many situations. Many such traits are so constant in members of a given culture that members of that culture take those traits for granted and only notice their absence when they meet people brought up in cultures different from their own. The behavior of such people may seem, therefore, baffling and even perverse. It should be noted in this connection that "culture" as used by anthropologists signifies not only science, art, philosophy, and religion, but also such practices of daily life as ways of preparing and eating food, of sleeping, of taking care of children, of personal cleanliness, and so on. Thus it comes about that millions of people in the Middle East feel disgusted at the very thought of eating pork, while millions of people in the West consider ham or pork sausage a delicacy. There is no need to

2. M. J. Herskovits, *Man and His Works,* New York, Knopf, 1948. Ed.

assume the existence of genes which would make one like or dislike pork; it is all a matter of conditioning or habit. Similarly, people who are used to sleeping in beds feel most uncomfortable sleeping in hammocks as people in tropical America do, and vice versa. Again this is a matter of training, not of genes. In some of these cultural traits retraining is possible, with some effort. At the cost of some uncomfortable nights, this writer has now learned to sleep very comfortably either in a bed or a hammock.

❂   ❂   ❂

No correlation has ever been established between the contents of a culture and any physical or physiological characteristics of its possessors. Quite the contrary, cultures have again and again been successfully taken over, as a whole or in part, by people of quite different racial stocks. And conversely, the same population can and has often changed its culture drastically during the course of its history with or without outside influence. After 1852, Japan rapidly absorbed much of European culture, and quite successfully blended it with its indigenous cultural pattern, so successfully in fact that it caused considerable difficulty to the original possessors of this culture in the years immediately following Pearl Harbor. But the Japanese have not become members of the Caucasian race—even though Hitler did propose to make them "honorary Aryans."

Less than two thousand years ago, the ancestors of most modern Americans and Europeans were barbarians eking out a rough and precarious existence in the forests and swamps of northern Europe. But these barbarians responded magnificently when they were given an opportunity to borrow foreign cultures developed by the rather different peoples who inhabited the lands around the eastern part of the Mediterranean Sea. Successful cultural reconditioning can now be observed any day in the large universities in such centers as New York, London, Paris, or Moscow, which attract students from all over the world. At least some of these students become culturally more similar to each other and to their hosts than they are to some of their own biological relatives who have stayed at home. Evidence that cultural patterns are determined by the genes is utterly lacking. What evidence there is rather contradicts this hypothesis.

❂   ❂   ❂

PROPAGATION OF CULTURE

Cultural traits can be transmitted potentially to any number of persons regardless of descent relationships. Language is the most ancient and powerful means of progagating culture. The invention of writing, and

later of printing, has made the dissemination of culture vastly more efficient. Transmission of culture can now occur through space, between persons who live in different parts of the world. It can occur through time,
from persons long dead to living ones. Founders of religions, philosophers, poets, scientists, artists, inventors, intellectual and political leaders
exert strong influence, for good or for evil, centuries and millennia after
their deaths. Most of mankind today are directly or indirectly influenced
by Christian civilization, and particularly by its branch known as Western
civilization. These civilizations perpetuate systems of ideas and of ideals
arrived at some two thousand years ago in Palestine, in Greece, Rome,
and in other parts of the Mediterranean world.

Acquired cultural characteristics are transmitted; acquired bodily characteristics are not. Acquired bodily characteristics die with the body
which acquired them. The body is a by-product of the self-reproduction
of genes, not the other way round as wrongly supposed by Lamarck and
his followers. There is no inherent mechanism of replication or reproduction in culture traits. This makes culture both more easily transmittable
and more easily modifiable than is biological heredity. All the inventions,
scientific findings, musical compositions, and literary works which mankind now living treasures and enjoys were created by relatively few
individuals through their personal efforts. Galileo, Leonardo, Newton,
Beethoven, Faraday, Shakespeare, Darwin, Dostoevsky, to name only a
handful of the more recent originators of our precious cultural heritage,
exert great and enduring influence on millions of people. In cultural fields,
more than anywhere else, the many owe much to the few. The splendor
of the arts and the advances of science and technology are often due to
astonishingly small numbers of persons. Biological evolution is, in this
sense, a far less efficient process.

Considered biologically, culture is, of course, a part of the environment
in which the development of a person is taking place. Indeed the "environment" consists not only of physical variables such as temperature,
humidity, light, quantity and quality of the food, but includes also the
interrelations which are established between individuals of one species
and of other species living in the same habitat. Culture is, however, a
nearly exclusively human phenomenon, and as such it deserves to be regarded as the third group of determinants of human personality, along
with heredity and environment.

VARIATIONS WITHIN A CULTURE

The processes of socialization and acculturation render it certain that
every normal person can function as a member of the community into

which he is born. As Benedict wrote in her *Patterns of Culture:* "From the moment of his [the child's] birth the customs into which he is born shape his experience and behavior. By the time he can talk, he is the little creature of his culture, and by the time he is grown and able to take part in its activities, its habits are his habits, its beliefs his beliefs, its impossibilities his impossibilities." [3] Every American or European wears clothing of a certain type, and regards appearing naked in public places as indecent. But an Amazonian Indian goes naked, and to him clothing, other than a loincloth, is a nuisance. Every American learns to use a fork and a spoon for eating; every Chinese is adept with chopsticks. In most cultures eating human flesh is a loathsome crime, but in some it is a heroic act and a source of power and authority.

❖   ❖   ❖

The way of life of a society shows an internal cohesion. Comparative studies of many kinds of societies, both nonliterate and advanced, have disclosed that the beliefs, customs, and practices of a group of people usually hold together as related parts of a whole. A durable culture shows a certain patterning or integration of its components. A change in one part of a culture may, then, require correlated changes in the whole pattern to make the system workable. The importance of this integration has been shown in a series of unpremeditated experiments. The contacts of primitive societies with Western civilization have in many instances resulted in a breakdown of the former and even in the extinction of the tribes which formed these societies. The breakdown was caused by attempts, frequently honest and well-meaning, on the part of the Western "civilizers" to alter drastically some aspects of the way of life of the subject populations without harmonizing these alterations with other aspects. The mental health and happiness of a people, both as individuals and as a group, is often dependent on the possession of a coherent cultural pattern.

❖   ❖   ❖

But it should be self-evident that within the bounds of each cultural pattern different individuals are not merely allowed but indeed are forced to develop variations in personality and behavior. Kluckhohn and Murray distinguish four groups of personality determinants: [4] constitutional, group membership, role, and situational. Constitutional determinants include, apart from sex and age, the individual genotype which makes one

3. Ruth Benedict, *Patterns of Culture,* New York, New American Library, 1934. Ed.
4. Clyde Kluckhohn and Henry Murray, editors, *Personality in Nature, Society and Culture,* New York, Knopf, 1948. Ed.

individual react differently from another in similar socializing influences. Group membership refers to the fact that most persons live amongst, and hence identify themselves with, either Hottentots or Eskimoes, Americans or Russians, or whatever the case may be. Accordingly, they are expected to behave and even to feel like a Hottentot, an Eskimo, an American, or a Russian usually does. Within each group, different individuals choose, or are assigned to play, different roles. There are craftsmen, farmers, aristocrats, tinkers, tailors, soldiers, and sailors. This again imposes or favors certain behavior patterns, and often leads to identification with and loyalty to the chosen or assigned role. And, finally, every person finds himself in a situation that is, to some extent, unlike that of any other person, and these unique situations come and go as long as life endures. Hence no two persons, not even members of a pair of identical twins, are ever exactly alike.

The existence of intracultural variations is clearly of the greatest importance for the success of a culture as a whole and for the welfare of the society which is the possessor of this culture. In fact, some nonliterate cultures permit a relative uniformity of personalities among their members, while in advanced cultures the variety of functions to be performed or roles to be played is very great. In bringing about intracultural diversity of personality and behavior, constitutional or genetic variables are most important, and yet they are given a minimum of, if any, attention by some anthropologists and sociologists. Assertions have even been made that the existence of genetically controlled personality variations is doubtful, or not proven, or of no particular importance compared to environmental and cultural influences. Such assertions can only be explained by the general lack of comprehension of the true nature of heredity. The problem here involved would appear less intricate, and certainly would provoke less passionate partisanship, if it was more generally realized that genetic conditioning does not necessarily imply a fixity of any trait. Still less does it exclude the influence of environment on the trait concerned, or dash the hope of being able to control human phenotypes.

❉   ❉   ❉

ENVIRONMENT AND THE MANIFESTATION OF HEREDITY

Heredity, the constellation of genes, that is, the genotype, brings about the development of the organism by imposing specific patterns on the assimilated food materials derived from the environment. The interaction of heredity and environment makes the development of the organism pass through a succession of stages, from fertilization, through growth, reproduction, and old age, to death. At any stage the outcome of the development depends on the genotype and on the succession of environments which the developing organism has encountered up to that stage. The appearance, the structure, and the functional state which the body has at a given moment constitute its phenotype. The phenotype of a person changes continuously through life as his development proceeds. The changes in the manifestation of heredity in the phenotype contrast with the relative stability of the genotype. But it should be understood that the relative stability of the genotype is a dynamic one—the genes are "stable" not because they are isolated from the environment, which they are not, but because they reproduce themselves faithfully.

The phenotype of a person is at any given moment determined by his genotype plus his life history. The question of whether a given "trait" of an individual, such as his eye color, state of health, musical ability, or good or bad behavior, is determined by heredity or by environment is, thus, pointless. The question so stated results from the old confusion of biological heredity with legal inheritance. All traits arise in the process of living, which implies a heredity and a succession of environments. An eye color presupposes an eye, well-being or disease have no meaning without an organism which can or cannot contract certain diseases or suffer certain breakdowns, and benevolent and malevolent behavior imply the existence of a person who can discriminate between good and evil.

The so-called "nature nurture," or heredity-environment, problem must be stated very carefully to be meaningful. The problem concerns *differences* between individuals or populations. Some people have blue and others brown eyes, some enjoy good health while others suffer from various infirmities, some can create or enjoy good music and others feel no interest in it, some are desirable members of society and others are outlaws. It is reasonable to inquire to what extent the observed *differences* in phenotypes are due to the existing variety of human genotypes and to the diversity of the environments in which people live.

There is not one nature-nurture problem, but many. To put it more precisely, what must be studied is the variation observed in a given characteristic among human beings in general, or among members of certain human populations. The part played in this variation by the existence of a diversity of human genotypes may then be evaluated and compared with the part played by the existing diversity of environments. The results obtained will be valid only for the characteristic or trait studied, for the population examined, and for the time and place when and where the examination is made.

For example, the relative importance of heredity is, taking mankind as a whole, probably greater in the variability of eye colors and skin colors than in the variability of height or weight. But it does not follow that skin color is always more strictly hereditary than height. Indeed, skin colors vary widely among the inhabitants of Virginia—from very pale to black. These wide variations are largely due to heredity, to the presence of genes of European and of African origin among the inhabitants of this state. The degree of variation in skin color is relatively much smaller among the inhabitants of a village in Norway, or of a village in central Africa, the populations of which are of more uniform racial origins. In this lesser variation the relative importance of the environmental component, the tanning effect of exposure to sunlight, is much greater than it is in Virginia. In general, if the environment in which a population lives is made constant, the phenotypic variation observed will be due to genotypic diversity. Conversely, making the population genotypically uniform results in the observed variation being environmental in origin.

CONTROLLING THE PHENOTYPE

All human traits and qualities result from the interaction of heredity and environment in the process of living. This statement applies with equal force to the so-called physical traits and to the so-called psychic, behavioral, or cultural traits of persons and populations. Its applicability to the latter category of traits is frequently and vehemently challenged. To many people the idea that heredity enters as one of the determinants of the intellectual and emotional life of human beings is intensely distasteful. This distaste has two causes. First, there is no question that overvaluation of heredity and undervaluation of environment in human development has been, and is being used as a specious reason for bias, oppression, and cruelty of man to man. Suffice it to recall that it was this kind of prostitution of biology which was employed by the Nazis as justification for their crimes. Second, to many people heredity seems a sinister force which suddenly emerges mysteriously to inflict blows of physical or mental illness on the innocent. Along with this idea goes the equally mistaken one that a hereditary disease is an incurable disease, that hereditary traits cannot be influenced or modified by the environment, and therefore that traits which are susceptible of being modified or improved are, hence, not hereditary. All these misconceptions give rise to the notion that a biological inheritance of human psychic traits is incompatible with personal freedom.

Modern biology would have earned its keep if it did nothing else but show that these fears are unfounded, and biology certainly has done this. There is no such thing as a purely inherited or a purely environmental trait, since all traits arise during the process of development of a person who has a certain genotype *and* who lives in a certain environment. The degree to which the phenotypic manifestation of human heredity, or of the heredity of any biological species, can be predicted and controlled depends to a considerable extent upon our knowledge and understanding of the developmental processes involved.

How vast the possibilities in this field are can be seen from the steady progress of medicine and hygiene, the applied biological sciences that devise methods for the management of the human phenotype.

❋   ❋   ❋

Man is a social animal who lives in organized groups with other men, and who cannot live otherwise for more than a generation—and rarely that long. Man is also the biological species which possesses the capacity to create and to transmit to succeeding generations "the social legacy the individual acquires from his group," i.e., the ability to create and to transmit culture. He is the only species which has developed culture, and the emergence of such a species has been a unique event in the history of all life on this earth, and perhaps also in the history of the Cosmos. He is, however, by no means the only social animal, and his society is by no means the only one known among animals. Being a social animal does not mean that one is necessarily a cultural animal, although the development of the capacity to have a culture is possible presumably only in a social animal. The genetic basis of culture is, then, different from that of social habits, and so is the adaptive function of culture different from that of society. A comparison of human and animal societies may throw light on the nature of both.

HIGHLY ORGANIZED INSECT SOCIETIES

Allee [5] has characterized societies as "all groupings of individuals which are sufficiently integrated so that natural selection can act on them as units." If the degree of integration is taken as a measure of the perfection of a society, then the most advanced social organizations are unquestionably found among social insects, and particularly among ants and termites. Here, the division of labor is carried to the point where biologically the most important function of any organism, i.e., procreation, is borne by a specialized caste, while most of the other individuals who compose the society remain sexually undeveloped. In some termite and ant nests, containing thousands or tens of thousands of individuals, there may exist only a single fertile female. A "queen" in some termite species may lay several thousand eggs in every twenty-four hours, and may continue doing so

5. W. C. Allee, *Cooperation Among Animals*, New York, Schumann, 1951. Ed.

without interruption for several years. Her abdomen is so distended by enormous ovaries that she can neither procure food nor move by herself. She is, in effect, a living egg factory, immured in a special cell in the interior of the nest, and conscientiously and continuously tended by numerous workers. In the honeybee and in some ants, the queen mates with a male only once in her lifetime, during the mating flight which usually precedes the foundation of the colony.

In addition to one or more "queens" and males, an ant or termite colony may contain numerous individuals belonging to "worker" castes. In some species, all workers carry on different functions as the occasion requires. In others, there is a worker caste proper, which takes care of procuring food and building the nest and keeping it in good repair. A soldier caste defends the colony from intruders. A curious caste are the "repletes" in honey ants, the function of whom is to serve as storage barrels for food (stowed away in their inflated bellies). Among other curiosities in ants are individuals with "phragmotic" heads, which fit like plugs into the entrance holes leading into the nest. These individuals function as "doormen," since they withdraw and permit the passage of individuals of their own colony, but not of outsiders.

It is well established by suitable experiments that members of an ant or a termite colony recognize each other and differentiate between nest-mates and strangers. Among ants, not only the nest itself but also a certain amount of the surrounding territory is defended against all comers. The owners of the territory frequently fight and may kill the intruders, including individuals from other nests of the same species. It appears that, at least in some ants, the sense used to discriminate between nest-mates and outsiders is that of smell. Every member of a colony seems to share one common family odor, and any individual who does not have that odor is treated as an enemy. Conversely, every member of a colony shows a complete "devotion" to the commonwealth. Readiness to sacrifice life itself is certainly the supreme test of the devotion an individual feels for another individual or for the community, and members of an insect society are often subjected to this trial. Members of the soldier caste, where such is present, or the undifferentiated workers in species which have no soldiers, attack the enemy, even one vastly stronger than themselves, completely heedless of danger. The sting of a honeybee worker has reversed barbs, and is often left in the flesh of her victim; the bee who loses her sting is, however, mortally wounded in the process, and it dies soon thereafter. A bee attacking an enemy with her sting is thus committing suicide.

Many insects perform wonderfully complex, coordinated operations which, in the environments where the species normally lives, promote the survival and reproduction of the individual or benefit its progeny. Among social insects, members of a colony may carry out intricate cooperative undertakings which aid the commonwealth. Thus, some species of termites and ants build nests of a kind of elaborate architecture that is strictly characteristic of their own species and is different in different species. The sauva ants (*Atta*) feed chiefly on special fungi which they cultivate in their nests. Probably everybody who has had a chance to observe life in the forests of tropical America has seen long files of sauva workers carrying to their subterranean nests portions of leaves which they had cut from a large variety of plants. In the nest, other workers chew the leaves to a spongy pulp which is stored in special chambers and infected with spores of a fungus. The mycelial growth and the fruiting bodies of the fungus are then used to feed all members of the colony. It is interesting to note that the sexually developed individuals which establish new colonies transport a pellet of the fungus with them, and feed the first crop of new workers on fungi grown on their own excrements. When the workers appear, they immediately take to leaf cutting and to the construction of fungus "gardens" for the new nest.

Perhaps no other feature of the biology of ants has impressed observers as much as the ability of certain species to wage organized warfare against other species, and to secure for themselves and to profit by the work of "slaves." From time to time the workers of the ant *Polyergus* form raiding parties which surge from their own nest to invade that of another species of ants in the neighborhood. The raiders overwhelm the resistance of the owners, kill many of the defenders, enter the nest, seize the larvae and pupae of the victims, and carry them off to their own nest. Curiously enough, the adult workers which hatch from these larvae and pupae become faithful "slaves" who serve their captors as they would their own colonies. Some slave-making species of ants are unable to secure their own food and to maintain their own nests, and are entirely dependent on slave labor for their existence.

ANTHROPOMORPHIC FALLACIES

It was to be expected that some scientists were overly inspired by these observations of social insects, and were so carried away by their enthusiasm as to suggest that man may well learn from insect societies how to organize or improve his own societies. Surely, there is little to admire in slavery and in slave raids, but, on the other hand, the absolute and selfless "devotion" of an individual bee, ant, or termite to the common weal of the colony are admirable. They seem to show a high degree of integrity and heroism which we rightly admire in some of our fellow men. The whole life of these insects appears to display unflinching purposefulness, which all too often we seek in vain in humans.

It is really unfortunate that a closer scrutiny makes the whole pretty picture disappear like a dream. Finding altruism and other human qualities in insects proves to be a delusion. Ants and termites are neither heroes when they defend their own nests, nor villains when they rob those of their neighbors. They are devoid of virtues and vices because they lack the freedom to decide between possible alternative courses of action. They act as they do because their behavior represents an innate response to a given environmental situation. Insect behavior is, then, not reducible to a common ethical measure with human actions. Praise and blame have meaning only in connection with acts in which the individual is at least to some extent a free agent. Although they show some interesting parallels, human societies and insect societies are distinct phenomena.

The insect is born fully equipped to perform the often very complex work by means of which its species secures its livelihood and provides for its offspring. A larva of a caddis fly just hatched from its egg proceeds to build its house in a characteristic way and from certain materials although it has never seen another larva building a house. An ant is ready to participate in the activities of its nest-mates soon after it hatches from its pupa. It does not undergo a period of training or apprenticeship.

This does not mean that the insect is an automaton able to do just one thing. Far from this. A bee worker hatched from the pupa spends her first two or three days incubating the brood and preparing brood cells. Then follow about three days during which the worker feeds the older larvae with honey and pollen, and four to nine days in which she feeds the younger larvae with brood food. A two-weeks-old bee now takes her first trial flight outside the hive, and for the next ten days or so she works as a "house bee," receiving and storing the food brought by other bees, cleaning the hive, and guarding its entrance. Only after that period does she become a field worker, foraging for pollen and nectar. There is even conclusive evidence to show that insects are capable of some simple learning —such as associating the presence of food with marks of certain colors and shapes. But, as Haskins puts it: "It is as though the invertebrate mind functioned like a shallow vessel which, once filled, cannot be emptied. By contrast, the vertebrate mind is a deeper pitcher. It may be partially emptied many times, and as often refilled with liquids of another dye. As the mammalian mind grows older, it grows wiser. Faced with a rapidly changing environment, the invertebrate mind merely grows less well adapted and more thoroughly confused." [6]

6. C. P. Haskins, *Of Societies and Men,* London, Allen & Unwin, 1951.

ADAPTIVE ADVANTAGES OF INNATE AND LEARNED BEHAVIOR

The great advantages and the equally considerable limitations of innate, or instinctive, behavior are obvious. Every individual of a species (except rare and pathological deviants) is born adapted to the environment which he is likely to encounter. An individual needs no training or preparation for the business of living. He is born fully prepared by his genotype. And the genotype is, in turn, molded in the evolutionary process and controlled by natural selection. As shown earlier, the adaptation to the environment by means of genotypic specialization is advantageous in a relatively constant environment. The "wisdom" of instinct is a distillation of the experience of the species in the environments which it has met regularly, or at least repeatedly, during its evolution. The adaptation reached by a genotypic fixation of behavior may be extremely precise and subtle. The individual does exactly what is best for himself and for his species. He does exactly what he should do at precisely the right time.

The drawback of genotypic specialization and fixation is that the possibilities of adaptation to environmental changes become severely limited. Numerous and ingenious experiments have shown that when an animal is placed in novel environments, its innate behavior loses its "wisdom." The animal is likely to do exactly the wrong thing, damaging its own chance of survival or that of its offspring. A commonplace example of such grossly unadaptive behavior is the attraction of many species of insects to light, resulting in their death in countless numbers near electric and other lighting fixtures. More refined experiments regarding instincts have been made by Fabre, on many insects, and by Lorenze and Tinbergen, chiefly on vertebrates. They agree in showing that innate behavior tends to be fixed, predictable, uniform, and adaptive under normal living conditions. But this behavior shows relatively little capacity for adaptive modification when confronted with unusual circumstances.

Conversely the capacity to learn causes the animal to modify its behavior in accordance with the circumstances of its individual life experience. Learned behavior is not separate from or independent of innate behavior; it is, rather, a modification of the latter. Tinbergen defines learning as "a central nervous process causing more or less lasting changes in the innate behavioral mechanisms under the influence of the outer world." Some capacity to learn is present in most animals, but the relative importance of learned behavior is greater in higher than in lower animals. Expressed in genetic terms, an individual's capacity to learn permits this individual to modify his behavioral phenotype. The adaptive significance of learning ability is that it brings about a plasticity of this phenotype.

The behavior is no longer determined by the animal's inherited nature within fairly narrow limits; it is also determined by the experience of living.

❋   ❋   ❋

It has been said that "Man was formed for society." The seclusion of an anchorite is assuredly neither the usual nor the normal human environment. The environment of a member of our species is set by the society of which he is a member. He is influenced by his relations to other persons in the same society; in fact, interpersonal relations constitute the most important aspect of human environment. These relations often determine the ability of an individual to survive and to leave progeny. The interpersonal relations which prevail in a society or in a group within a society affect its chances for survival and perpetuation. These simple considerations have led many thinkers, from Darwin to our day, to suppose that man's behavior as a member of society is, like the structure and the physiology of his body, molded in the evolutionary process and controlled by natural selection. The basic assumption to this view has been stated most concisely by Leake as follows: "The probability of survival of a relationship between individual humans or groups of humans increases with the extent to which that relationship is mutually satisfying." [7]

The greatest, although by no means the only, difficulty for this utilitarian explanation of ethics lies in understanding the nature and origin of the moral sense which every man, to some degree, has. Admiration of the good and disapprobation of the bad is ineradicable from the human mind. Regardless of whether a person does or does not live up to the standards of ethics accepted in his social environment, he usually feels that these standards have some validity which he cannot gainsay. Yet the kinds of behavior which many ethical codes hold most praiseworthy do not always promote the survival and welfare of the person who adopts them. Indeed, the selfish often prosper more than the generous, the cunning more than the truthful, and the cowards more than the brave. The rule that whatever promotes the survival and welfare of the individual is good, and whatever puts them in jeopardy is bad, is no safe guide in ethics. Yet such a rule would be expected to hold if ethics were a product of what nineteenth-century evolutionists called the survival of the fittest in the struggle for existence. There is a glaring conflict between the "gladiatorial theory of existence," seemingly implied in the evolutionary theory, and the Golden Rule, which, to most people, still stands as the most trenchant statement of the guiding principle of ethics. In 1893, T. H.

7. C. D. Leake and P. Romanel, *Can We Agree?* Austin, Texas, University of Texas Press, 1950, Ed.

Huxley admitted with admirable courage and sincerity that he was thwarted by the contradiction.

Ways towards a resolution of this apparent conflict were not opened up until it was realized that the "gladiatorial theory" is not only not a necessary part of the theory of natural selection, let alone a part of the theory of evolution, but is, in fact, invalid on purely biological grounds. Biological fitness is by no means always promoted by the ability to win in combat. It is much more likely to be furthered by the inclination to avoid combat, and in any case, it is measured in terms of reproductive success rather than in terms of the numbers of enemies destroyed. Moreover, not only individuals, but also groups of individuals, such as tribes and races, are units of natural selection. It is, then, at least conceivable that evolutionary processes may promote the formation of codes of ethics which, under some conditions, may operate against the interests of a few individuals but which favor the group to which these individuals belong. Biological analogies are readily available. Among ants, termites, and other social insects, an individual often sacrifices his life for the sake of the colony. In social insects most individuals do not reproduce, and natural selection favors the forms of behavior which increase the chances of survival of the colony, and particularly of its sexual members.

This, certainly, does not dispose of all the difficulties in understanding ethics on an evolutionary basis. In man, codes of morality and ethics vary from group to group. In a given group they may undergo changes with time, as history abundantly shows. Are we to conclude that, in man, natural selection favors the ethical codes which benefit the group at the expense of the individual? Such a view would leave unresolved the ethical paradox of conflicting interests between the individual and the society to which he belongs. We saw that the behavior of an ant is largely determined by its biological heredity, and that the ant has no choice but to act as it does. On the other hand, man is free to choose between various courses of action, and feels himself responsible for the choices he makes. He can choose to work for or against the society. Should he always sacrifice himself to the interest of his group, and does the group always have the right to expect its members to do so? This is, of course, one of the greatest problems facing mankind.

The Evolution of Altruism *

by Hans Selye (1956)

THE EVOLUTION OF INTERCELLULAR ALTRUISM

The most characteristic feature of life is egotism; it is also its most an-cient and essential property. As soon as the first amebalike cell was cre-ated in the primeval ocean it had to shift for itself or die. Unlike all the inanimate substances around it, it was not indifferent to its own fate; it lived and it wanted to go on living. To achieve its aim this cell did not have to compete with other living beings, for it alone was alive and the inanimate did not "mind" being exploited.

But another characteristic of life is to multiply and to develop: soon there were two cells, then millions, each looking out for itself first—and often necessarily at the expense of the others. This led to clashes in which the strongest cells won out.

Then, sometime, somewhere, a few cells stuck together and formed a colony with a community of interests, a sort of *collective egotism*. At this moment altruism was born. With regard to other living beings, the com-munity still behaved like an egotist; but it was in the interest of each cell within this colony that the other members also had to strive, because the strength of the whole depended upon all its parts. From any one cell's point of view, altruism toward other cells of the same colony became a form of egotism.

The biologic efficiency of this collective life proved to be so great that, in the course of the struggle for survival, cells found it useful not only to stay together but to rely ever more upon each other. Eventually large numbers of them learned even to share a single life; they came to form *a single living being*. Man himself is such a multicellular organism. The cells of his body are so strictly interdependent that they could not sur-vive in separation. Some of them have become indispensable to feed the whole body, others to move it, yet others to coordinate its manifold ac-tivities; but, in acquiring this high specialization for their respective tasks, the parts have given up their capacity for independent existence. The choice between egotism and altruism does not arise between the cells of a single multicellular body; within its confines, there is no motive for competitive fight.

When any kind of foreign living body enters into our tissues, there is considerable wear and tear: that is, local stress, due to a clash of interests between the invader and the invaded. This is evidenced by an essentially inflammatory defensive response. Only our own organs (for instance, nerves, blood vessels) can penetrate our tissues without causing much stress. Here, no protective barricade is built to fend off the invader; in fact, he is welcome: even the dense tissue of bone melts away under the soft pressure of an invading blood vessel which brings it nourishment.

It took countless generations before single cells evolved the art of peaceful interdependence to avoid internal stress. (Even now a revolt can break out occasionally in a part which forgets the principle of collective altruism. This is what we call *cancer*. It kills the whole as well as itself by its own unrestrained expansion.)

In short, we may look upon the evolution of complex living beings as essentially a process permitting many cells to develop in harmony, with a minimum of stress between them, so as best to serve the interests of the whole community.

THE EVOLUTION OF INTERPERSONAL ALTRUISM

Now let us consider relationships between various multicellular living beings, for instance, between men. Here there has also evolved an inter-dependence between individuals who have specialized for diverse tasks. Some raise food, others provide transportation, and others administer and coordinate the activities of the community. But, at present, teamwork be-tween them is still far less satisfactory and far more conducive to stress than between the various organs of one person.

The Origin of Ethics *

by Edmund W. Sinnott (1957)

It thus seems plausible that a natural propensity for some of the moral values, particularly as they concern the immediate group, may have been instilled into man through a long selective period in which the most extreme and selfish individualists were less likely to survive because they lacked the protection given by their fellows. Something like a social instinct may therefore have been developed which became the forerunner of conscience. It should be remembered, however, that this could have been only rudimentary and that animals and primitive man himself have but the germs of altruism. In a state of nature, selfishness is still the rule and not the exception. It seems most unlikely that from a process that is directed ultimately to survival there should have arisen the high selflessness that distinguishes the moral codes of our great religions and the lives of those who almost everywhere are recognized as saintly.

To lift to such a level the primitive social sense exhibited by some of the higher animals and by early man, something more seems to be needed than mere survival value. Here again, as in the case of beauty and other values of the spirit, man's character apparently undergoes an elevation in its moral goals as he grows further out from barbarism. Just as his appreciation of beauty rises, so does his moral sensitivity and the sort of behavior he regards as right. If one takes the long view and compares man's moral codes as the centuries have succeeded each other he will be convinced, I think, that progress has been made, bleak as the moral climate of our time may seem. Among enlightened peoples men are no longer held in slavery, children employed for long hours, petty thieveries punished by death or bears baited for sport. To account for these things on any materialistic theory is very difficult. Such a theory, says President Conant: "fails to accommodate what I regard as highly significant facts, not facts of science but facts of human history. These are the unselfish ways in which human beings often act with compassion, love, friendliness, self-sacrifice, the desire to mitigate human suffering. In short, it is the problem of 'good,' not 'evil,' that requires some other formulation of human personality than that provided by the usual naturalistic moralist." [1]

The right may exist in many forms or levels, as does the beautiful, and in each of them the role of man's spiritual monitor is to distinguish the better from the worse. Conscience in this way may be a useful moral guide without setting up standards of conduct in detail. The almost universal judgment of mankind that unselfishness and a willingness to serve one's fellows is right and much to be encouraged suggests that it is a major goal of man's nature. It will express itself in various ways in different situations but is the broad foundation for *all* ethical behavior. If one loves his neighbor as himself he does not need to be specifically enjoined from

1. James B. Conant, *Modern Science and Modern Man* (New York: Columbia University Press, 1952), p. 99.

murder, theft or covetousness. All the minor details of human behavior can be spelled out in various ways, but this one admonition underlies them all. To set any behavior against the background of this basic standard and thus judge whether it is right or not is the task of moral character. Just as man tries to distinguish what is beautiful from what is not, at a given aesthetic level, so he attempts to decide what behavior conforms to this basic moral goal and what does not, whatever the particular code is under which he lives. To this extent, at least, man seems not to be empty but to come into the world with an inborn feeling that love for his fellows is a high and worthy thing. Only the germ of this has been attained through natural selection. It is not implanted by conditioning, though this may greatly aid—or hinder—its expression. It is, I believe, a natural development of man's questing, aspiring, creative nature, the ultimate goal of his spiritual life. "The kelson of the creation is love."

This is opposed to the idea of moral relativism, so often supported by anthropologists; the idea that any set of customs and codes is as valid as any other and that there are no moral absolutes at all. This opinion is by no means universal today, however, even among these scientists. In a recent paper Professor Clyde Kluckhohn emphasizes fundamental human similarities in ethical standards. Says he: "If, in spite of biological variation and historical and environmental diversities, we find these congruences, is there not a presumptive likelihood that these moral principles somehow correspond to inevitabilities, given the nature of the human organism and of the human situation? . . . Moral behavior in specific instances and in all its details must be judged within a wide context *but with reference to principles which are not relative. . . .* What is right for Hindus in 1955 may not be precisely the same as what is right for Americans in 1955 but it will be of the same *kind. . . .* Some needs and motives are so deep and so generic that they are beyond the reach of argument: pan-human morality expresses and supports them. . . . Principles as well as contexts must be taken into account." [2] . . .

Man has certain basic values, better developed in some individuals than in others but present in everyone. These are part of his inborn nature, and as he grows in spiritual stature he becomes more sensitive to them. Whatever he does that helps to realize them and thus brings final satisfaction we may call right, and anything that thwarts them and thus leads at last to unhappiness is wrong. This measuring stick is often very difficult to use, for the results of an act may be far-reaching and are often long deferred; but this test provides a foundation for morals which in a sense is absolute, since it is grounded on the character of life's goals and thus of

2. Clyde Kluckhohn, "Ethical Relativity: sic et non," *Jour. Philosophy,* 52:663-677, 1955.

life itself. Conscience is the inner sense of what these goals are, and pain and pleasure result from the degree to which we are successful in attaining them. Thus we may define the right as *whatever is in harmony with the basic goals of human life.*

Essentially this conclusion is well stated from the viewpoint of a psychologist by Professor Hadley Cantril. "No matter how diverse the social and ethical standards may be that mankind has developed through the ages and in various cultural groups and sub-groups, they are all ultimately attempts to increase the possibility of gaining greater satisfaction in living. And human living—as such—has, I believe, similar psychological aspects due to the similarities we have as human beings." [3] Jennings carries this idea still further: "It is the promotion of life, of its fulness and adequacy, that is the right; it is the degradation or destruction of life that is the wrong." [4]

What we have called the goals of the spirit are still entangled with a host of lower and purely bodily ones in man's animal nature. He is by no means compounded all of sweetness and light. A little lower than the angels he may be, and god-like in many of his attributes, but an animal he still remains. What makes him so exciting and interesting, so significant in the order of nature, is not only that he is the place where matter and spirit meet but that he is the battleground where a higher nature is struggling to emerge from a lower one; where those bodily goals, so necessary for individual survival in his long ascent up the evolutionary pathway, are being challenged for supremacy by spiritual ones that are beginning to emerge. Two moral codes are fighting in him, the code of the jungle and the code of the good society. The self, as we have seen, is a basic biological fact and therefore *self*ishness is the motive for most animal behavior—if not selfishness for the individual, as it oftenest is, then selfishness for the family or herd.

As man developed more highly organized societies these have required as a condition for their success that he give up much of the grasping, selfish behavior so necessary in his earlier course and devote himself to the welfare of his group, since only thus can his own welfare be assured. A society has often been likened to an organism since it is an organized system, not of cells but of individual human beings. The parts of an organism are differentiated, each performing its particular activity but all so interrelated that the welfare of the whole is maintained. In much the same way there is division of labor in a human social organism, each indi-

3. Hadley Cantril, "Ethical Relativity from the Transactional Point of View," *Jour. Philosophy*, 52:677-687, 1955.
4. H. S. Jennings, *The Universe and Life* (New Haven: Yale University Press, 1933), p. 74.

vidual's activity being so related to those of others that the society as a whole maintains its health and vigor. If any part or organ of the body fails to perform its normal function, the goal of the whole will not be reached and there are illness and suffering. Unless by homeostasis the normal state can be again restored, death may ensue. In somewhat the same way, antisocial behavior by an individual will injure the social group and cause unhappiness and pain not only to himself but to many others. Conduct that violates the goals of society or of the individual is wrong conduct. The concept of self must thus be expanded to include the greater self of society and finally of all mankind. Thou shalt love thy neighbor as thy*self*. It is out of this expanded self that love is born. This new relationship, grown from its simple beginnings in the world of beasts, has helped lift man's goals and aspirations higher than they ever could have reached if his activities were still centered in himself. It is an important step in the expansion of his spiritual life.

Is There a Genetic Basis for Altruism? *

by J. B. S. Haldane (1932)

It can be shown mathematically that in general qualities which are valuable to society but usually shorten the lives of their individual possessors tend to be extinguished by natural selection in large societies unless these possess the type of reproductive specialisation found in social insects. This goes a long way to account for the much completer subordination of the individual to society which characterises insect as compared to mammalian communities. Of course on Lamarckian principles one would expect exactly the opposite effect. The worker bees are descended from queens and drones, none of which have worked for very many generations, probably some million. One would expect the complex instincts of the worker to be gradually lost by disuse in these circumstances. They are not. Man on the other hand is, on the whole, induced by society to behave better than he would if left to his own devices. On Lamarckian principles he ought to be getting innately better in each generation. There is, unfortunately, no evidence for this view.

What is more . . . I doubt if man contains many genes making for altruism of a general kind, though we do probably possess an innate predis-

* From J. B. S. Haldane, *The Causes of Evolution*, Harper & Row Publishers, Inc., New York, 1932. Reprinted by permission.

position for family life. But psychologists are perhaps right in regarding social life as an extension of family life, and theologians can use no more vivid metaphors than the fatherhood of God and the brotherhood of man. For in so far as it makes for the survival of one's descendants and near relations, altruistic behaviour is a kind of Darwinian fitness, and may be expected to spread as the result of natural selection.

But the altruism of the social insects is more thoroughgoing. That is why moralists tell us to imitate them. But it is hard to see how such behaviour could become congenitally fixed in a species which did not practise reproductive specialisation. It may be that, as Hudson [1] suggested, man will adopt this practice, but the first steps to it, as pointed out by my wife,[2] would involve a very drastic interference with our present moral code, and would be most violently opposed by the moralists who would like us to imitate the insects in other respects.

1. Hudson (1919), *A Crystal Age,* London.
2. Haldane, Charlotte (1926), *Man's World,* London.

Society as a Superorganism

Editor's Note. Experiments with highly organized colonies of insects have shown that the differentiation of the specialized members is not inherited by the individual insect but appears to be chemically controlled by the whole colony in the course of its development in very much the same way as the growth and development of a single organism is controlled. These discoveries raise some interesting questions. Does natural selection operate primarily on the colony as a unit? Is a highly integrated group of specialized organisms in the process of evolving into a superorganism?

The Termite and the Cell *

by Martin Lüscher (1953)

Termites, closely related to the roaches, are relatively primitive insects which have nevertheless developed extremely elaborate social habits. There are no solitary termites. Of about 2,000 known species, all have at least three structurally different forms, which live together as castes with different functions in the colony. Some have as many as six or seven easily distinguishable castes.

Each caste generally possesses its own special instincts. The king and queen, which develop from winged forms that lose their wings after the nuptial flight, start the colony. They are extremely well adapted to their single occupation of reproduction: some tropical queens lay as many as 20,000 eggs a day. When they die, nymphs in the colony develop into a special caste to take over their procreative function. These nymphs grow reproductive organs, molt and become fertile. They are called supplementary reproductives.

All other castes in the colony are sterile, though they contain both sexes. There are soldier castes equipped with enormous jaws to defend the colony against predators. Many species also have worker castes, which build the nests and runways, collect food to feed the colony, and take care of the eggs. Termite species of the temperate regions usually have no true worker caste; their growing nymphs do the work.

The caste arrangement among the termites has attracted investigators for a long time. Early authorities thought that the differences among the castes were hereditary, but it is now known that they develop according to the needs of the colony, just as the embryonic cells of a single animal differentiate into blood cells, bone cells, muscle cells and so on. Thus a colony of termites is a kind of superorganism in which we can study differentiation in a nice, convenient way.

It was the late S. F. Light and his colleagues at the University of California who demonstrated that the caste of a termite is not fixed by its genes. They discovered that any nymph in a colony of termites could develop into either a soldier or a king or queen of the supplementary type, depending on what the situation demanded. For example, in the species of damp-wood termites they studied, a new colony always has one soldier, and only one, in the first brood. When they removed the soldier, another always developed, from a nymph which would not normally have become a soldier. This occurred up to as many as six times; each time a soldier developed to replace the one removed. Likewise the elimination of reproductive termites caused replacements to develop.

Light's group concluded that the soldier and reproductive termites already present in a colony inhibited the development of additional ones, and that they exercised this inhibition by means of some kind of "social hormone." To test this theory the experimenters fed extracts from reproductive termites to developing nymphs. There was a slight inhibition in the production of new kings and queens and a delay in egg-laying, but the effect was not sufficiently clear-cut to prove the social hormone theory. Recent experiments, however, have given the theory more definite support.

If a termite colony is considered as a superorganism, then caste differentiation may be looked upon as an embryological problem. The differentiation of cells in an embryo is always initiated at a certain critical stage in their development; in other words, a given group of cells cannot change until it is ready. One may assume that a nymph in a termite colony similarly can be induced to differentiate only at a critical stage of readiness.

We have studied the critical period in the European dry-wood termite. We marked the nymphs individually with colored spots and kept them under observation daily for two years in flat nests. They went through several molts. As individuals molted they were measured and marked again, since they lost the spots with the castoff skin. From time to time the king and queen were removed so that new reproductive termites would develop from the nymphs.

This investigation showed that there is a certain critical period during the molting interval when a nymph has the capacity to differentiate into a special adult caste. But whether it will do so depends on the make-up of the colony's population. Most of the nymphs never differentiate, though they may go on molting at regular intervals. These arrested nymphs function as workers in this species, which has no true worker caste. We call them "pseudo-workers." A pseudo-worker may develop wing pads and become a winged adult.

The pseudo-workers are the undifferentiated elements of a colony, like the undifferentiated cells of an organism. What makes them convenient for study of differentiation is that they can be followed individually, unlike the cells in an organism.

It is easy to produce new adults of the reproductive caste experimentally. All one has to do is to remove the king or queen or both from the colony. Within 10 to 20 days several supplementary reproductive adults of both sexes appear. But the colony will tolerate only one king and one queen; it promptly kills and eats the excess.

So long as the reproductive pair is present, no others are produced. Yet if the pair is removed for only 24 hours, new candidates at once begin to develop as their replacements. In that 24-hour period the differentiation of the candidate nymphs is irreversibly determined. Even though the original king and queen are returned to the colony, the nymphs that have started to differentiate will go on developing and become reproductive adults in 8 to 14 days.

The capacity of a given nymph to change into a reproductive adult depends on the stage of its molting cycle. Any nymph that has just molted when the king and queen are taken away is certain to change. This capacity decreases exponentially with time after the molt: 19 days after the molt it has dropped to 50 per cent (*i.e.*, only half of the nymphs at this stage will change), and 38 days after the molt it is only 25 per cent. The curve showing the decline in the capacity to change resembles a radioactive decay curve (See Figure 6–2.)—the half-life period in this case being 19 days (at a temperature of 80 degrees Fahrenheit). That is, after 19 days half of the originally capable termites have lost their competence to change. As in the disintegration of radioactive atoms, we cannot predict just which individuals will lose the capacity in 19 days; all we can say is that each has a 50 per cent chance of losing it by that time.

From the fact that the loss of competence is an exponential function we

Figure 6–1. *Developmental possibilities of a nymph of the termite* Kalotermes flavicollis *are shown in this drawing. From the eggs at the bottom of the drawing hatch the young nymphs. They molt five to seven times until they reach the pseudergate stage, represented by the termite in the middle of the drawing. At this stage the termite can molt many times without growing, and under environmental influences can change into a supplementary reproductive (left side of drawing), or, by way of a soldier nymph stage, into a soldier (right). At the pseudergate stage the nymph can also change, by way of two wingpadded nymphs, into a winged form (top). Reproductives and soldiers may also arise from younger or wing-padded nymphs.*

350

may conclude that it is due in each individual to a single biochemical event, possibly the disintegration or synthesis of a specific molecule.

All this bears a striking resemblance to an embryo cell's loss of competence to change. The University of Rochester embryologist Johannes Holtfreter has made a series of transplantations of ectoderm (outer) tissue from a young amphibian embryo into the site where nerve tissue

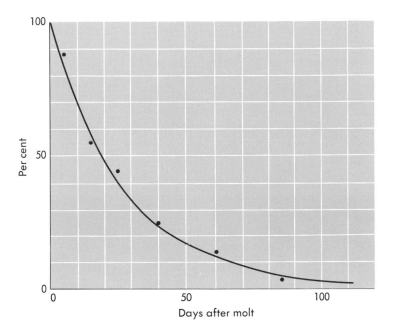

Figure 6–2. Probability of change in a nymph decreases with time after molt. With the proper stimuli all nymphs that have just molted will change. Of the nymphs that molted 19 days earlier, only 50 per cent will change.

develops in another embryo. He found that the tissue he took from the donor gradually lost its competence to change into nerve tissue as he kept it in artificial culture longer and longer before implanting it. For the tissue as a whole the decline in ability to differentiate was not exponential, but for the individual cells in the tissue it very likely is, as a single cell must change or not change: there are no gradations in between. The changes in the transplanted tissue ranged from complete alteration to a spinal cord to development of only a few neural cells.

We must of course be cautious about applying the conclusions about the termite "superorganism" to an actual organism, but it seems reasonable to assume, as a working hypothesis, that the loss of ability to differentiate in cells, as in individual insects in a termite colony, is effected by a single event, possibly a change in a single molecule in each cell.

In a termite colony, as we have seen, differentiation depends on two things: (1) the nymph's competence to change, and (2) inhibition, or lack of inhibition, by the colony. How is this inhibition exerted: by some active substance, a social hormone as Light suggested, or merely by a scent or other sensory warning given off by the king and queen?

To test these possibilities we observed two colonies of termites separated by a fine metal screen through which the colonies could maintain contract by rubbing antennae. One colony had a king and queen, the other was orphaned. The termites showed a strong tendency to touch antennae through the screen. Now the experiment had a curiously mixed result. The contact through the screen did not prevent the groups from behaving, in one sense, like two separate colonies: the orphaned colony always proceeded to produce its own king and queen. But in many instances it promptly killed them, as if it perceived the king and queen on

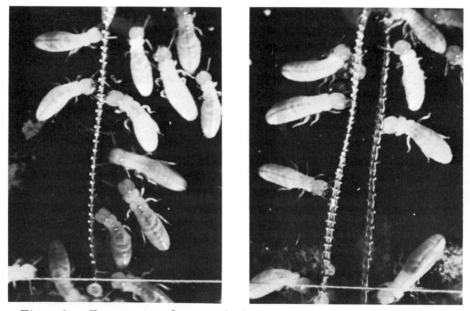

Figure 6–3. Two termite colonies at the left are separated by wire mesh. When the king and queen are removed from one, they are regenerated but then killed. At the right the colonies are separated by a double screen, which keeps the antennae of their members from touching. These colonies behave as though they were isolated. (Photographs by M. Lüscher, University of Bern. Reprinted by permission.)

the other side of the screen and considered its own surplus. This confusion progressed so far that the colony went on producing and killing kings and queens until it had almost annihilated itself.

When the colonies were separated by a screen which prevented any contact, the orphaned colony produced and supported its own king and queen in the normal way. On the whole the experiment upheld the idea that an active principle transmitted by contact, that is, a social hormone, is responsible for the suppression of differentiation.

It seems likely that production of the soldier caste also is regulated by an inhibitory substance. But Frances Weesner of the University of California has recently discovered that in at least one species of termite the production of soldiers is controlled by a promoting factor as well as an inhibitory one.

The upshot of all this work on termites is that the balance of castes in a termite society seems to be maintained by means of certain special hormones, produced by the differentiated adults and acting upon the undifferentiated nymphs. A similar theory would explain many facts in cell differentiation. In an organism the various kinds of differentiated cells are always kept in constant proportions. By analogy with the termite colony, we might assume that each type of cell produces a specific hormone which inhibits the production of an excessive number of cells of the same type. Some evidence for this was recently gained by S. Meryl Rose

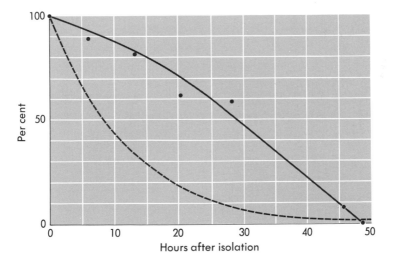

Figure 6–4. Embryonic tissue also loses its capacity for change with time after its isolation. The solid line shows the percentage of transplants that changed; the dotted line, the theoretical decrease in this capacity for single cells.

of the University of Illinois, who showed that in the presence of an amphibian animal's adult tissues the differentiation of an embryo's cells into the same tissues is inhibited.

There is another striking parallel between a termite society and an embryonic organism. Among the termites winged adults develop only in colonies with a population of at least 100. When nymphs with incipient wings are transferred from such a large colony to a small one before a critical stage of their molting cycle, they lose their wing pads at the next molt. Similarly, at one stage in the development of an embryo, differentiation occurs only if the embryo contains a certain minimum number of undifferentiated cells. The same is true in some tissue cultures: differentiation can take place only when the culture has grown to a certain size.

The study of differentiation in termite societies has discovered so many analogies to cell differentiation that it is reasonable to expect more. The superorganism idea promises to be immensely helpful, for we can isolate and experiment on individual termites as we cannot on single cells in an embryo. We must always keep in mind, however, that an analogy can do no more than suggest a working hypothesis, which is not to be taken too seriously until it is experimentally confirmed.

The Beginnings of Multicellular Organization [*]

by Maurice Sussman (1964)

Living organisms are generally classified as unicellular (the individual is a single cell) or multicellular (the individual is an organized collection of specialized cells). Although such categories imply rigid distinctions, we must remember that discrete boundaries do not exist in nature but only in the minds of biologists. In reality, there is a middle ground occupied by organisms that are neither wholly unicellular nor wholly multicellular. For example, many protozoa, algae, fungi, and bacteria are transiently colonial; they come together for a time as loose clusters of quasi-independent cells with no organization or specialization. Others attain a primitive level of organization in which some cells of the colony become specialized for feeding, others for sexual reproduction, etc. A highly complicated level of organization is attained by the cellular slime molds.

[*] From Maurice Sussman, *Growth and Development*, 2nd edition, Prentice-Hall, Inc., Englewood Cliffs, N.J. Copyright © 1964 by Prentice-Hall, Inc. Reprinted by permission.

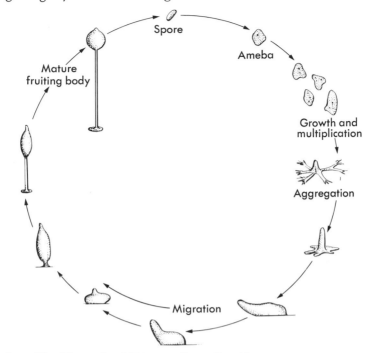

Figure 6–5. The life cycle of Dictyostelium discoideum.

Figure 6–5 summarizes the life cycle of one species called *Dictyostelium discoideum*. The cycle is divided into four stages: growth, aggregation, migration, and fruiting-body construction. It starts with the germination of spores to yield ameboid cells. The amebae live in soil or on the agar surface of a Petri dish, feed upon bacteria, and reproduce by binary fission. When the supply of bacteria is exhausted, the cells stop growing and enter the aggregation stage. The amebae stream radially toward central collecting points, and, as they reach the center, a conical mound of cells is built up. When all the cells have aggregated, each conical mound falls over on its side and is transformed into a worm-like slug up to a few millimeters in length. The slug migrates over the surface and finally comes to rest. It then proceeds to construct a fruiting body with a round mass of spores at the top, a stalk enclosed in a cellulose sheath below, and a basal disc at the bottom. These constitute three clearly different cell groups. Ultimately the spores are cast off to repeat the cycle while the stalk cells and basal disc cells desiccate and die. Figure 6–6 shows photographs of an aggregation sequence, migrating slugs, and the construction of a fruiting body.

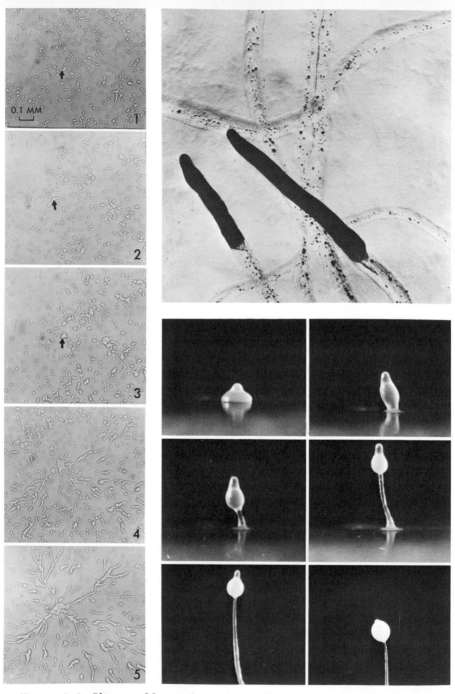

Figure 6–6. Slime mold morphogenesis. Left, an aggregation sequence (M. Sussman); top right, migrating slugs (From J. T. Bonner, The Cellular Slime Molds, Princeton: Princeton University Press, 1959, Plate II. Reprinted by permission.); bottom right, the stages of fruiting body construction—the photographs were taken approximately 1½ hours apart (Also from Bonner, Plate III).

CELLULAR DIFFERENTIATION

As we have mentioned, the fruiting body consists of three distinct cell groups: spores, stalk cells, and basal disc cells. These have different sizes, shapes, constitutions, and functions—as do the various tissue cells of a frog embryo. We can ask: What determines whether any given cell in the assembly will become a spore, a stalk cell, or a basal disc cell? This appears to depend on the order in which it entered the aggregate (Figure 6–7). Cells that enter the aggregate first become the leading element of the migrating slug and ultimately the lower stalk of the fruit. Later arrivals take up progressively more posterior positions in the slug and become upper stalk cells and spores. Those entering last constitute the tail

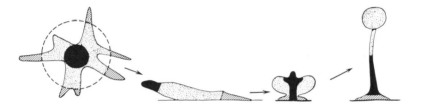

Figure 6–7. The fate of the cells depends on the order in which they enter the aggregate.

end of the slug and ultimately the basal disc of the fruit. One can tentatively imagine that the position of a cell in the aggregate and in the slug subjects it to a specific set of environmental conditions and of chemical signals imposed by its neighbors that forces it to become a spore or stalk cell or basal disc cell. These differences are reversible, at least up to the slug stage. Slugs have been cut into head and tail segments. Both could construct fruits separately. Although not completely normal, both contained all three types of cell even though the head segment would ordinarily have given rise to stalk only, while the tail segment would have yielded spores and basal disc cells but not stalk. . . .

You can also ask whether any cell at any stage of the morphogenetic [1] sequence *can give rise to progeny* capable of producing normal fruiting bodies. To answer the question, cells from aggregates, slugs, and immature fruiting bodies have been dispersed and grown separately. The colo-

1. *Morphogenetic* pertains to the structural development of an organism or part. *Morphology* is that branch of biology which deals with the form and structure of an organism as a whole. Ed.

nies derived from each could, without exception, aggregate and form normal fruiting bodies with viable spores. In the mature fruiting bodies it is impossible to free most of the stalk cells from their cellulose sheath but at least some cells coming from this region, if collected early enough, live and yield progeny that can make complete fruiting bodies. Thus, when a cell engages in fruit construction its genetic constitution remains intact, i.e., it can pass on to its descendants the ability to produce complete, normal fruiting bodies.

Cell differences can be observed even during aggregation. Thus, under proper circumstances we can demonstrate that some cells have a great capacity to affect their neighbors in such a way as to initiate centers of aggregation (such as the one seen in Figure 6–6) whereas others have very little initiative capacity. Specialization is also apparent in the migrating slugs. Thus, if the front end (about 10 per cent of the cells) is separated from the rear, the former continues to migrate while the latter stops in its tracks. If the front end is replaced, the two parts fuse and migration of the whole slug proceeds once again in normal fashion. This suggests that two "cell types" are present—leader cells and follower cells—and that the behavior of the slug results from the organized interaction of the two. Finally, we can observe the differentiated synthesis of special biochemical products. . . . These products arise at particular stages of the morphogenetic sequence, some of them being confined to special locations in the fruiting body.

The details described above suggest the following conclusions:

1. The aggregate, slug, and fruiting body are organized entities in which different cells do different things both biochemically and morphologically in order to complete the organization of the whole.

2. These specialized activities are not caused by nor do they lead to changes in the genetic material of the cell. Furthermore, at least up to the slug stage, a cell fated to engage in one kind of activity, can by a change in position be made to engage in another.

3. The underlying biochemical activities are subject to rather strict control with respect to time and place of occurrence.

CELL INTERACTIONS DURING DEVELOPMENT

The development of a multicellular system is accompanied, indeed is largely directed, by a hierarchy of cell interactions. One cell or one tissue may stimulate or inhibit the development of another by exchange of appropriate chemical agents. It is this matrix of interactions that ensures the harmonious organization of the whole, i.e., that the components will develop at the right time, in the right place, and in the right amount. . . . the use of cell interactions reaches its highest degree of complexity and

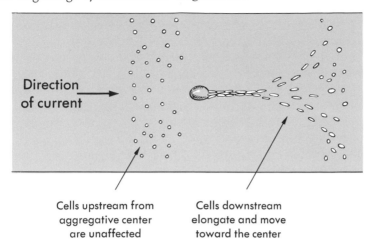

Direction
of current

Cells upstream from
aggregative center
are unaffected

Cells downstream
elongate and move
toward the center

Figure 6–8. Myxamebae aggregate under a flowing stream.

refinement in vertebrate embryos. However, the beginnings of these control mechanisms operate in primitive organisms such as the slime molds.

A case in point is the way in which the cells aggregate. A large body of evidence has been accumulated to show that outlying cells are attracted and caused to aggregate by specific chemical agents produced at the center. (Attraction of cells by a chemical compound is called *chemotaxis*.) For example, if the amebae are permitted to aggregate under a flowing stream of water (see Figure 6–8), a center forms, but the cells upstream are unaffected. Only those downstream are attracted by and move toward the center. This is just what you would expect if the center were producing some chemical agent that was being carried downstream by the water current.

REGULATION

The term regulation refers to the ability of a morphogenetic system to turn out normal products of different absolute sizes. For example, individual frog embryos or tadpoles or adult frogs of a given species can vary markedly in size and yet be perfectly proportioned. That is, the sizes of the various parts (arms, legs, head, etc.) at each developmental stage bear precise relationships to the total length or total mass. Furthermore, if cells are surgically removed from a very young frog embryo, the remainder will develop into a very tiny but nonetheless normal tadpole, again demonstrating the same relationships of parts to whole.

Slime molds also display regulatory capacities. For example, myxame-

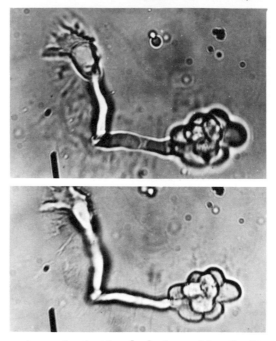

Figure 6–9. *Two views of a fruiting body formed by the "fruity" mutant of* D. *discoideum. This highly organized multicellular structure consists of 9 spores, 2 stalk cells, and 1 basal cell.* (*M. Sussman.*)

bae packed together in a solid mass produce relatively large migrating slugs and fruiting bodies; some are as much as 5 mm in length and contain hundreds of thousands of cells. If sparsely distributed, the myxamebae form tiny slugs and fruits a few tenths of a millimeter in length that contain a few hundred cells. Yet the gross proportions and the relative amounts of cells comprising the various parts of the fruiting body remain the same. This regulatory capacity is exhibited at a truly startling level by a mutant of *Dictyostelium discoideum* that can form normal fruits containing hundreds of thousands of cells or as little as 10 to 12 cells. Figure 6–9 shows an example of the latter, a fruiting body with 12 cells—9 spores, 2 stalk cells, and 1 basal disc cell. This exquisite miniature contains all the elements of a morphogenetic system—cellular differentiation, cell interactions, and regulation—but in a tiny packet of cells.

MUTANT STRAINS

A large number of mutants have been isolated that display aberrant morphogenetic capacities. Some of these can form fruiting bodies with viable spores but with fruits whose shapes and sizes are greatly different from those of the parent strains; such mutants are given names like fruity, curly, glassy, and forked. Other mutants are *morphogenetically deficient*, i.e., they cannot complete the normal morphogenetic sequence; instead they stop at various stages along the way. Thus, some cannot aggregate at all and remain as separate cells; others aggregate but cannot develop any further.

RÉSUMÉ

In summary, these are some of the lessons learned thus far from the study of slime molds:

1. Just as in the development of higher forms (vertebrate embryos, etc.), slime-mold development involves the appearance of new cell types that play a causal role in the construction of the multicellular whole. We would like to know how and when they arise and in what numbers, and what physiological mechanisms are responsible for the roles they play in morphogenesis.

2. Here, as in other developmental systems, a matrix of cellular interactions helps to regulate the process. Chemical signals are passed during aggregation, and during migration of the slug. Finally, chemical signals tell some cells to become spores, others to become stalk cells, and still others to become basal disc cells in the fruiting body. We want to know what are these signals and how they act.

3. Normal development need not involve participation of huge numbers of cells. Instead, even as few as 10 cells can provide the components and the chemical interactions needed to construct a perfect fruiting body.

A Society Is an Organism *

by Herbert Spencer (1882)

When we say that growth is common to social aggregates and organic aggregates, we do not thus entirely exclude community with inorganic aggregates: some of these, as crystals, grow in a visible manner; and all of them, on the hypothesis of evolution, are concluded to have arisen by integration at some time or other. Nevertheless, compared with things

* From Herbert Spencer, *Principles of Sociology*, D. Appleton and Company, New York, 1882, Vol. 1.

we call inanimate, living bodies and societies so conspicuously exhibit augmentation of mass, that we may fairly regard this as characteristic of them both. Many organisms grow throughout their lives; and the rest grow throughout considerable parts of their lives. Social growth usually continues either up to times when the societies divide, or up to times when they are overwhelmed.

Here, then, is the first trait by which societies ally themselves with the organic world and substantially distinguish themselves from the inorganic world.

It is also a character of social bodies, as of living bodies, that while they increase in size they increase in structure. A low animal, or the embryo of a high one, has few distinguishable parts; but along with its acquirement of greater mass, its parts multiply and simultaneously differentiate. It is thus with a society. . . .

The multiplying divisions, primary, secondary, and tertiary, which arise in a developing animal, do not assume their major and minor unlikenesses to no purpose. Along with diversities in their shapes and compositions there go diversities in the actions they perform: they grow into unlike organs having unlike duties. Assuming the entire function of absorbing nutriment at the same time that it takes on its structural characters, the alimentary system becomes gradually marked off into contrasted portions; each of which has a special function forming part of the general function. A limb, instrumental to locomotion or prehension, acquires divisions and sub-divisions which perform their leading and their subsidiary shares in this office. So is it with the parts into which a society divides. A dominant class arising does not simply become unlike the rest, but assumes control over the rest; and when this class separates into the more and the less dominant, these, again, begin to discharge distinct parts of the entire control. With the classes whose actions are controlled it is the same. The various groups into which they fall have various occupations: each of such groups also, within itself, acquiring minor contrasts of parts along with minor contrasts of duties. . . .

This division of labour, first dwelt on by political economists as a social phenomenon, and thereupon recognized by biologists as a phenomenon of living bodies, which they called the "physiological division of labour," is that which in the society, as in the animal, makes it a living whole. Scarcely can I emphasize sufficiently the truth that in respect of this fundamental trait, a social organism and an individual organism are entirely alike. When we see that in a mammal, arresting the lungs quickly brings the heart to a stand; that if the stomach fails absolutely in its office all other parts by-and-by cease to act; that paralysis of its limbs entails on the body at large death from want of food or inability to escape; that loss of even such small organs as the eyes, deprives the rest of a service essential to their preservation; we cannot but admit that mutual dependence of parts is an essential characteristic. And when, in a society, we see that the workers in iron stop if the miners do not supply materials; that makers of clothes cannot carry on their business in the absence of those who spin and weave textile fabrics; that the manufacturing community will cease to act unless the food-producing and food-distributing agencies are acting; that the controlling powers, governments, bureaux, judicial officers, police, must fail to keep order when the necessaries of life are not supplied to them by the parts kept in order; we are obliged to say that this mutual dependence of parts is similarly rigorous. Unlike as the two kinds of aggregates are in sundry respects, they are alike in respect of this fundamental character, and the characters implied by it.

How the combined actions of mutually-dependent parts constitute life of the whole, and how there hence results a parallelism between national life and individual life, we see still more clearly on learning that the life of every visible organism is constituted by the lives of units too minute to be seen by the unaided eye.

An undeniable illustration is furnished us by the strange order *Myxomycetes.* The spores or germs produced by one of these forms, become ciliated monads which, after a time of active locomotion, change into shapes like those of amœbæ, move about, take in nutriment, grow, multiply by fission. Then these amœba-form individuals swarm together, begin to coalesce into groups, and these groups to coalesce with one another: making a mass sometimes barely visible, sometimes as big as the hand. This *plasmodium,* irregular, mostly reticulated, and in substance gelatinous, itself exhibits movements of its parts like those of a gigantic rhizopod; creeping slowly over surfaces of decaying matters and even up the stems of plants. Here, then, union of many minute living individuals to form a relatively vast aggregate in which their individualities are apparently lost, but the life of which results from combination of their lives, is demonstrable. . . .

On thus seeing that an ordinary living organism may be regarded as a nation of units that live individually, and have many of them considerable degrees of independence, we shall perceive how truly a nation of human beings may be regarded as an organism.

The relation between the lives of the units and the life of the aggregate, has a further character common to the two cases. By a catastrophe the life of the aggregate may be destroyed without immediately destroying the lives of all its units; while, on the other hand, if no catastrophe abridges it, the life of the aggregate immensely exceeds in length the lives of its units. . . .

From these likenesses between the social organism and the individual organism, we must now turn to an extreme unlikeness. The parts of an animal form a concrete whole; but the parts of a society form a whole that is discrete. While the living units composing the one are bound together in close contact, the living units composing the other are free, not in contact, and more or less widely dispersed. . . .

Though coherence among its parts is a prerequisite to that co-operation by which the life of an individual organism is carried on; and though the members of a social organism, not forming a concrete whole, cannot maintain co-operation by means of physical influences directly propagated from part to part; yet they can and do maintain co-operation by another agency. Not in contact, they nevertheless affect one another through intervening spaces, both by emotional language, and by the language, oral and written, of the intellect. . . .

The mutual dependence of parts which constitutes organization is thus effectually established. Though discrete instead of concrete, the social aggregate is rendered a living whole.

But now, on pursuing the course of thought opened by this objection and the answer to it, we arrive at an implied contrast of great significance —a contrast fundamentally affecting our idea of the ends to be achieved by social life.

Though the discreteness of a social organism does not prevent subdivision of functions and mutual dependence of parts, yet it does prevent that differentiation by which one part becomes an organ of feeling and thought, while other parts become insensitive. . . .

Hence, then, a cardinal difference in the two kinds of organisms. In the one, consciousness is concentrated in a small part of the aggregate. In the other, it is diffused throughout the aggregate: all the units possess the capacities for happiness and misery, if not in equal degrees, still in degrees that approximate. As, then, there is no social sensorium, it results that the welfare of the aggregate, considered apart from that of the units, is not an end to be sought. The society exists for the benefit of its members; not its members for the benefit of the society. It has ever to be remembered that great as may be the efforts made for the prosperity of the body politic, yet the claims of the body politic are nothing in themselves, and become something only in so far as they embody the claims of its component individuals.

Is Mankind Developing into a Superorganism? *

by René Dubos (1962)

The large cities of the modern world remind one of beehives and of ant hills. Each individual person in them has a specialized function and returns to rest at a particular place—as if he were but one among so many other interchangeable units in an immense colony. As human societies become larger, older, and more dependent on technology, the colonial organization becomes more intricate and less flexible. The formal resemblance between human institutions and the colonies of social insects is indeed so striking that one might conclude that modern man is doomed to become a specialized worker or soldier in a stereotyped social machine. Fortunately, the formal resemblances are misleading, because human societies differ in a fundamental way both in origin and in structure from all known insect colonies. Whereas each of the latter represents a gigantic single family, human societies in contrast represent groupings of many families. Whatever their degree of specialization, all the members of a given insect colony are but as many brothers and sisters from the same parents. The members of human groups are bound by social ties, but they differ in genetic make-up.

The great genetic diversity of human beings in most parts of the world provides an immense range of potentialities for evolutionary development. Furthermore, as we have seen, evolution in man is now primarily of a psychosocial nature, and the great density of human populations increases enormously the rate of change by facilitating contacts and cross-fertilization of ideas. We have seen also that two types of changes, which appear at first sight contradictory, seem to be occurring at the same time. On the one hand, individual human beings become more and more differentiated; on the other hand, all of them become increasingly integrated into complex social structures. For this reason, human societies are not likely to achieve the state of almost complete stability which is made possible by the rule of instincts in insect colonies.

The Fallacy of Aggregation Ethics *

by George Gaylord Simpson (1949)

These proposed systems, which might collectively be called aggregation ethics, see as ethically good the increased aggregation of organic units into higher levels of organization. One form of the argument runs more or less as follows: Evolution has involved a succession of organic levels, with progressive complication and perfection of coordinated structure and function on each level. The protozoans and other one-celled (or noncellular, or cellularly undivided) organisms represent a lowest level. The next level is that of multicellular individuals, metazoans, with increasing differentiation of the cells and their grouping into organs with increasingly specialized functions. There follows another level, the highest, in which metazoan individuals were aggregated into hyperzoan organisms. The hyperzoan or "epi-organism" is society, into which individuals are to merge and be integrated as subordinate parts of a higher whole. It is, in fact, the so-called organic state, considered as having an individuality and life of its own. In such a state individuals, or "persons," as some adherents of this view prefer to call them, exist for the state, the good and rights of which are separate from and superior to that of the individuals composing it.[1]

The whole argument and its ethical implications among which may be support of authoritarian and totalitarian ideologies as biologically and ethically right, are thoroughly erroneous. When the state or any other social structure is called an "organism," the word is being used in a way fundamentally different from its use for a biological organism such as an

* From George Gaylord Simpson, *The Meaning of Evolution.* Yale University Press, New Haven and London. Copyright 1949 by Yale University Press. Reprinted by permission.

1. The organismic theory of the state was pre-evolutionary and in its modern form seems to stem directly from Comte. It was left for some modern biologists to "discover," wholly falsely, as I believe, that the trend of evolution confirms the organic nature of the state, makes this biologically "right," and establishes as the basis for ethics the promotion of the state as an organism. Among those who have most insistently and fully expressed this idea is the physiologist R. W. Gerard, for instance in "A Biological Basis for Ethics" (*Philosophy of Science,* 9 [1942], 92–120). The whole complicated subject of organic levels was rather fully treated from many points of view, including that of Gerard and the "epi-organism school," in R. Redfield, *Levels of Integration in Biological and Social Systems* (Lancaster, Pennsylvania, Jacques Cattell, 1942).

ameba, a tree, or you. The state is not an individual or a person in anything like the same sense that these organisms are individuals. The parts that compose it retain, and indeed intensify, their organic individuality. Their relationship to the social unit is entirely different from the relationship of cells or organs to a metazoan individual. Calling a state an "organism" and concluding that it is therefore comparable with a metazoan organism is a glaring example of the fallacy of the shifting middle term. Use of the comparison as an analogy provides an interesting descriptive metaphor, but its use in support of an aggregation ethic is a particularly egregious misuse of analogy, confusing it with equivalence and extending it as an interpretive principle far beyond the point to which it is valid even as metaphorical description.

Furthermore, if the fallacious use of "organism" as a shifting term is avoided, it is quite evident that merging of the individual into a higher organic unit is not a common trend in evolution and, specifically, is not at all a trend in human evolution. The trend in human evolution and in many other evolutionary sequences has been, on the contrary, toward greater individualization. The particular type of social organization characteristic of man, as opposed, say, to that in an anthill, has been based on high individualization of its members and has intensified this.

The Collective Issue *

by Pierre Teilhard de Chardin (1955)

The still unnamed Thing which the gradual combination of individuals, peoples and races will bring into existence, must needs be *supra-physical,* not *infra-physical,* if it is to be coherent with the rest. Deeper than the common act in which it expresses itself, more important than the common power of action from which it emerges by a sort of self-birth, lies reality itself, constituted by the living reunion of reflective particles.

And what does that amount to if not (and it is quite credible) that the stuff of the universe, by becoming thinking, has not yet completed its evolutionary cycle, and that we are therefore moving forward towards some new critical point that lies ahead. . . .

We are faced with a harmonised collectivity of consciousnesses equivalent to a sort of super-consciousness. The idea is that of the earth not only becoming covered by myriads of grains of thought, but becoming enclosed in a single thinking envelope so as to form, functionally, no more than a single vast grain of thought on the sidereal scale, the plurality of individual reflections grouping themselves together and reinforcing one another in the act of a single unanimous reflection.

This is the general form in which, by analogy and in symmetry with the past, we are led scientifically to envisage the future of mankind, without whom no terrestrial issue is open to the terrestrial demands of our action.

To the common sense of the "man in the street" and even to a certain philosophy of the world to which nothing is possible save what has always been, perspectives such as these will seem highly improbable. But to a mind become familiar with the fantastic dimensions of the universe they will, on the contrary, seem quite natural, because they are directly proportionate with astronomical immensities.

In the direction of thought, could the universe terminate with anything less than the measureless—any more than it could in the direction of time and space?

We as yet understand nothing of the way in which our conscious selves are related to the separate lives of the billions of cells of which the body of each of us is composed. We only know that the cells form a vast nation, some members of which are always dying and others growing to supply their places, and that the continual sequence of these multitudes of little lives has its outcome in the larger and conscious life of the man as a whole. Our part in the universe may possibly in some distant way be analogous to that of the cells in an organised body, and our personalities may be the transient but essential elements of an immortal and cosmic mind.

Francis Galton
Inquiries into Human Faculty and Its Development, 1883

Suggestions for Further Reading

* Bonner, John Tyler: *The Ideas of Biology,* Harper Modern Science Series, edited by James R. Newman, Harper & Brothers, 1962. Chapters 4 and 5 on the development of multicellular organisms and colonies of organisms are especially interesting in connection with the readings in this Part.

* Cassirer, Ernst: *An Essay on Man,* A Doubleday Anchor Book, Doubleday & Company, 1953. An introduction to a philosophy of human culture which reveals an underlying unity in all man's creative activities.

* Paperback edition

* Sinnott, Edmund W.: *The Biology of the Spirit*, The Viking Press, 1955. Chapter II contains very interesting descriptions of the phenomenon of integrated self-regulation in living organisms.

* Sussman, Maurice: *Animal Growth and Development*, Prentice-Hall Foundations of Modern Biology Series, Prentice-Hall, Inc., 1961. Readers who would like to know more about the process of morphogenesis (multicellular organization described in physical and chemical terms) will find the whole of Sussman's book very rewarding. It is, however, somewhat more technical than most of the books recommended here.

* Teilhard de Chardin, Pierre: *The Phenomenon of Man*, Harper Torchbook, Harper & Brothers, 1961. Teilhard's thesis of an emerging collective soul and mind of humanity can only be fully appreciated by reading the whole of his very original and provocative book.

* Whyte, Lancelot Law: *The Next Development in Man*, A Mentor Book, The New American Library of World Literature, 1949. The author presents an original view of the relationship of man to society, a view based upon the principle of unitary thought. He suggests the kind of civilization man must build in order to fulfill his destiny.

7

How Successfully Is Man Controlling His Environment?

The problems brought about by man's misuse and dissipation of natural resources and by the creation of an artificial environment are discussed in these selections. Will technology be able to make good these losses or are we creating a less favorable environment for future generations?

How Successfully Is Man Controlling His Environment?

Introduction

HOMO SAPIENS is the only species that has been able to adapt the environment to its own needs, instead of being adapted by the slow process of natural selection to fit the environment. This rapid adaptation has been achieved through innovations such as housing, clothing, agriculture, but with these changes in the environment, the balance of nature has been upset. Essential materials are not being recycled, and resources are being used more rapidly than they are being replaced. These facts raise the question of whether man is creating a favorable environment for himself today at the expense of making a less favorable one for future generations.

Actually the whole matter of the use of natural resources is so vast that only a small sampling of the various aspects can be presented here. Considerable attention is given to the problem of soil erosion on a world-wide basis. Some mention is made of air and water pollution and of the problems arising from the construction and design of cities. For instance, by building highways and city streets we have waterproofed a considerable percentage of the earth's surface, changed the distribution of rainfall, and even affected the climate of the earth. In the cities the air is drier and the nights are warmer. The frost-free season in Chicago is a whole month longer than in the surrounding countryside. Dust in the air has affected the amount of rainfall and the amount of sunshine reaching the earth's surface. These differences are not just confined to city areas. The observed dustiness of the air over the United States, outside of cities, doubled from 1957 to 1963. Even in Hawaii the turbidity of the air has increased by 30 per cent in the last ten years. And average temperatures over the earth have dropped markedly since 1950.[1]

Edward Hall presents the view that modern cities are also creating unfavorable cultural environments for many people, exposing them to stressful conditions which may precipitate psychological problems. Stress and its effects on biological communities will be studied in more detail in Part 8. Here the main emphasis is on the use of natural resources. Living space, for instance, is a fundamental resource and it is becoming increasingly scarce.

Many people believe that the source of the cure for these ills is identical with their cause—more technology applied to each specific aspect of the

1. These facts and figures are taken from "The Research Frontier" by Reid A. Bryson, *Saturday Review*, April 1, 1967.

problem. Land erosion, for example, can be counteracted by better farming techniques, better fertilizers, better hybrid grains. Just recently an improved form of rice has been developed which is reported in some areas to yield ten times as much as the older types of rice. The threat of food shortage can also be met by less conventional solutions, such as making a palatable human food from algae.

Water supplies could be dramatically increased by the desalination of sea water. Several techniques are now available to do this, although they are too expensive to be used on a large scale. But more efficient methods may soon be perfected.

Our dwindling energy sources, now obtained largely from fossil fuels, will become almost limitless when power can be obtained from atomic fusion. Energy is one of the key natural resources because it can help solve other scarcity problems. For instance, the utilization of lower grade metallic ores is feasible, but it requires the expenditure of more energy to mine and refine them.

Again, the whole subject of the technological possibilities for the future is much too large to be dealt with in any detail. The reader is undoubtedly already familiar with many of the possibilities as they are presented in current magazines and newspapers.

Several of the authors of these readings point to the dangers of relying upon a more and more complicated technology to solve these fundamental problems. Even today, as Norbert Wiener tells us, a disruption in the water supply for the city of New York for a period of five or six hours would cause death and disease. In the not too distant future, when all the readily available natural resources are expended, technology will become dependent on complex machinery to extract low-grade ores, desalinate sea water, and unlock energy from the atom. Then any major catastrophe that destroyed the machinery would cause the whole complex to grind to a halt. "And it is difficult to see how man would again be able to start along the path of industrialization with the resources that would then be available."[2]

Ironically, the inventive mind that gave man his unusual adaptability, and resulted in his dominance, could be leading him into the evolutionary cul-de-sac of overspecialization. Other organisms that become overspecialized have not survived an unexpected change in the environment.

Furthermore, a highly technological society creates an increasing demand for integration and efficiency which may threaten man's freedom. "It seems clear," says Harrison Brown, "that the first major penalty man will have to pay for his rapid consumption of the earth's non-renewable resources will be that of having to live in a world where his thoughts and

2. See Harrison Brown selection, p. 414.

actions are ever more strongly limited, where social organization has become all-pervasive, complex, and inflexible, and where the state completely dominates the actions of the individual." [3]

Is it possible to avoid these pitfalls and still continue to expand our scientific knowledge? Is a complete rejection of science and a "return to nature" the only solution? René Dubos suggests that man can solve his problems only by turning his attention to the human aspects. "Unless the science of material things becomes equalled by the science of humanity," he says, "I fear that man cannot survive indefinitely in the artificial environment he is creating." [4]

The idea of an artificial environment is highly distasteful to many people but is it really so bad in itself? We already live in a somewhat artificial environment and some aspects of it are beautiful and satisfying. Handsome cities have been built, swamps have been drained, gardens and parks have been created. A fertile farm landscape with well-tended fields of grain is a delightful sight. In these cases man has achieved a harmonious relationship with nature.

On the other hand, it is frightening to hear city planners talk about the megalopolis of the future which will extend over the whole land so that from birth to death we may live in a completely man-made environment, isolated from many other species whose presence we now enjoy. We are told, "We shall learn to get along without birds if necessary." [5] But, as René Dubos points out, some association with other living things may be psychologically necessary to our well-being: "Living is everywhere a collective enterprise. However primitive or complex, all organisms in nature spend much of their existence in the company of several representatives of their own kind. Even more interestingly, they always occur in intimate and lasting associations with other forms of life not genetically related to them. In fact, most living things soon die of starvation or disease when separated from the other species with which they exist in partnership under natural conditions." [6]

The complexity and interrelatedness of the living world is a subject which is just beginning to be appreciated and studied scientifically. New sciences have been springing up in the past century: ecology, ethology, cultural anthropology. These sciences approach their subject matter in a different manner from the older physical sciences. They strive for a total picture of the many contributing factors which maintain a state of dynamic equilibrium in the living world.

3. From Harrison Brown, *The Challenge of Man's Future,* The Viking Press, New York, 1954, p. 219.

4. From René Dubos, *Man Adapting,* Yale University Press, 1965.

5. See quote on page 429.

6. From René Dubos, *Man Adapting,* p. 90.

The method of the physical sciences has traditionally been to simplify the problem to be studied, to isolate the phenomenon in question, and to control all the other variables. This method does not work in studying the relationship between living things and their environment. The factors are so interwoven that, as in a plate of spaghetti, one strand cannot be removed without disturbing all the others.

Several of the authors suggest that if we are ever to create an environment which satisfies all man's needs, we must stop thinking in terms of solving isolated mechanical problems and more in terms of understanding the total biotope. This may involve training minds that are capable of evaluating many aspects simultaneously and making judgments based on this broad view. Specialists are necessary, of course, but the absence of any programs for training generalists, for training minds that are capable of synthesizing knowledge from different fields, may be another symptom of the trend toward overspecialization. The new sciences of life, on the other hand, are a hopeful sign that ways can be found of using science and rationality for the creation of what Dubos calls "a science of humanity," a science that will be capable of designing an environment suitable for the maximum enjoyment and development of human beings.

Destruction of the Natural Environment

Man—man has the world in the hollow of his hand. He is a standing refutation of an old superstition like predestination—or a new one like determinism. His chances seem all but boundless, and boundless might be his optimism if he had not already thrown away so many of his opportunities. That very marsh was the home of waterfowl as valuable as they were beautiful. Now they must die, because in this world all breeding grounds are already crammed full. When he slays the birds, he lets loose their prey, and his worst enemy, the insects. He wastes his forests faster than he replaces them, and slaughters the mink and the beaver and the seal. He devours his limited coal supply ever faster; he fouls the rivers, invents poison gases and turns his destruction even on his own kind. And in the end he may present the spectacle of some Brobdingnagian spoiled baby, gulping down his cake and howling for it too.

Donald Culross Peattie
Almanac for Moderns (1935)

Our Plundered Planet *
by Fairfield Osborn (1948)

In the parlance of the biologist, a "generalized" animal is one which has adhered to the general, standard form of its ancestors as if guided by the determination to progress conservatively and without distortion, or even not at all.

In contrast, "specialized" animals are those that are strongly developed in one or more particular ways. Figuratively speaking, they "choose" a special way of living and cast their fate and pin their hopes for survival on the expectation that the conditions surrounding them will remain approximately as they are. The animal world is rich with examples, some of which border on the grotesque, such as the anteater, the sloth, the giraffe and the hippopotamus. Although the advantages of specialization are clear enough, the disadvantages are very substantial. A basic one is that, as a general rule, the process of evolution is irreversible. While there are exceptions to this rule it is usually the case that once the trend is well under way there can be no withdrawal. In effect, this spells a commitment to a special environment.

Many living things have characteristics which permit them to be classified as either generalized or specialized types. This distinction can be applied, more or less appropriately, not only to the higher forms of animal life but downward through the scale even to plant life.

Man is one of the best examples of the generalized type. Early man, as observed above, lacked notable or specialized physical endowments. He excelled at nothing except in the use of his wits and in the capacity of doing things with his hands. His survival was due principally to his use of his growing intelligence in a scene where many competing forms of life were either stronger or swifter or endowed with other physical characteristics, of a specialized nature, which protected them. From a physical point of view man is still a generalized animal, only three characteristics being specialized. The first is the human eye, which is intricate and highly developed and capable of seeing in three dimensions as well as in color. Another of man's specialized characteristics is the foot. This is a wonderful mechanism, in that the elemental primate foot has been "made over" into an arched platform, able to apply a powerful force at the ball, and consequently the only foot that can take a human step. Because of this man is able to move with facility in an upright position, an accomplishment possessed by no anthropoid apes. Finally, there is the human brain, undoubtedly one of the most extraordinary cases of specialization to be found in nature. Can it prove an instrument of foresight and wisdom as well as one of amazing ingenuity?

A survey of the innumerable and various kinds of living things that have existed on this earth since the beginning of the Paleozoic era, approximately 500,000,000 years ago, indicates that the major tendency of evolution is towards specialization, which proves successful as long as conditions do not change. But in studying the present and in looking backward over the vast panorama of cycles and ages of this earth's history, one thing becomes perfectly clear and that is that conditions are constantly changing, even though, as a rule, almost imperceptibly. Faced with changing conditions, life must move to similar conditions, or change to meet changed conditions; failing to move or change it perishes. As a rule the choice was not made, or there was an incapacity to make it. More kinds of living things have become extinct than exist today. The price of survival is high.

It will be immediately apparent that, judged by his physical make-up, man is extremely generalized, which contributes to his capacity of adapting himself with readiness to the extremes of physical environment. It is likely that no other form of living thing, certainly no other mammal, has ever developed an equal capacity. Man can exist in the extreme cold of the polar regions, in the attenuated atmosphere of the earth's highest mountains, in desert heat or in the pervasive humidity of the tropics. Does this amazing capacity insure his survival? Perhaps, if there were not other far more compelling circumstances to consider.

The essence of man's situation is slowly becoming obvious. His physical adaptability, in the pattern of biological history, provided, until recently, its own guarantee of his survival. The characterizing of man as a generalized type, and therefore as one most capable of adaptation to changing conditions, seems illogical now—outdated by the course of events of even the last few decades. Today one cannot think of man as detached from the environment that he himself has created. True, one never actually was justified in doing so. Yet even as recently as the latter years of the last century, the projections of man's mind in the form of the physical changes he was effecting on the earth itself were not of sufficient extent to be recognized as a new and profound change in the evolution and even in the destiny of mankind. The groundwork had been laid in earlier centuries. The explosion, world-shaking, has occurred in this one. The mechanical, chemical and electrical sciences, man's mind-extensions, are changing the earth. A concept, recently expressed, speaks of man as now becoming for the first time a *large-scale geological force.* The effects upon man's social and political relationships are not within the immediate scope of this book, although the present world-wide disturbances in human civilization can at least partially be accounted for by the havoc, described in subsequent pages, that man is working upon his natural environment. These disturbances will unquestionably increase in violence, even to the point of social disintegration, if the present velocity of destruction of the earth's living resources continues. Man has it in his power to stop this havoc. He also still has it in his power to remedy enough of the damage that he has caused to permit the survival of his civilization. The question is, Will he do it and will he do it in time?

Man, then, has exchanged the safety and flexibility of generalized characteristics, which since his primitive days have largely contributed to his survival, for extreme specialization. Through the development of the physical sciences, funneled into vast industrial systems, he has created and continues to create new environments, new conditions. These extensions of his mind-fertility and his mind-restlessness are superimposed, like crusts, on the face of the earth, choking his life sources. The conditions under which he must live are constantly changing, he himself being the cause of the changes. In this metamorphosis he has almost lost sight of the fact that the living resources of his life are derived from his earth-home and not from his mind-power. With one hand he harnesses great waters, with the other he dries up the water sources. He must change with changing conditions or perish. He *conquers* a continent and within a century lays much of it into barren waste. He must move to find a new and un-spoiled land. He must, he must—but where? His numbers are increasing, starvation taunts him—even after his wars too many are left alive. He causes the life-giving soils for his crops to wash into the oceans. He falls back on palliatives and calls upon a host of chemists to invent substitutes for the organized processes of nature. Can they do this? Can his chemists dismiss nature and take over the operation of the earth? He hopes so. Hope turns to conviction—they *must*, or else he perishes. Is he not na-ture's "crowning glory"? Can he not turn away from his creator? Who has a better right? He has seemingly "discovered" the secrets of the universe. What need, then, to live by its principles!

❊   ❊   ❊

THE FLATTERY OF SCIENCE

Many people have the notion that lands that have been misused and that have become sterile can be restored to fertility by the use of chemi-cals. Nowadays we are so impressed by the "marvels of modern science" that we are apt to consider it capable of any accomplishment—even of patching up nature. We move, live, and have our being in a world of gad-gets and inventions. That is, about a billion of us do—the other billion live on the land trying to produce a subsistence for themselves and suffi-cient products of the land for the rest of us who live in towns and cities. It is realized of course—when one stops to think of it—that forests once de-stroyed take a long time to grow again, but it is apt to be forgotten that, for example, in some of the desolated countries around the Mediterranean the forests have never reappeared—nor could they, even with the most careful nurturing, within many, many lifetimes because the land has been denuded of its soil. There is a general impression that all that is needed to make such land areas productive is to apply fertilizers. There could be no greater illusion.

The treatment of soil by chemicals and fertilizers is, it is true, of consid-erable help in preventing land from becoming sterile. Under existing methods of rapid crop production, the mineral elements that nourish plant growth are removed from the soil faster than they can be replaced by natural processes, and consequently fertilization of the soil with such chemicals as lime, phosphate and potash is an accepted practice in agri-culture today. Further, the organic matter of which soils have been de-pleted can be restored in large measure by the application of animal manure or by means of so-called "green manuring," which consists of plowing leguminous plants and grasses into the surface. There can be no argument concerning the fact that manures and chemical fertilizers are necessary aids in maintaining soil fertility, but at best fertilizers are corrective supplements. In no sense should they, especially chemical fertilizers alone, be thought of as substitutes for the natural processes that account for the fertility of the earth. In the long run life cannot be supported, so far as our present knowledge goes, by artificial processes. The deterioration of the life-giving elements of the earth, that is proceeding at a constantly accelerating velocity, may be checked but cannot be cured by man-applied chemistry. As to the speed of this deterioration, it has recently been estimated that there has been a greater loss of productive soil in the last few decades than the accumulated loss in all previous time. While this is solely an estimate and not substantiable by actual figures, it is nevertheless one made by able men who have attempted to present the case of erosion throughout the world with the greatest accuracy at their command.

There are two reasons why artificial processes, unless they are recognized as complements to natural processes, will fail to provide the solution to the present crisis. The first is concerned with the actual nature of productive soil. The second reason is a practical one and hinges upon the difficulty, if not impossibility, of instructing great numbers of people who work on the land regarding the extremely complicated techniques that need to be applied to produce even a reasonable degree of fertility by artificial methods.

As to the first reason, it is necessary to bear in mind that the body of the soil itself must be stable in the sense that it becomes impracticable to maintain good soil conditions if land is in movement because of abnormally active processes of erosion by either wind or water. Consequently, the first problem under any circumstances continues to be the prevention of continuing erosion which, as we shall see, has already so greatly devastated so many areas on the earth where inadequate methods of agriculture were employed. Further, it is above all necessary to keep the fact in mind that fertile soil is *alive* in the sense that it harbors many different kinds of living organisms that function in relationship to one another and provide, in effect, the health and productivity of the soil itself. *Ingenious as man is he cannot create life.* Stated simply, soil is fertile largely because of the living organisms that are within it, in combination, of course, with its mineral nutrients. Nowhere on this earth has nature provided a more intricate pattern of interrelated life than in the soil. There is an immense variety of animal life, ranging in size and kind from burrowing rodents, insects, earthworms, down through the scale to animals and life forms of microscopic size such as protozoa and bacteria. These living things make two basic contributions to soil fertility, one as important as the other. The first is soil cultivation, including the letting in of air and water through the soil made by the passage of their bodies, whether large or small. In this job countless living things are at work—among them various kinds of insects and other invertebrates running in numbers to millions per acre. The earthworm is the most familiar but by no means the most important of nature's soil farmers.

The second contribution comes through the fact that animal life and living bacteria are the media by which organic remains are mixed with the minerals in the soil. These living elements—innumerable hosts of them, invisible to the eye—are, in effect, the soil chemists. The scope and complexity of the work of the bacteria alone are almost beyond definition. Even a summary glance at some of their functions will illustrate the extremely complex nature of productive soil and the reasons why the successful application of man-made processes cannot effectively be substituted for the processes of nature. One of the activities, for example, of bacteria is gathering nitrogen from the air, combining it in forms which in turn can be made into protein by plants. Free nitrogen in the air cannot be used directly by plants, and it will be remembered that all living things —plant and animal—from virus to man, are essentially protein organisms. Another group of bacteria decomposes the protein existing in dead animal and plant materials in the soil and changes them to ammonia. Ammonia contains nitrogen, but most plants cannot use the ammonia as such since the nitrogen must be available in the form of nitrates. The transformation from ammonium salts to nitrates is brought about by two other groups of bacteria. The first group changes the ammonia to nitrites, which also are unusable by plants. But a second group of bacteria is standing by whose function is that of converting the nitrites to nitrates, in which form nitrogen is readily absorbed and used by the plants to build more protein, and proteins mean growth. The reverse process also takes place in that some bacteria change nitrates to nitrites and finally to free nitrogen, which goes back to the air. Here it is picked up by the nitrogen-gathering bacteria and again transformed by them into usable protein-building compounds.

There is an almost endless number of other characteristics of fertile soil. For instance, iron is a necessary element for living cells; no matter how minute the quantity, it is required for the growth and well-being of all plants and animals. In the complete absence of iron, green plants will fade to a yellow and finally die. If such a condition develops it may indicate a slow or improper bacterial activity of the soil. Certain soil bacteria are able to take up iron and accumulate it on the surface of their cells, where it is quickly changed to some more soluble form of iron compound which is then taken up by the plants.

Another important constituent of living matter is phosphorus. This element is also tied up with proteins. The bony framework of man and animals consists largely of calcium phosphate. This material is insoluble, but despite this fact it is converted to a soluble form of phosphate by the action of soil bacteria, the solvent action being largely due to carbonic acid produced by these bacteria.

In addition to the four major chemicals of soil, namely nitrates, lime, phosphates and potash, there are a number of other essential elements such as copper, manganese, zinc and boron, known as "minor" or "trace" elements. It is only in recent years that the essential nature of these constituents has become known. Previously it was erroneously assumed that the amounts used by plants were too small to be important. There is no room for doubt that most of them, perhaps all of them, are vital to health and strength, even though they are supplied by nature in minute volume. The mere fact that they exist and appear to have a definite part in the life scheme emphasizes the extreme complexity of earth fertility. The relationship between land health and the health of human beings, as well as other animals . . . is actually no more than another aspect of the delicate and complex relationship of all life. How in the face of these things can we accept the idea that "science" is capable of providing for the continuity of human life by substituting its methods for those of nature?

The second reason why undue reliance should not be placed upon soil chemistry to produce sufficient crops and fibers is purely a realistic one. In theory the answer might be found in that direction. In measuring the possibilities from a practical point of view, however, one would have to assume that workers on the land could be so instructed that they would be capable of putting into practice methods of agriculture that in themselves are extremely complicated and would be able to adopt techniques that are not even as yet fully developed. It is one matter to get fairly satisfactory results on an experimental farm—it is obviously quite another to hope for such results where millions of workers on the land are involved, many of them bound by unfavorable customs of land use, especially when we are speaking of large numbers of people in many parts of the earth who, unfortunately, are completely illiterate, so that education in highly technical methods of land use would have to be conducted by oral instruction or by visual demonstration—at best a virtually insuperable task.

Another recourse that is sometimes blandly proposed is the establishment of great central "food factories" where edible plants shall be grown in solutions of chemicals and water. The culture of plants by this method, known as hydroponics, is not a new art, having been started in England in a crude way as long ago as the eighteenth century. In more recent times it has been used extensively by research workers in many countries as a tool to arrive at facts regarding the growth of plants. So far the production of plant growth on any large scale through this method has not been accomplished, and those who are conversant with its complexities strongly question its practicability. Also, there is a good deal of glib talk that we shall be able to derive a large proportion of our basic subsistence from marine resources, including plant life. Far be it from anyone to laugh off the time and effort which are being devoted to studying the possibilities of filling human needs by these expedients. Maybe the day will actually come when hard-pressed and diminishing human populations will be grateful for being able to fall back upon such resources. All life once came from the water and perhaps, who knows, we shall have to return to it for survival. But one can only wonder whether the proponents of such ideas have thought the situation through to its ultimate implications. Such theories of action would bring with them a social revolution of such magnitude that the whole structure of human society would be torn apart. People who live on the soil and produce their living from it constitute more than one half the total earth population.

The issue might as well be clearly drawn. The problem is how to conserve the remaining good natural soils that exist on the earth, together with the complementary resources of forests, water sources and the myriads of beneficial forms of animal life. There is no other problem. If that is not solved the threat to human life will grow in intensity and the present conditions of starvation that are already apparent in various parts of the earth will seem as nothing in the years that lie ahead.

❊   ❊   ❊

THE PLUNDERER

Man's misuse of the land is very old, going back thousands of years even to the earliest periods of human history. It can be read in the despairing chronicle of ruins buried in sand, of rivers running in channels high above their surrounding landscapes, of ever-spreading deltas, of fallen terraces which once held productive fields or rich gardens. It can be seen in man-made deserts and in immense reaches of bare stone from which the once fertile soils have been washed and blown away. Occasionally in some wasted part of the earth a trace may be found of what was once there before the vast destruction. Perhaps there will be a terrace wall that did not fall away, but held its portion of soil where still are growing vines or olive trees that flourish in spite of the surrounding desolation, or a grove of ancient cedars in some protected place where all but their own small plot of land has been destroyed around them.

Erosion and its fatal consequences have often been attributed to a gradual change in the climate of a region or, more especially, to the fact that certain regions have, over long periods of time, suffered a marked diminishment in rainfall. Recent research does not, however, support the earlier theory that unfavorable trends in weather conditions can be held responsible. There are no real evidences of the geologic desiccation of a region, namely from natural causes, within historic times. China, where many regions are severely wasted, has not been subjected to climatic changes of any moment for more than three thousand years. Palestine has today the same general weather conditions that it had in Biblical times. A small stand of cedars of Lebanon, untouched for many centuries because it was considered a sacred grove and was protected by a wall which kept out goats, supports the opinion that weather was not responsible for the loss of all the immense forests of cedar which existed within historic times. And in North Africa an olive grove, standing since Roman days, gives evidence that there too the weather cannot be solely blamed for dead cities buried under drifts of sand. Natural forces have played their part but man himself is intimately involved.

How did man become a land destroyer? It is easy to understand in present times, with the world so crowded and in need of food, how any overpopulated country, such as China or India, might deplete its land in a desperate effort to feed its crowding millions. It is difficult for people, under the urgency of immediate need, to take thought for the future of unborn generations, or of the earth that must support them. Indeed they have neither the time nor the strength nor generally the knowledge to do so; the best they can hope for is to keep alive, whatever the cost. But this was not always so. Once, as has already been told, man's numbers were limited, and up to historic times he had plenty of land to support him adequately without allowing that land to become depleted.

And yet, even in the earliest periods of history, man frequently suffered the feeling of being crowded, though the world population was but a small fraction of its present number. This sense of *crowdedness* arose from the concentration of people upon those lands which appeared most desirable and from a reluctance to strike out and develop new areas. Only under various kinds of pressure do people leave their home country. Even within recent times, and despite the ease of modern transportation and methods of communication, this reluctance of people to leave the place they are accustomed to is evident. The colonial empire that was Germany's prior to the First World War had all told a mere 24,000 or so German inhabitants, and Italy was never successful in placing more than 10,000 of her people in all her African colonies. The inclination of people to concentrate is age-old. It is one of the major causes for the long and consistent record of land damage by the human race and occurred throughout the earliest periods of human history.

How human beings first developed their methods and customs for gaining subsistence from the earth is largely a matter of conjecture. Cereals were probably cultivated in Asia as long ago as 8000 B.C., and it appears that grains were grown in Turkestan and northern Persia before animals were domesticated.

One of the theories concerning the first domestication of animals holds that it was not done by the hunters or nomads but by the settled tillers of the soil and, at that, more or less by accident. Farmers seeing the seasons change around them, and the weather bring them good or bad crops, began to believe in beings beyond themselves who had control of natural forces. Thus they developed a religious sense connected with the natural world, and associated the wild animals of their own region with the gods controlling their destiny. The practice of propitiating various deities by sacrificing to them appropriate animals did, it is true, spring up in many parts of the world. Since the animals were hard to come by, it is thought that farmers of long ago built stockades of fairly large dimension, driving wild cattle and sheep into them yet not confining them so closely that they would not breed. So, in time, they were easy to take for religious purposes, and also gradually became used to people and thus domesticated.

All the while the earlier practice of living by the chase survived, which tended to divide early human societies into two groups, the hunters or nomads on the one hand and the tillers of the soil and townspeople on the other. After animals had become domesticated many of the former group took to the keeping of herds and flocks, moving from one grazing region to another and continuing the life of nomadism. There are many evidences of long-continued and bitter feuds between these groups throughout this period of history. These early feuds have their counterparts in other countries in subsequent times, as, for example, the dominance of the Mesta in Spain in the fifteenth century and the land-grabbing tactics of the cattle and sheep men in the United States at this very moment. At any period, whether then or now, the eventual result is inevitably the same—severe and frequently permanent damage to the fertility of the land.

In the region that lies between the Tigris and Euphrates Rivers there once was a land suggestive of the Garden of Eden, a rich land whose people lived well, built flourishing cities, established governments and developed the arts. Advanced methods of agriculture were developed including the building of a complex and extensive system of irrigation works during the reign of Hammurabi, about 2000 B.C., by which the waters of the two great rivers were drawn off to increase the fertility of the land. Gradually great changes took place and the whole region deteriorated. This may have been because of the cutting of forests outside the cities, the exposing of land to eroding rains at certain seasons, and to the quick runoff which must always mean a dearth of water from natural sources later on. Also it may have been because of overgrazing on grasslands, which would have a similar effect. Eventually enemies seemed to have caused the final undoing through wrecking or blocking the ditches and canals which were the life streams of the settled populations. In this way they could plunder, graze in the fields and gardens and cut down trees for firewood, and in the end impoverish or destroy the people. These cities with all their elaborate civilization are today lost under the sand.

❊   ❊   ❊

The pattern of land use still follows that of the ancient populations— "*cut, burn, plant, destroy, move on.*" It is a method known as *milpa*, common not only in Mexico but in most other countries in Latin America and, for that matter, elsewhere in the world. Such a system is possible where there are limited numbers of people who have plenty of room to move on from one place to another, leaving the wounds on the land behind them for time and nature to heal. Under the pressure of tremendously increased populations, and with the growth of cities and towns and the disappearance of new and unspoiled land, the eventual results of the system are fatal.

❊   ❊   ❊

And now to come to the United States, the country of the great illusion, the country that "can feed the world." What has happened there during its turbulent and inconceivable period of development and what is happening there today?

The story of our nation in the last century as regards the use of forests, grasslands, wildlife and water sources is the most violent and the most destructive of any written in the long history of civilization. The velocity of events is unparalleled and we today are still so near to it that it is almost impossible to realize what has happened, or, far more important, what is still happening. Actually it is the story of human energy unthinking and uncontrolled. No wonder there is this new concept of man as a large-scale geological force, mentioned on an earlier page.

In the attempt to gain at least some perspective let us review a little. Our people came to a country of unique natural advantages, of varying yet favorable climates, where the earth's resources were apparently limitless. Incredible energy marked the effort of a young nation to hack new homes for freedom-loving people out of the vast wilderness of forests that extended interminably to the grassland areas of the Midwest. Inevitably the quickest methods were used in putting the land to cultivation, not the desirable methods. Great areas of forest were completely denuded by ax or fire, without thought of the relationship of forests to water sources, or to the soil itself. Constantly there was the rising pressure for cultivable land caused by the rapid inpouring of new settlers. By about 1830 most of the better land east of the Mississippi was occupied. In that year there were approximately 13,000,000 people in this country, or less than one tenth of the present population. In the meanwhile the land in the South, long occupied and part of the original colonies, was being devoted more and more extensively to cotton, highly profitable as export to the looms in England, and tobacco, for which there was a growing world market. These are known as clean-tilled crops, meaning that the earth is left completely bare except for the plantings and is a type of land use most susceptible to loss of topsoil by erosion. Today a large proportion, in many areas from one third to one half, of the land originally put to productive use for the growing of cotton and tobacco has become wasteland and has had to be abandoned. It is not unusual for Southerners to blame the Civil War and its aftereffects for their impoverishment. There are other reasons.

There is no particular point in tracing the westward surge of settlers over the great grass plains that lay beyond the Mississippi and on to the vast forested slopes bordering the Pacific. Everyone knows the story. It is significant, however, that the movement, dramatic as any incident in human history, was symbolized by the phrases "subjugating the land" and "conquering the continent." It was a positive conquest in terms of human fortitude and energy. It was a destructive conquest, and still continues to be one, in terms of human understanding that nature is an ally and not an enemy. . . .

A detailed presentation of what has happened area by area would fill many volumes. A large amount of precise information has been gathered together by various governmental services, by other conservation agencies, and by a handful of individuals whose perception has led them to give attention to an unfolding drama that is as yet visible to so few.

The submission of the following general facts may serve to throw light on what has happened to our land since those bright days when we began to "conquer the continent."

The land area of the United States amounts to approximately one billion nine hundred million acres. In its original or natural state about 40 per cent was primeval forest, nearly an equal amount was grass or range lands, the remainder being natural desert or extremely mountainous.

Today the primeval or virgin forest has been so reduced that it covers less than 7 per cent of our entire land area. If to this there are added other forested areas consisting of stands of second- or even third-growth forests, many of which are in poor condition, and if scattered farm woodlands are also included, it is found that the forested areas now aggregate only slightly more than 20 per cent of the total land area of our country. If urban lands, desert and wastelands, and mountaintop areas, are subtracted there is left somewhat over one billion acres which can be roughly divided into three categories: farm croplands, farm pasture lands and range-grazing lands.

❋   ❋   ❋

A consideration of the situation of land resources in our country shows that other than forests there are, as mentioned above, about a billion acres that fall into the three categories of farm croplands, farm pasture lands and open-range grazing lands. Of these, farm croplands are the largest in area, running to approximately 460,000,000 acres. What has happened in regard to these resources and what is going on now?

The most recent report of the Soil Conservation Service of the government contains a number of pertinent statements. They are a factual recital. They point to a velocity of loss of the basic living elements of our country which, if continued, will bring upon us a national catastrophe. Already every American is beginning to be affected in one way or another by what is happening. This report indicates that of the above billion acres considerably more than one quarter have now been ruined or severely impoverished, and that the remainder are damaged in varying degrees. Furthermore, the damage is continuing on all kinds of land—cropland, grazing land, and pasture land. Here are other highlights in the report:

The loss we sustain by this continuing erosion is staggering. Careful estimates based on actual measurements indicate that soil losses by erosion from all lands in the United States total 5,400,000,000 tons annually. From farm lands alone, the annual loss is about 3 billion tons, enough to fill a freight train which would girdle the globe 18 times. If these losses were to go on unchecked, the results would be tragic for America and for the world.

The results would not only be disastrous—they already are far too costly for the country to continue to bear. For example, in a normal production year, erosion by wind and water removes 21 times as much plant food from the soil as is removed in the crops sold off this land.

Nor is loss of plant food our only expense from erosion. The total annual cost to the United States as a result of uncontrolled erosion and water runoff is estimated at $3,844,000,000. This includes the value of the eroded soil material and the plant nutrients it contains, the direct loss sustained by farmers, and damages caused by floods and erosion to highways, railroads, waterways, and other facilities and resources.

The loss in the productive capacity of our farms cannot be figured so easily, but it is plain that farm lands which have lost so much topsoil and plant nutrients cannot produce as bountifully as they did before they were slashed and impoverished by erosion.

In that fact lies the significance of America's erosion problem for America's citizens. We do not have too much good cropland available for production of our essential food and fiber crops in the future. If we do not protect what we have, and rebuild the land which can still be restored for productive use, the time inevitably will come—as it already has come to some areas of the world—when United States farm lands cannot produce enough for us and our descendants to eat and to wear.

The Soil Conservation Service has only been in existence since 1935. It was created by Congress in that year not so much as a result of the government's vision or strategy but principally because the people of this country had been struck with dread by the revulsion of nature against man that was evidenced by the Dust Bowl incident on May 12 of the previous year. On that day, it will be recalled, the sun was darkened from the Rocky Mountains to the Atlantic by vast clouds of soil particles borne by the wind from the Great Plains lying in western Kansas, Texas, Oklahoma and eastern New Mexico and Colorado—once an area of fertile grasslands but now denuded by misuse, much of it to the point of permanent desolation. In the years since its inception this government service has gained extraordinary results in advancing the science of proper land use and in assisting soil conservation districts, set up under state law, and encouraging voluntary and co-operative action among farmers. These conservation districts now exist in all the states and have been the medium through which better methods of land use have been adopted in many of the farming regions. At best, however, this vital program—one of the most elemental that affects the lives of the people of our country—is only well started.

*   *   *

Action of government, in the last analysis, rests mainly, of course, on the point of view of the people. The fact that more than 55 per cent [1] of our population live in cities and towns results inevitably in our detachment from the land and apathy as to how our living resources are treated. As a result the majority of voters in the United States at this time neither know nor care about the problem that is facing us. As a consequence, elected representatives of the majority of our people are likewise apathetic.

1. The estimate today is that approximately two out of three people in the United States live in metropolitan areas. It is expected that in another ten years this figure will rise to three out of four. Ed.

This great section of our people, as well as large portions of our rural population, either do not realize what is going on or are lulled into a false sense of security by misleading reports regarding the status of our life-supporting resources that are inspired by groups having special interests in such properties as timberlands, cattle and water rights. Probably, however, the most potent soporific affecting popular opinion comes from the belief we all innately share these days to the effect that the marvels of modern technology can solve any of the riddles of life. The miraculous succession of modern inventions has so profoundly affected our thinking as well as our everyday life that it is difficult for us to conceive that the ingenuity of man will not be able to solve the final riddle—that of gaining a subsistence from the earth. The grand and ultimate illusion would be that man could provide a substitute for the elemental workings of nature.

It will be recalled that the country as a whole became seriously alarmed, and converted alarm into action, after that dramatic day when the dust clouds from the Far West hid the sun from the Capitol in Washington and darkened the Eastern cities. Do we need another catastrophic warning from nature to stir us to further action, or can we not now accept the many evidences of approaching crisis and take steps to ward it off?

There are mountains in Attica which can now keep nothing but bees, but which were clothed, not so very long ago, with fine trees producing timber suitable for roofing the largest buildings, and roofs hewn from this timber are still in existence. There were also many lofty cultivated trees.

The annual supply of rainfall was not lost, as it is at present, through being allowed to flow over a denuded surface to the sea, but was received by the country, in all its abundance—stored in impervious potter's earth—and so was able to discharge the drainage of the heights into the hollows in the form of springs and rivers with an abundant volume and wide territorial distribution. The shrines that survive to the present day on the sites of extinct water supplies are evidence for the correctness of my present hypothesis.

Plato, *Critias*

The Web of Life *
by John H. Storer (1953)

EROSION

There is a spot in the woodlands of southeastern Tennessee that can never be forgotten by one who has seen it. To reach it, one may travel for a hundred miles through forest-covered hills, rich with laurel, azalea, and rhododendron, and along springs and brooks and ravines which sometimes open up into green meadows where cattle graze.

Suddenly this green world disappears. The forest gives way to a hundred square miles of desert as dead as the Sahara. The rolling hills are cut into rows of low, steep-sided ridges, sterile and bare of any life. The soil is dry, the springs and brooks are gone. In this area the annual rainfall is less than in the surrounding country. The winds are stronger. It is hotter in summer and colder in winter. Here and there on this desert there stand in rows the dead skeletons of small trees, planted by people who hoped to start a new forest.

Figure 7–1. Erosion at Copper Basin of Tennessee. This man-made desert was once covered by a forest. When the forest was killed by acid from a copper smelter, the soil lost its power to store water and support life. For many years, attempts to replant this land failed. (Photograph courtesy of the U. S. Forest Service.)

The soil in the nearby woodland is dark, rich, and spongelike. That on the desert is coarse, hard, and yellow. This desert was once covered by a forest and by rich forest soil. But today that soil lies five miles down the valley at the bottom of a reservoir and the shoals of coarse desert soil grow deeper, year by year, as every rain washes its fresh quota down to the reservoir.

This all happened because, many years ago, a copper smelter was built here, and the fumes from the smelter killed the surrounding trees, thereby setting in motion a train of events that finally produced the desert. The owners of the smelter have long since learned to control these fumes, which no longer poison the air so seriously; but the harm has been done.

After the fumes were controlled, many attempts were made to restore the forest. Desert grasses were planted, in the hope of furnishing some green cover to hold the soil in place; for rich soil cannot exist without the help of plants to build and protect it. But with the killing of the forest, the living soil that gave it life had also been killed. There were no roots to hold the soil in place, no litter to absorb the rain, and the grass seeds washed away, the seedling trees withered, and the dead soil continues to this day to wash from the desert and drift down to fill the reservoir.

At prohibitive cost, this desert land could be restored to life. Modern machinery might partly substitute for some of nature's processes. It could fill the eroding gullies and build level ridges along the hillsides to stabilize the soil, hold the rain, and so give moisture to the earth. Then seeds of grass and trees could take root and find a chance for life. Given such a start, nature could take over and slowly rebuild the ruined soil, organizing once again the community of living things that makes life possible for the forest.[1]

1. Since the publication of this selection, reforestation efforts in this area have been partially successful. Ed.

THE MAKING OF TOP SOIL

Figure 7–2. Soil building—the first step. The slopes below the mountain peaks are composed of rock fragments broken from the face of the cliffs by the forces of erosion. Slowly this broken rock will be moved down into the valleys by water, wind and gravity. On the journey the pieces will become finer and finer until small enough to form the parent material of soil. Before it can support life, soil must be anchored into place, enriched and organized into a living entity by the action of plants and animals. (Photograph courtesy of the U. S. Forest Service.)

Figure 7–3. Life makes its start. Lichens, moss and ferns on rock. The pioneer lichens can exist on bare rocks with very little moisture. They secrete acid which dissolves minerals from the rocks, thus creating a seed bed which can absorb moisture and provide food and a foothold for higher plants. The plant roots spread out into the surrounding soil tying it into place. From this foundation of raw soil the plant community develops. (Photograph courtesy of the U. S. Department of Agriculture.)

Figure 7–4. Plants build the soil. The chlorophyll in green leaves is the only substance in nature that has the power to harness the sun's energy and combine it with elements from air, water and rock into living tissue. This plant tissue becomes the food that supports all animal life, and it is the basis for the organic matter that is an essential part of productive soil. Thus, the creative work of plants is the essential first step toward building fertility into soil. By a maze of rootlets and fungus-fibres, most of them invisible to the eye, plants tie the soil together. When these die, they make a sponge of absorbent rotting vegetation. Insects, worms and other animals burrow to reach this organic food, mixing it thoroughly in the process and enriching it with their own remains. (Photograph by Larry Pringle. Reprinted by permission of the National Audubon Society.)

Figure 7–5. *Nature plows the soil. Today's catastrophe prepares riches for tomorrow. In a mature forest the soil may be corrugated with ridges and hollows which hold rain and melting snow and that show where trees have fallen and mixed soil with vegetation. (Photograph courtesy of the U. S. Forest Service.)*

Figure 7–6. *Small animals play a useful role as nature's farmers, plowing the soil. Some, like this prairie dog, dig in it to make their homes; others, like the shrews and moles, tunnel through it in their search for insects and worms, thus helping to make it rich and absorbent. (Photograph courtesy of the Soil Conservation Service.)*

Figure 7–7. *Insects and worms do their share in building soil, plowing and mixing it with the vegetation and enriching it with their remains as they burrow to seek shelter and food among the plant roots. This tunnel was dug by a cicada wasp to make a home for its young. (Photograph by Hal H. Harrison. Reprinted by permission of the National Audubon Society.)*

Figure 7–8. The finished product—a soil profile. The light lower layer is the parent soil composed of rock particles. The dark upper layer is topsoil, built by the action of plants and animals. It may be as much as 40% organic matter—i.e., living and dead plant and animal life. This soil has the power to absorb water and the ability to produce crops. (Photograph courtesy of the Soil Conservation Service.)

ARTERIES OF LIFE

Of all the many things that go into the making of life, water is the one most prodigally used. It not only makes up a big proportion of all living substance, but it is steadily used up to maintain the life process and must be as constantly renewed. The liquid that an animal drinks is only a small fraction of the water that goes into the maintenance of its life.

It is hard to make an accurate measurement of all the water used. But . . . it is estimated for example that it takes about 900 pounds of water to produce a single pound of dried alfalfa, and about 500 pounds to produce one pound of wheat. Translate the plant into meat, and it takes about 5,000 pounds of water to produce one pound of beef.

The supply of water directly controls the kinds of plants and animals that can live in any environment, and so in any community the source of its water supply is a most fundamental part of the environment.

Many parts of the earth's surface receive but little rain. In some places the loss of moisture through evaporation is greater than the entire rainfall. So, existence here is totally dependent on the arteries that bring this life blood from the community's heart in the distant mountains and distribute it to actively functioning members in the dry lowlands.

When we speak of nature's arteries, it would be natural to think of rivers. Actually, as far as dry-land communities are concerned, by the time water has reached a river its value is largely gone. The river is carrying the water back to the ocean just as fast as it can get it there. To be useful on dry land, water must be delayed on its journey to the sea, and distributed so as to render the greatest amount of service.

When rain strikes a bare slope its natural course is to go straight downhill by the shortest way, and into the nearest river. It may leave a little moisture along the way, but the land is soon as dry as if it had never rained. There is no water left to support life, and except for primitive forms, there is no life to use water when it does rain. We have seen how such a condition developed on the man-made desert in Tennessee. We have seen, too, how nature goes about building a protective sponge on the earth's surface to control run-off. We will now see how the life of a community in the lowland depends on the effective working of this sponge in the higher country that supplies the water.

Our western mountains receive a good part of their water in the form of snow. High in the mountains of Colorado, at the headwaters of the Colorado River, a snowflake coming to rest on a spruce branch may lie for days where it falls, buried among its neighboring flakes. As it melts, some of its moisture evaporates back into the air, for water evaporates fast in the thin, decompressed air of the high mountains. The rest of the melting flake eventually flows down the tree trunk, or drips off onto the snow below. Here, shaded from the sun, the melting of the banked snow may take weeks or months. The water reaches the soil slowly. An insulating blanket of spruce needles protects the soil from freezing. Thus its open pores are always ready to receive water, which spreads out to fill every crevice between the soil particles and to filter into the humus, which holds it like a sponge.

As it fills the soil, part of this moisture adheres to the surfaces of every particle of mineral matter and humus, while the rest continues to sink slowly, following the pull of gravity.

Thus it is that at last the hillside becomes filled with a huge store of water, part of it anchored in place, the balance moving slowly downward, a great, leisurely, underground river as the unhurried melting of the snow sends in fresh supplies from above to maintain its flow.

The capacity of the ground to hold this vast moving reservoir depends chiefly on the quality and depth of the soil, for a good soil can hold the equivalent of more than a third of its bulk in water.

A good share of this stored water will go into maintaining the organization: nourishing the vegetation that makes the reservoir possible, building and protecting the soil, helping to grow timber and grass.

The remainder continues on its way downward, into the flat dry lands below the mountains. Here there is little rain or snow to moisten the surface, but moisture from the moving water in the ground rises by capillary action to supply thirsty vegetation, and where the water sinks too deep for this, longer-rooted plants reach far down for it.

As the earth in the valley fills with water, some of it may break out on the surface in springs to join the brooks that carry the overflow from the mountains. Many of these springs rise unseen, to flow into the stream beneath the surface. Thus slowly the stream swells as it travels, until at last it joins the Colorado River, a boiling turmoil of flowing silt mixed with water, "too thin to plow, too thick to drink," carrying the run-off from nearly a quarter of a million square miles of land.

This mixture of silt and water might tell us an interesting story about the land it comes from. It carries to its destination in Lake Mead, above Hoover Dam, an average of nearly seven hundred thousand tons of silt each day throughout the year. All of this great mountain of dirt has been scoured off the hillsides, covering thousands of miles of watershed, and each ton represents a bit of land that might have served as a reservoir to store the water and control its flow.

The flow of the river may rise to more than sixty million gallons a minute in summer when the snow is melting on the mountains, and may sink to about two million gallons in the winter when the surplus water has run out. This flow has always varied with the seasons, and it has always carried great quantities of silt. But the load of silt has greatly increased in recent years with destruction by fire and overgrazing of nature's protective cover on the hillsides. And with the increasing erosion of the natural storage reservoir, the extremes of flood and drought have greatly increased. So the life-giving potentialities of the river in the lowlands are greatly affected by conditions on the watersheds, many hundred of miles away.

DESTRUCTION OF A FOREST

Figure 7–9. The destruction of a forest may spell disaster to communities a thousand miles away. These trees have been blown down by a windstorm. With broken roots and weakened sap flow they can no longer withstand the attacks of bark beetles. Under a tangle of branches they are inaccessible to the woodpeckers which ordinarily catch most of the beetles that do gain entrance. (Photograph courtesy of the U. S. Forest Service.)

Forest 7–10. *The beetles, thus protected, multiply like an explosion, spreading out to attack and kill the healthy trees in the surrounding forest. (Photograph courtesy of the U. S. Forest Service.)*

Figure 7–11. *The dead trees dry out and turn to tinder, becoming far more vulnerable to fire than a normal forest. The catastrophe started by a local windstorm may spread out to ruin the forests on an entire watershed. The same thing may happen when man improperly cuts or grazes a forest. (Photograph courtesy of the U. S. Department of Agriculture.)*

Figure 7–12. After the fire, this land is dead, and from it death will spread for a thousand miles over the country below, carried there by floods and droughts. (Photograph courtesy of the U. S. Forest Service.)

Figure 7–13. When raindrops strike unprotected soil, the fine soil particles spatter into the air. Falling back they fill the crevices between the larger particles, making the surface waterproof. Instead of being stored in the earth, the water now runs off in a flood, tearing away the unprotected soil as it goes. (Photograph courtesy of the U. S. Department of Agriculture.)

Figure 7–14. Erosion caused by first rain following fire, Los Padres National Forest, California. This rain was approximately one inch. (Photograph courtesy of the U. S. Forest Service.)

Figure 7–15. Two farms buried in a single grave. Across the center of this picture runs a dark layer of rich topsoil that once formed the fertile surface of a farm. Above it lie the layers of sterile sand and gravel washed by floods from other farms on the slopes above. The topsoil from these upper farms has been carried away, or so mixed with the gravel as to be useless. More recent floods have started to cut a new channel across the land, exposing the layers that we see here carrying the soil down to bury other farms, or to drop it in the ocean. Total depth, 13 feet. (Photograph courtesy of the U. S. Department of Agriculture.)

MAN IN THE WEB

We must remember that one of water's chief uses to man is as a solvent, cleaner, and conveyor of wastes.

As the early cities grew along the river courses, pollution was not a serious problem, for the wastes from each city were diluted by the flowing water, oxidized by the bacteria, used as fertilizer by the water plants, and filtered through the river sands and gravel, so as to reach the next user in fairly clean condition. But with the multiplication of cities and their discharges, the water became filled with an unsupportable load of poisons from the factories, offal from the slaughterhouses, raw sewage from the homes. These killed the cleansing plants, used up the purifying oxygen in the water, and clogged the filtering gravels with filth.

And so, today, the water supply for many of our cities enters the city water system as a dark chocolate-colored fluid, straight out of the sewers and factories of its neighbors upstream. In one midwestern city some tests of the water showed that, during the period of low water in the winter, it was one-half straight sewage. Later in the season, when the river filled with run-off from the spring rains, it was a thin, liquid mud, made from good topsoil washed off the improperly cultivated farm lands above.

By thorough and expensive treatments, the mud and the visible sewage can be removed from the water and its remaining load of bacteria killed by disinfectants before it is distributed to citizens for drinking, washing, and cooking. But then it is discharged back into the river, boiling out from the city sewer as foul and offensive as when it entered. The sewers of the city have become an integral part of the watershed that supplies the cities downstream.

This pollution of the rivers is as truly a destruction of a basic natural resource as is the overcutting of a forest or the wrong management of good land. Like the land and the forest, the water can be restored to usefulness at great expense if it is not too far gone. But, as in their case, this destruction is unnecessary. It should be cleaned by its users before its return to the river. This can be done by proper treatment. The wastes and chemicals thus recovered are valuable for fertilizer and other uses. But it is an expensive process, beyond the immediate means of many communities and factories, so the nationwide drive toward improvement is slow and painful. . . .

There is much still unknown about the proper management of watersheds, and since the conditions on each will vary with soil and slope, climate and vegetation, the management required on them may differ widely. But in every case the first requirement is to protect the soil, which is nature's reservoir for storing water where it falls.

So, we see the river and the lands of its watershed as a great living organism, with its heart in the mountains that supply its life blood. This blood flows out through the streams that form the arteries above ground and below, coming down from a hundred thousand hidden sources—the mountain springs and meadows, the patches of moist woodland with the porous soil beneath them, the shaded snow banks and the afternoon thunderstorms, the flow of every raindrop held back by the delaying stems of grass and flowers, absorbed by bits of rotting wood, filtering into the soil through a million root tunnels and worm holes, delayed, but slowly moving down the hillside through the soil, to bring a steady, even flow of life to the great functioning body of civilization in the valley.

Every action that affects the lands of the watershed has its direct influence on the functioning of the whole organism. The growing leaf that shades the snow to delay its melting is doing its microscopic share to give an even flow of water through the summer. Combining its influence with a hundred billion other leaves, it may determine the success or failure of the harvest in the valley.

We may see and appreciate the devastation caused on the watershed by a carelessly dropped cigarette. But what of the sheep whose hoof compresses a worm hole or destroys the protecting grass roots on the hillside? Nature may repair these tiny bits of damage if they are kept within limits with which she can cope. But when too many sheep combine to lay a hillside bare, they set the stage for rivers filled with silt, and for next season's floods and droughts over huge areas of distant lands.

The people who use the watershed hold in their hands the lives and well-being of the millions who depend on it. Every action that affects the health of the watershed becomes a matter of vital concern to those millions, and to all the communities of plants and animals that together make up the whole.

And so we see that every community may be divided into four parts: first, its members with their immediate environment; second, the distant and unknown lands that send out their influence by stream and wind, by wing and padded foot to affect the local environment; third, the actions of men whose influence spreads out to affect in some way nearly every community living on the earth; and last, but most important, those influences that mold the minds of men, giving them the incentives to wise or unwise actions; for in the end these lie at the very center of earth's great web of life. To support this web, the soil must be maintained alive and functioning.

Technology Can Solve the Problem *
by Colin Clark (1958, 1960)

The physical resources of the world are capable of yielding an immensely increased food supply if we only make the effort to make use of them; the real trouble is that we don't yet really try. In an article in *Nature* (3rd May, 1958), and in more detailed statements elsewhere (and no geographer or agricultural scientist has published any criticism of my figures), I have stated that the agricultural resources of the world would suffice for ten times the present world population.

❋   ❋   ❋

The world's total land area (excluding ice and tundra) is 123 million sq. km., from which we exclude most of the 42½ million sq. km. of steppe or arid lands, discount anything up to half the area of certain cold or sub-humid lands, but could double 10 million sq. km. of tropical land capable of bearing two crops per year. We conclude that the world possesses the equivalent of 77 million sq. km. of good temperate agricultural land. We may take as our standard that of the most productive farmers in Europe, the Dutch, who feed 385 people (at Dutch standards of diet, which give them one of the best health records in the world) per sq. km. of farm land, or 365 if we allow for the land required to produce their timber (in the most economic manner, in warm climates—pulp requirements can be obtained from sugar cane waste). Applying these standards throughout the world, as they could be with adequate skill and use of fertilizers, we find the world capable of supporting 28 billion people, or ten times its present population. This leaves us a very ample margin for land which we wish to set aside for recreation or other purposes. Even these high Dutch standards of productivity are improving at a rate of 2 per cent per annum.

❋   ❋   ❋

Even these figures have not taken any account of further improvements in agricultural and biological techniques, which will almost certainly take

* From Colin Clark, "Overpopulation—Is Birth Control the Answer?" *Family Planning*, April 1960, and "World Population," *Nature*, 181, 1958. Reprinted by permission of Macmillan (Journals) Ltd., and the author.

place, nor of any food which our descendants may obtain from the sea. And if we really wish to look several centuries ahead and to predict a world population so large that it will have outrun even these resources, we can safely say that by that time our descendants will so far surpass us in wealth and in skill that they will be quite capable of building themselves large, artificial satellites, on which they can dwell in the sunny climates of outer space.

By the End of the Twentieth Century *
by David Sarnoff (1964)

Science and technology will advance more in the next thirty-six years than in all the millennia since man's creation. By the century's end, man will have achieved a growing ascendancy over his physical being, his earth, and his planetary environs.

The primary reason is man's increasing mastery of the electron and the atom from which it springs. Through this knowledge he is capable of transforming everything within his reach, from the infinitesimally small to the infinitely large. He is removing the fetters that for more than a million years have chained him to the earth, limited his hegemony over nature, and left him prey to biological infirmities.

By the year 2000 A.D. I believe our descendants will have the technological capacity to make obsolete starvation, to lengthen appreciably the Biblical life-span, and to change hereditary traits. They will have a limitless abundance of energy sources and raw materials. They will bring the moon and other parts of the solar system within the human domain. They will endow machines with the capacity to multiply thought and logic a millionfold.

FOOD

The Western nations by the turn of the century will be able to produce twice as much food as they consume, and—if political conditions permit—advanced food production and conservation techniques could be extended to the overpopulated and undernourished areas. New approaches —the protein enrichment of foods, genetic alteration of plants and ani-

---

* From David Sarnoff, "By the End of the Twentieth Century," *Fortune,* May 1964. Copyright © 1964 by Time Inc. Reprinted by permission of Fortune and RCA.

mals, accelerated germination and growth by electronic means—will be widely used. The desalinization of ocean waters and tapping of vast underground fresh-water lakes, such as the one some believe to underlie the Sahara, can turn millions of desert acres to bloom. The ocean itself, covering seven-tenths of the earth's surface, will systematically be cultivated for all kinds of plant crops and fish—floating sea farms. Man's essential nutrients are reducible to chemical formula, and ultimately the laboratory will create highly nutritive synthetic foods, equaling in palatability and price the products of the land. As that happens, his total dependence upon the products of the soil will terminate.

RAW MATERIALS

Technology will find ways of replenishing or replacing the world's industrial materials. The ocean depths will be mined for nickel, cobalt, copper, manganese, and other vital ores. Chemistry will create further substitutes for existing materials, transmute others into new forms and substances, and find hitherto unsuspected uses for the nearly 2,000 recognized minerals that lie within the earth's surface. Oil and coal will be used increasingly as the basis for synthetics. Long before we have exhausted the existing mineral resources, the world will have developed extraction and processing techniques to keep its industries going largely on raw materials provided by the ocean waters and floor, the surface rocks, and the surrounding air. The rocks that crust the earth contain potentially extractable quantities of such basic metals as iron, copper, aluminum, and lead. The ocean abounds in a variety of chemicals.

ENERGY

The energy at man's disposal is potentially without limit. One pound of fissionable uranium the size of a golf ball has the potential energy of nearly 1,500 tons of coal, and the supply of nuclear resources is greater than all the reserves of coal, oil, and gas. Increasingly, electric-power plants will be nuclear, and atomic energy will be a major power source particularly in the underdeveloped areas. Small atomic generators will operate remote installations for years without refueling. Electronic generators, converting energy directly to electricity, will light, heat, and cool homes, as will solar energy. Many areas of the world also may draw power from thermal gases and fluids within the earth's crust. Ultimately, even more powerful energy sources will be developed—thermonuclear fusion, the tapping of heat from deep rock layers, the mutual annihilation of matter and anti-matter.

HEALTH

Science will find increasingly effective ways of deferring death. In this country, technology will advance average life expectancy from the Biblical three score and ten toward the five-score mark, and it will be a healthier, more vigorous, and more useful existence. The electron has become the wonder weapon of the assault on disease and disability. Ultraminiature electronic devices implanted in the body will regulate human organs whose functions have become impaired—the lungs, kidney, heart —or replace them entirely. Electronics will replace defective nerve circuits, and substitute for sight, speech, and touch. Chemistry will help to regenerate muscles and tissues. Laser beams—highly concentrated light pulses—operating inside the body within needle-thin tubes will perform swift, bloodless surgery. By the end of the century, medical diagnosis and treatment will be indicated by computers, assembling and analyzing the latest medical information for use by doctors anywhere in the world.

GENETICS

Before the century ends, it will be possible to introduce or eliminate, enhance or diminish, specific hereditary qualities in the living cell— whether viral, microbic, plant, animal, or human. Science will unravel the genetic code which determines the characteristics that pass from parent to child. Science will also take an inanimate grouping of simple chemicals and breathe into it the spark of elementary life, growth, and reproduction. When this occurs, man will have extended his authority over nature to include the creative processes of life. New and healthier strains of plants and animals will be developed. Transmitted defects of the mind or body will be corrected in the gene before life is conceived. If cancer proves to be genetic or viral, it may be destroyed at the source. There appear, in fact, to be few ultimate limits to man's capacity to modify many forms of living species.

COMMUNICATIONS

Through communication satellites, laser beams, and ultraminiaturization, it will be possible by the end of the century to communicate with anyone, anywhere, at any time, by voice, sight, or written message. Satellites weighing several hundred tons will route telephone, radio, and television, and other communication from country to country, continent to continent, and between earth and space vehicles and the planets beyond.

Participants will be in full sight and hearing of one another through small desk instruments and three-dimensional color-TV screens on the wall. Ultimately, individuals equipped with miniature TV transmitter-receivers will communicate with one another via radio, switchboard, and satellite, using personal channels similar to today's telephone number. Overseas mail will be transmitted via satellite by means of facsimile reproduction. Satellite television will transmit on a worldwide basis directly to the home, and a billion people may be watching the same program with automatic language translation for instant comprehension. Newspaper copy, originating on one continent, will be transmitted and set in type instantly on another. Indeed, by the year 2000 key newspapers will appear in simultaneous editions around the world.

TRAVEL

From techniques developed for lunar travel and other purposes, new forms of terrestrial transport will emerge. Earth vehicles riding on air cushions and powered by nuclear energy or fuel cells will traverse any terrain and skim across water. Forms of personal transportation will include such devices as a rocket belt to carry individuals through the air for short distances. Cities at opposite points of the globe will be no more than three to four hours apart in travel time, and individuals will breakfast in New York, lunch in Buenos Aires, and be back in New York for dinner. Indeed, the greatest problem will be the adjustment of time and habit to the tremendous acceleration in the speed of travel. As rocket systems are perfected and costs are reduced, it is possible to foresee the transport of cargo across continents and oceans in tens of minutes. Within and among cities and countries, the foundation will be laid for the movement of freight through undergound tubes, automatically routed to its destination by computers. . . .

AIR AND SPACE

Around earth, a network of weather satellites will predict with increasing accuracy next season's floods and droughts, extremes of heat and cold. It will note the beginnings of typhoons, tornadoes, and hurricanes in time for the disturbances to be diverted or dissipated before they reach dangerous intensity. Ultimately, the development of worldwide, long-range meteorological theory may lead to the control of weather and climate. Space will become hospitable to sustained human habitation. Manned laboratories will operate for extended periods in space, expanding our basic and practical knowledge about the nature of the universe, the planets, and earth. Permanent bases will be established on the more habitable planetary neighbors, and from these a stream of televised reports and radioed data, inanimate and conceivably also living matter, will flow to earth.

Despite these enormous changes, the machine in the year 2000 will still be the servant of man. The real promise of technology is that it will release man from routine drudgeries of mind and body and will remove the final imprint of the cave. In doing so, science will give new validity to Alfred North Whitehead's profound observation that civilization advances "by extending the number of important operations we can perform without thinking about them." Man's mind will then be free for the creative thinking that must be done if the impact of science is to be harmonized with man's enduring spiritual, social, and political needs.

The Dangers of Relying on Invention *

by Norbert Wiener (1954)

In depending on the future of invention to extricate us from the situations into which the squandering of our natural resources has brought us we are manifesting our national love for gambling and our national worship of the gambler, but in circumstances under which no intelligent gambler would care to make a bet. Whatever skills your successful poker player must have, he must at the very least know the values of his hands. In this gamble on the future of inventions, nobody knows the value of a hand. . . .

If the food supply is falling short, or a new disease threatens us, inventions to relieve it must be made before famine and pestilence have done their work. Now, we are far nearer to famine and pestilence than we like to think. Let there be an interruption of the water supply of New York for six hours, and it will show in the death rate. Let the usual trains bringing supplies into the city be interrupted for forty-eight hours, and some people will die of hunger. Every engineer who has to deal with the administration of the public facilities of a great city has been struck with terror at the risks which people are willing to undergo and must undergo every day, and at the complacent ignorance of these risks on the part of his charges. . . .

The very increase of commerce and the unification of humanity render the risks of fluctuation ever more deadly.

Patterns of the Future *

by Harrison Brown (1954)

Within a period of time which is very short compared with the total span of human history, supplies of fossil fuels will almost certainly be exhausted. This loss will make man completely dependent upon water-power, atomic energy, and solar energy—including that made available by burning vegetation—for driving his machines. There are no fundamental physical laws which prevent such a transition, and it is quite possible that society will be able to make the change smoothly. But it is a transition that will happen only once during the lifetime of the human species. We are quickly approaching the point where, if machine civilization should, because of some catastrophe, stop functioning, it will probably never again come into existence.

It is not difficult to see why this should be so if we compare the resources and procedures of the past with those of the present.

Our ancestors had available large resources of high-grade ores and fuels that could be processed by the most primitive technology—crystals of copper and pieces of coal that lay on the surface of the earth, easily mined iron, and petroleum in generous pools reached by shallow drilling. Now we must dig huge caverns and follow seams ever further underground, drill oil wells thousands of feet deep, many of them under the bed of the ocean, and find ways of extracting elements from the leanest of ores—procedures that are possible only because of our highly complex modern techniques, and practical only to an intricately mechanized culture which could not have been developed without the high-grade resources that are so rapidly vanishing.

As our dependence shifts to such resources as low-grade ores, rock, sea-water, and the sun, the conversion of energy into useful work will require ever more intricate technical activity, which would be impossible in the absence of a variety of complex machines and their products—all of which are the result of our intricate industrial civilization, and which would be impossible without it. Thus, if a machine civilization were to

* From Harrison Brown, *The Challenge of Man's Future*. The Viking Press, New York. Copyright 1954 by Harrison Brown. Reprinted by permission.

stop functioning as the result of some catastrophe, it is difficult to see how man would again be able to start along the path of industrialization with the resources that would then be available to him.

The situation is a little like that of a child who has been given a set of simple blocks—all the blocks of one type which exist—with which to learn to build, and to make the foundation for a structure, the upper reaches of which must consist of more intricate, more difficult-to-handle forms, themselves quite unsuited for the base. If, when the foundation was built, he conserved it, he could go on building. But if he wasted and destroyed the foundation blocks, he would have "had it," as the British Royal Air Force would say. His one chance would have been wasted, his structure of the future would be a vanished dream, because there would be nothing left with which to rebuild the foundation.

Our present industrialization, itself the result of a combination of no longer existent circumstances, is the only foundation on which it seems possible that a future civilization capable of utilizing the vast resources of energy now hidden in rocks and seawater, and unutilized in the sun, can be built. If this foundation is destroyed, in all probability the human race has "had it." Perhaps there is possible a sort of halfway station, in which retrogression stops short of a complete extinction of civilization, but even this is not pleasant to contemplate.

Once a machine civilization has been in operation for some time, the lives of the people within the society become dependent upon the machines. The vast interlocking industrial network provides them with food, vaccines, antibiotics, and hospitals. If such a population should suddenly be deprived of a substantial fraction of its machines and forced to revert to an agrarian society, the resultant havoc would be enormous. . . .

Should a great catastrophe strike mankind, the agrarian cultures which exist at the time will clearly stand the greatest chance of survival and will probably inherit the earth. Indeed, the less a given society has been influenced by machine civilization, the greater will be the probability of its survival.

The Man-Made Environment

A World of Cities *

by Kevin Lynch and Lloyd Rodwin (1960)

Although the city itself is five thousand years old, the metropolis is a new phenomenon, dating from a mere hundred years ago. Its scale alone differentiates it from any older type of urban settlement. Even ancient Rome, with its million inhabitants, was in visible relation to its surrounding countryside. One could easily walk from one district to another or from the central to the rural area. In the metropolis this is hardly possible; even in a car it may take hours to move from center to periphery. Thus the city has swollen to a vast organism whose scale far transcends individual control.

Throughout the metropolis the environment is man-made—even its plants and trees are there by man's agency. Yet the density of its population in the outer parts, at least, is much lower than in the traditional city, so that we observe single dwellings and factories dispersed among gardens, parks, small woods, and open spaces. In the suburb, city and country fuse, and their long rivalry may here find its resolution.

It appears that these metropolitan complexes will become the dominant environment, at least in the most highly developed regions of the world, for they contain the bulk of the population, and they produce and consume most of the goods. The living space will become a set of such areas, at times separated by areas of low population that provide raw materials. The metropolitan regions may be very large, even grouped in chains several hundred miles long; but they will be more or less continuously urbanized, however low their suburban densities may be. Each one will display a continuous dense web of facilities for communication, within which the population will be highly interdependent on a daily basis in terms of supply, of information, and of commutation to and from work.

* From Kevin Lynch and Lloyd Rodwin, "A World of Cities," *The Future Metropolis,* George Braziller, New York. Copyright © 1961 by the American Academy of Arts and Sciences. Reprinted by permission. Originally published in *Daedalus,* Winter 1961.

Such a society will be intricate and its means of communication will be increasingly mechanized and depersonalized. Most likely it will also be mobile and egalitarian. This type of mass habitation and this kind of inhabitant may well become a world-wide pattern. . . . assume that, in significant areas of the world and within a period as short as fifty years, the majority of the world's population will be accommodated in vast metropolitan complexes, each on a scale of twenty million people or more. Staggering as this figure may seem, it is by no means unreasonable, and even if exaggerated, it sets the stage for an evaluation of this new phenomenon in the history of society.

Human Needs and Inhuman Cities *

by Edward T. Hall (1968)

One does not shape and maintain the physical environment independent of the spatial experience which may range from the most intimate—like the physical touching of bodies—to reveries of future excursions to lakes and mountain streams. The sense of crowding—or its obverse—is so subtle and so complex that I am continually amazed by some newly-identified facet of the subject. I was once discussing spatial needs with a psychiatric colleague, and he pointed out the relationship of unused distant space to one's immediate feelings about life in the city. He phrased it as follows: "If I were to hear that the people of the United States no longer had access to our national parks—the forests and the mountains and the great outdoors—even though I might never visit them again, my anxiety rate would be very considerably increased. The important thing is to know that they are there and available." When someone says that his anxiety rate has risen, this is only another way of saying that he is under stress and that his adrenals are working overtime. There are, of course, a great many individuals who, having lived all of their lives in cities, have no conscious need for open space and, in fact, may even be quite threatened by the outdoors. There is great danger, however, in allowing a large segment of our population to grow up without ever experiencing the complex relationships that make up ecological systems: we could develop a citizenry so uninformed as to permit the destruction of its own biotope. I'm not certain that we have not already done so. As recently as five years ago, it was difficult to interest people in the significance of John Calhoun's work with rats or John Christian's studies of the consequences of animal crowding. Today, one can hardly pick up a newspaper without reading about a new study on the effects of crowding. This recent preoccupation with crowding is symptomatic of our times, and our growing awareness of its consequences is extremely encouraging. There is also an increased awareness of the fact that each species inhabits its own biotope, and that each biotope constitutes part of a complex of interrelated ecosystems. All too many Americans, however, still demand specific answers to specific problems. In considering the human environment we are not dealing with the kind of phenomena in which one can identify specifics. Rather, we are dealing with systems that require a holistic approach, a point made repeatedly and elegantly by René Dubos. The ethological evidence is overwhelming: as populations build up, so does stress. Eventually, the animals' capacity to withstand stress begins to diminish, and the population collapses for a variety of reasons.[1]

Man has confounded the issue considerably and in a number of different ways. Not only has he produced a variety of extensions—like automobiles and factories—but his evolution is now primarily *by means* of his extensions. Man has conditioned himself to respond in quite different ways to the environmental pressures of crowding. In a sense, people of different ethnic groups respond differently to crowding because they perceive space differently. Man, as a culture-producing and culture-internalizing organism, has created a number of different types of biotopes for himself. There is no standard way of measuring spatial adequacy for man. What is crowded to an Englishman would be impossibly controlled or understimulating to an Arab. Because of such cultural differences, it is

1. This process has been described in detail by John Christian (The Pathology of Overpopulation; *Military Medicine*, no. 7, pp. 571–603, July 1963) and V. C. Wynne-Edwards (Self-regulatory Systems in Populations of Animals; *Science*, vol. 147, pp. 1543–1548, March 1965: *Animal Dispersion in Relation to Social Behavior*, New York, 1962). Highly relevant here is the work of Wilhelm Schäfer, director of the Frankfort Natural History Museum. Schäfer's 1956 monograph, *Der Kritische Raum und die Kritische Situation in der Tierischen Sozietät* (*Critical Room and Critical Situations in Animal Societies*), deals with the added dimension of pollution and how it has been handled in the past on simpler organizational levels. Schäfer's point is that as population builds up, a critical situation is reached when a population contaminates its environment with its own waste products. If this crisis is not solved, the population dies rapidly and the process is irreversible. Each solution to a crisis has in it the seeds of new crises because solutions permit larger and larger concentrations of organisms, hence greater volume of pollutants which, in turn, demands more drastic solutions.

necessary to introduce an added dimension that will make it possible to equate space needs of different cultures according to some sort of self-anchoring scale. We have this instrument in the various theories of sensory deprivation and information overload. Man can suffer from both, and an excess of either can destroy him.[2] An adequate environment balances sensory inputs and provides a mix that is congenial as well as consistent with man's culturally conditioned needs. This principle applies to housing standards for our own socioeconomic groups in the United States. I mention this by way of introduction because the hidden structure of culture is one of the most consistently ignored features of our 20th-century life.

What are we facing today in our cities? In the summer of 1966, Joseph Alsop devoted three columns to cities. In his first column he wrote: "The problem of the cities, in the form it is now assuming, is the most urgent, the most difficult, and the most frightening American domestic problem that has emerged in all the years of this country's history since the Civil War." [3] In a recent *Fortune* article, Edmund Faltermayer states: "Two things are common to all the great cities: an exciting and beautifully designed downtown section, and a large middle-class population living close to the central core. Most American cities fail miserably on both counts. Only New York, Chicago, and San Francisco can even make claims to greatness." [4] Elsewhere I have stated: "the implosion of the world populations into cities everywhere is creating a series of destructive behavioral sinks more lethal than the hydrogen bomb. Man is faced with a chain reaction and practically no knowledge of the structure and the cultural atoms producing it." [5] The current urban crisis is sufficiently acute that its seriousness is widely recognized by layman and specialist alike.

In the last thirty years, the character of our major cities in the United States has changed radically.[6] Our cities have been bombed by urban re-

---

2. In a paper presented at the 133d annual convention of the American Association for the Advancement of Science in Washington, D.C., on Dec. 27, 1966, A. S. Welch and B. L. Welch reported that small laboratory animals went into a stupor when crowded (overstimulated). The specific agent was an overabundance of three chemicals—norepinephrine, dopamine, and serotinin—which carry messages across the synapses separating nerve cells in the brain. See also: A. S. Welch and B. L. Welch, "Differential Effect of Chronic Grouping and Isolation on the Metabolism of Brain Biogenic Amines," *The ASB Bulletin,* vol. 13, no. 2, p. 48 (April 1966); B. L. Welch, "Psychophysical Responses to the Mean Level of Environmental Stimulation: a Theory of Environmental Integration," Symposium of Medical Aspects of Stress in the Military Climate, Washington, D.C., April 1964 (sponsored by Walter Reed Army Institute of Research).

3. Joseph Alsop, "Matter of Fact," *The Washington Post,* Aug. 1, 3, and 5, 1966.

4. Edmund K. Faltermayer, "What it takes to make great cities," *Fortune,* vol. LXXV, pp. 118–123, 146–151, January 1967.

5. Edward T. Hall, *The Hidden Dimension,* Garden City, N.Y., 1966. See, in particular, chap. XII, p. 155.

6. This has been extensively reported on by: Charles Abrams, *The City is the Frontier,* New York, 1965; Constantinos A. Doxiadis, *Urban Renewal and the Future*

newal in as devastating a way—and just as effectively—as though they had been bombed by an outside enemy. The difference is that if the bombing had come from an enemy, we would have recognized the crisis and would have done something about it. As it is, there is no coherent program for our cities. People are moved about, neighborhoods destroyed, whole viable social groups disrupted, and the fabric of life disintegrated—all in the name of progress. Fried's and Gleicher's studies of the Boston West End are particularly relevant.[7] For the first time we have documentation on an impressive scale of what happens when the members of a dominant ethnic group and class look at the biotope of another group and fail to see its structure. Seeing no structure, they then proceed to destroy. The same thing has happened in many other American cities.

Lewis Mumford accustomed us to thinking of the city as a process. He showed the city as the focus of the activities that make up what we have come to know as civilization. It is not too farfetched to think of a society as a living organism. Using a physiological analog, the city can be compared to various parts of the nervous system. If we were, therefore, to perceive in a human body the wild and uncontrolled growth of cells in the central nervous system or in an important ganglion of nerves controlling the functions, we would know that there was a serious crisis and act accordingly. In considering the plight of our cities, I do not detect signs of the kind of urgent action that the symptoms warrant. . . .

In our economic development programs overseas, we recognize that there is very little that can be done about underdeveloped countries until a large, stable, middle class has been established. These countries that have been recipients of so much foreign aid are by and large characterized by a massive, undereducated, underskilled working or peasant class. Our own major cities are now taking on many of the population characteristics of the so-called underdeveloped countries. As the poor move into cities and the middle-class flees to the suburbs, the situation in the cities becomes increasingly unstable. Cities today are principally for the very rich and the very poor.

In New York City, traffic is at a standstill because we cannot bring ourselves to do the obvious: discourage private vehicles from entering Manhattan by imposing high tolls, and restrict the time when trucks can use

of the American City, 1966; *Architecture in Transition*, New York, 1963; Marc Fried, "Grieving for a lost home," in Leonard J. Duhl's *The Urban Condition*, New York, 1963; Herbert Gans, *The Urban Villagers*, Cambridge, Mass., 1960; Nathan Glazer and Daniel P. Moynihan, *Beyond the Melting Pot*, Cambridge, Mass., 1963; Victor Gruen, *The Heart of Our Cities*, New York, 1964; E. H. Gutkind, *The Twilight of Cities*, New York, 1962; Jane Jacobs, *The Death and Life of Great American Cities*, New York, 1961; Barbara Ward, "The Menace of Urban Explosion," *The Listener*, vol. 70, no. 1807, pp. 785–787, November 14, 1963 (BBC, London).

7. Marc Fried and Peggy Gleicher, "Some Sources of Residential Satisfaction in an Urban Slum," *Journal of the American Institute of Planners*, vol. 27, 1961.

the streets. . . . A city that is impossible to get out of without going through the added tension of traffic jams has a higher stress factor than one with decent, pleasant public transportation. There is an added element of the irrational here for, as Alsop says in the columns referred to above, we are willing to spend billions upon billions for freeways and expressways that pour automobiles into our cities, and virtually nothing for schools. Yet our schools represent one of the single most important features of our environment: people will make all kinds of sacrifices to move into an area where there are good schools, and almost nothing will hold them in an area where the schools are inferior.

When Europeans first settled North America, man was presented with an entire continent rich in resources. He had the tools, the know-how, and the energy to exploit it. No such situation has ever existed before, nor will it exist again on our planet. In the four generations since the Civil War, when this country was forming itself, we acquired some very bad habits which permitted every man to do as he pleased. Now we have suddenly run out of frontier, and we have produced an economy which, like a chain reaction, is self-sustaining. Our prosperity could ruin us, not because it is bad to be prosperous, but simply because we don't know how to plan for the added dwellings, automobiles, boats, and airplanes. Nor have we learned how to dispose of the resulting pollutants, or how to design the spaces for the masses of people that are moving in and out of our cities each day.

In the United States we allow individuals to do virtually anything: pollute the lakes, contaminate the atmosphere, build a high-rise next door that makes our own living space uninhabitable because it shuts off the view, create walled-in slums in public housing high-rise, transform a potential recreation area on a lake into a run-down industrial waste, plow up the countryside, bulldoze trees, and build thousands of identical prefabricated bungalows in open country. Peter Blake in his book, *God's Own Junkyard,* has documented this aspect of our anarchic and anomic approach to planning. I have discovered (to my sorrow) that in building a house, plumbers and electricians often make important decisions overruling the owner and the architect: they change walls with abandon, run pipes where they should never be, and arrange interior spaces at will. Similarly, important decisions on the national scene are often made by officials, both public and private, who have little or no knowledge of the consequences of their actions.

If our historical traditions have been no help to us in the face of our environmental crisis, the state of present day knowledge about man also leaves much to be desired. We need to learn much more about man's basic nature and his requirements as a biological organism. Also lacking is basic data on what constitutes optimum conditions for man's social and cultural development. In recent years I have been conducting research on one aspect of man's requirements: his spatial needs. In the course of my work on proxemics—man's perception and use of space—I've observed that man shares with lower life forms certain basic needs for territory. Each man has around him an invisible series of space bubbles that expand and contract, depending upon his emotional state, his culture, his activities, and his status in the social system. People of different ethnic origins need different kinds of spaces, for there are those who like to touch and those who do not. There are those who want to be auditorially involved with everybody else (like the Italians), and those who depend upon architecture to screen them from the rest of the world (like the Germans).

❊   ❊   ❊

Culture transcends man in both time and space; i.e., it is "supraorganic," which tends to divert attention from the fact that culture, on the individual, behavioral, objective, or manifest level, is also rooted in, or can be traced back to, biological activities. Man's cultural elaborations of his handling of space illustrate this point. However, man has so developed his use of space in any given culture, as well as the great diversity of space patterns between cultures, that it is fruitful to consider lower life forms as a means of clarifying the phylogenetic base.

Systematic territorial studies begin with the publication of Howard's *Territory in Bird Life*, a scant 41 years ago.[8] I would predict, however, that as the understanding of the relationship of organisms to their setting increases, we will begin to find that the lines between physiology and ecology will not be so sharply drawn as they were at first.

For example, space needs may be as basic as the need for food. Not only does everything occur in a time-space plane (largely taken for granted), but the handling of space can be, and often is, a life and death matter for many organisms. For example, small mammals which have not yet established a territory are much more vulnerable to predation than those which have.[9] Hediger [10] reports that wild animals, in captivity, may not even survive being moved from zoo to zoo, or even from pen to

8. See Howard, H. E.: *Territory in Bird Life*. London: Murray, 1920.

9. See Errington, P.: "The Great Horned Owl as an Indicator of Vulnerability in the Prey Populations." *J. of Wildlife Management*, 2:190, 1938.

10. See Hediger, H.: Bemerkungen Zum Raum-Zeit System der Tiere, *Schweiz. Z. f. Psychol.*, 5:241–267, 1946.

————: *Wild Animals in Captivity*. London: Butterworths, 1950.

————: *Studies of the Psychology and Behavior of Captive Animals in Zoos and Circuses*. London: Butterworths, 1955.

pen. He also states that animals that are restricted in their use of space become stupid.

It is difficult to pinpoint precisely at what time in the phylogenetic scale territoriality first became established; each expert seems to have his own idea about it. It appears that it is present in all of the vertebrates, in some form or another [11] although Washburn and DeVore [12] state that baboons in Africa have no territoriality, but a "home range" instead. Hediger notes that animals often use space in ways thought to be of strictly human invention. There are homes and yards, eating and drinking places, sleeping and bathing places, places for food storage, lavatories, sun-bathing places, terraces, nurseries, and places for sex. Hediger also introduces some new concepts to the study of space; these are: flight distances, social distances and personal distances.

Flight distance is that distance to which you can approach an animal before it will flee. This varies from species to species, and can also vary within a species, according to the situation. As with territoriality, there is a relation between the size of the animal and the amount of space used. Flight distance is shorter with small animals than with large ones. An antelope has a flight distance of approximately 500 yards, while a lizard can be approached within six feet.

Social distance is a highly predictable distance, beyond which some vertebrates lose contact with the herd, the flock or their mates. They will not normally exceed the limits placed on them by social distance.

Personal distance is the normal spacing that animals maintain between themselves and their fellows. Personal distance can be likened to a bubble that surrounds the organism. Outside this bubble, two organisms are not immediately involved with each other. Inside it, they are. Anyone who has observed birds sitting on a wire can quickly spot the very regular spacing which is a function of personal distance. [See Figures 7–16, 7–17, 7–18.]

Even from this all too brief review, it can be seen that most animals are restricted to their own territory or home range, and have limits placed on them as to how close they can come to their fellows, as well as how far they can stray before losing contact with the group.

Man has elaborated his use of space to such a degree that it is difficult to determine just how space-bound he is. There are indications, however, that man, like the other vertebrates, moves within the framework of highly-patterned spatial systems.

11. See Burt, W. H.: "Territoriality and Home Range Concept as Applied to Mammals." *J. of Mammal.*, 33:139–159, 1943. Also Kalela, O.: "Range Possession as a Factor of Population Ecology in Birds and Mammals." *Vanamo*, 16:1–48, 1954.

12. See Washburn, S. L. and DeVore, I.: "The Social Life of Baboons." *Scientific American*, 204:62–71, 1961.

Figure 7–16. Some species huddle together and require physical contact with each other. Others completely avoid touching. No apparent logic governs the category into which a species falls. Contact creatures include the walrus, the hippopotamus, the pig, the brown bat, the parakeet, and the hedgehog among many other species. The horse, the dog, the cat, the rat, the muskrat, the hawk, and the blackheaded gull are non-contact species. Seals along the Pacific coast give a perfect example of contact behavior. (Photograph by LeBoeuf, University of California, Santa Cruz. Reprinted by permission. The editor derives the ideas for Figures 7–16, 17, and 18 from Plates III, I, and II of Edward T. Hall, Hidden Dimension, *Doubleday & Co., 1966.)*

Figure 7–17. Non-contact species, such as these swans, avoid touching. (Reprinted courtesy of the American Museum of Natural History.)

Figure 7–18. Personal distance is the term applied by the animal psychologist, H. Hediger, to the normal spacing that non-contact animals maintain between themselves and their fellows. These swallows demonstrate this natural grouping. (Reprinted courtesy of the American Museum of Natural History.)

Personal distance in man varies from culture to culture, and is cause for considerable marginally-felt discomfort, irritation, and some misunderstanding between people. People reared in cultures where the distance is shortest will be perceived as "pushy" by those with a longer personal distance. On the other hand, people with a long personal distance will be seen as cold, aloof, withdrawn or standoffish by the individual with a short personal distance, simply because they cannot be approached closely enough for him to become involved with them.

❁   ❁   ❁

A more detailed description of cross cultural differences in the perception and use of space can be found in my book on proxemics, *The Hidden Dimension.*

In addition to ethnic differences in spatial requirements there are differences in people's tolerance for change and mobility within their own culture. Studying the effects of urban renewal on slum dwellers, I've found evidence that indicates that very poor uneducated people have a much lower tolerance for being displaced than people of the middle class. Even a move across the street can be traumatic because it alters the pattern of social relationships. This is one of many reasons why the "Instant Rehab" experiments in New York are of great importance. These experiments involve a procedure for stripping the interior of an old building down to the brick or masonry wall, cutting a large hole in the roof and lowering a core of preassembled utility systems through the roof. Using this technique, trained crews of workmen have completely rehabilitated a small building in less than two days. For this short period, the building's occupants can be moved to a hotel and then moved back to their old home with no disruption of their relationships with friends and neighbors.

For several years I have been investigating the relationship of the working class urbanite to his environment. It is a widely known fact that high-rise public housing has not been popular with high-rise occupants, particularly those families who have recently come to the city from rural areas. My studies indicate that the high-rise not only structures relationships between the occupants in specific ways, but it removes many of the challenges and the reasons for neighborhood organizations that, in turn, perform important social functions. The high-rise has proved to be a new source of anomie in ghetto life.

To further complicate the problem of providing adequate public housing for our urban populations, Glazer and Moynihan have documented the long suspected fact that the United States is not the melting pot that we once believed it to be. Ethnicity is much more tenacious than had originally been supposed. Yet, nowhere do I find evidence that plans for public housing recognize the existence of different needs for different ethnic groups. Recent growing concern about air pollution and water pollution, sparked by newspaper and TV publicity, has helped alert the country to these dangers. Hopefully we will soon have national pollution controls. Another grave danger, but one that is not widely understood, is the very real danger we face from overcrowding in our urban centers.

It seems obvious to say that our total environment must be protected and enhanced by the establishment of national and local planning authorities using the best available professional advisors. Two key disciplines needed for successful environmental planning are architecture and city planning. Unfortunately, neither discipline is presently prepared for this responsibility. The education of architects and planners is, with few exceptions, extraordinarily limited. Conspicuously absent from their curricula are the social sciences and humanities, to say nothing of ecology. Training in even rudimentary research procedures is rare. In the last seven years, I have addressed at least three thousand architects, published in the principal architectural journals, taught architecture students, and interviewed dozens of architects. In all this time I have failed to find a single instance of a systematic recording of feedback from users which could later be incorporated in the program or design phase of a new building. All over the United States, rooms have been getting steadily smaller, ceilings lower, walls more transparent to sound, views cut off, and what do we know about the consequences? What hard data is there? What feedback from users? Every year the Government spends millions for building, yet nowhere is there a requirement for impartial reporting on the degree to which these buildings serve the needs of their occupants. For example, FHA requirements for housing, promulgated with the best of intentions, bear little or no relation to man's sense of crowding and often place a straitjacket on all but the most pedestrian designers. Even the Pueblo Indians in New Mexico, if they are to qualify for FHA loans, must choose from among FHA designs (White man's designs). Pueblo designs do not meet FHA standards, yet they are vastly more suited to the needs of the Indians than any design produced by FHA.

The professions of architecture and city planning suffer from serious deficiencies which must be corrected if this nation is to plan intelligently to improve the total environment. I refer to: first, the widespread practice of architects and planners to pay close attention to a narrow reference group (other architects), awarding prizes on the basis of two-dimensional representations of structures often *before they are built!* Few designers see *people* in their projects, only prizes from other architects and planners. Second, there is a failure to get feedback on the consequences of their design decisions. Third, there are no plans for making available the results of studies that could be used in plans for future building: there is no clearinghouse for architectural programs or designs by geographic area, class, or ethnic group. The result is that even when architects and planners want such data they have virtually no way of finding it.

The extreme seriousness of our environmental crisis demands immediate and total response from this nation with the full participation of business, Government, and universities. By and large our universities have failed to become involved in environmental problems until very recently, and most of them are still only peripherally involved. It is no wonder that students all over the country are setting up their own underground colleges to meet today's problems. I would suggest that the university, as we know it, should be stood on its head. It is not enough to set up centers for "urban studies," which are usually only a sop to the urban crisis and a means to tap Government and foundation funds. Instead, the entire university should be *involved* in the urban process.

In recent years I have had the good fortune to work with design students. As a group they are the most responsive students I have ever taught. If they are properly trained, they question the basic design of virtually everything. They examine the processes at work in a given activity and they question assumptions endlessly. Because they are already trained to look for new ideas and new solutions, I have found that I can train these students in half the time required for liberal arts students. Tómas Maldonado has suggested that design education be integrated into the university curriculum on a level with the divisions of physical science, behavioral science, and the humanities. This idea should certainly be tried. I can think of nothing more important than teaching young people about environmental design, using data from such disciplines as ecology, psychology, and anthropology. Particular attention should be focused on research into man's spatial requirements. I am confident that such research and training would produce students who could discover better answers to man's needs than are currently being provided.

In addition to training and research programs, I would hope that some of our universities would begin to produce some working models of man's biotope. These would enable us to make accelerated studies of the total environment including our cities. At the moment we have no working model for the city.

In closing I would like to stress the importance of considering this small planet as an entire ecological system. This concept is basic to any planning for the improvement of our environment. We can regard our cities either as disaster areas beyond remedial action or as opportunities to learn more about man and his relationships to his environment. If we choose the latter course we will find that most of our urban problems merely reflect basic inadequacies in our total environment; inadequacies which will have to be remedied if man is going to persist on this planet.

Men-Like-Ants? *

by Joseph Wood Krutch (1962)

It is true, of course, that man became man rather than simply a member of the animal kingdom when he ceased merely to accept and submit to the conditions of the natural world. But it is also true that for many thousands of years his resistance to the laws of animal nature and his modifications of his environment were so minor that they did not seriously interfere with natural law and required no such elaborate management of compensating adjustments as became necessary as soon as his intentions, desires, and will became effective enough to interfere with the scheme of nature.

It was not until well into the nineteenth century that his interferences did become extensive enough to force a dawning realization of the fact that you cannot "control nature" at one point without taking steps to readjust at another the balance which has been upset. Improved methods of agriculture exhaust the soil unless artificial steps are taken to conserve and renew it. You cannot destroy all the vermin without risking the destruction of useful animals. You cannot, as we are just discovering, poison noxious insects without risking the extinction of birds who are an even more effective control. It is not that we should not interfere with nature, but that we must face the consequences of this interference and counteract or ameliorate them by other interferences. You dare not, to put it as simply as possible, attempt to manage at one point and to let nature take her course at another.

❖   ❖   ❖

Let us suppose for a moment that those are in the right who say that the context of nature has ceased to be the most desirable context for civilized life, that man can live in a wholly man-made world and that he will in time forget all that he once drew from his contemplation of that world

* From Joseph Wood Krutch, "A Naturalist Looks at Overpopulation," *Our Crowded Planet: Essays on the Pressures of Population*, edited by Fairfield Osborn. Doubleday and Company, Garden City. Copyright © 1962 by Doubleday and Company. Reprinted by permission.

of which he has ceased to be a part. Let us suppose further that his increase in numbers stopped before space itself gave out, and that he has reached what some seem to think of as the ideal state, i.e., living in cities which are almost coextensive with the surface of the earth, nourishing himself on products of laboratories rather than farms, and dealing only with either other men or the machines they have created.

What will he then have become? Will he not have become a creature whose whole being has ceased to resemble Homo sapiens as we in our history have known him? He will have ceased to be consciously a part of that nature from which he sprang. He will no longer have, as he now does, the companionship of other creatures who share with him the mysterious privilege of being alive. The emotions which have inspired a large part of all our literature, music, and art will no longer be meaningful to him. No flower will suggest thoughts too deep for tears. No bird song will remind him of the kind of joy he no longer knows. Will the human race have then become men-like-gods, or only men-like-ants?

Man: The Lethal Factor *

by Loren Eiseley (1963)

The great life web which man has increasingly plucked with an abruptness unusual in nature shows signs of "violence in the return," to use a phrase of Francis Bacon's. The juvenile optimism about progress which characterized our first scientific years was beginning to be replaced early in this century by doubts which the widely circulated *Decline of the West* by Oswald Spengler documents only too well. . . . In the scores of books analyzing these facts, . . . [it is not] easy to find a word spared to indicate concern for the falling sparrow, the ruined forest, the contaminated spring—all, in short, that spells a life in nature still to man.

As one of these technicians wrote in another connection involving the mere use of insecticides, and which I here shorten and paraphrase: "Balance of nature? An outmoded biological concept. There is no room for sentiment in modern science. We shall learn to get along without birds if necessary. After all, the dinosaurs disappeared. Man merely makes the process go faster. Everything changes with time." And so it does. But let

* From Loren Eiseley, "Man: The Lethal Factor," *The Key Reporter*, Spring 1963. Reprinted by permission of Phi Beta Kappa. Originally a joint Sigma Xi-Phi Beta Kappa lecture which was printed in both *American Scientist*, Vol. 51, No. 1, and *The Key Reporter*.

us be as realistic as the gentleman would wish. It may be we who go. I am just primitive enough to hope that somehow, somewhere, a cardinal may still be whistling on a green bush when the last man goes blind before his man-made sun. If it should turn out that we have mishandled our own lives as several civilizations before us have done, it seems a pity that we should involve the violet and the tree frog in our departure.

To perpetrate this final act of malice seems somehow disproportionate, beyond endurance. It is like tampering with the secret purposes of the universe itself and involving not just man but life in the final holocaust— an act of petulant deliberate blasphemy.

Suggestions for Further Reading

Calhoun, John B.: "Population Density and Social Pathology," *Scientific American,* February 1962. Experiments on populations of rats under various conditions of crowding throw light on the relationships of space, stress, and social pathology.

* Carson, Rachel: *Silent Spring,* A Crest Reprint, Fawcett Publications, 1964. A dynamic indictment of man's misuse of his environment with special emphasis on the use of insecticides and other chemicals.

Hall, Edward T.: *The Hidden Dimension,* Doubleday & Company, 1966. The author proposes many interesting thoughts on the use of space in animal and human societies in addition to the ones presented in the selection in this Part.

"Man and His Habitat," *Bulletin of the Atomic Scientists,* October and November 1961. A symposium on the interaction of man and his environment; contains articles on pollution, climate control, and natural resources.

* Sears, Paul: *Where There Is Life,* Dell Publishing Co., Inc., 1962. An excellent introduction to ecology and the evolution of natural history. The author explores the fitness of our environment and the place of individual organisms in nature.

Selye, Hans: *The Stress of Life,* McGraw-Hill Book Company, Inc., 1956. A somewhat technical book but very pertinent in connection with these readings.

* Storer, John H.: *The Web of Life,* Signet Science Library, The New American Library of World Literature, 1953. The intricate interrelationship of living things can be best appreciated by reading this book in its entirety.

* Thomson, George: *The Foreseeable Future,* Viking Explorer Books, The Viking Press, 1960. Describes what technology will probably be able to achieve for us in the next hundred years. Energy, transport, communications, weather control, food, and medicine are among the subjects considered.

* Paperback edition

8

Is Civilization Retarding the Evolution of Man?

Improved standards of living and advances in modern medicine have changed both the death rate and the birth rate of mankind. These changes have precipitated the population crisis and may also be interfering with the process of natural selection. Is there any evidence that these factors are adversely affecting human evolution?

Is Civilization Retarding the Evolution of Man?

Introduction

CIVILIZATION has brought improved health and better living conditions to the majority of men. It has also brought new techniques for preserving life and for controlling the size of the human family. It is a strange irony that these benefits are accompanied by serious new threats to the development and quality of human life. The change in the natural balance between the death rate and the birth rate has produced a rising tide of humanity which threatens to engulf the advances that society has made.

Population has gone through surges of growth at other periods in history. Each important increase followed a major invention that made more resources available to mankind: the invention of agriculture, the harnessing of non-human power, the opening up of new lands, the organization of people into urban centers. In all cases the population grew until it had absorbed the new supply of food, power, and usable space. Then a new balance was automatically struck between population and the means of subsistence.

The present surge in population however, results from a different kind of discovery: the control of death by modern medicine and sanitation. Death control has upset nature's automatic balance by allowing more people to live while not providing, in itself, new sources of subsistence. Fortunately, technology during this same era has been making available more food and energy. However, these factors are not linked in the simple cause and effect relationship that automatically maintained the balance before. The result is that the population of the world is growing at the rate of about 190,000 people every 24 hours. In effect, we are adding enough people each month to make a new city the size of Chicago.

Such a rate of growth precipitates many problems. It makes economic growth doubly difficult for the developing nations so that the gap between the have and have-not nations is an ever-widening rift.

Crowded conditions in cities are causing stress which may be partly responsible for rising delinquency and mental illness. Hudson Hoagland describes experiments with animal communities which show that overcrowding causes animals to die—not from starvation—but from overactivity of the pituitary and adrenal glands, a condition that is typical of acute stress. Crowding also brings on various types of pathological behavior. In one experiment with Norway rats, patterns of social behavior were badly deranged at twice normal crowding.

Overpopulation also threatens basic human liberties. The more densely occupied living area offers less freedom of activity for each individual. A high degree of organization and regimentation must be maintained to prevent the members from interfering with each other. As Paul Sears puts it, ". . . indefinite increase brings into play an inexorable physical principle that applies to any dynamic particle, from molecule to man. Whenever such units are confined within a finite space—bottle or continent as the case may be—any increase in numbers brings about a decrease in mean free path. Up to a certain point this may even have advantages when man happens to be the dynamic particle. Beyond that point it increasingly limits his freedom." [1]

Some scientists are concerned that by keeping alive the unfit who carry deleterious genes we are thwarting the process of natural selection and may be undermining the biological fitness of man. The living conditions in civilized societies not only decrease the weeding-out of harmful genes but also increase the rate of mutation. Exposure to radiation (both from medical practice and from atomic testing) and the use of a number of mutagenic drugs are undoubtedly increasing the mutation load of mankind. Discussion of the radiation hazard is not included in these readings. It is a complicated issue in itself and has been dealt with in detail earlier in this series.[2]

Most geneticists believe that any increase in the genetic load will inevitably cause early death and suffering to more people. On the other hand, P. B. Medawar points out that although many mutated genes are unconditionally detrimental (whether they appear in homozygous or heterozygous form), some genes are injurious only in the homozygous form. They may even be beneficial and confer hybrid vigor in the heterozygous individual. These differences are still not fully understood.

The deleterious genes that do appear tend to be protected from natural selection by the practice of modern medicine and will, therefore, become more widely distributed throughout the population. However, as Dobzhansky says, natural selection in itself is no guarantee of biological safety. "The geological strata of the earth's crust contain fossilized remains of countless thousands of species which became extinct without issue. Yet the evolution of these species was controlled by natural selection. They became extinct mostly because natural selection made them too specialized to live in environments which were only temporary." [3]

1. From Paul Sears, "The Perspective of Time," *Bulletin of the Atomic Scientists,* October 1961.
2. See *The Mystery of Matter,* by Louise B. Young, ed., Oxford University Press, New York, 1965, Part 9.
3. From Theodosius Dobzhansky, *The Biological Basis of Human Freedom,* 1956, p. 84.

Adaptability is the quality that is lost in overspecialization and this is the quality, Dubos insists, that we should be most concerned to cultivate in man. "A certain amount of stress, strain, and risk," he says, "seems essential to the full development of the individual." [4]

Another problem that has worried geneticists in recent years has been brought about by the use of contraceptives to control birth. These methods were used first and most extensively by the best educated segment of the population with the result that the reproductive rate of this segment declined relative to the birth rate of the disadvantaged. A number of theoretical studies were made which showed that such a differential birth rate could lead to an alarmingly rapid decline in intelligence. However, the concern on this point has been somewhat eased by the fact that no definite evidence has been found to demonstrate this drop in I.Q. In fact, a study made of school children in Scotland showed a slight rise in average I.Q. over a fifteen-year period. Scientists disagree over the meaning of these results. Does the rise in I.Q. indicate a real genetic improvement in intellectual endowment or are other factors covering a real decline?

The reader will be interested to observe how widely different conclusions can be drawn from one set of experimental data. As we have seen before, science is not always a cut-and-dried method of finding the facts and drawing a unique conclusion from them. The interpretation is often complex and the conclusion depends partly on what the scientist expects to find.

In the past two decades changes have occurred in the differential birth rate which have led some writers to hope that it was a transitional phenomenon that may correct itself when knowledge about birth control becomes universal. The fertility of the underprivileged has fallen more rapidly than the fertility of the well-to-do.[5] Furthermore, in the United States there seems to be a preference for a moderate number of children among those who consciously choose the size of their families. "The change represents neither a return to the very large family of the last century nor the abandonment of effective contraception, but does indicate a preference for more children than characterized the families of the 1920's and 1930's." [6]

The discussion about the differential birth rate raises another question. Should men and women with unusual talents devote their main time and effort to the expression of these talents rather than assume the burdens of a large family? Darwin expressed this opinion: "Great law-givers, the

4. See René Dubos selection, p. 47.
5. See *Determinants and Consequences of Population Trends,* United Nations, Department of Social Affairs, Population Division, New York, 1953, pp. 85–86.
6. From *The Growth of U.S. Population,* National Academy of Sciences, National Research Council, Committee on Population, 1965, pp. 5–6.

founders of beneficent religions, great philosophers and discoverers in science, aid the progress of mankind in a far higher degree by their works than by leaving a numerous progeny." [7]

The cultural evolution of man depends on the intellectual achievements of a relatively small number of people and this cultural evolution transcends the biological in rapidity and significance. It is possible that some of the conditions of civilized societies that are detrimental to biological evolution are conducive to cultural advance and, therefore, may do more good than harm in the total process of human evolution.

7. From Charles Darwin, *The Descent of Man,* 1871, p. 229.

Death and Birth Control

Some of mankind's "well-wishers" are thrown into a panic by the sharp decline of the death rates which, they contend, suspends the action of natural selection in man. This is, however, not necessarily valid. A far greater and more present danger comes from the uncontrolled growth of human populations or, as the fashionable expression goes, from "the population explosion." Indeed, never before has mankind been faced with such a cruel paradox. Darwinian fitness is reproductive fitness; mankind has inherited the earth; having become so widespread and abundant the human species is an unprecedented biological success; and yet it is this very success which threatens to choke the advance of his culture and reduce him to starvation and misery.

Theodosius Dobzhansky
Mankind Evolving (1962)

Population *

by Kingsley Davis (1963)

For hundreds of millenniums *Homo sapiens* was a sparsely distributed animal. As long as this held true man could enjoy a low mortality in comparison to other species and could thus breed slowly in relation to his size. Under primitive conditions, however, crowding tended to raise the death rates from famine, disease and warfare. Yet man's fellow mammals even then might well have voted him the animal most likely to succeed. He had certain traits that portended future dominance: a wide global dispersion, a tolerance for a large variety of foods (assisted by his early adoption of cooking) and a reliance on group co-operation and socially transmitted techniques. It was only a matter of time before he and his kind would learn how to live together in communities without paying the penalty of high death rates.

Man remained sparsely distributed during the neolithic revolution, in spite of such advances as the domestication of plants and animals and the invention of textiles and pottery. Epidemics and pillage still held him back, and new kinds of man-made disasters arose from erosion, flooding and crop failure. Indeed, the rate of growth of the world population remained low right up to the 16th and 17th centuries.

Then came a spectacular quickening of the earth's human increase. Between 1650 and 1850 the annual rate of increase doubled, and by the 1920's it had doubled again. After World War II, in the decade from 1950 to 1960, it took another big jump (see middle illustration Figure 8–1). The human population is now growing at a rate that is impossible to sustain for more than a moment of geologic time.

Since 1940 the world population has grown from about 2.5 billion to 3.2 billion.[1] This increase, within 23 years, is more than the *total* estimated population of the earth in 1800. If the human population were to continue to grow at the rate of the past decade, within 100 years it would be multiplied sixfold.

Projections indicate that in the next four decades the growth will be even more rapid. The United Nations' "medium" projections give a rate during the closing decades of this century high enough, if continued, to multiply the world population sevenfold in 100 years. These projections are based on the assumption that the changes in mortality and fertility in regions in various stages of development will be roughly like those of the recent past. They do not, of course, forecast the actual population, which may turn out to be a billion or two greater than that projected for the year 2000 or to be virtually nil. So far the UN projections, like most others in recent decades, are proving conservative. In 1960 the world population was 75 million greater than the figure given by the UN's "high" projection (published in 1958 and based on data up to 1955).

In order to understand why the revolutionary rise of world population has occurred, we cannot confine ourselves to the global trend, because this trend is a summation of what is happening in regions that are at any one time quite different with respect to their stage of development. For instance, the first step in the demographic evolution of modern nations—a decline in the death rate—began in northwestern Europe long before it started elsewhere. As a result, although population growth is now slower in this area than in the rest of the world, it was here that the unprecedented upsurge in human numbers began. Being most advanced in demographic development, northwestern Europe is a good place to start in our analysis of modern population dynamics.

In the late medieval period the average life expectancy in England, according to life tables compiled by the historian J. C. Russell, was about 27 years. At the end of the 17th century and during most of the 18th it was

1. The estimate in 1967 was 3.42 billion. Ed.

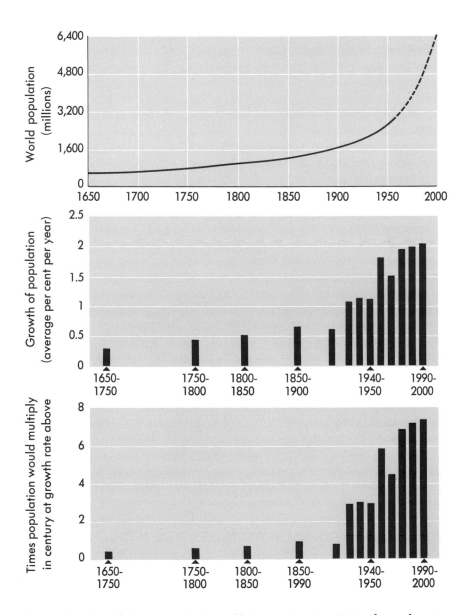

Figure 8–1. Population growth of world from 1650 to 1960 is shown by curve at top of page, projected to the year 2000. The middle chart shows the rate of growth, and the bottom chart the number of times the population would multiply in 100 years at that growth rate for various periods.

about 31 in England, France and Sweden, and in the first half of the 19th century it advanced to 41.

The old but reliable vital statistics from Denmark, Norway and Sweden show that the death rate declined erratically up to 1790, then steadily and more rapidly. Meanwhile the birth rate remained remarkably stable (until the latter part of the 19th century). The result was a marked increase in the excess of births over deaths, or what demographers call "natural increase" (see Figure 8–2). In the century from about 1815 until World War I the average annual increase in the three Scandinavian countries was 11.8 per 1,000—nearly five times what it had been in the middle of the 18th century, and sufficient to triple the population in 100 years.

For a long time the population of northwestern Europe showed little reaction to this rapid natural increase. But when it came, the reaction was emphatic; a wide variety of responses occurred, all of which tended to reduce the growth of the population. For example, in the latter part of the 19th century people began to emigrate from Europe by the millions, mainly to America, Australia and South Africa. Between 1846 and 1932 an estimated 27 million people emigrated overseas from Europe's 10 most advanced countries. The three Scandinavian countries alone sent out 2.4 million, so that in 1915 their combined population was 11.1 million instead of the 14.2 million it would otherwise have been.

In addition to this unprecedented exodus there were other responses, all of which tended to reduce the birth rate. In spite of opposition from church and state, agitation for birth control began and induced abortions became common. The age at marriage rose. Childlessness became frequent. The result was a decline in the birth rate that eventually overtook the continuing decline in the death rate. By the 1930's most of the industrial European countries had age-specific fertility rates so low that, if the rates had continued at that level, the population would eventually have ceased to replace itself.

In explaining this vigorous reaction one gets little help from two popular clichés. One of these—that population growth is good for business— would hardly explain why Europeans were so bent on stopping population growth. The other—that numerical limitation comes from the threat of poverty because "population always presses on the means of subsistence"—is factually untrue. In every one of the industrializing countries of Europe economic growth outpaced population growth. In the United Kingdom, for example, the real per capita income increased 2.3 times between the periods 1855–1859 and 1910–1914. In Denmark from 1770 to 1914 the rise of the net domestic product in constant prices was two and a half times the natural increase rate; in Norway and Sweden from the

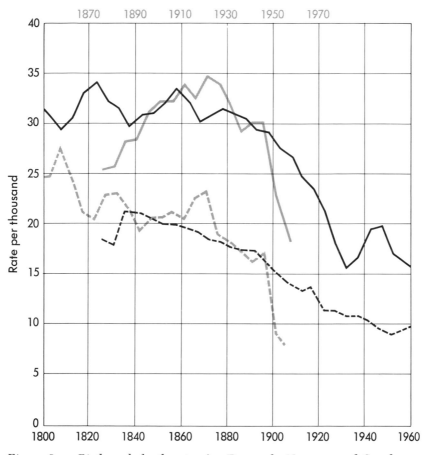

Figure 8–2. Birth and death rates for Denmark, Norway, and Sweden combined (black lines and dates) are compared with Japanese rates (gray lines and dates) of 50 years later. Japan has been passing through a population change similar to that which occurred earlier in Scandinavia. Area between respective birth-rate curves (solid lines) and death-rate curves (broken lines) shows natural increase, or population growth that would have occurred without migration. In past few years both Japanese rates have dropped extremely rapidly.

1860's to 1914 it was respectively 1.4 and 2.7 times the natural increase rate. Clearly the strenuous efforts to lessen population growth were due to some stimulus other than poverty.

The stimulus, in my view, arose from the clash between new opportunities on the one hand and larger families on the other. The modernizing society of northwestern Europe necessarily offered new opportunities to people of all classes: new ways of gaining wealth, new means of rising socially, new symbols of status. In order to take advantage of those opportunities, however, the individual and his children required education, special skills, capital and mobility—none of which was facilitated by an improvident marriage or a large family. Yet because mortality was being reduced (and reduced more successfully in the childhood than in the adult ages) the size of families had become potentially larger than before. In Sweden, for instance, the mortality of the period 1755–1775 allowed only 6.1 out of every 10 children born to reach the age of 10, whereas the mortality of 1901–1910 allowed 8.5 to survive to that age. In order to avoid the threat of a large family to his own and his children's socioeconomic position, the individual tended to postpone or avoid marriage and to limit reproduction within marriage by every means available. Urban residents had to contend particularly with the cost and inconvenience of young children in the city. Rural families had to adjust to the lack of enough land to provide for new marriages when the children reached marriageable age. Land had become less available not only because of the plethora of families with numerous youths but also because, with modernization, more capital was needed per farm and because the old folks, living longer, held on to the property. As a result farm youths postponed marriage, flocked to the cities or went overseas.

In such terms we can account for the paradox that as the progressive European nations became richer, their population growth slowed down. The process of economic development itself provided the motives for curtailment of reproduction, as the British sociologist J. A. Banks has made clear in his book *Prosperity and Parenthood.* We can see now that in all modern nations the long-run trend is one of low mortality, a relatively modest rate of reproduction and slow population growth. This is an efficient demographic system that allows such countries, in spite of their "maturity," to continue to advance economically at an impressive speed.

Naturally the countries of northwestern Europe did not all follow an identical pattern. Their stages differed somewhat in timing and in the pattern of preference among the various means of population control. France, for example, never attained as high a natural increase as Britain or Scandinavia did. This was not due solely to an earlier decline in the birth rate, as is often assumed, but also to a slower decline in the death rate. If we historically substitute the Swedish death rate for the French, we revise the natural increase upward by almost the same amount as we do by substituting the Swedish birth rate. In accounting for the early and easy drop in French fertility one recalls that France, already crowded in the 18th century and in the van of intellectual radicalism and sophistication, was likely to have a low threshold for the adoption of abortion and contraception. The death rate, however, remained comparatively high because France did not keep economic pace with her more rapidly industrializing neighbors. As a result the relatively small gap between births and deaths gave France a slower growth in population and a lesser rate of emigration.

Ireland also has its own demographic history, but like France it differs from the other countries in emphasis rather than in kind. The emphasis in Ireland's escape from human inflation was on emigration, late marriage and permanent celibacy. By 1891 the median age at which Irish girls married was 28 (compared with 22 in the U.S. at that date); nearly a fourth of the Irish women did not marry at all, and approximately a third of all Irish-born people lived outside of Ireland. These adjustments, begun with the famine of the 1840's and continuing with slight modifications until today, were so drastic that they made Ireland the only modern nation to experience an absolute decline in population. The total of 8.2 million in 1841 was reduced by 1901 to 4.5 million.

The Irish preferences among the means of population limitation seem to come from the island's position as a rural region participating only indirectly in the industrial revolution. For most of the Irish, land remained the basis for respectable matrimony. As land became inaccessible to young people they postponed marriage. In doing so they were not discouraged by their parents, who wished to keep control of the land, or by their religion. Their Catholicism, which they embraced with exceptional vigor both because they were rural and because it was a rallying point for Irish nationalism as against the Protestant English, placed a high value on celibacy. The clergy, furthermore, were powerful enough to exercise strict control over courtship and thus to curtail illicit pregnancy and romance as factors leading to marriage. They were also able to exercise exceptional restraint on abortion and contraception. Although birth control was practiced to some extent, as evidenced by a decline of fertility within marriage, its influence was so small as to make early marriage synonymous with a large family and therefore to be avoided. Marriage was also discouraged by the ban on divorce and by the lowest participation of married women in the labor force to be found in Europe. The country's failure to industrialize meant that the normal exodus from farms to cities was at the same time an exodus from Ireland itself.

Ireland and France illustrate contrasting variations on a common theme. Throughout northwestern Europe the population upsurge resulting from the fall in death rates brought about a multiphasic reaction that eventually reduced the population growth to a modest pace. The main force behind this response was not poverty or hunger but the desire of the people involved to preserve or improve their social standing by grasping the opportunities offered by the newly emerging industrial society.

Is this an interpretation applicable to the history of any industrialized country, regardless of traditional culture? According to the evidence the answer is yes. We might expect it to be true, as it currently is, of the countries of southern and eastern Europe that are finally industrializing. The crucial test is offered by the only nation outside the European tradition to become industrialized: Japan. How closely does Japan's demographic evolution parallel that of northwestern Europe?

If we superpose Japan's vital-rate curves on those of Scandinavia half a century earlier (see Figure 8–2), we see a basically similar, although more rapid, development. The reported statistics, questionable up to 1920 but good after that, show a rapidly declining death rate as industrialization took hold after World War I. The rate of natural increase during the period from 1900 to 1940 was almost exactly the same as Scandinavia's between 1850 and 1920, averaging 12.1 per 1,000 population per year compared with Scandinavia's 12.3. And Japan's birth rate, like Europe's, began to dip until it was falling faster than the death rate, as it did in Europe. After the usual baby boom following World War II the decline in births was precipitous, amounting to 50 per cent from 1948 to 1960—perhaps the swiftest drop in reproduction that has ever occurred in an entire nation. The rates of childbearing for women in various ages are so low that, if they continued indefinitely, they would not enable the Japanese population to replace itself.

In thus slowing their population growth have the Japanese used the same means as the peoples of northwestern Europe did? Again, yes. Taboo-ridden Westerners have given disproportionate attention to two features of the change—the active role played by the Japanese government and the widespread resort to abortion—but neither of these disproves the similarity. It is true that since the war the Japanese government has pursued a birth-control policy more energetically than any government ever has before. It is also clear, however, that the Japanese people would have reduced their childbearing of their own accord. A marked decline in the reproduction rate had already set in by 1920, long before there was a government policy favoring this trend.

As for abortion, the Japanese are unusual only in admitting its extent. Less superstitious than Europeans about this subject, they keep reasonably good records of abortions, whereas most of the other countries have no accurate data. According to the Japanese records, registered abortions rose from 11.8 per 1,000 women of childbearing age in 1949 to a peak of 50.2 per 1,000 in 1955. We have no reliable historical information from Western countries, but we do know from many indirect indications that induced abortion played a tremendous role in the reduction of the birth rate in western Europe from 1900 to 1940, and that it still plays a considerable role. Furthermore, Christopher Tietze of the National Committee for Maternal Health has assembled records that show that in five eastern European countries where abortion has been legal for some time [2] the rate has shot up recently in a manner strikingly similar to Japan's experience. In 1960–1961 there were 139 abortions to every 100 births in Hungary, 58 per 100 births in Bulgaria, 54 in Czechoslovakia and 34 in Poland. The countries of eastern Europe are in a developmental stage comparable to that of northwestern Europe earlier in the century.

Abortion is by no means the sole factor in the decline of Japan's birth rate. Surveys made since 1950 show the use of contraception before that date, and increasing use thereafter. There is also a rising frequency of sterilization. Furthermore, as in Europe earlier, the Japanese are postponing marriage. The proportion of girls under 20 who have ever married fell from 17.7 per cent in 1920 to 1.8 per cent in 1955. In 1959 only about 5 per cent of the Japanese girls marrying for the first time were under 20, whereas in the U.S. almost half the new brides (48.5 per cent in the registration area) were that young.

Finally, Japan went through the same experience as western Europe in another respect—massive emigration. Up to World War II Japan sent millions of emigrants to various regions of Asia, Oceania and the Americas.

In short, in response to a high rate of natural increase brought by declining mortality, Japan reacted in the same ways as the countries of northwestern Europe did at a similar stage. Like the Europeans, the Japanese limited their population growth in their own private interest and that of their children in a developing society, rather than from any fear of absolute privation or any concern with overpopulation in their homeland. The nation's average 5.4 per cent annual growth in industrial output from 1913 to 1958 exceeded the performance of European countries at a similar stage.

As our final class of industrialized countries we must now consider the frontier group—the U.S., Canada, Australia, New Zealand, South Africa and Russia. These countries are distinguished from those of northwestern Europe and Japan by their vast wealth of natural resources in relation to their populations; they are the genuinely affluent nations. They might be

2. In January 1967 the Romanian government banned abortions and imposed additional income taxes upon childless couples. Ed.

expected to show a demographic history somewhat different from that of Europe. In certain particulars they do, yet the general pattern is still much the same.

One of the differences is that the riches offered by their untapped resources invited immigration. All the frontier industrial countries except Russia received massive waves of emigrants from Europe. They therefore had a more rapid population growth than their industrializing predecessors had experienced. As frontier countries with great room for expansion, however, they were also characterized by considerable internal migration and continuing new opportunities. As a result their birth rates remained comparatively high. In the decade from 1950 to 1960, with continued immigration, these countries grew in population at an average rate of 2.13 per cent a year, compared with 1.76 per cent for the rest of the world. It was the four countries with the sparsest settlement (Canada, Australia, New Zealand and South Africa), however, that accounted for this high rate; in the U.S. and the U.S.S.R. the growth rate was lower—1.67 per cent per year.[3]

Apparently, then, in pioneer industrial countries with an abundance of resources population growth holds up at a higher level than in Japan or northwestern Europe because the average individual feels it is easier for himself and his children to achieve a respectable place in the social scale. The immigrants attracted by the various opportunities normally begin at a low level and thus make the status of natives relatively better. People marry earlier and have slightly larger families. But this departure from the general pattern for industrial countries appears to be only temporary.

In the advanced frontier nations, as in northwestern Europe, the birth rate began to fall sharply after 1880, and during the depression of the 1930's it was only about 10 per cent higher than in Europe. Although the postwar baby boom has lasted longer than in other advanced countries, it is evidently starting to subside now, and the rate of immigration has diminished. There are factors at work in these affluent nations that will likely limit their population growth. They are among the most urbanized countries in the world, in spite of their low average population density. Their birth rates are extremely sensitive to business fluctuations and social changes. Furthermore, having in general the world's highest living standards, their demand for resources, already staggering, will become fantastic if both population and per capita consumption continue to rise rapidly, and their privileged position in the world may become less tolerated.

3. During the 1960's this rate has been declining in the United States. The per cent rate of growth for 1963–66 was 1.3. The estimate for the year 1967–68 was 1.0 per cent. See the *Statistical Yearbook,* United Nations, 1968, and *Current Population Reports,* U.S. Department of Commerce, August 13, 1968. Ed.

Let us shift now to the other side of the population picture: the nonindustrial, or underdeveloped, countries.

As a class the nonindustrial nations since 1930 have been growing in population about twice as fast as the industrial ones. This fact is so familiar and so taken for granted that its irony tends to escape us. When we think of it, it is astonishing that the world's most impoverished nations, many of them already overcrowded by any standard, should be generating additions to the population at the highest rate.

The underdeveloped countries have about 69 per cent of the earth's adults—and some 80 per cent of the world's children. Hence the demographic situation itself tends to make the world constantly more underdeveloped, or impoverished, a fact that makes economic growth doubly difficult.

How can we account for the paradox that the world's poorest regions are producing the most people? One is tempted to believe that the underdeveloped countries are simply repeating history: that they are in the same phase of rapid growth the West experienced when it began to industrialize and its death rates fell. If that is so, then sooner or later the developing areas will limit their population growth as the West did.

It is possible that this may prove to be true in the long run. But before we accept the comforting thought we should take a close look at the facts as they are.

In actuality the demography of the nonindustrial countries today differs in essential respects from the early history of the present industrial nations. Most striking is the fact that their rate of human multiplication is far higher than the West's ever was. The peak of the industrial nations' natural increase rarely rose above 15 per 1,000 population per year; the highest rate in Scandinavia was 13, in England and Wales 14, and even in Japan it was slightly less than 15. True, the U.S. may have hit a figure of 30 per 1,000 in the early 19th century, but if so it was with the help of heavy immigration of young people (who swelled the births but not the deaths) and with the encouragement of an empty continent waiting for exploitation.

In contrast, in the present underdeveloped but often crowded countries the natural increase per 1,000 population is everywhere extreme. In the decade from 1950 to 1960 it averaged 31.4 per year in Taiwan, 26.8 in Ceylon, 32.1 in Malaya, 26.7 in Mauritius, 27.7 in Albania, 31.8 in Mexico, 33.9 in El Salvador and 37.3 in Costa Rica. These are not birth rates; they are the *excess* of births over deaths! At an annual natural increase of 30 per 1,000 a population will double itself in 23 years.

The population upsurge in the backward nations is apparently taking place at an earlier stage of development—or perhaps we should say *unde-*

velopment—than it did in the now industrialized nations. In Britain, for instance, the peak of human multiplication came when the country was already highly industrialized and urbanized, with only a fifth of its working males in agriculture. Comparing four industrial countries at the peak of their natural increase in the 19th century (14.1 per 1,000 per year) with five nonindustrial countries during their rapid growth in the 1950's (32.2 per 1,000 per year), I find that the industrial countries were 38.5 per cent urbanized and had 27.9 per cent of their labor force in manufacturing, whereas now the nonindustrial countries are 29.4 per cent urbanized and have only 15.1 per cent of their people in manufacturing. In short, today's nonindustrial populations are growing faster and at an earlier stage than was the case in the demographic cycle that accompanied industrialization in the 19th century.

As in the industrial nations, the main generator of the population upsurge in the underdeveloped countries has been a fall in the death rate. But their resulting excess of births over deaths has proceeded faster and farther, as a comparison of Ceylon in recent decades with Sweden in the 1800's shows. (See Figure 8–3.)

In most of the underdeveloped nations the death rate has dropped with record speed. For example, the sugar-growing island of Mauritius in the Indian Ocean within an eight-year period after the war raised its average life expectancy from 33 to 51—a gain that took Sweden 130 years to achieve. Taiwan within two decades has increased its life expectancy from 43 to 63; it took the U.S. some 80 years to make this improvement for its white population. According to the records in 18 underdeveloped countries, the crude death rate has dropped substantially in each decade since 1930; it fell some 6 per cent in the 1930's and nearly 20 per cent in the 1950's, and according to the most recent available figures the decline in deaths is still accelerating.

The reasons for this sharp drop in mortality are in much dispute. There are two opposing theories. Many give the credit to modern medicine and public health measures. On the other hand, the public health spokesmen, rejecting the accusation of complicity in the world's population crisis, belittle their own role and maintain that the chief factor in the improvement of the death rate has been economic progress.

Those in the latter camp point out that the decline in the death rate in northwestern Europe followed a steadily rising standard of living. Improvements in diet, clothing, housing and working conditions raised the population's resistance to disease. As a result many dangerous ailments disappeared or subsided without specific medical attack. The same pro-

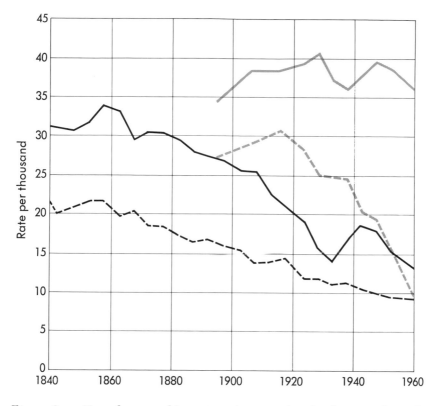

Figure 8–3. New demographic pattern is appearing in the nonindustrial nations. The birth rate (solid line) has not been falling significantly, whereas the death rate (broken line) has dropped precipitously, as illustrated by Ceylon (gray). The spread between the two rates has widened. In nations such as Sweden (black), however, the birth rate dropped during development long before the death rate was as low as in most underdeveloped countries today.

cess, say the public health people, is now at work in the developing countries.

On the other side, most demographers and economists believe that economic conditions are no longer as important as they once were in strengthening a community's health. The development of medical science has provided lifesaving techniques and medicines that can be transported overnight to the most backward areas. A Stone Age people can be endowed with a low 20th-century death rate within a few years, without waiting for the slow process of economic development or social change.

International agencies and the governments of the affluent nations have been delighted to act as good Samaritans and send out public health missionaries to push disease-fighting programs for the less developed countries.

The debate between the two views is hard to settle. Such evidence as we have indicates that there is truth on both sides. Certainly the newly evolving countries have made economic progress. Their economic advance, however, is not nearly rapid enough to account for the very swift decline in their death rates, nor do they show any clear correlation between economic growth and improvement in life expectancy. For example, in Mauritius during the five-year period from 1953 to 1958 the per capita income fell by 13 per cent, yet notwithstanding this there was a 36 per cent drop in the death rate. On the other hand, in the period between 1945 and 1960 Costa Rica had a 64 per cent increase in the per capita gross national product and a 55 per cent decline in the death rate. There seems to be no consistency—no significant correlation between the two trends when we look at the figures country by country. In 15 underdeveloped countries for which such figures are available we find that the decline in death rate in the 1950's was strikingly uniform (about 4 per cent per year), although the nations varied greatly in economic progress—from no improvement to a 6 per cent annual growth in per capita income.

Our tentative conclusion must be, therefore, that the public health people are more efficient than they admit. The billions of dollars spent in public health work for underdeveloped areas has brought down death rates, irrespective of local economic conditions in these areas. The programs instituted by outsiders to control cholera, malaria, plague and other diseases in these countries have succeeded. This does not mean that death control in underdeveloped countries has become wholly or permanently independent of economic development but that it has become temporarily so to an amazing degree.

Accordingly the unprecedented population growth in these countries bears little relation to their economic condition. The British economist Colin G. Clark has contended that rapid population growth stimulates economic progress. This idea acquires plausibility from the association between human increase and industrialization in the past and from the fact that in advanced countries today the birth rate (but not the death rate) tends to fluctuate with business conditions. In today's underdeveloped countries, however, there seems to be little or no visible connection between economics and demography.

In these countries neither births nor deaths have been particularly responsive to economic change. Some of the highest rates of population growth ever known are occurring in areas that show no commensurate economic advance. In 34 such countries for which we have data, the correlation between population growth and economic gain during the 1950's was negligible, and the slight edge was on the negative side: —.2. In 20 Latin-American countries during the period from 1954 to 1959, while the

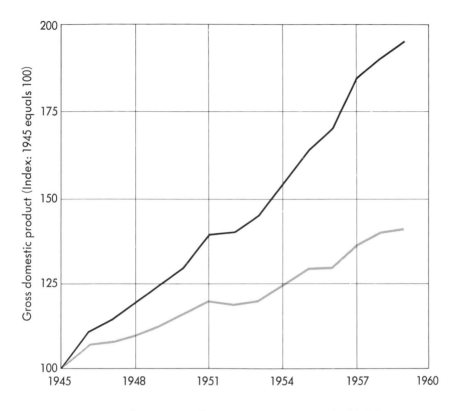

Figure 8–4. Gross domestic product of Latin America doubled between 1945 and 1959 (black line) but population growth held down the increase in per capita product (gray line).

annual gain in per capita gross domestic product fell from an average of 2 per cent to 1.3 per cent, the population growth rate *rose* from 2.5 to 2.7 per cent per year.

All the evidence indicates that the population upsurge in the underdeveloped countries is not helping them to advance economically. On the contrary, it may well be interfering with their economic growth. A surplus of labor on the farms holds back the mechanization of agriculture. A rapid rise in the number of people to be maintained uses up income that might otherwise be utilized for long-term investment in education, equipment and other capital needs. To put it in concrete terms, it is difficult to give a child the basic education he needs to become an engineer when he is one of eight children of an illiterate farmer who must support the family with the produce of two acres of ground.

By definition economic advance means an increase in the amount of product per unit of human labor. This calls for investment in technology, in improvement of the skills of the labor force and in administrative organization and planning. An economy that must spend a disproportionate share of its income in supporting the consumption needs of a growing population—and at a low level of consumption at that—finds growth difficult because it lacks capital for improvements.

A further complication lies in the process of urbanization. The shifts from villages and farmsteads to cities is seemingly an unavoidable and at best a painful part of economic development. It is most painful when the total population is skyrocketing; then the cities are bursting both from their own multiplication and from the stream of migrants from the villages. The latter do not move to cities because of the opportunities there. The opportunities are few and unemployment is prevalent. The migrants come, rather, because they are impelled by the lack of opportunity in the crowded rural areas. In the cities they hope to get something—a menial job, government relief, charities of the rich. I have recently estimated that if the population of India increases at the rate projected for it by the UN, the net number of migrants to cities between 1960 and 2000 will be of the order of 99 to 201 million, and in 2000 the largest city will contain between 36 and 66 million inhabitants. One of the greatest problems now facing the governments of underdeveloped countries is what to do with these millions of penniless refugees from the excessively populated countryside.

Economic growth is not easy to achieve. So far, in spite of all the talk and the earnest efforts of underdeveloped nations, only one country outside the northwestern European tradition has done so: Japan. The others are struggling with the handicap of a population growth greater than any industrializing country had to contend with in the past. A number of them now realize that this is a primary problem, and their governments are pursuing or contemplating large-scale programs of birth limitation. They are receiving little help in this matter, however, from the industrial nations, which have so willingly helped them to lower their death rates.

The Christian nations withhold this help because of their official taboos against some of the means of birth limitation (although their own people privately use all these means). The Communist nations withhold it because limitation of population growth conflicts with official Marxist dogma (but Soviet citizens control births just as capitalist citizens do, and China is officially pursuing policies calculated to reduce the birth rate).

The West's preoccupation with the technology of contraception seems unjustified in view of its own history. The peoples of northwestern Europe utilized all the available means of birth limitation once they had strong motives for such limitation. The main question, then, is whether or not the peoples of the present underdeveloped countries are likely to acquire such motivation in the near future. There are signs that they will.

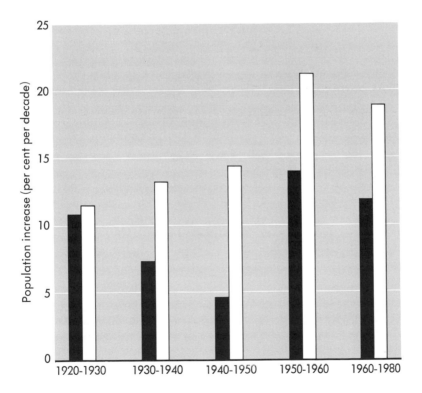

Figure 8–5. Differential population growth in underdeveloped regions (white bars) and developed regions (black bars) is plotted. The 1960–80 projections may turn out to be low.

Surveys in India, Jamaica and certain other areas give evidence of a growing desire among the people to reduce the size of their families. Furthermore, circumstances in the underdeveloped nations today are working
more strongly in this direction than they did in northwestern Europe in
the 19th century.

As in that earlier day, poverty and deprivation alone are not likely to
generate a slowdown of the birth rate. But personal aspirations are. The
agrarian peoples of the backward countries now look to the industrialized,
affluent fourth of the world. They nourish aspirations that come directly
from New York, Paris and Moscow. No more inclined to be satisfied with
a bare subsistence than their wealthier fellows would be, they are demanding more goods, education, opportunity and influence. And they are
beginning to see that many of their desires are incompatible with the enlarged families that low mortality and customary reproduction are giving
them.

They live amid a population density far greater than existed in 19th-
century Europe. They have no place to which to emigrate, no beckoning
continents to colonize. They have rich utopias to look at and industrial
models to emulate, whereas the Europeans of the early 1800's did not
know where they were going. The peoples of the underdeveloped, overpopulated countries therefore seem likely to start soon a multiphasic limitation of births such as began to sweep through Europe a century ago.
Their governments appear ready to help them. Government policy in
these countries is not quibbling over means or confining itself to birth-
control technology; its primary task is to strengthen and accelerate the
peoples' motivation for reproductive restraint.

Meanwhile the industrial countries also seem destined to apply brakes
to their population growth. The steadily rising level of living, multiplied
by the still growing numbers of people, is engendering a dizzying rate of
consumption. It is beginning to produce painful scarcities of space, of
clean water, of clean air and of quietness. All of this may prompt more
demographic moderation than these countries have already exercised.

Editor's Note. "The ultimate check to population," said Thomas Malthus, "appears to be a want of food, arising necessarily from the different ratios according to which population and food increase. But this ultimate check is never
the immediate check, except in cases of actual famine. The immediate check
may be stated to consist in all those customs, and all those diseases, which
seem to be generated by a scarcity of the means of subsistence; and all those
causes, independent of this scarcity, whether of a moral or physical nature,
which tend prematurely to weaken and destroy the human frame. These
checks to population . . . are constantly operating with more or less force in

every society, and keep down the number to the level of the means of subsistence . . ." [1]

Both Malthus himself and many of his followers went on to point out the
natural corollary to this law of population. If we increase the means of subsistence of poor and starving people, we will at the same time cause their
population to increase and thereby reduce them to a poorer state then they
had been in before.

Today the achievements of modern medicine have made it possible to control many diseases and reduce the death rate dramatically. The result has been
an alarming increase in world population. The Malthusian dilemma is thus
restated in slightly different form.

The Dangerous Doctor *
by William Vogt (1948)

The modern medical profession . . . continues to believe it has a duty
to keep alive as many people as possible. In many parts of the world doctors apply their intelligence to one aspect of man's welfare—survival—
and deny their moral right to apply it to the problem as a whole. Through
medical care and improved sanitation they are responsible for more millions living more years in increasing misery. Their refusal to consider their
responsibility in these matters does not seem to them to compromise their
intellectual integrity. . . . They set the stage for disaster; then, like Pi-
late, they wash their hands of the consequences.

The Ethical Dilemma of Science †
by A. V. Hill (1952)

The dilemma is this. All the impulses of decent humanity, all the dic-
tates of religion and all the traditions of medicine insist that suffering
should be relieved, curable diseases cured, preventable disease prevented.

* From William Vogt, *Road to Survival,* William Sloane Associates, Inc., New
York. Copyright 1948 by William Vogt. Reprinted by permission.

† From A. V. Hill, "The Ethical Dilemma of Science," *Nature,* 170, 1952. Reprinted
by permission of Macmillan (Journals) Ltd. and the author.

1. From Thomas R. Malthus, *Essay on the Principle of Population,* 6th edition,
1816, J. M. Dent & Sons Ltd.

The obligation is regarded as unconditional: it is not permitted to argue that the suffering is due to folly, that the children are not wanted, that the patient's family would be happier if he died. All that may be so; but to accept it as a guide to action would lead to a degradation of standards of humanity by which civilization would be permanently and indefinitely poorer. . . .

Some might [take] the purely biological view that if men will breed like rabbits they must be allowed to die like rabbits. . . . Most people would still say no. But suppose it were certain now that the pressure of increasing population, uncontrolled by disease, would lead not only to widespread exhaustion of the soil and of other capital resources but also to continuing and increasing international tension and disorder, making it hard for civilization itself to survive: Would the majority of humane and reasonable people then change their minds? If ethical principles deny our right to do evil in order that good may come, are we justified in doing good when the foreseeable consequence is evil?

Cybernetics of Population Control *

by Hudson Hoagland (1964)

There is an ambivalence about many scientific discoveries, and it is ironical that our best humanitarian motives in medicine and public health are primarily responsible for the grave dangers of the population explosion. I would like to consider some of the ways in which nature deals with overcrowding in other animal societies, since this may shed some light on our own population problems.

In multiplying cultures of microorganisms, the growth rate accelerates exponentially; but as toxic metabolic products such as acids or alcohols accumulate, the rate declines and the curve describing numbers of organisms as a function of time ultimately flattens off.

Insect populations are regulated in various ways. The fruit fly, *Drosophila,* above certain population densities, decreases its egg laying in an amount proportional to the density. In flour beetles, below a fixed number of grains of flour per beetle, cannibalism occurs in some species and egg production drops off; in one species, crowding results in females punctur-

* From Hudson Hoagland, "Cybernetics of Population," *Bulletin of the Atomic Scientists,* February 1964. Copyright 1964 by the Educational Foundation for Nuclear Science. Reprinted by permission.

ing and destroying some of the eggs they have produced. Frequency of copulation also declines with crowding. There are some species of flour beetles with glands that produce a gas, the release of which is increased with crowding. This gas is lethal to larvae and acts as an antaphrodisiac at high densities of population. Flour contaminated with beetle excrement inhibits egg production of another species; mixing this contaminated flour with fresh flour decreases the rate of population growth, despite the ample food supply.

Among mammals other than man, it was long thought that food and predators were the controlling factors in limiting populations. It was thought, for example, that the four-year cycles of buildup and decline of lemming populations, terminating in their suicidal migrations, were due to an increase in predators accompanying population growth which ultimately caused the panic and decline. But the migrations and deaths appear now not to be caused by the predators. Rather, the predators appear to multiply in response to the multiplying prey. While the lemming cycles have not been studied as systematically as those of other species, it seems likely that these four-year fluctuations in population densities are determined by factors now known to regulate cycles in other species.

Minnesota jack rabbit populations rise and fall through cycles of several years' duration. There is a buildup followed by a dying off. It was observed that when the animals died off there was usually plenty of food —they didn't starve. There was no evidence of excessive predators. Furthermore, the bodies showed no sign of any specific epidemic that killed them. To quote from a 1939 study of the dead animals: "This syndrome was characterized primarily by fatty degeneration and atrophy of the liver with a coincident striking decrease in liver glycogen and a hypoglycemia preceding death. Petechial or ecchymotic brain hemorrhages and congestion and hemorrhage of the adrenals, thyroid, and kidneys were frequent findings in a smaller number of animals. The hares characteristically died in convulsive seizures with sudden onset, running movements, hind-leg extension, retraction of the head and neck, and sudden leaps with clonic seizures upon alighting. Other animals were typically lethargic or comatose." The adrenals were hypertrophied in some cases and atrophied in others. Such signs—liver disease, hypertension, atherosclerosis, and adrenal deterioration—are typical of the acute stress syndrome that results from overactivity of the pituitary-adrenal axis.

Studies of rodents showed that, after the severe stress of winter crowding in burrows when population densities were high, there was much fighting among the males, sex drives were at a low ebb, the young were often eaten, and the females produced premature births. There was also increased susceptibility to nonspecific infections—another byproduct of excessive production of adrenal corticoids. After such a colony has been depleted in numbers through effects of the stress syndrome, the colony then tends to build up again, and so it goes through repeated cycles of growth and decline.

A pair of deer were put on a small island of about 150 acres in Chesapeake Bay about forty years ago. The deer were kept well supplied with food. It was found that the colony grew until it reached a density of about one deer per acre. Then the animals began to die off despite adequate food and care. Examination of the dead animals again showed marked evidence of the adrenal stress syndrome. In studies of crowding in the Philadelphia zoo, it was found that in some species of animals there was a tenfold increase in atherosclerosis under conditions of severe crowding, and there were many other symptoms characteristic of stress. John Christian of the Naval Medical Research Institute made population studies in relation to crowding of mice. In his 1950 paper in the *Journal of Mammalogy*, "The Adrenopituitary System and Population Cycles in Mammals," he wrote: "We now have a working hypothesis for the die-off terminating a cycle. Exhaustion of the adrenopituitary system resulting from increased stresses inherent in a high population, especially in winter, plus the late winter demands of the reproductive system, due to increased light or other factors, precipitates population-wide death with the symptoms of adrenal insufficiency and hypoglycemic convulsions."

OVERCROWDING IN RAT COLONIES

John Calhoun of the National Institutes of Health, investigating rats kept at critical levels of crowding, found high infant mortality, high abortion rates, and failures of mothers to build nests. The rats showed evidence of the stress syndrome, and Calhoun speaks of "pathological togetherness." He has reviewed this work in "Population Density and Social Pathology" (*Scientific American,* February 1962), a summary of which follows.

Calhoun confined wild Norway rats in a one-quarter acre enclosure with plenty of food and water. At the end of 27 months the population stabilized itself at 150 adults. One would expect from the very low adult mortality rate in uncrowded conditions a population of 5,000, not 150 rats. But infant mortality was extremely high. The stress from social interaction disrupted maternal behavior so that only a few of the young survived. Calhoun later studied groups of domesticated white rats confined indoors in observation rooms under better controlled conditions. Six different populations were examined. Each group was allowed to increase to twice the number that his earlier experience indicated could occupy the space allotted with only moderate stress. Pathological behavior was most marked in the females. Pregnancies were often not full term. There were many abortions and many maternal deaths. The mothers often could not nurse or care for their young. Among the males there was much sexual deviation, cannibalism, and other abnormal behavior ranging from frenetic overactivity to pathological withdrawal in which some males emerged from their nests only to eat and drink. Patterns of social behavior were thus badly deranged at twice normal crowding.

The experiments took place in four interconnecting pens, each six by six feet in area. Each was a complete dwelling unit with a drinking fountain, a food hopper, and an elevated artificial burrow containing five nest boxes. There was comfortable space for 12 adult rats in each pen. The setup should thus have been able to support 48 rats comfortably. At the stabilized number of 80 to 100—double the comfortable population—which they were allowed to reach by breeding, an even distribution would have been 20 to 25 adults in each pen. But the rats did not dispose themselves in this way.

Biasing factors were introduced in the following fashion. Ramps were arranged enabling the animals to get from one pen to another and so to traverse the entire four pens in the room. However, the two end pens, numbers 1 and 4, each had only one ramp, which connected them with pens 2 and 3, respectively, while the middle pens had two ramps each, i.e., ramps connecting pen 2 with pen 3 and with pen 1, and ramps connecting pen 3 with pen 2 and pen 4. This arrangement immediately skewed the probabilities in favor of a higher density in the two middle pens, since pens 2 and 3 could be reached by two ramps whereas pens 1 and 4 by only one ramp each. But with the passage of time strange aspects of the behavior of the group skewed the distribution in an unexpected way, and also resulted in an unexpected arrangement of the sex ratios. The females distributed themselves about equally in the four pens, but the male population was concentrated almost overwhelmingly in the middle pens. One reason for this was the status struggle among the males. Shortly after six months of age, each male enters into a round-robin of fights that eventually fixes his position in the social hierarchy. Such fights took place in all of the pens, but in the end pens it became possible for a single dominant male to take over the area as his territory.

Calhoun describes how this came about. The subordinate males in all pens adopted the habit of arising early. This enabled them to eat and drink in peace. Rats generally eat in the course of their normal wanderings, and the subordinate residents of the end pens, having been defeated by the dominant male in these pens, were likely to feed in one of the middle pens where they had not had as yet to fight for status. When, after feeding, they wanted to return to their original quarters, they found it very difficult to do so. By this time the dominant male in the end pen would have awakened and he would engage the subordinates in fights as they tried to come down the single ramp to the pen. For a while the subordinate would continue his efforts to return to what had been his home pen, but after a succession of defeats he would become so conditioned that he would not even make the attempt. In essence, Calhoun points out that the dominant male established his territory of domination in the end pens and his control over a harem of females—not by driving the other males out but by preventing their return over the one ramp leading to the end pen. While the dominant male in an end pen slept a good part of the time, he made his sleeping quarters at the base of the ramp. He was therefore on perpetual guard, awakening as soon as another male appeared at the head of the ramp. He usually only had to open his eyes for the invader to wheel around and return to the adjoining pen. Since there were two ramps for pens 2 and 3, no one male could thus dominate both of them. The dominant males in pens 1 or 4 would sleep calmly through all the comings and goings of his harem. Seemingly he did not even hear them. His condition during his waking hours reflected his dominant status. He would move about in a casual and deliberate fashion, occasionally inspecting the burrow and nests of his harem. But he would rarely enter a burrow, as some other males did in the middle pens, merely to ferret out the females. A dominant male might tolerate other males in his domain, provided they were phlegmatic and made themselves scarce. Most of the time these subordinate males would hide in the burrows with the adult females, and only come out onto the floor to eat and drink. These subordinate males never tried to engage in sex activity with the females.

In the end pens, where population density was thus kept low, the mortality rate among infants and females was also low. Of the various social environments that developed, the breed pens—as the end pens were called—were the only healthy ones. The harem females generally made good mothers and protected their pups from harm. The pregnancy rates of the females in the middle pens were the same as those in the end pens, but a very much lower percentage of their pregnancies terminated in live births. In one series of experiments 99 per cent of the young born in pens 2 and 3 perished before weaning.

The females that lived in the densely populated middle pens became progressively less adapted to building adequate nests. Normally, this is an undertaking that involves repeated periods of sustained activity with the searching out of appropriate materials, such as the strips of paper which were made available to them, and transporting the strips to a nest which they arrange in a cup-like form. In the crowded middle pens, however, the females began merely to pile the strips in heaps, sometimes trampling them into a pad, showing little signs of cup formation. Later they would bring fewer and fewer strips to the nesting site, and in the midst of transporting a bit of material would drop it and engage in some other activity, occasioned by contact and interaction with other rats met on the way. In the extreme disruption of their behavior during the later months of the population's history, they would build no nests at all, but would bear their litters on the sawdust in the burrow box. The females also lost the ability to transport their litters from one place to another—a thing they would normally do with skill. If they tried to move a litter, they would drop individuals and scatter them about on the floor. The infants thus abandoned throughout the pens were seldom nursed. They would die where they were dropped and were thereupon eaten by the adults. In the middle pens, when a female would come into heat, she would be relentlessly pursued by all of the males until she was exhausted. This caused the death of 25 per cent of the females in a relatively short time in the crowded pens; in contrast, only 15 per cent of the adult males died over the same time in the middle pens. In the end brood pens, however, this sort of thing didn't happen. The females in these pens would retire to bear their young in nests they made in a normal fashion, and were protected from the excessive attention of the other males by the dominant male.

In the middle pens, a great deal of fighting among the males went on, with now one and now another assuming the dominant position. In contrast, in the end brood pens, one male predominated and peace generally reigned. These dominant males took care of the females and of the juveniles, never bothering them in any way.

Among the subordinate males there was much abnormal behavior. For instance, there was a group of homosexuals. They were really pansexual animals, and apparently could not discriminate between sex partners. They made sexual advances to males, juveniles, and females that were not in estrus. They were frequently attacked by their more dominant associates, but they very rarely contended for status.

Another type of male emerged in the crowded pens. This was essentially a very passive type that moved a good deal like a somnambulist. These rats ignored the others of both sexes, and all the other rats ignored them. Even when the females were in estrus, these passive animals made no advances to them, and only very rarely did other males approach them for any kind of play. They were healthy, attractive, and sleek, but were simply zombies in their conduct.

The strangest of all the abnormal male types was what Calhoun called the probers. These animals, which always lived in the middle pens, took no part at all in the status struggle. Nevertheless they were the most active of all the males, and persisted in their activities in spite of attacks by the dominant animals. In addition to being hyperactive, the probers were hypersexual, and in time many of them became cannibalistic. They were always chasing females about, entering their nests, having intercourse with them in the nest, a thing that the normal rats would never do. These probers conducted their pursuits of estrus females in a very abnormal manner, abandoning all the courtship ritual that is characteristic of normal mating rats.

These experiments of John Calhoun show the development of serious pathology in a society which is directly attributable to overcrowding at only twice the number of rats per unit area required for a healthy society.

STRESS IN HUMAN POPULATIONS

A question of immediate interest is to what extent the stress syndrome may be a factor in reducing the growth rate of human populations. As far as I know, there are no adequate data to answer this question. Studies from a number of laboratories including our own have demonstrated that the human pituitary adrenal system responds under stress in a way similar to that of other mammals. There is indirect evidence that inmates of concentration camps experienced acute forms of the stress syndrome that may have accounted for many deaths. Concentration camps are more appropriately compared with highly congested animal populations than are city slums, since even in very crowded cities, the poor do have some mobility. They can escape from their immediate congestion on streets and associate with other segments of the population. The incidence of street gangs and juvenile delinquency is especially characteristic of overcrowded city areas and constitutes a form of social pathology. Several studies have also indicated a higher incidence of schizophrenia and of other psychotic and neurotic behavior in congested urban areas than in more spacious environments, but other factors may be involved here. The increased incidence of atherosclerosis and other cardiovascular pathology associated with urban living and its competitive stresses may also be enhanced by crowding, although direct evidence for this is lacking. In underdeveloped countries with high birth rates and recently lowered death rates, producing population growth of two to four per cent per year, any growth-retarding effect of the stress syndrome is masked by the use of health measures that are enhancing life expectancies.

V. C. Wynne-Edwards has published an important book called *Animal Dispersion in Relation to Social Behavior* (New York: Hafner Press, 1962). He points out that in nature, a failure in food supplies results in a local emergency of overpopulation, which, especially among highly mobile flying animals such as insects and birds, can be immediately relieved by emigration. The social hierarchy and code of behavior of the animals are devices to force out the surplus. The exiles may be condemned to perish or, under other circumstances, may set up new populations in other places. In the cold boreal and arctic regions, most of the new species are birds. Many birds are adapted to exploit intermittent or undependable food crops and are to a large extent nomadic in their search for regions where supplies are temporarily plentiful. Two distinct biological functions appear to be served by emigration. One is that of a safety valve to give immediate relief to overpopulation, and the other is a pioneering function to expand and replenish the range of the species as a whole, and provide for gene exchange. Emigration of the safety valve kind is associated with stress and with quickly deteriorating economic conditions (locusts). On the other hand, providing pioneers is something that can be afforded only when conditions are good. In more variable environments, both functions tend to become important and to be exercised on a large scale. The individuals expelled are in either case usually the junior fraction of the hierarchy.

In this connection I am reminded of human colonizing activities—the Greek city-states that sent young people off to colonize the Mediterranean area—Italy, Sicily, Africa, and the lands around the Aegean and the Black Sea. The younger sons of families in Britain, Portugal, Spain, and France established colonies in the western hemisphere and relieved population pressures at home. Like many animal colonies, Australia was originally colonized by a group very low in the British pecking order, i.e. prisoners. . . . But today, human emigration on a significant scale has ceased everywhere. The frontier is no more and there are few areas that welcome immigrants in our world of intensified nationalisms.

Wynne-Edwards points out that the same general social machinery that controls safety-valve emigrations is involved in regulating seasonal redispersions of animals, such as the annual two-way migrations of birds. Many studies, other than those already mentioned, of mortality promoted by stress have been made. The white stork has been intensively studied. Nestling mortality is often very heavy under crowded conditions and individual chicks may be deliberately killed and sometimes eaten by one of their parents, usually the father. This is most likely to happen where the parents are beginners or young adults, and presumably of lower social status in the pecking hierarchy. The killing off of the young under prolific breeding conditions is characteristic of a great many birds and mammals, and is a direct result of social stress. The killing of the young and cannibalism are known to occur quite widely in mammals; for instance, in rodents, lions, and also in primitive man. Cases of cannibalism are found in fish, spider crabs, and spiders and of fractricide in various insect larvae. *In all cases experimentally investigated, the mortality is found to be dependent on population density and to cease below a certain critical population density.*

Mortality from predation has also been examined. This appears to be density-dependent to the extent that the prey cooperates by making its surplus members especially vulnerable to predators. The density-dependent elements in predation thus seem to arise on the part of the prey and not on that of the predators. Because of lowered resistance to infective agents following prolonged stress, disease as a form of predation may effectively reduce excessive population. In this case a surplus of individuals predisposed to injury by their dominant fellows naturally experiences a variable amount of uncontrolled mortality; this tends to fall most heavily on the young, which are as yet unprotected by acquired immunity from bacterial and viral infections. But social stress can lead to casualties at all ages, both through direct and mortal combat and through stress-induced disease. The victim of severe stress is likely to develop physiological disorders affecting many organs, especially the lymphatic apparatus, including the spleen and thymus, and also the nervous system, circulatory, digestive, and generative organs, and the endocrine glands, especially the adrenal cortex, which serves an intermediary role between the stressor and the organs responding to adrenal cortical hormones. Social stress is sometimes partly physical, as when the exercise of pecking order rights leads to the infliction of wounds or to withholding food and shelter. But, as Wynne-Edwards points out, it may be largely mental, just as we find that man, in his simpler-minded states, may die from the conviction that he has been bewitched. Cases are known of birds, mammals, and amphibians similarly dying from nonspecific injuries apparently induced by social stress.

THE POPULATION EXPLOSION

What about man? What can we do about the world population explosion? We could, of course, do nothing and just wait for the stress syndrome or a new virus to do its work. It has been said that until recently our politicians had washed their hands of the population problem but are now wringing their hands over it. We can leave the "solution" to some trigger-happy dictator with a suitable stockpile of nuclear weapons, or perhaps we can finally decide on an optimal population for the world and, by education and social pressure, try to see that it is not exceeded. At the present average growth rate of two per cent per year, there will be one square yard of earth per person in 600 years. Population growth depends only on the difference between birth rate and death rate. Man is the only animal that can direct its own evolution. Which of these two variables will he manipulate?

Prejudice, indifference, and hostility are the major blocks to population limitation. We know many methods of birth control: coitus interruptus, jellies, douches, diaphragms, condoms, surgical procedures, and "the pill," and ongoing research will give us more and better methods. None of these are of value if people refuse their use. For the very poor and illiterate, the cost and the difficulties of using contraceptives demand massive financial, social, and educational government aid. Prudery and politics, myth, superstition, and tradition, have so far rendered birth control ineffective in countries most in need of it.

Grenville Clark in a recent paper has argued that the population explosion probably cannot be controlled until the world has acquired universal and complete disarmamant under world law, at which time a substantial part of the $120 billion now being spent on weaponry might, if we are wise enough, be used to raise the living standards of have-not peoples. He bases this view on the often demonstrated fact that birth control procedures are used extensively only by literate and prosperous people with hope and ambition for bettering their own lots and those of their children. The take-off point for family planning and limitation requires a critical level of education and prosperity not now found in the very poor countries.

A still more gloomy view might be that if we do not manage to disarm in a decade or so we shall probably solve the population problem by nuclear extermination. In any case, the two major problems of our time— nuclear war and the population explosion—are closely linked together. Physics and the medical sciences have given all of mankind these two worldwide challenges never dreamed of by previous generations. Only by fundamental changes in our ways of thinking can these problems be solved.

The Dysgenic Aspects of Modern Civilization

Medical Utopias *
by René Dubos (1961)

One needs only to recall the toll exacted by infectious diseases in the past to appreciate the magnitude of the contributions made to human health by sanitation and drug therapy. But even this achievement may turn out to be a not unmixed blessing. . . .

The availability of techniques capable of postponing death in every age group and arresting almost any type of disease will increasingly present to the medical conscience difficult alternatives. For example, to save the life of a child suffering from some hereditary defect is a humane act and the source of professional gratification, but the long-range consequences of this achievement mean magnified medical problems for the following generations. Likewise, prolonging the life of an aged and ailing person must be weighed against the consequences for the individual himself and also for the community of which he is a part. These ethical difficulties are not new, of course, but in the past they rarely presented issues to the medical conscience because the physician's power of action was so limited. Soon, however, ethical difficulties are bound to become larger as the physician becomes better able to prolong biological life in individuals who cannot derive either profit or pleasure from existence, and whose survival creates painful burdens for the community. Increasing numbers of these persons cannot pull their full weight in society and require constant medical supervision and economic assistance. They constitute a social burden which is likely to grow heavier with time, precisely as a result of medical advances.

Even more important than these economic considerations, however, is the fact that many of the biologically defective individuals who are saved from death transfer to their progeny the genetic basis of their deficiencies. Speaking of our "load of mutations," Professor H. J. Muller has repeatedly emphasized that, as medical science becomes more effective in prolonging

* From René Dubos, *The Dreams of Reason*, Columbia University Press, New York. Copyright © 1961 by Columbia University Press. Reprinted by permission.

survival, there will be an increase in the frequency of detrimental genes allowed to accumulate in our communities. A continuation of this trend would mean that, in Professor H. J. Muller's words,

Instead of people's time and energy being mainly spent in the struggle with external enemies of a primitive kind such as famine, climatic difficulties, and wild beasts, they would be devoted chiefly to the effort to live carefully, to spare and to prop up their own feeblenesses, to soothe their inner disharmonies and, in general, to doctor themselves as effectively as possible. For everyone would be an invalid, with his own special familial twists.

In a recent essay on "The Control of Evolution in Man," the English geneticist C. D. Darlington expressed tersely the same thought:

Those who were saved as children return to the same hospital with their children to be saved. In consequence, each generation of a stable society will become more dependent on medical treatment for its ability to survive and reproduce.

Let us hasten to say that not all geneticists take such a dark view of our biological future in the Western World. But even the most optimistic probably recognize that there is much truth in a quatrain published shortly after World War I in the London *Spectator:*

> Science finds out ingenious ways to kill
> Strong men, and keep alive the weak and ill—
> That these a sickly progeny may breed,
> Too poor to tax, too numerous to feed.

While these problems are becoming more apparent in modern societies, they are not entirely new. In ancient civilizations and among primitive peoples one commonly finds social customs and taboos which represent efforts to deal with the difficulties created by the existence of individuals who are economically or biologically deficient. Among Eskimos and certain Indian tribes aged people were expected to abandon the camp and die once they had become too much of a burden for the group. For different but related reasons, unwanted children in Sparta were left to die from exposure.

Similar practices which seem cruel to us have been defended by some of the most idealistic social philosophers in the past. Suicide of the sick was encouraged by the Stoics and other sects, and even by Plato. At Marseilles in Roman times poison was kept in the city for those who could present to the Council a reason for wishing to rid themselves of life. More surprisingly, a similar attitude was taken in the seventeenth century by

Sir Thomas More. In More's Utopia those who were incurably ill and in continual pain were urged by the priests and magistrates to kill themselves.

Yf the dysease be not onelye vncurable, but also full of contynuall payne and anguyshe, then the priestes and the magistrates exhort the man, seynge he ys not able to doo annye dewtye of lyffe, and by ouerlyuing hys owne deaths is noysome and yrkesome to other, and greuous to hymself; that he wyll determyne with hymselfe no longer to cheryshe that pestilent and peynefull dysease: and, seynge hys lyfe ys to hym but a tourmente, that he wyll nott be vnwyllynge too dye.

In this respect at least, the ethics of the modern world have grown loftier since Plato and More. All decent men now regard human life as sacred, worth preserving at whatever cost to the individual or to society. Moreover, experience has taught us that some of the greatest achievements of the human race have come from individuals suffering from handicaps which rendered their physical lives miserable and would have prevented their survival in an unsheltered environment. Thus, selfish motives agree with modern ethics and religious ideals in making the preservation and prolongation of human life the ultimate goal of medicine.

There is, furthermore, biological justification for this attitude, for it is misleading to speak of biological defectives without regard to the environment in which these individuals live and function. Medical techniques can make up for genetic and other deficiencies that would be lethal in the wilderness. While it is certain that the physically handicapped could not long survive under "natural" conditions, it is also true that medical and other social skills make it possible for men to live long and function effectively in the modern world even though they be tuberculous, diabetic, blind, crippled, or psychopathic. The reason is that fitness is not an absolute characteristic, but rather must be defined in terms of the total environment in which the individual has to spend his life.

It must be realized, however, that fitness achieved through constant medical care has grave social and economic implications which are commonly overlooked. We can expect that the cost of medical care will continue to soar because each new discovery calls into use more specialized skills and expensive items. Today medical care represents some eight per cent of the national income in the United States. There is certainly a limit to the percentage of society's resources that can be devoted to the maintenance of medical establishments, and a time may come when medical ethics and policies will have to be reconsidered in the harsh light of economics.

Furthermore, it must not be taken for granted that the power of science is limitless. Only during the past few decades and in but a few situations has medical treatment enabled the victims of genetic disabilities to survive and to reproduce on a large scale. If the numbers of biologically defective individuals continues to increase, therapy may not be able to keep pace with the new problems that will inevitably arise. Yet, failure to meet these problems might eventually lead to biological extinction. It is urgent, therefore, that we formulate medical policies compatible with our system of ethics, yet practical within the limitations imposed by economic and biological consequences.

For the reasons outlined above, and despite all we can do in the way of medical care, new problems of disease will endlessly arise and require ever increasing scientific and social efforts, making of medical Utopia a castle in the air that can exist only in the Erewhon of political Utopia. Yet, it is clear that the lay and paramedical organizations established during the past fifty years to deal with problems of health are based on the optimistic assumption that, given enough time and financial resources, science can develop techniques to prevent or cure most diseases, and that only social and economic limitations will in the future stand in the way of ideal health. Anyone who has dealt with congressional appropriation committees knows that their willingness to pour public money into medical research comes from the belief that if the program is pursued for a few more years, science will provide ways to eliminate disease.

The belief that disease can be conquered through the use of drugs deserves special mention here because it is so widely held. Its fallacy is that it fails to take into account the difficulties arising from the ecological complexity of human problems. Blind faith in drugs is an attitude comparable to the naïve cowboy philosophy that permeates the Wild West thriller. In the crime-ridden frontier town the hero single-handedly blasts out the desperadoes who have been running rampant through the settlement. The story ends on a happy note because it appears that peace has been restored. But in reality the death of the villains does not solve the fundamental problem, for the rotten social conditions which opened the town to the desperadoes will soon allow others to come in unless something is done to correct the primary source of trouble. The hero moves out of town without doing anything to solve this far more complex problem; in fact, he has no weapon to deal with it and is not even aware of its existence.

Similarly, the accounts of miraculous cures rarely make clear that arresting an acute episode does not solve the problem of disease in the social body—or even in the individual concerned. . . .

To state it bluntly once more, my personal view is that the burden of disease is not likely to decrease in the future, whatever the progress of medical research and whatever the skill of social organizations in applying new discoveries. While methods of control can and will be found for almost any given pathological state, we can take it for granted that disease will change its manifestations according to social circumstances. Threats to health are inescapable accompaniments of life.

Health is an expression of ability to cope with the various factors of the total environment, and fitness is achieved through countless genotypic and phenotypic adaptations to these factors. Any change in the environment demands new adaptive reactions, and disease is the consequence of inadequacies in these adaptive responses. The more rapid and profound the environmental changes, the larger the number of individuals who cannot adapt to them rapidly enough to maintain an adequate state of fitness and who therefore develop some type of organic or psychotic disease. "It is changes that are chiefly responsible for diseases," wrote Hippocrates in *Humours*, "especially the great changes, the violent alterations both in seasons and in other things." And he stated again in *Regimen in Acute Disease*, "The chief causes of disease are the most violent changes in what concerns our constitutions and habits."

A perfect policy of public health could be conceived for colonies of social ants or bees, whose habits have become stabilized by instincts. Likewise, it would be possible to devise for a herd of cows an ideal system of husbandry with the proper combination of stables and pastures. But unless men become robots, their behavior and environment fully controllable and predictable, no formula can ever give them permanently the health and happiness symbolized by the contented cow. Free men will develop new urges, and these will give rise to new habits and new problems, which will require ever new solutions. New environmental factors are introduced by technological innovations, by the constant flux of taste, habits, and mores, and by the profound disturbances that culture and ethics create in the normal play of biological processes. It is because of this instability of the physical and social environment that the pattern of disease changes with each phase of civilization, and that medical research and medical services cannot be self-limiting. Science provides methods of control for the problems inherited from past generations, but it cannot prepare solutions for the specific problems of tomorrow because it does not know what these problems will be. Physicians and public health officials, like soldiers, are always equipped to fight the last war. . . . who could have dreamed a generation ago that hypervitaminoses would become a common form of nutritional disease in the Western World; that the cigarette industry, air pollutants, and the use of radiations would be held responsible for the increase in certain types of cancer; that the introduction of detergents and various synthetics would increase the incidence of allergies; that advances in chemotherapy and other therapeutic procedures would create a new staphylococcus pathology; that alcoholics and patients with various forms of iatrogenic diseases would occupy such a large number of beds in the modern hospital? . . .

What may be worth asking is whether medical science can help the individual and society to develop a greater ability to meet successfully the unpredictable problems of tomorrow. This is an ill-defined task for which there is hardly any background of knowledge. Traditionally, medicine is concerned with retarding death and also with preventing pain and minimizing effort. Its achievements in this field have added greatly to the duration, safety, and charm of individual existence. While scientific medicine has continued to emphasize the detailed study of particular diseases and specific remedies, it has placed less emphasis on the nonspecific mechanisms by which the body and soul deal with the constant and multifarious threats to survival. The question is whether it is possible to increase the ability of the individual and of the social body to meet the stresses and strains of adversity. In this regard it may be worth considering that preoccupation with the avoidance of threats and dangers does not have the creative quality of goal-seeking. It is at best a negative attitude, one that does not contribute to growth, physical or mental. In our obsession with comfort and security we have given little heed to the future, and this negligence may be fatal to society and, indeed, to the race.

Whatever the theories of physicians, laboratory scientists, and sociologists, it is of course society that must decide on the types of threats it is most anxious to avoid and on the kind of health it wants—whether it prizes security more than adventure, whether it is willing to jeopardize the future for the sake of present-day comfort. But the decision might be and should be influenced by knowledge derived from a study of the manner in which different ways of life can affect the future of the individual and of society. Although this knowledge does not yet exist, a few general remarks appear justified.

It is a matter of common experience that, while man's physical and mental resources cannot develop to the full under conditions of extreme adversity, nevertheless a certain amount of stress, strain, and risk seems essential to the full development of the individual. Normal healthy human beings have long known, and physiologists are beginning to rediscover, that too low a level of sensory stimulation may lead to psychotic disorders, and that man functions best when a sufficient number of his neurons are active. Analogous considerations seem to be valid for the lower levels of biological functions, and recent studies illustrate that at least some of the mechanisms involved in training and in adaptability are not beyond experimental analysis.

It has been shown by Dr. Curt P. Richter and his associates that the domesticated laboratory rat differs from its wild ancestor, the Norway rat, in many anatomic and physiologic characteristics that can be measured by objective tests. As a result of selection and of life in the sheltered environment of the laboratory, the domesticated rat has lost most of its wild ancestor's ability to provide for itself, to fight, and to resist fatigue as well as toxic substances and microbial diseases. The domesticated rat has become less aggressive in behavior but also less able to meet successfully the strains and stresses of life, and therefore it could hardly survive competition in the free state. As a result of domestication, in Dr. Richter's words:

(1) the adrenal glands, the organs most involved in reactions to stress and fatigue, and in providing protection from a number of diseases, have become smaller, less effective . . . (2) the thyroid—the organ that helps to regulate metabolism, has become less active . . . (3) the gonads, the organs responsible for sex activity and fertility, develop earlier, function with greater regularity, bring about a much greater fertility . . . The finding of a smaller weight of the brain and a greater susceptibility to audiogenic and other types of fits, would indicate that the brain likewise has become less effective.

It must be pointed out, on the other hand, that the domesticated rat is better adapted than its wild ancestor to laboratory life and to many of the artificial tricks and stresses devised by the experimenter.

While some of these changes may be phenotypic, it is probable that most of them are the expression of mutations selected by life in the laboratory. But, whatever their mechanism, the effects of domestication on the wild rat are not without relevance to the future of mankind. Human societies made up of well-domesticated citizens, comfort-loving and submissive, may not be the ones most likely to survive.

The study of so-called germ-free animals has revealed other aspects of this problem. Animals born and raised in an environment free of detectable microorganisms can grow to a normal size and are capable of reproducing themselves for several generations, but they exhibit a high susceptibility to infection, even to the most common types of microorganisms that would be innocuous for animals raised in a normal, exposed environment. Furthermore, germ-free animals produce only small amounts of lymphoid tissue, and their plasma is extremely low in gamma globulin—deficiencies which may be of little consequence in the protected environment of the germ-free chamber, but which are great handicaps under normal conditions of life.

Experimental situations of these types illustrate the fact that a sheltered life alters in many ways the ability of the organism to cope with the stresses of life. "Let a man either avoid the occasion altogether, or put himself often to it, that he may be little moved with it," Bacon wrote in his essay "Of Nature in Men." While Bacon's aphorism is a picturesque statement of an important sociomedical problem, the solution that it offers hardly fits the modern world. Man cannot "put himself often to threats" whose nature he cannot anticipate. But he can perhaps cultivate the biological mechanisms that will enable him to respond effectively when the time of danger comes. . . .

I should like to emphasize the fact that during the past century mankind in the Western World has become adapted to very new conditions of its own making, and that even more drastic alterations in the ways of life will soon demand further adaptations. One century is a short time on the evolutionary scale for man, and for this reason genetic changes cannot possibly account for the adaptations that have occurred during the past few decades and that will be needed in the near future. On the other hand, and fortunately, each individual has in reserve an enormous range of potential adaptive resources that can be called into play under demanding circumstances. These adaptive potentialities have made it possible for millions of people to move in one generation from life on isolated farms to the tensions of Broadway and Forty-second Street, and we shall have to depend on similar mechanisms to survive the even more drastic and more rapid changes which are in the offing.

For many millennia mankind has moved forward and upward, even though in an erratic and halting manner. It has taken all sorts of disasters and upheavals in its stride, often deriving from painful experiences the stimulus for a more brilliant performance. We cannot create for our descendants a world free of stresses—nor, in my opinion, should we do it if we could. . . .

In medicine as in other social pursuits the long-range welfare of individuals and of the community often makes it advisable to forgo some of the immediate comforts and pleasures for the sake of the future. Overemphasis on the avoidance of effort and pain is an attitude fraught with dangers for the individual and even more for society; indeed, it amounts to social suicide.

Thus, medical philosophy at a high level transcends the problems posed by the care of the sick patient and must take into consideration the philosophical meaning of human existence. If we believe that the individual is but a link in a long chain of human adventure and that the continuity of human life is our collective responsibility, then it is wrong to jeopardize the future of the group for the sake of today's comfort. Medicine is one of the highest forms of social philosophy because it must look beyond the individual patient to mankind as a whole. The more effective it becomes through scientific knowledge, the more it must concern itself with the long-range consequences of its practices for future generations.

The Future of Man *

by P. B. Medawar (1959)

The argument that advances in medicine and hygiene are undermining the overall fitness of mankind is based on the belief that there is a hereditary or genetic element in all human ills and disabilities, even if it amounts to no more than a predisposition. This is known to be true of some diseases and not known to be false of any, so there can be no disagreement here. In its simplest form, the argument then runs as follows: because of the discovery of insulin, antibiotics, and so on, we are preserving for life and reproduction people who even ten years ago might have died. We are therefore preserving the genetically ill-favoured, the hereditary weaklings, who can intermarry with and therefore undermine the constitution of normal people; and as a result of all this, mankind is going downhill.

If by "going downhill" is meant "declining in biological fitness," with the implication that mankind will probably die out, this argument is simply a museum of self-contradictions.[1] It is true that we preserve for life people who even ten years ago might have died; but then we do not live ten years ago: we live today. It is also true that if some disaster were to destroy the great pharmaceutical industries to which diabetics and the victims of Addison's disease literally owe their lives, then a great many of them might die; but what could be deduced from this, except the lunatic inference that people who might conceivably die tomorrow might just as well be dead today? So let me put the argument in a form in which it might be put by a humane and intelligent person. He might say something like this:

---

* From P. B. Medawar, *The Future of Man.* Basic Books, Inc., New York. © 1959 by P. B. Medawar. Reprinted by permission.

1. For if medical treatment confers fitness upon the unfit, there can be no fear of extinction; if it fails to do so, the fear of extinction does not arise. The "going downhill" argument seems to contemplate the predicament of modern man in primitive surroundings, without insulin, penicillin, central heating and other allegedly debilitating devices; but it is not clear why such an exercise should be supposed to be informative.

I live in a country with a National Health Service, and the effect of this is that, in a sense, I myself suffer from diabetes and rheumatoid arthritis, and so on—from mental deficiency too. Of course my sufferings are only economic, in the sense that it is my taxes that help to pay the bill; but, as a result of them, I can afford to have only two children, though I very much wanted three. Now I am a sound and healthy person, and though I'm all for helping other less lucky people, it is clear that what you call their "biological fitness" is being bought at the expense of mine.

There are two arguments here, and they cannot be considered apart. The first is that people of a genetically sound constitution are being crowded out by the inferior. My spokesman was too humane to resent the idea that the inferior should survive and have children, but he saw some danger in the fact that the population of the future would contain fewer of his descendants because it would contain more of theirs. The second point he makes is that inborn resistance to a disease can be taken as evidence of a *general* soundness of body, of fitness in some rounded and comprehensive sense; so that even if the people he described as unlucky could all be cured of their particular disabilities, there would still be a deep-seated, though hidden, deterioration of mankind.

The arguments I have just outlined are serious and respectable, but they are not generally valid; they may sometimes represent the very opposite of the truth. Consider one of the forms of inborn resistance to a very severe form of malaria, subtertian malaria. It is now known that people can enjoy a definite inborn resistance to subtertian malaria if their red blood corpuscles contain something between 30 and 40 per cent of an unusual form of haemoglobin, haemoglobin S as opposed to haemoglobin A. One of the consequences of possessing haemoglobin S is that the red blood cells tend to collapse if deprived of oxygen; they become sickle shaped instead of remaining rounded, and people whose blood behaves in this way are said to show the "sickle cell trait." Sickle cell trait can be found in parts of Africa, in some Mediterranean countries, and in parts of India—always in places where malaria has been or still is rife. It is not a disabling condition, so its victims should not be said to "suffer" from it; and, in any event, it confers a high degree of resistance to the multiplication of the malaria organism in the blood.

This sounds like a splendid example of Nature's ingenuity in coping with a particularly murderous disease, malaria, without the help of these new-fangled drugs; but until one knows the rest of the story one cannot appreciate how devilishly ingenious it is.

The formation of haemoglobin S instead of A is due to an inborn difference of a particularly uncompromising kind, in the sense that if a person is genetically qualified to produce haemoglobin S, by possessing the appropriate "gene" or genetic factor, then he surely will. People who show sickle cell trait do so because they have inherited the gene that changes haemoglobin A to S from one, and only one, of their parents. But when two such people marry and have children, one quarter of their children, on the average, will inherit that gene from both their parents; all their haemoglobin, instead of only part of it, will be of type S; and as a result of this they usually die early in life of a destructive disease of the blood known as "sickle cell anaemia." This highly successful form of inborn resistance to malaria therefore makes it certain that a number of children will die.

The situation as a whole can be set out in the form of a balance sheet or equation. In some parts of the world where malaria is rife, people with sickle cell trait are the fittest people. Alongside them are, on the one hand, normal people, whose haemoglobin is wholly of type A; and, on the other hand, the victims of sickle cell anaemia, whose haemoglobin is wholly of type S. The proportion in which these three classes occur adjusts itself automatically to a pattern in which the loss of life due to malaria and to sickle cell anaemia nicely counterbalances the gain that is due to the greater fitness of those with sickle cell trait. Nevertheless, in malarial regions, populations which possess this genetic structure are fitter than populations which do not.[2]

Essentially the same explanation will account for the widespread occurrence in certain parts of Italy of the disease known as Cooley's anaemia and, not impossibly, for the otherwise paradoxically high incidence of a certain fatal inborn disease of the pancreas[3] in Great Britain and elsewhere. In all such cases it may turn out that there is, or recently has been, some special advantage in having inherited from one parent, and one parent only, the genetical factor which produces such disastrous effects when it is inherited from both.

The moral of this story—though morality seems to have little to do with it—is that mankind will improve if we stamp out inborn resistance to malaria by stamping out malaria itself. Sickle cell anaemia is in fact disappearing from the Negro population of America at about the rate we should expect if malaria had ceased to be a scourge to it 200 or 300 years ago. So the only good thing about inborn resistance to malaria is—inborn

2. This interpretation of sickle cell trait is the outcome of a brilliant combined operation between geneticists, chemists and clinicians, among them J. V. Neel, E. A. Beet, L. Pauling, V. M. Ingram and A. C. Allison. . . .

3. Cystic disease of the pancreas, the frequency of which, in Great Britain, has been put at one in 2,000—a frequency much higher than mutation at known rates could account for. There has therefore grown up the uneasy suspicion that the carriers of the harmful gene may have, or may have had, some special advantage over normal people (see L. S. Penrose, "Mutation in Man," in *The Effect of Radiation on Human Heredity.* 101: WHO, Geneva, 1957).

resistance to malaria: it does *not* reveal any general soundness of constitution; and this is just the opposite to what my imaginary spokesman supposed. It is simply not true to say that advances in medicine and hygiene must cause a genetical deterioration of mankind. There is more to be feared from a slow decline of human intelligence, but that is a different matter: *if* it is happening, it is because the rather stupid are biologically fitter than those who are innately more intelligent, not because medicine is striving to raise the biological fitness of those who might otherwise be hopelessly unfit.

This question of a possible decline of intelligence is very important, and I shall devote my [next] lecture to it; but, having referred to mind, and defects of mind, it is essential to make this point. Some forms of idiocy and imbecility are congenital. "Congenital" is a vague word, but I use it here to refer to an idiocy which follows from an inborn defect of the genetic make-up as it was laid down at the moment of conception. This defect represents an inborn difference from other people, but it is no more a property of the genetic make-up *as a whole* than the inborn difference between people of blood-groups A and B. One particular form of imbecility, now known as phenylketonuria, is the effect of a single, particular, and accurately identified inborn error of metabolism. In point of inheritance it is essentially similar to another disturbance of metabolism, alkaptonuria, the most serious effect of which is usually no worse than a darkening of the urine after it is exposed to air. To suppose, then, that congenital imbecility pointed to some general inborn inadequacy or degeneracy is nonsense—ignorant and cruel nonsense, too. Our ambition should be to *cure* phenylketonuria, for it is an illusion to suppose that congenital afflictions are necessarily incurable; and if eventually we do cure phenylketonuria, we shall in no sense be conniving at a genetical degradation of mankind.

. . . One form of gross mental defect, mongolism, is by no means so simple in origin as phenylketonuria: it is the result of a damaging genetic accident involving a whole chromosome, and it is not at all easy to see how it might be cured. But it *is* an accident, a particular accident, one which happens more often to the children of older mothers; it is not to be thought of as an outward fulfilment of some inner degeneracy of a family stock.[4] When you come to think of it, all defects of the genetic constitution must have an accidental or unpremeditated or casually intrusive quality—"epiphenomenal" is the word; for it is impossible, indeed self-

---

4. The fact that mongolism can be traced back to a particular chromosomal abnormality was discovered by J. Lejeune, M. Gauthier and R. Turpin, *C.R. Acad. Sci., Paris,* **248,** 602, 1959. . . . In mongoloids there is an extra chromosome, apparently because one pair exists in triplicate instead of in duplicate. . . .

contradictory, that any animal should have evolved into the possession of some complex and nicely balanced genetic make-up which rendered it unfit. It is this fact that justifies our always hoping to find a cure.

❂   ❂   ❂

THE CONSEQUENCES OF BIRTH CONTROL

One of the questions I put in my first lecture was this: is the practice of birth control and family limitation so unnatural that it is bound to have evil consequences? This is a question of great practical importance, so let me spend the rest of this lecture discussing what the consequences of birth control might be.

Two distinct things are involved in a deliberate limitation of the size of families, and they must be kept apart. The first is a restriction of the total number of children born to a married couple; the second is a tendency which need not go with it, though it usually does—a tendency to complete a family earlier in life than would otherwise have been the case.[5] I shall discuss the second first.

The most general effect of an earlier completion of families will be to shorten the average gap between successive generations: one generation will follow another more quickly, so that whatever genetical changes are happening will happen faster in terms of calendar years. More particularly, we can look forward to a sharp decline in the numbers of newborn children afflicted by mongolian idiocy or by any other disease that increases in frequency with their mother's age. We can expect the sex ratio at birth to shift still further in favour of males, with the consequence that, for the first time perhaps, men will outnumber women in their marriageable years. . . .

So much seems fairly clear: the rest is a matter of guesswork. An earlier completion of families implies that a human pedigree will hereafter run through a succession of rather younger mothers—mothers much younger, on the average, than they would have been a hundred years ago. This may just conceivably be a good thing, for the following reason. It is now widely agreed that human egg-cells do not increase in number after birth: women make do with the entirely adequate number they have to begin with, and the egg-cells are used up progressively in course of life. This means that egg-cells are obliged to wait for years in the ovary before they

5. Cohort analysis makes it possible to work out what fraction of a whole family is completed in each year after the parents' marriage or the mother's birth. In this country (if we disregard the interruptions caused by two wars) there has been a slow progressive decline in the mean age at marriage and a slow increase in the proportion of the family completed by the fifth or tenth year after marriage.

are shed into the Fallopian tube, where they may or may not be fertilized. Too long a wait, repeated generation after generation, might just possibly be harmful. It would be very interesting to study the recorded pedigrees of noble families and trace the fate of the lineages that went through last daughters of last daughters of last daughters. I must tell you that when an experiment which reproduces this state of affairs is carried out on animals called *rotifers,* the lines of descent that pass repeatedly through older mothers invariably die out. But then rotifers are very lowly and highly aberrant pond animalcules; it would be most unsound to draw far-reaching conclusions from a comparison between rotifers and human beings of noble birth.[6]

A restriction of the total number of children has several genetic consequences, one of which is to make the work of human geneticists even more difficult than it already is. It will certainly reduce the frequency of all diseases of children—above all haemolytic disease of newborn children —which tend to take a more severe form or to occur more often in the later children of a family. . . .

But what people really fear when they talk about the biological evils of birth control is this. In many countries, families are deliberately restricted to two or three or four. The pressure of natural selection against low degrees of fertility will therefore be, to some extent, relaxed. I mean that if families now average only two or three children, there will no longer be the same sharp discrimination between married couples who *could* only have had two or three children and those who, had they wished it, could have had ten or twelve. Some discrimination there surely will be; but it is theoretically possible that in a matter of tens or hundreds of generations the proportion of innately very fertile men and women may go down.

If this were to happen, I think it would be looked back upon as an example—yet another example—of the way in which the level of fertility comes to be adapted to the prevailing circumstances. It is a fallacy to assume, as I fear some biologists still do assume, that the fertility of a species is a kind of primeval fixture—as if animals and plants were driven by some demon of fertility to have vastly more offspring than are needed. One can hear it said that the explanation of natural selection itself is that living things produce an allegedly "prodigious" number of offspring, of which only a chosen few are spared. But to say this is to forget that the level of fertility adopted by any species is just as much the *consequence* of natural selection as its cause. There is in fact no good reason to fear that an innate decline of human fertility must be a stage on the road to extinction or that we shall face a struggle to keep mankind alive.

6. The source of nearly all our actuarial knowledge of rotifers is A. I. Lansing: see A. Comfort, *The Biology of Senescence,* London, Routledge & Kegan Paul, 1956.

Of course, one can imagine circumstances in which a low level of fertility might be very disadvantageous. Some frightful disaster might oblige a handful of human beings to populate the entire world anew. But why worry about the imaginary dangers of a low level of fertility in the distant future when confronted by the real dangers of a high level of fertility as it affects so many countries of the world today? These are real and present worries; yet one of them at least we can spare ourselves: on present evidence, there is no reason to believe that the world-wide adoption of the practice of birth control would have biologically malign effects. On the contrary, there is every reason to believe that failure to adopt some measure of family limitation will lead, in the long run, to misery, privation, and economic distress.

INTELLIGENCE AND FERTILITY

. . . In this lecture I propose to discuss the possibility that in some countries, Great Britain among them, the average level of human intelligence is going down.

If we classify children of any chosen age by the scores they get in intelligence tests, and then make a diagram showing what number or proportion were awarded each possible score, we shall find that the diagram is smooth and pretty well symmetrical. The average score divides the children equally, and the most numerous single group is the one whose members have this average and middling score. Moreover, the number of children who exceed the average by a certain quantity is about equal to the number who fall short of the average by that quantity; and as we get further and further from the average in either direction, so the number of children to be counted gets less and less.

What I have said about the distribution of scores in intelligence tests applies to other characteristics of human beings—to the heights of adult men or women, for example. Within the whole range of heights and wits there are people who are exceptional in the sense of being in a minority— uncommonly good at intelligence tests, or uncommonly small; but to call them "abnormal" is a bit misleading, because it suggests that they are separated from the rest of us by sharp divisions or bold steps. It is true that some people do lie right outside the normal range of variation—are abnormally small or abnormally dull for reasons that call for special explanations. So it is with idiots and imbeciles: those who fall short of the average to the degree of utter incapability do seem to form a class apart.[7]

7. The distribution of scores in intelligence tests is, to a good approximation, normal (Gaussian): see J. A. Fraser Roberts on "The Genetics of Oligophrenia," *Congr. Int. Psychiatrie,* pp. 55–117 (Paris, Hermann, 1950). Fraser Roberts points

But it is no longer these unlucky people that biologists have in mind when they discuss the possibility that intelligence may be declining. Many years ago, to be sure, the rumour got around that mankind would lose its wits because idiots and imbeciles are riotously fertile![8] In fact they are nothing of the kind. Many are sterile; and in any event confinement to home or homes makes it impossible for most of them to have children . . . No: the problem arises over the greater fertility of those who are somewhat below the average of intelligence; and the fear is that their progeny are tending to crowd the rest of the population out. This might happen for one or both of two reasons: because, generation by generation, they tend to have larger families than the more intelligent; or because, generation by generation, they tend to have them earlier in life. For if the more intelligent parents start having children later and space them more widely apart, then, even if they end up with the same size of family in the long run, they are bound to be left behind.

All rational discussion of the possibility that intelligence may be declining starts from our knowledge of a certain association between the average performance of children in intelligence tests and the size of the families they belong to: in some countries, Great Britain among them, children who belong to small families are known to do better in intelligence tests than the children of larger families. The relationship between the average score of children and the number of their brothers and sisters is pretty consistent over the whole range of family sizes: taken by and large, children with x brothers and sisters do better than children with $x + 1$. A great mass of evidence points to a clear *negative correlation* between the size of a family and the average performance of its members.[9]

Before asking how this negative correlation is to be explained, and what its genetic implications may be, we must take some view about what is to be inferred from a score in an intelligence test. Some people speak with angry contempt of "*so-called* intelligence tests"; having satisfied themselves of the absurdity of claims which psychologists no longer make for them (and which the better psychologists never did make), they dismiss the entire subject from their minds. Others profess to attach no meaning to the word "intelligence"—but try calling them *un*intelligent and see how they react. At the risk of being peremptory, because time is

out that whereas the feeble-minded form part of the lower end of the normal distribution of intelligence, idiots and imbeciles form a group that lies outside it.

8. For the history of popular attitudes towards fertility and mental abnormality, see A. Lewis, *Eugen. Rev.*, **50**, 91, 1958.

9. The case for a decline in the average level of intelligence, based upon the negative correlation between intelligence and size of family, is argued by C. Burt, *Intelligence and Fertility* (Occasional Papers on Eugenics, No. 2: London, Eugenics Society, 1952), and G. Thomson, *The Trend of National Intelligence* (*ibid.*, No. 3, 1947).

short, I shall take the view that intelligence tests measure intellectual aptitudes which are important, though very far from all-important; and that
these aptitudes make up a significant fraction of what we all of us call
"intelligence" in everyday life. Only one disclaimer is important: intelligence tests can be valuable when they are applied to children still at
school and to feeble-minded adults; their application to adults in general
is very much more restricted in scope.

There are quite a number of possible explanations of the negative correlation between intelligence and family size. One possibility is that, for
some reason, a child's intelligence declines with his position in the family;
the first child being the most intelligent; the last, the least. The intelligence of each child might depend upon the age of his mother when she
bore him, for a mother must be older when she bears (say) her fourth
child than when she bears her third. This idea goes against all common
understanding, but for purely technical reasons it is rather difficult to test.
If we set aside certain forms of imbecility which are obviously exceptional, the most accurate tests show that matters to do with rank of birth
do *not* explain the relationship between intelligence and family size.[10]

A second possibility is that size of family can itself affect a child's proficiency in those intelligence tests which rely heavily upon some outward or
inward skill in the use of words. For one thing, a child in a large family
will listen and contribute much more to the unscholarly prattle of its
brothers and sisters than will a child in a family of two or three. There is
good evidence that inexperience in the use of words does play some part
in the negative correlation between scores in intelligence tests and family
size. It seems entirely reasonable that it should. Words are not merely the
vehicles in which thought is delivered: they are part of thinking; and lack
of experience in the use of words, even unspoken words, may well put a
child at a disadvantage in a test.[11]

There is a less direct way in which the size of the family he belongs to
might be related to a child's performance. On the whole children of large
families are not quite so tall, at any given age of childhood, as the children of smaller families—perhaps because, on the average, they have
been a little less well nourished. They grow more slowly, therefore,
though they might make up for that disadvantage by continuing their
growth a little longer, ending up no smaller than the better fed. But if, at
any chronological age, the children of large families are a little backward
physically, might they not be backward in mental growth as well? And

10. See J. A. Fraser Roberts, *Brit. J. Psychol.* (Statistical Section), October 1947,
p. 35.
11. The argument here is J. D. Nisbet's, *Family Environment* (Occasional Papers
on Eugenics, No. 8: London, Eugenics Society, 1953).

may they not eventually catch up with the others, given a little time? Mental and physical growth are not exactly in gear, so backwardness in size can by no means be construed as backwardness of mind; but so far as our meagre evidence goes, there is some small but definite connexion between the intelligence of children and their size at any given age of childhood; and, with some reservations, this might account for a certain small part of the negative correlation between intelligence and family size.[12]

Another possibility is that a lowly score in an intelligence test is part of a child's inheritance from its parents, though not an inheritance in the technical or genetic sense. Unintelligent parents, we might reason, have large families because they have neither the skill nor the will to have smaller families; and, being unintelligent, their conversation and precepts will tend to have a rudely pragmatical character, and their houses to be bare of books. The nature of the home he comes from is known to affect a child's performance in verbal tests of intelligence; but there is no suggestion here that a child of unintelligent parents would be at any disadvantage if he were to be brought up in a more educated home.

Yet another possibility is that the children of less intelligent parents do *not* start on the same footing as the children of the more intelligent; that their lack of intelligence is something which good upbringing can palliate but cannot completely cure; that differences of intelligence are inherited in the technical or genetic sense.

At this point I shall ask you to assume (what I think no one denies) that differences of intelligence *are* to some degree inborn. There are certain obstinate and persistent correlations of intelligence between parents and their children and between the children of a family among themselves—correlations that do not disappear when as much allowance as possible is made for differences of upbringing, environment, and family size. The study of these correlations in the population generally, and, in particular, of the values they take in foster-children and in identical twins who have been reared apart, suggests that not less than half of the observed variation of intelligence is an inborn variation. For many environments, it may be a good deal more than half. It does not do to be more particular, because the concept of an inborn variation in characters greatly affected by the environment is very complex, and the people who know most about it are the least inclined to express it in a numerically exact form.[13]

12. See J. D. Turner, *Growth at Adolescence* (Oxford, Blackwell, 1955).
13. For a cogent discussion of the theory underlying the attempt to discriminate between genetic and environmental influences, see L. Hogben, *Nature and Nurture* (London, Allen & Unwin, 1945).

Let us agree, then, that differences of intelligence are strongly inherited. We must now ask, why do less intelligent parents run to larger families? Teachers, demographers and social workers incline to believe that the answer is mainly this. Less intelligent parents have larger families because they are less well informed about birth control or less skilful in its practice; because they are less well able to see the material disadvantages of having more children than can be well provided for, or the more than material advantages of having the children one really wants. I do not like to put it this way because it seems to import a moral judgment which, valid or not, has no bearing on the argument. Someone might insist that it was *right* for all parents to have all the children they were capable of having, and that the unintelligent live up to that precept because, being more innocent than learned but worldly people, they have a clearer perception of what is right or wrong. All this is beside the point. The point is that they have more children, and are unintelligent, whether that does them credit or not; and if they do have more children, there is a certain presumption that innate intelligence in the population at large will decline. It should not decline at anything like the speed suggested by the boldly negative correlation between intelligence and size of family, because, as we have agreed, some part of that correlation can be traced to causes in which inborn differences of intelligence need play no part.

I say there is a *presumption* that the average level of intelligence will decline. It is not a certainty. In the first place, one highly important piece of information is missing. What about the intelligence of married couples who have no children, or of people who never marry at all? Those who are oppressed by the possibility of a decline of intelligence point, with some reason, at the many highly learned people who are childless; and they remind us that when a population is classified by the occupations of its members, something is to be learned from the fact that manual labourers are much more fertile than those who live mainly on their wits. But those who think that the dangers of a decline are greatly exaggerated point out that idiots and imbeciles, and some of the feeble minded, are very infertile too. Our uncertainty about the intelligence of those who have no children is awkward because it means that we cannot give a very confident answer to a very important question: to what extent are the parents of each successive generation a representative sample of the population of which they form a part?

A second reason for saying that the decline of intelligence is no more than a presumption is that there can obviously be no certainty in the matter until we know exactly how differences of intelligence are inherited. The argument for a decline is based on the belief that differences of intelligence are under the control of a multitude of genes, no one of which can be recognized individually; and it is assumed that the contributions of these genes to intelligence are *additive* in a certain technical sense. Are these assumptions justifiable? There is no reason at all to doubt that inborn differences of intelligence over the normal range of variation are under the control of a very large number of genes; but the idea that their contributions are additive requires a little consideration.

The word "additive" refers to a particular pattern of co-operation of interaction between genes. An additive pattern of interaction implies (amongst other things) that there will be no such thing as hybrid vigour in respect of intelligence; it implies that a person who is mainly heterozygous or hybrid in his make-up with respect to the many genes that control intelligence will lie somewhere between the extremes of brightness or dullness that correspond to those genes in their similar or homozygous forms.[14] If, on the contrary, the genes that controlled differences of intelligence were all to exert their greatest effect in the hybrid or heterozygous state, then there would be no correlation between the intelligences of parents and their children: the children of parents who both had the high intelligence conferred by a hybrid make-up would as often as not be of low intelligence, and parents of low intelligence would as often as not give birth to children much more intelligent than themselves. But in actual fact the correlation of intelligence between parents and children is just as great as the correlation between the children of a family among themselves, so there is no good reason to doubt that the genes interact in the manner which I have described as additive. Nor, as I say, is there any reason whatsoever to doubt that a great many genes are at work. Both assumptions I have made are therefore justifiable, and there is a fair case for the belief that intelligence is declining. There is an equally good case for the belief that the decline could not go on indefinitely, but this, for the moment, I shall defer.

Is there any *direct* evidence that intelligence is declining? In 1932, a grand survey was made of the scores in intelligence tests of 90,000 Scottish schoolchildren whose eleventh birthdays fell within that year. Fifteen years later, in 1947, a very similar test was carried out on some 70,000 children of the same ages. No decline was apparent: the boys' scores had improved slightly; the performance of girls had risen even more. Taken together, the children of 1947 were two or three months ahead of the children of 1932 in terms of mental age.[15]

14. Not all geneticists use the word "additive" in quite this sense: some confine it to interactions between alleles at different loci.
15. *The Trend of Scottish Intelligence,* Publ. Scottish Council for Research in Education, XXX (University of London Press, 1949). For the style of test used in 1947, see L. A. Terman and M. Merrill, *Measuring Intelligence* (London, Harrap, 1937).

At first sight these results were immensely reassuring. A decline of intelligence on the scale we now fear might not have been shown up by tests only fifteen years apart; but an increase was more than most people had dared to hope for. But could anything have happened to conceal a genuinely innate decline? Unhappily it could. The children of 1947 were, on the average, an inch and a half taller than their predecessors of 1932. They were ahead in physical as well as in mental age; [16] but this does not imply that they were bound to end up taller and brighter when they reached adult stature of body and mind. Again, the pattern of family sizes might have altered in those fifteen years. If it did, the alteration would surely have taken the form of a decline in the proportion of the largest families, and this alone would account for a certain rise of average score. There is much else besides. The children of 1932 were taught by methods that may not have prepared them so well for intelligence tests; they grew up when a wireless set had not yet become a voluble piece of ordinary household furniture, and they lacked, then, whatever experience in "verbalization" (if you will pardon the expression) comes from a constant familiarity with the spoken word.

But still: I gave you theoretical reasons for thinking that there might be a slow decline of intelligence, and direct evidence which, taken at its face value, shows that no such decline occurred. Can the theory itself be incomplete or wrong?

Some geneticists believe that it is incomplete, and I should like to explain their reasons. I asked a moment ago why the less intelligent should run to larger families, and gave the answer that has seemed reasonable to most of us: they have larger families for reasons connected with their lack of intelligence. But some geneticists look to another explanation: that people of mediocre or rather lowly intelligence are intrinsically more fertile, innately more capable of having children, than people of very high or very low intelligence.

To accept this interpretation is by no means to deny that inborn differences of intelligence are controlled by a multitude of genes, or that inborn variation of intelligence is mainly additive in character. A new point is being made: that people of mainly heterozygous make-up are innately

16. The figure I mention—1½ inches—is arrived at by interpolation between two figures given to the Scottish Council by the Education Health Service of Glasgow: over the period 1932–47, the average height of nine-year-olds increased by 1.3 inches and of thirteen-year-olds by 1.7 inches: see *Social Implications of the 1947 Scottish Mental Survey,* Publ. Scottish Council for Research in Education, XXXV (University of London Press, 1953), the authors of which find it hard to believe that so great an increase of height (with all that it implies of better upbringing) should not be associated with an improved performance in intelligence tests. Dr. J. M. Tanner has made the same point.

more fertile—are innately *fitter*, as biologists use that word. When people of mediocre intelligence marry and have children, then, in the simplest possible case, some half of their children will grow up to be like themselves; the other half will consist of relatively infertile people of very high and of very low intelligence in about equal numbers. The children of each successive generation will therefore be recruited mainly from parents of mediocre intelligence, but they will always include among them the very bright and the very dull. It is possible, as a theoretical exercise, to construct a balance sheet of intelligence in which gains and losses cancel each other out: a reservoir of parents of mediocre or even lowly intelligence maintains a natural and stable equilibrium in the population, for, among their children, the all but complete sterility of low-grade mental defectives will cancel with the lesser fertility of the very bright.[17]

There is one respect, I think, in which this argument carries a lot of weight. It sets a natural limit to any likely rise or fall of intelligence. If a tyrant were to carry out an experiment on human selection, in an attempt to raise the intelligence of all of us to its present maximum, or to degrade it to somewhere near the minimum that now prevails, then I feel sure that his attempts would be self-defeating: the population would dwindle in numbers and, in the extreme case, might die out. In the long run, the superior fitness of heterozygotes would frustrate his dastardly schemes. This is a cheering hypothesis, but it does not imply, I fear, that our population is already in a state of equilibrium; that the average level of intelligence may not fall a good deal further yet. It does not imply that we have already used up all the resources of additive genetic variation that can be called upon before natural selection intervenes. Nothing could be more unrealistic than to suppose that our population is already in a state of natural and stable equilibrium, with a nice balance between gain of intelligence and loss. We cannot disregard the purely arbitrary element in whatever it is that decides the size of a family—disregard the massive evidence of the Royal Commission on Population on the spread of the practice of birth control.[18] Nor can we neglect the fact that habits of fertility keep changing rapidly. The census of 1911 revealed a sharp increase in the difference between the sizes of families born to labourers and to professional men, but there are hints in the census of 1951 that the difference may since have declined.[19] There is no need to assume that professional

17. See L. S. Penrose, *Lancet*, **2**, 425, 1950; *Brit. J. Psychol.*, **40**, 128, 1950; *Proc. Roy. Soc.*, B 144, 203, 1955.

18. *Report of an Enquiry into Family Limitation* by E. Lewis-Faning; Papers of the Royal Commission on Population, Vol. 1, London, HMSO, 1949.

19. This trend toward a declining differential has been verified by more recent statistics as quoted in the two selections that follow. Ed. See also R. Freedman, P. K. Whelpton, and A. A. Campbell, *Family Planning, Sterility and Population Growth* (New York, McGraw-Hill, 1959).

men are innately more intelligent than labourers; the argument would be equally valid if for professional men and labourers we were to substitute the people who do or do not believe that intelligence will decline; I am saying that there has been a *change* in the habits of fertility, and that when such changes are in progress, the idea of a natural equilibrium must be set aside.

Much else could be said to the same effect. For example, it is not true that the most highly educated people are the least fertile. Apart from imbeciles and idiots, the least fertile members of our population in terms of educational standing are those whose schooling stopped short of university but went beyond what is legally required.[20] I feel that the members of this group are less fertile because they *choose* to have fewer children; I am not inclined to believe that they are either unusually intelligent or unusually stupid; so far as innate intelligence goes, they may be a perfectly fair sample of the population as a whole. Nor do I think that some subconscious premonition of infertility directs them towards occupations which they merely appear to choose. But they are a numerous class, and the least fertile; what they contribute to our understanding of a natural balance of fertility is evidence that no such balance exists.

Again, the pattern of mortality as it falls upon the large families of poorer people has changed dramatically in the past fifty years. The death-rate of children within a week or so of birth has fallen rapidly, and begins to compare with mortality in the children of the better off; but deaths during the first year of life—deaths due mainly to infectious diseases—have not yet fallen so far, and do not compare so well.[21] This is but a fragment of the evidence that must turn our thoughts away from the idea that we are in a state of equilibrium. At one time, I suppose, there may have been some natural equilibrium between intelligence and fertility—adjusted, perhaps, to an average family size of eight or ten. Perhaps matters were so adjusted that the brightest and dullest of our forebears were incapable of having families larger than four or five. But the families we have in mind today belong to the lower half of what is possible in the way of human fertility, and it is hard to believe that a new equilibrium could have grown up around families with an average size of two or three.

Let me now summarize this long and complicated argument. It is a fair assumption that a child's performance in an intelligence test—imperfect as such tests are—gives one *some* indication of its wits. It is a fact that the average performance of children in intelligence tests is related to the size

20. See the *Fertility Report* (London, HMSO, 1959) on the General Census of 1951.

21. These remarks should be qualified by more recent evidence: see J. N. Morris, *Lancet*, 1, 303, 1959.

of the family they belong to: the larger the number of their brothers and sisters, the lower, on the average, will be their scores. Part of this negative correlation between intelligence and size of family can be traced to causes which have no genetic implications, whether for good or ill. But differences of intelligence are strongly inherited, and in a manner which, in general terms, we think we understand. If innately unintelligent people tend to have larger families, then, with some qualifications, we can infer that the average level of intelligence will decline. There are good reasons for supposing that intelligence could not continue to decline indefinitely, but equally good reasons for thinking that it may have some way yet to go. In any event, the decline will be a slow one—much slower than the boldly negative correlation between intelligence and size of family might tempt one to suppose. All our conclusions on the matter fall very far short of certainty: there are serious weaknesses in our methods of analysis, and grave gaps in our knowledge which, it is to be hoped, someone will repair.

Profound changes in habits of fertility have been taking place over the past fifty or hundred years; and they are not yet complete. The decline of intelligence (if indeed it is declining) may be a purely temporary phenomenon—a short-lived episode marking the slow transition from free reproduction accompanied by high mortality to restricted reproduction accompanied by low mortality. But even if the decline looked as if it might be long lasting, it would not be irremediable. Changes in the structure of taxation and in the award of family allowances and educational grants may already have removed some of the factors which have discouraged the more intelligent from having larger families; and in twenty-five years' time we may be laughing at our present misgivings. I do not, however, think that there is anything very much to be amused about just at present.

The Threat of Genetic Deterioration [*]

by Marston Bates (1955)

The people who are alarmed about the present threat of genetic deterioration in man often sound as though this were something new, a side product of our present industrial civilization in which the ordinary force

of "natural selection" in keeping the race up to the mark was being nullified. I think this is questionable.

They speak of various "dysgenic" effects, which, on the one hand, tend
to promote the survival and even encourage the reproduction of the unfit;
and which, on the other hand, tend to impede the reproduction of many
desirable groups and individuals. Their picture of present conditions is
certainly plausible enough, and I do not want to argue that man tends to
act in his best interests, from the biological point of view. My only argument is that this behavior does not seem to me particularly new—that
man has been acting in ways contrary to his best interests for a very long
time. This may very well prove disastrous for the species in the near future. The puzzle, really, is how he has been able to get along so well so
far.

War is clearly a dysgenic force in that it gathers together the finest
physical and mental specimens of a people at the time of the height of
their reproductive powers and arbitrarily kills off large numbers of them,
and prevents normal family relations among the rest for appreciable periods of time. Meanwhile, those rejected on physical or mental grounds are
left at home to perpetuate the race.

These dysgenic influences have been systematized in modern times
through the system of conscript armies and through the increasing size of
armies in relation to noncombatant populations. This may represent an in-
tensification of the dysgenic force, but still the general principle of war
acting as an agent of unnatural selection has been in force for a very long
time.

Maybe way back in the Pleistocene, war, or the general homicidal
tendency, was a force of "natural" selection. One can visualize the most
intelligent and the most agile men succeeding in preserving their own
heads while they garnered heads from their tribal enemies; conversely,
the dull and stupid would be eliminated in their first adolescent skir-
mishes. This sort of thing might well favor bloodthirstiness as well as intelligence, and may account for much that is deplorable in human nature.

But these possibly favorable effects of homicide began to disappear
way back in the Neolithic when war presumably started to become systematized and when survival value would presumably go to the Spartan
tradition. For a very long time now war has not tended to promote the
survival of physical or intellectual types that we would regard as highly
desirable.

Differential fertility, whereby the better-educated and the more prosperous tend to have fewer children, is, in its contemporary form, a new
phenomenon. It may also be a transient phenomenon. The Royal Commission that looked into these matters in England found that the small-

family pattern was tending to spread to the lower income groups as knowledge of contraception and awareness of the difficulty of properly raising many children increased. At the same time, the upper income groups were tending to have more children, tending to have as many children as they thought they could afford. This may be a response to a feeling that a family of several children provides a "healthier" living environment than the one-child or two-child family at one time so prevalent. Most of us who were "only" children feel that we missed something important, and we want to give our own children the benefit of fraternal companionship (and rivalry)—if we can afford it.

The alarmists are worried about the dysgenic effect of this new and possibly transient differential family pattern. Relatively little attention, however, has been paid to a similar force of negative selection that has been in operation for a long time—the celibacy of the priesthood. One would think that this, over the last 1,900 years in Europe, would have had an appreciable effect on the genetic constitution of several European populations, if any eugenic program can be effective. For generation after generation the people with inclinations for scholarship, for quiet and peace, for all sorts of pursuits which most of us would take to indicate "favorable" traits, entered monasteries and nunneries and were removed from the reproducing population. The effect of this, particularly during the Middle Ages, must have been enormous—despite the occasional leakage of genes from the monastic population reflected in some of the more scandalous stories of the times.

There may have been a real genetic effect from this ideal of celibacy. I have heard one friend express the opinion that the decline of intellectual leadership in Italy after the Renaissance may have been a result of this long-term negative selection, though probably few people would agree with such an explanation.

The Christian ideal of celibacy is not unique as a dysgenic force, and one could probably collect many examples of man behaving in a manner that, biologically, would seem irrational. Infanticide is sometimes quoted as an example of ancient eugenics, and certainly the congenitally deformed were destroyed at birth in many cultures. But it would be very difficult to assess the potentialities of a newborn infant except for obvious deformities that would probably, in any case, prevent the individual from growing up to form part of the reproductive population, so that I would doubt whether infanticide has ever really been effective as a eugenic force. It may more frequently have been dysgenic when human sacrifice was involved, because the gods would hardly be appeased by an offering of sickly and obviously undesirable victims.

There are many aspects of the present condition of man that seem to me alarming, but the immediate danger of genetic deterioration is not one of them. We are constantly learning more and more about human heredity. . . . The human genetic make-up is so diverse, and the various desirable and undesirable (by whatever definition) genes are so broadly scattered through the population that the effect of any eugenic program would be very slow—and conversely, the effects of failing to apply any eugenic program would be equally slow. In a world full of immediate threats, the threat of genetic deterioration of man seems remote.

Evolution in Process *

by Theodosius Dobzhansky (1962)

Man has not only evolved; for better or for worse, he *is* evolving. Our not very remote ancestors were animals, not men; the transition from animal to man is, on the evolutionary time scale, rather recent. But the newcomer, the human species, proved fit when tested in the crucible of natural selection; this high fitness is a product of the genetic equipment which made culture possible. Has the development of culture nullified the genes? Nothing could be more false. Culture is built on a shifting genetic foundation. It is fairly generally admitted that genetic changes in the human species are influenced by culture. But many people are reluctant to credit that genetic changes may influence culture. The reluctance comes from an almost obsessive fear that biological influences on culture are somehow incompatible with democratic ideals; social sciences must be guarded against the encroachment of biology. Admittedly, most of the biologists' forays into the realm of sociology warranted distrust. But the estrangement must be overcome. Man's future inexorably depends on the interactions of biological and social forces. Understanding these forces and their interactions may, in the fullness of time, prove to be the main achievement of science.

ARE PEOPLE BECOMING LESS INTELLIGENT?

The somber prognosis of declining intelligence has called forth a number of investigations designed to test its validity. By far the most significant of these are the surveys conducted by the Scottish Council for Research in Education [1] [described on p. 485] . . . The fifteen-year interval between 1932 and 1947 corresponds to at least half of the average length of a human generation. Has, then, the intelligence of the Scottish children dwindled during this time? Far from it, the average score has slightly but significantly increased! A similar situation was found in the United States: the soldiers drafted for the Second World War scored on the average higher than those in the First World War. . . .[2]

This result causes some embarrassment to the prophets of doom. . . . The investigators who conducted the surveys honestly admit that they expected a decline, not a rise. Among the explanations offered are improved health of the children and increasing "test sophistication," i.e., a greater familiarity with the kinds of tasks employed in the tests. It is amusing that some writers use these explanations to argue that the Scottish surveys have borne out the predictions that the intelligence is declining! Truly, they let no inconvenient fact interfere with their predilections. . . .

A lucid analysis of the changing patterns of differential fertility in the United States is given in a short paper by Kirk.[3] His data come mainly from a study of the families of men in *Who's Who*, which aims "to include the names, not necessarily of the best, but rather of the best known, men and women in all lines of useful and reputable achievement." These men have had smaller families than the general population of their generation. However, "the difference is progressively narrowing, and the younger men listed in *Who's Who* may approximate or exceed the national average in completed family size for their age groups. Men in *Who's Who* marry later, but more of them marry and they have fewer childless marriages than the general population of comparable age."

Kirk finds, in agreement with investigators of other population groups, that fertility is negatively related to social mobility. Those who have an inherited social status have more children than do the "self-made" men who had to struggle for comparable status. Large families may obviously act as impediments in such a struggle. "If present trends continue, the genetic qualities of men in *Who's Who* will be biologically perpetuated in the future at least in their numerical proportion to the general population. This is in marked contrast to the situation prevailing for at least two generations in the past." This again agrees with a more general trend. The

1. *The Trend of Scottish Intelligence*, London, University of London Press, 1949; and *Social Implications of the 1947 Scottish Mental Survey*, London, University of London Press, 1953.
2. See R. D. Tuddenham, "Soldier Intelligence in World Wars I and II," *Am. Psychol.*, 3, 1948.
3. D. Kirk, "The Fertility of a Gifted Group: A Study of the Number of Children of Men in *Who's Who*," *Milbank Mem. Fund*, 1956 Ann. Conf., 78–98.

birth rates in the United States have increased during the postwar period in all social strata; however, the greatest relative increases have been in those socially favored groups which had been characterized earlier by deficient fertility.

It may well be that the situation in which the economically more successful and, also, the more intelligent (which is certainly not the same thing) strata of the population failed to produce their proportional quota of children was only a temporary one. It may have arisen, especially in the West, because some people became familiar with efficient methods of progeny limitation before others. It is interesting in this connection that the patterns of differential fertility in at least some underdeveloped and non-Western countries favored greater families in economically more prosperous strata. This seems to have been the situation in China.[4]

The direction of natural selection in the human species has certainly been shifting in the environments created by cultural changes. Have these changes and shifts been on the whole beneficial or injurious? Asking this is really another way of posing the question whether man must accept the "natural" drift of evolution as something preordained and inevitable. The alternative to such acceptance is pitting the forces of man's knowledge and wisdom against the forces of nature.

4. See Ping-ti Ho, *Studies on the Population of China, 1368–1953*, Cambridge, Harvard University Press, 1959. Also F. L. K. Hsu, "The Family in China: The Classical Form," in Ruth N. Anshen, *The Family. Its Functions and Destiny*, New York, 1959.

Suggestions for Further Reading

Bates, Marston: *The Prevalence of People*, Charles Scribner's Sons, 1955. A good general book on the population problem. The chapters on the role of war, famine, and disease in controlling population are especially recommended.

* Brown, Harrison: *The Challenge of Man's Future*, Compass Books Edition, The Viking Press, 1956. A stimulating book which places special emphasis on the relationship of natural energy resources and food supply to population.

Calder, Ritchie: *Commonsense About a Starving World*, The Macmillan Company, 1962. Deals with the pressure of the increasing world population on the food supply.

* Dobzhansky, Theodosius: *Heredity and the Nature of Man*, Signet Science Library, The New American Library of World Literature, 1963. The last two chapters discuss the genetic load, the radiation hazard, and the implications of relaxed natural selection.

* Dubos, René: *Man Adapting*, Yale University Press, 1965. In the author's own words the dominant theme of this book "is that the states of health or disease are the expression of the success or failure experienced by the orga-

* Paperback edition

nism in its efforts to respond adaptively to environmental challenges." See especially the chapters on the population problem, disease, and medical science.

Hauser, Philip, ed.: *The Population Dilemma,* Prentice-Hall, Inc., 1963. A series of papers prepared for a meeting of the American Assembly at Columbia University, 1963.

Osborn, Fairfield, ed.: *Our Crowded Planet,* Doubleday & Company, 1962. A group of essays on the population problem by a number of noted writers and scientists.

* Young, Louise B., ed.: *Population in Perspective,* Oxford University Press, 1968. A comprehensive anthology of readings on the population problem; provides historical perspective as well as breadth of coverage in cultural, economic, political, and religious aspects.

9

Should Man Control His Own Evolution?

Man now has the scientific knowledge to take an active part in guiding his own evolution. These readings discuss the various proposals for improving the human species by controlling the heredity as well as the environment. Does the present state of our knowledge justify our tampering with the genetic constitution of man or should we still assume that "Nature knows best"?

Should Man Control His Own Evolution?

Introduction

SCIENTISTS BELIEVE that it will soon be possible for mankind to undertake conscious control of both of the major factors that influence his evolution: his cultural environment and his genetic endowment.

The cultural environment has, of course, been under some conscious control for many years. Education, social reform, and religious ethics all have as their primary objective an improvement in the performance and nature of mankind. There may be disagreement about the effectiveness of some of these methods, but the principle of a conscious control of the cultural environment is accepted as being not only moral but necessary. On the other hand, conscious control of the human genetic endowment has never been attempted on any scale and is considered by many people to be both dangerous and immoral.

The history of eugenics throws some light on this negative response. The idea of a planned differential reproduction that would favor the most desirable segment of the population was suggested very early by a number of philosophers. Plato, for instance, in *The Republic,* outlined the following program:

It follows from what has been already granted, that the best of both sexes ought to be brought together as often as possible, and the worst as seldom as possible, and that the issue of the former unions ought to be reared, and that of the latter abandoned, if the flock is to attain to first-rate excellence; and these proceedings ought to be kept a secret from all but the magistrates themselves, if the herd of guardians is also to be as free as possible from internal strife. . . .

We must therefore contrive an ingenious system of lots, I fancy, in order that those inferior persons, of whom I spoke, may impute the manner in which couples are united, to chance, and not to the magistrates. . . . And those of our young men who distinguish themselves in the field or elsewhere, will receive, along with other privileges and rewards, more liberal permission to associate with the women, in order that, under colour of this pretext, the greatest number of children may be the issue of such parents.[1]

However, it was not until the late nineteenth century that Charles Darwin's cousin, Francis Galton, initiated the science which he christened "eugenics," from the Greek word meaning well-born. Galton's first book, *Hereditary Genius,* described a statistical study he had made of the oc-

1. *The Republic of Plato,* translated by John Llewelyn Davies and David James Vaughan, New York, A. L. Burt, Publisher. Book V, p. 183.

currence of certain traits in family lines. In later books he also discussed
the methods for improving the human stock through the encouragement
of proper matings. He suggested that the study of twins would prove a
useful method for estimating the relative importance of environment and
heredity.

Other scientists took up this idea, and many studies were made of identical and fraternal twins who had been raised in different environments.
These studies showed a strong correlation between certain physical and
mental endowments in closely related individuals. The results of these
studies were interpreted quite differently by sociologists and geneticists.
Heated discussion generated around two opposing points of view: heredity determines the character and fate of each individual, and man is a
tabula rasa at birth—his character is completely shaped by his education
and environment.

Today it is generally believed that both heredity and environment are
important in the development of the human phenotype. While it is true
that certain traits are more easily influenced by environmental factors
than other traits, the total personality results from the interaction of the
genotype with the individual's life experiences.

During the 1930's and early 1940's, the young science of eugenics fell on
evil days when it was perverted by the Nazis to support racism. It was
used to justify the extermination of the Jews and other excesses that
shocked the rest of the world. Two conflicting opinions on intermarriage
are presented in these readings by Adolf Hitler and Bernard Shaw. Hitler
maintains that any cross-breeding leads to mediocrity and degeneration.
Shaw advocates promoting diversity by breaking down the class and
racial distinctions which tend to set up small inbreeding groups. The
genetic disadvantages of either of these two extremes are discussed by
George Gaylord Simpson. He maintains that the situation which now exists with many distinctive sub-groups that occasionally cross-breed produces maximum diversity.[2]

The recent advances in biology have made possible a number of new
techniques that can be brought to bear on eugenics. H. J. Muller advocates the widespread use of artificial insemination. By this method the
sperm of the most talented and healthiest men can be used to father many
children. It is now possible to deep-freeze sperm and preserve them *in
vitro* over a long period of time. Soon it may be possible to use the same
technique with ova. It is interesting to speculate whether these discoveries
might lead to a form of reproductive specialization similar to that practiced by insect societies.

2. In small isolated populations genetic drift and natural selection tend to
produce distinctive characteristics. Refer back to Part 2, page 117 n 1.

Even more revolutionary proposals for positive eugenics are described by René Dubos and Joshua Lederberg. Biochemists are learning to control genes at the molecular level and someday they may be able to replace deleterious genes with normal genes in the human chromosomes. If these transplantations can be achieved, scientists may be able to create tailor-made people for special functions in society (again we are reminded of the specialized individuals in the insect colonies.) Perhaps the most revolutionary suggestion is Lederberg's proposal that vegetative reproduction could be applied to man by causing an undifferentiated human cell to divide without fertilization. In this way perfect copies of superior individuals could be produced.

These proposals, interesting and exciting as they may be, raise doubts in the minds of many of the investigators. As we have seen, maximum diversity in the array of human genotypes helps to maintain adaptability in the species. Many of the proposals would have the effect of reducing genetic diversity.

Many writers also fear that the eugenic programs threaten basic human rights such as the freedom to choose a mate and the right to have children. Some of the suggestions are based on voluntary systems but many of them would need the intervention of the state to make them effective.

Another more general objection is that we do not know exactly what changes would be the most effective in promoting human evolution. "The fundamental difficulty," as René Dubos puts it, ". . . is that we do not know what we are nor where we are, what we want to become nor where we want to go." [3]

In the meantime science marches on. Techniques for manipulating the genetic endowment of the human species are rapidly becoming more available. Will these techniques be put to use before we know where we are going? Should scientists be free to experiment with human genetic material?

We saw in studying man's control of the physical environment that the absence of an over-all plan and the lack of appreciation for the complexity of the various factors, resulted in waste and destruction of our natural resources. If each separate human problem were "corrected" at the molecular level without regard to any general policy, could we not expect human waste and suffering to result?

It has been suggested, for example, that we could relieve the population pressure by breeding smaller people.[4] But would decreased size affect intelligence, rate of maturation, length of life? Would decreased

3. See René Dubos selection, p. 547.
4. See "Where Is Science Taking Us?" by Robert J. Hansen and Myle J. Holley, Jr., *Saturday Review,* December 2, 1968, originally published in *Technology Review.*

size actually relieve stress from overcrowding? Or is the significant factor the number of individual personalities in a given living space?

There are so many aspects to consider and our knowledge is so fragmentary, we wonder whether it would not be best to let nature take its own course and refrain from interference until we understand more about what we are doing.

On the other hand, a very good case can be made for the point of view that man should apply his intelligence to take an active part in the evolutionary process. This may be the only practical way to counteract the degenerative effects of civilization on the human stock. "It is a profound truth," says P. B. Medawar, ". . . that nature does *not* know best; that genetical evolution, if we choose to look at it liverishly instead of with fatuous good humor, is a story of waste, makeshift, compromise, and blunder." [5]

"To put it in another way," said Julian Huxley, "progress has hitherto been a rare and fitful by-product of evolution. Man has the possibility of making it the main feature of his own future evolution, and of guiding its course in relation to a deliberate aim." [6] Evolution has invented mind; from this point on rational purpose emerges as the driving force that will carry us forward into an ever-open future.

5. From P. B. Medawar, *The Future of Man,* Basic Books, Inc., New York, 1959, p. 95.
   6. See Julian Huxley selection, Part 4, p. 248.

Nature and Nurture

The Observed Order of Events *

by Francis Galton (1883)

We find ourselves face to face with two great and indisputable facts that everywhere force themselves on the attention and compel consideration. The one is that the whole of the living world moves steadily and continuously towards the evolution of races that are progressively more and more adapted to their complicated mutual needs and to their external circumstances. The other is that the process of evolution has been hitherto apparently carried out with, what we should reckon in our ways of carrying out projects, great waste of opportunity and of life, and with little if any consideration for individual mischance. Measured by our criterion of intelligence and mercy, which consists in the achievement of result without waste of time or opportunity, without unnecessary pain, and with equitable allowance for pure mistake, the process of evolution on this earth, so far as we can judge, has been carried out neither with intelligence nor ruth, but entirely through the routine of various sequences, commonly called "laws," established or necessitated we know not how.

An incalculable amount of lower life has been certainly passed through before that human organisation was attained, of which we and our generation are for the time the holders and transmitters. This is no mean heritage, and I think it should be considered as a sacred trust, for, together with man, intelligence of a sufficiently high order to produce great results appears, so far as we can infer from the varied records of the prehistoric past, to have first dawned upon the tenantry of the earth. Man has already shown his large power in the modifications he has made on the surface of the globe, and in the distribution of plants and animals. He has cleared such vast regions of forest that his work that way in North America alone, during the past half century, would be visable to an observer as far off as the moon. He has dug and drained; he has exterminated plants and animals that were mischievous to him; he has domesticated those that serve his purpose, and transplanted them to great distances from their

* From Francis Galton, *Inquiries into Human Faculty and Its Development.* J. M. Dent & Co., London, and E. P. Dutton & Co., New York, 1883.

native places. Now that this new animal, man, finds himself somehow in existence, endowed with a little power and intelligence, he ought, I submit, to awake to a fuller knowledge of his relatively great position, and begin to assume a deliberate part in furthering the great work of evolution. He may infer the course it is bound to pursue, from his observation of that which it has already followed, and he might devote his modicum of power, intelligence, and kindly feeling to render its future progress less slow and painful. Man has already furthered evolution very considerably, half unconsciously, and for his own personal advantages, but he has not yet risen to the conviction that it is his religious duty to do so deliberately and systematically.

❀   ❀   ❀

We cannot but recognise the vast variety of natural faculty, useful and harmful, in members of the same race, and much more in the human family at large, all of which tend to be transmitted by inheritance. Neither can we fail to observe that the faculties of men generally, are unequal to the requirements of a high and growing civilisation. This is principally owing to their entire ancestry having lived up to recent times under very uncivilised conditions, and to the somewhat capricious distribution in late times of inherited wealth, which affords various degrees of immunity from the usual selective agencies.

In solution of the question whether a continual improvement in education might not compensate for a stationary or even retrograde condition of natural gifts, I made inquiry into the life history of twins, which resulted in proving the vastly preponderating effects of nature over nurture.

Variety of Human Natures *

by Theodosius Dobzhansky (1964)

GENOTYPE AND PHENOTYPE

Everybody . . . except identical twins, has a nature different from everybody else's. A human being unquestionably is not born a *tabula rasa*, a blank slate. This biological fact does not contradict the principle "that all men are created equal." Arguments will be presented . . . to show that equality should not be confused with identity, or genetic diversity with inequality. People need not be identical twins to be equal before God, before the law, and in their rights to equality of opportunity.

Just how different human beings are made by their different natures must be investigated. Differences between individuals arise owing to their different natures, or heredities, as well as to their different environments, or nurtures. Heredity does not necessarily decree a person's fate. It may be only a conditioning, a bias, a proclivity with which a human individual enters this world. The Danish geneticist W. Johannsen suggested in 1911 that the genotype of an individual be distinguished from his phenotype, and this distinction remains basic for clear thinking about the relations between heredity and environment. In a nutshell, the genotype is the sum total of the heredity the individual has received, mainly, as shown [previously] . . . , in the form of DNA in the chromosomes of the sex cells. The cytoplasm may also contain some heredity determinants; if so, they are likewise constituents of the genotype. The phenotype is the appearance of the individual—the structure and functions of his body. The concept of the phenotype subsumes, of course, not only the external appearance, but also the physiological, normal and pathological, psychological, sociocultural, and all other characteristics of the individual.

It is often stated that the genotype of an individual is set at fertilization and does not change during an individual's lifetime. It is also said that the genotype is somehow isolated from the environment. These are shorthand metaphors which are liable to be misunderstood. The genes an individual received from his parents in the sex cells have many times replicated themselves as the fertilized egg became two, four, eight, and finally billions of cells. Surely, then, I do not have the genes I inherited from my parents; I do, however, have numerous true copies of the genes with which I started my existence. To replicate themselves, the genes must certainly interact with their environment—the copies of the genes can only be built from materials taken from the environment. The interaction is, however, a circular process—genes make more of themselves.

The phenotype, in contrast to the genotype, changes all the time, and its changes are directional rather than cyclic. I am certainly different at present from what I was in the embryonic state, or in infancy, or as a youth; if I live longer, I shall change further. A human egg cell weighs roughly one twenty-millionth of an ounce; a spermatozoon weighs much less; an adult person weighs, let us say, 160 pounds, or some fifty billion times more than an egg cell. The material for this growth comes evidently from the environment. In a broad sense, this is the food the organism consumes and transforms into constituents of the body. A human body transforms its food in ways somewhat different from a dog's or a frog's or a fly's body. The transformations occur according to the instructions emanating from the genes. . . . The outcome of the transformation depends, however, not only on the genes, but also on the materials to be transformed, that is, on the kind of food the organism consumes and on the conditions under which it develops.

The phenotype is, then, a result of interactions between the genotype and the sequence of the environments in which the individual lives. It is also said that the genotype determines the "norm of reaction" of the organism. This should not be understood to mean that some reactions of the organism are somehow intrinsically normal and others abnormal; the "norm of reaction" is simply the sum total of developmental responses which a carrier of a certain genotype might give in all possible environments, natural or artificial, favorable or unfavorable. My present phenotype, what I am as an organism and as a human being, has been determined by my genotype and by the whole succession of the environments I have encountered in my life. The "environments" include, of course, everything that can influence man in any way. They include the physical environment—climate, soil, nutrition—and, most important in human development, the cultural environment—all that a person learns, gains, or suffers in his relations with other people in the family, community, and society to which he belongs.

GENETICS OF SOCIAL INEQUALITY

The existence of glaring social inequalities is often felt to be in need of justification. In some societies with such class or caste inequalities a belief is widely held, by the privileged as well as by the underprivileged, that the social stratification reflects biological, genetic, differences in human quality. The class privilege is hereditary, so the argument goes, because human ability is also hereditary. A scientific sanction for this argument becomes desirable, especially when the established order begins to be questioned by those lacking the favored status. It is probably not accidental that the pioneering studies on human heredity were undertaken in Victorian England, well in advance of the development of Mendelian genetics and of the formulation of the genotype and phenotype concepts by Johannsen. The pioneer was Sir Francis Galton, a cousin and one of the early followers of Darwin.

In 1869, Galton published a book entitled *Hereditary Genius*. The category of "genius" was made rather generously inclusive. Galton assembled a large collection of pedigrees of persons who had achieved "eminence" in various fields of activity—statesmen, jurists, scientists, poets, clergymen, military and naval commanders, and so on. Everywhere he found that the incidence of eminent people among the ancestors and relatives of these persons was far greater than in the general population of England. The conclusion drawn by Galton was that, in the development of human abilities, the nature (which is what we would now call the genotype) was far more important than the nurture (environment).

In the United States, the problem assumed in Galton's time a rather different color. Here many self-made and self-reliant men were achieving ever-greater independence, respectability, and prosperity. The optimistic creed of the Age of Enlightenment, that men in general are good by nature and need only to have their good potentialities evoked by good opportunity, seemed to be borne out by the experiences of the European white man transplanted to the wide-open spaces of the New World. But there were some annoying exceptions. Some people living by crime and in poverty and degradation failed to respond to the stimuli of opportunity. Were these dregs of society the genetic counterparts, in contrast to Galton's eminent men? The 1875 study by R. L. Dugdale seemed to suggest exactly that. He attempted to trace the pedigrees of a group of families to whom he gave the pseudonym Jukes. The ancestor of the Jukes was a certain Max, who lived in the eastern United States in colonial days. Among some 709 of his descendants, Dugdale found 76 convicted criminals, 128 prostitutes, 18 brothel keepers, and over 200 paupers. In 1912 came H. H. Goddard's similar study on the group of families given the pseudonym Kallikak. These families were traced to a person called the "Old Horror," born during the American Revolution. Among his 480 known descendants, there were 143 feeble-minded, 24 alcoholics, 26 illegitimate children, 3 criminals, 33 prostitutes and other sexually immoral individuals.

The ghastly stories of the Jukes and Kallikaks, even more than Galton's accounts of the families of notables, made a great impression on the public in America and in Europe. The closing decades of the nineteenth and approximately the first quarter or third of the twentieth centuries were the periods of the greatest vogue of social Darwinism—a social philosophy for which Darwin was hardly responsible, but which claimed to be simply a straightforward application of Darwin's theory of biological evolution to human society. Darwin saw the biological evolution promoted by the "survival of the fittest" in the "struggle for existence." . . . Darwin used these expressions largely as metaphors, but there was nothing metaphorical about them to social Darwinists. Human society is a field of remorseless struggle, in which the best win, the worst lose, and the result is progress.

Herbert Spencer (1820–1903) thought that, with men or with animals, "If they are sufficiently complete to live, they do live, and it is well they should live. If they are not sufficiently complete to live, they die, and it is best they should die." The Yale University sociologist William Graham Sumner (1840–1910) thought that "poverty belongs to the struggle for existence, and we are all born into that struggle." The gist of his book entitled *What Social Classes Owe to Each Other* can be summed up simply —nothing at all! To a social Darwinist it is plain that social problems such as crime, poverty, and ignorance stem not from any defects of social organization but simply from defective genes. This view pleased greatly, and still continues to please, holders of stoutly conservative political opinions, down to Barry Goldwater.

Liberals and radicals recoiled from social Darwinism in horror. They did not, however, realize that the scientific foundations of social Darwinism fall somewhat short of imposing, and that the view which regards social problems as reducible to genetics is far from proved. They wanted the environment and social reform to be all-powerful, and they came to distrust all genetic conditioning of human traits. The result was that the stand a person took on the nature-nurture problem merely reflected his political convictions. This is surely contrary to the most elementary principle of scientific method—that any investigator who makes the same observations and experiments under similar conditions will obtain the same results, and consequently will have to reach the same conclusions.

PARTITIONING THE GENETIC AND ENVIRONMENTAL EFFECTS

The development of genetics has ushered in more critical methods and attitudes toward studies on human heredity. Similarity between members of a family in certain characteristics is no proof that this characteristic is inherited. It is certainly helpful to have a general or an admiral for a father or a relative if one chooses a military or a naval career, or an influential politician to help a career in politics. It does not take much discernment to see that growing up in surroundings of affluence is more propitious for the development of different tastes and attitudes than growing up among paupers. This simple consideration certainly did not escape Galton's attention; Galton was, however, firmly convinced that had his eminent personalities been born among paupers they would have struggled successfully to reach their eminence. Regrettably, this conviction is simply not subject to verification.

If everybody were growing up and living in a uniform environment, under conditions of perfect equality of opportunity, then, and only then, could one be certain that the differences among people would be due to their hereditary natures. Surely, this condition is nowhere fulfilled in reality. A descendant of the Jukes or the Kallikaks, growing as a youngster among paupers and criminals, is likely to be more susceptible to the

lures of violence and ill-gotten money, and less prone to seek to ameliorate his station by honest toil, than somebody who is brought up to abhor
wrongdoing and vice.

Furthermore, it gradually came to be recognized that the question
whether the nature or the nurture, the genotype or the environment, is
more important in shaping man's physique and his personality is simply
fallacious and misleading. The genotype and the environment are equally
important, because both are indispensable. There is no organism without
genes, and any genotype can act only in some environment. The nature-
nurture problem is nevertheless far from meaningless. Asking right questions is, in science, often a large step toward obtaining right answers. The
question about the roles of the genotype and the environment in human
development must be posed thus: To what extent are the *differences* observed among people conditioned by the differences of their genotypes
and by the differences between the environments in which people were
born, grew, and were brought up?

The insistence on stating the nature-nurture problem in just this way is
neither pedantry nor hairsplitting. The dichotomy of genetic and environmental traits is a false one because any trait is both genetic and environmental. I choose deliberately two farfetched examples to illustrate this
proposition. Surely, the presence in man of two eyes, instead of only one,
is a "hereditary" trait—it is a characteristic of the human species, of the
class of mammals, and even of most representatives of vertebrate animals.
Yet cases are recorded of the birth of so-called cyclopic monsters, with
only one eye. Developmental accidents can cause the appearance of this
monstrosity, and the cyclopic monsters have been induced experimentally
in animal materials. Consider now the ability to play the game of chess.
Surely it is due to human genes, since no animal other than man plays
chess. Just as surely, it is environmental, since nobody plays this game
in countries where it has not been introduced. However, where chess is
known, some people play it well and others do not play, or play it badly.
It is not pointless to ask to what extent these *differences* are due to genetic
predisposition and to environmental opportunity. (This question has not,
to my knowledge, been investigated.)

If the nature-nurture problem is stated correctly, the answer turns out
to depend on which differences, in what particular features or characters
or traits, are considered. For example, there are four blood types among
people, called types O, A, B, and AB. A competent technician can take
a drop of a person's blood and in a few minutes make the tests needed
to determine the blood type (Figure 9–1). Knowing one's blood type
may be important if a blood transfusion is needed. Giving A, B, or AB
blood to an O blood recipient may lead to grave or even fatal conse-

Anti-B serum   Anti-A serum

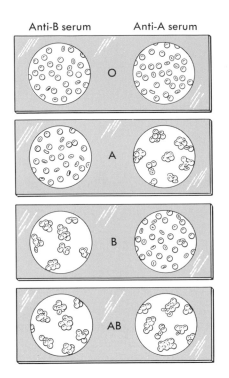

Figure 9–1. The tests to determine the blood type of a given person. A small amount of blood is mixed with each of two sera, the anti-A and the anti-B. The red blood cells either do, or do not, become clumped (agglutinated), depending on the blood type to which the person tested belongs.

quences. Now, the blood type of an individual is already determinable in a fetus, starting with the second or third month of pregnancy; it persists unchanged during the whole life, regardless of age, state of health, or conditions of living. The blood type is inherited according to Mendel's law. There is no known way of changing the blood type with which a person is born. It is fair to say that the differences among people in their blood types are rigidly determined by their genes.

By contrast, the language a person speaks comes from his environment, not from his genes. Note, however, that the ability to learn a language, any language, is genetic; some genetic defects make a child unable to learn to speak any language. Most human differences lie between the extremes of the rigid genetic and a purely environmental causation. To estimate the relative magnitudes of the genetic and environmental influences more precisely often turns out to be exceedingly difficult. This matter obviously needs much further research.

Let us make it very clear where the difficulty lies. With plants, or animals, or microorganisms, the problem can be solved relatively more easily than it can be in man. Different individuals, or their progenies, are raised in as uniform an environment as can be achieved, controlling the temperature, light, humidity, food, or soil in which the experimental organism grows. The differences observed between the phenotypes of the individuals thus obtained depend on their genotypes being different. Or vice versa, one can take individuals having the same or very similar genotypes, and have them develop in different environments. This will yield information concerning the degree of the developmental plasticity of the organism or character in question—how great a variety of phenotypes can be obtained by having the carriers of similar genotypes brought up in different environments.

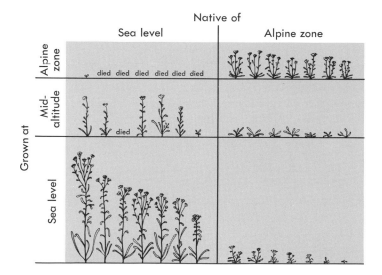

Figure 9–2. The yarrow plants (Achillea) *native at sea level and at high elevations in the Sierra Nevada Mountains of California. The sea-level form dies if planted at high elevations in the Alpine Zone, and develops poorly if planted at intermediate elevations in the mountains. The Alpine Zone form does not grow well at sea level or at intermediate elevations. The differences between the varieties native at low and at high elevations are owing to genes, those between the individuals of the same variety planted at different elevations are environmental.*

Beautiful experiments of the above kind were carried out on several species of plants by a group of investigators working at the Department of Plant Biology of the Carnegie Institution of Washington (J. Clausen, D. D. Keck, and W. M. Hiesey). Figure 9–2 summarizes the results of experiments on the yarrow (*Achillea*). In California, this plant grows from sea level up to the alpine zone of the Sierra Nevada Mountains. The race growing at sea level grows much taller (lower left in Figure 9–2) than the race native to the alpine zone (upper right in Figure 9–2). The young plants can be cut in several parts, and the cuttings replanted in locations of the experimenter's choice. The parts of the body of the same individual have presumably the same genotype. Now, the sea-level race planted in the alpine zone fails to survive the rigor of the climate; planted in a locality at an intermediate altitude (4,800 feet) it usually survives but does more poorly than in its native habitat. The alpine race, on the contrary, develops better in its native environment than lower down in the mountains or at sea level. In Figure 9–2 the vertical rows thus show the phenotypes that the same genotype engenders in different environments, while the horizontal rows show the reactions of different genotypes to similar environments.

TWINS

Experiments of the above sort, quite obviously, cannot be done with humans. It is extremely difficult even to arrange for two or several persons to be brought up from infancy to maturity in identical, or even in very similar, environments. All this does not prevent some people from holding very firmly set opinions regarding what heredity or environment can or cannot do. Fortunately, the problem, though certainly difficult, need not remain insoluble forever. Nature has provided an opportunity that can be utilized to obtain some accurately verifiable information.

Identical, or monozygotic, twins arise through the division of a single fertilized egg into parts, which develop into separate individuals. [See Figure 9–3.] Identical twins are, of course, always of the same sex; barring mutation, and possibly an unequal distribution of hereditary determinants in the egg cytoplasm, they have identical genotypes. Fraternal, or dizygotic, twins, come from two simultaneously maturing egg cells fertilized by two different spermatozoa. They may be either of the same or of different sexes, and they are genotypically as different on the average as brothers and sisters who are not twins. [See Figure 9–4.] The differences that can be observed between growing or grown-up identical twins are due largely to environmental influences. The differences between fraternal twins of the same sex provide a control experiment, showing the combined effects of the genotype plus the environmental influences. Particularly valuable, but regrettably rare from a scientific point of view, are instances of twins separated at as early an age as possible and brought up apart from each other in different environments. Such twins reared apart, and especially if reared in substantially different environments, would be analogous to the yarrow plants in the experiments described above; the study of such twins could furnish invaluable information on just how plastic the human development really is.

Many observations on twins have accumulated in the genetic literature, and yet, it must be pointed out again, further research in this field is necessary if the problem of the gene-environment interaction in man is to be advanced towards a more adequate understanding. For the time being one can draw only tentative conclusions, which will almost certainly stand in need of correction when better data become available. As a broad generalization, the study of twins shows that human variation in almost every character is due in part to a genetic diversity and in part to environmental dissimilarity. Different traits are, however, quite different in this respect. The genetic component is responsible for the human diversity in some traits to a much greater extent than in other traits, and the same is true of the environmental component.

By and large, such traits as the facial features, eye and hair colors and forms, and other external traits by which we usually recognize and distinguish persons whom we meet are strongly influenced by the genotype. This is the reason why monozygotic twins are so astonishingly similar in appearance as to deserve the name "identical." Countless stories (some fictional, others quite genuine) are told about twins being misidentified, or substituted one for the other; a girl who did not know that she had a twin sister was startled to meet her double, and so on. Fraternal twins differ in facial features and other external bodily traits about as much as do brothers and sisters who are not twins.

The suceptibility to some infectious diseases is genetically conditioned. Tuberculosis (consumption) is due to the invasion of the organism by certain bacteria (Koch's bacilli). It is, then, an "environmental," not a "hereditary" disease. An individual cannot contract it unless exposed to infection. However, especially in the slums of crowded cities, everybody is exposed to some degree. Whether a person, under these conditions, develops a clinical case of tuberculosis turns out to be influenced by his heredity. F. J. Kallman and D. Reisner studied seventy-eight monozygotic twin pairs one member of which was known to have tuberculosis, and found that forty-eight of the co-twins were also tubercular. This is usually expressed by saying that 61.5 per cent (forty-eight out of seventy-eight) of the twin pairs were "concordant." Among dizygotic twins about 18 per

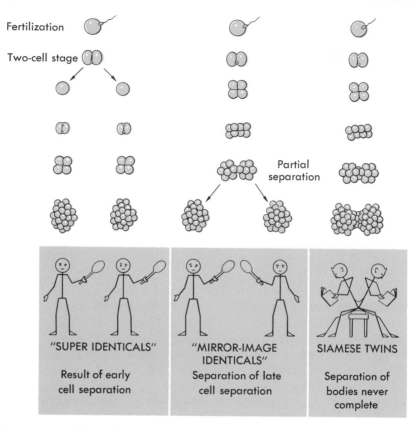

Figure 9–3. *Identical or monozygotic twins are one genetic person in two bodies. Identical twins begin life with the fertilization of a single egg cell by a single sperm cell. Early in its development the embryo splits into two (or more) distinct islands of growth, each of which goes on to develop into a complete human being.* (From Robert C. Cook, Human Fertility: The Modern Dilemma, *William Sloane Associates, Publishers, New York. Copyright 1951 by Robert C. Cook, p. 205. Reprinted by permission.*)

cent, among full siblings 19 per cent, and among half-siblings 9.5 per cent were concordant. The incidence of tuberculosis in the population from which the twins came was between 1 and 1.5 per cent.

Malaria is another infectious disease, caused by a parasite infecting the red blood cells. In many tropical and subtropical countries, certain districts are, or until recently were, heavily malarial, so much so that practically the entire population was infected at some period of their lives. Several genetic conditions . . . have recently been proved, or suspected, to

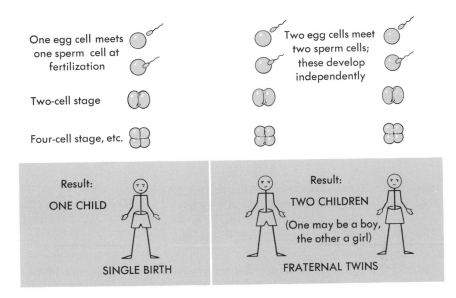

Figure 9–4. Single and multiple birth. Fraternal or dizygotic twins, or multiplets, arise from the germination of two or more separate egg cells by separate sperm cells. (From Robert C. Cook, Human Fertility: The Modern Dilemma, *William Sloane Associates, Publishers, New York. Copyright 1951 by Robert C. Cook, p. 206. Reprinted by permission.)*

confer an at least relative resistance to certain species of the malarial parasite.

Much discussion and polemics have centered on the problem of the inheritance of intelligence. Psychologists and educationists have worked out a technique of estimation of the so-called intelligence quotient, or I.Q. What is measured by the I.Q. is not necessarily the same as what is referred to in everyday language as intelligence, cleverness, aptitude, or wit. Still less does the I.Q. give an estimate of the value or worth of the person. The I.Q. as administered to school children is regarded as a measure of their ability to handle verbal symbols. This ability shows a fairly high correlation with scholastic success, and this is what makes the I.Q. measurements useful, but at the same time suggests the limitations of their usefulness. The I.Q. is certainly not independent of the environment, of the family background, schooling, and the circumstances under which the test is administered.

Identical twins perform appreciably more similarly in intelligence tests than do fraternal twins, and the latter only slightly more similarly than do siblings who are not twins. L. Erlenmeyer-Kimling and L. F. Jarvik have recently reviewed and summarized the data obtained in fifty-two different studies by different investigators. The mean correlation coefficient for the I.Q. scores of identical twins reared together is 0.87, and for those reared apart 0.75. A correlation coefficient of 1.00 would mean that the I.Q. measurement was absolutely precise and that the variations observed are entirely genotypic and not environmental. This is obviously not so, and particularly the fact that the twins reared apart show a lower correlation than those reared together is evidence that the I.Q. depends in part on environment. However, the fraternal twins give a correlation of only 0.53, the value being almost the same for twins reared together and those reared apart. This value, 0.53, is only slightly greater than the 0.49 for siblings, brothers or sisters who are not twins, and significantly lower than the values for identical twins reared either together or apart. Important additional evidence comes from studies on the I.Q.'s of genetically unrelated persons reared together, such as children brought up in the same orphanage or foster home. Here the correlation falls to 0.23. The correlation between foster children and foster parents is only 0.20. The value of 0 would indicate, of course, no correlation at all.

The conclusion is inevitable that the performance on I.Q. tests is determined jointly by genotypic and environmental variables. The genetic component is, for this trait, appreciably greater than the environmental component in the materials covered by the studies. This last qualification is necessary and important. The point is this: Every one of the fifty-odd studies concerned itself with twins, or siblings, or foster children in the same country and usually within a rather limited section of the same country. The range of the environments to which the individuals studied were exposed was accordingly a limited one, much more limited than it could be if the persons reared apart were brought up in, for example, a Western country, China, or an Indian tribe, or at least in different social classes and different economic and educational groups. It is, indeed, evident that the more uniform the environment is the greater, relatively, is the role of genotype differences, and vice versa.

Intelligence, or whatever it is that the I.Q. scores measure, is an important but certainly not the only quality in which we are interested in our fellow men. Temperamental, emotional, and other personality traits may for some purposes and under some circumstances be equally or even more important. The handicap that has impeded scientific study of personality traits is the difficulty of reliably measuring them. Psychologists have, however, been making progress with such measurements, by means of a variety of questionnaires, personality inventories, projective tests, and so on. It is too early at present to reach firm conclusions in this difficult field. It is nevertheless interesting to quote some of the results of the recent work of I. I. Gottesman (1963), who has administered the so-called Minnesota Multiphasic Personality Inventory (MMPI) and Cattell's High School Personality Questionnaire (HSPQ) to thirty-four pairs of monozygotic and an equal number of dizygotic teen-age twins. Gottesman divides the "factors" and "test dimensions" he has measured into the following groups, according to whether the observed variation is due chiefly to genotype or to environmental causes:

I. Equal Contributions of Heredity and Environment
   Confident Adequacy versus Guilt Proneness
   Depression
   Psychopathic Deviate

II. Heredity Predominant
   Social Introversion

III. Environment Predominant
   Submissiveness versus Dominance
   Shy, Sensitive versus Adventurous
   Liking Group Action versus Fastidiously
   Individualistic Psychasthenia

One can only agree with Gottesman's following conclusion: "Granting that the difficulties of accurately assessing the contribution of heredity to variation of socially important behavior are great, such efforts will not have been in vain if they contribute to a greater understanding of the sources of individual differences. The provision of an optimum environment for the optimum development of the various aspects of human behavior should follow such increased understanding."

WHAT THE DATA ON TWINS DO NOT MEAN

Notwithstanding all their weaknesses and incompleteness, the studies of twins have laid a firm foundation on which to build an understanding of the roles of genotypic and environmental variables in human development. It is fair, I think, to say that, as a general rule, whenever a variable human trait, whether structural or physiological or psychological, was at all adequately studied, both genotypic and environmental causes proved to be involved to some extent. Any way you look at it, man is a creature of both his nature and his nurture. It is folly to disregard either.

Many social scientists, and also followers of some schools of psychoanalysis, feel nevertheless a compulsive distrust of any genetic determinism in man. When the evidence becomes overwhelming they try to patch things up by saying that the genetic differences between persons are, after all, very small. For example: "In a consideration of behavioral differences among people, therefore, we may regard the biological factor as a constant, and hence eliminate it from our calculations" (L. White). I can understand (and perhaps forgive) statements of this sort only as a reaction against the excesses of social Darwinists; scientists should, however, resist the temptation to oppose exaggerations by making exaggerations opposite in sign. It has, to be sure, been claimed that the studies on identical and fraternal twins prove that what man is or can become is settled and predestined by his heredity. Environment matters little; heredity is the "dice of destiny."

To bring the issue to a focus, let us consider two examples of studies of twins; one of these studies has actually been misinterpreted, while the other is easily susceptible to misinterpretation. Several investigators sought to locate in prisons individuals sentenced for a variety of offenses and who had twin brothers. It did not matter what sort of crime the individual had committed. The question asked was simply whether the twin brother did or did not have a criminal record. When each twin had a criminal record, they were called "concordant"; when the brother of the imprisoned twin was found not to have an official criminal record, the twins were "discordant." Some of the twins were monozygotic and others dizygotic; the crucial fact ascertained by the investigation was that the monozygotic twins were concordant more often than the dizygotic ones, and the difference was large enough to be statistically assured. The other study arose as a by-product of the observation that lung cancer is found more often among cigarette smokers than among nonsmokers. The suggestion was then made by an eminent geneticist that the tendency to smoke may be genetically conditioned, and that the proneness to lung cancer may be a by-product of the same genotypes. An attempt was accordingly made to examine the tobacco-smoking habits of some pairs of monozygotic and dizygotic twins. With respect to this characteristic, too, the monozygotic twins were concordant more often than the dizygotic ones.

One of the investigators of the criminality in twins, J. Lange, has published his work in a book suggestively entitled *Crime and Destiny*.[1] We are entitled, however, to ask the question whether the information collected warrants the conclusion that some people are destined by their heredities to be criminals and others law-abiding, some to be smokers and others nonsmokers. Surely, nobody can be a tobacco smoker if he has no

1. Johannes Lange, *Crime and Destiny*, translated by Charlotte Haldane, New York, C. Boni, 1930. Ed.

access to tobacco. On the other hand, some people become smokers because among their friends and acquaintances it is considered smart or "masculine" to smoke. Whether or not a person becomes a criminal, and, if so, what kind of crime he commits, and if he commits a crime whether or not he is apprehended, all this evidently depends on the kind of laws the society in which he lives has instituted, and on the strictness of the law enforcement.

Nothing whatever in the data on the "criminal" twins proves that these people would have been criminals if they had been brought up differently. For all that the data tell us, these criminals might have been Galton's eminent men, righteous pillars of society and even prosecutors of criminals. And there is nothing to show that the present pillars of society could not have been like the Jukes or Kallikaks under other circumstances. However, the data are far from meaningless; they do show that in some environments, and presumably in any environment, persons with similar genotypes are likely to be on the average more similar in their behavior than persons with different hereditary endowments. Heredity is not the destiny that foreordains that a person will behave in a certain way regardless of circumstances. Heredity does predispose him to behave this way, rather than that way, under a given set of circumstances. It is a conditioning that to a certain extent biases man's choices and efforts of will.

As stated previously, certain people like to exaggerate the deterministic role of heredity and to underestimate that of environment. Others are loath to believe that heredity can have any influence, at least on socially significant human qualities, such as intelligence and character. The anti-hereditarians fear that if the genes are shown to have anything to do with man's behavior, this will deprive us of our freedom, make us mere automata, and render futile all attempts to improve man by education and social betterment. These fears go hand in hand with a misunderstanding of what heredity really "determines." It has been said before, but it will bear repetition, that heredity, the genotype, the genes do not determine "characters," such as proneness to criminality or smoking habits; the genes determine the reactions of the organism to its environment.

ENVIRONMENTAL ENGINEERING

Any trait or character, external or internal, physiological or psychological, is, at least in principle, modifiable and governable by genetic as well as by environmental influences. To be sure, one cannot, at least at present, change the blood type that one's genes have determined. There are many hereditary diseases for which no remedies are known. It does not, however, follow that a hereditary, gene-controlled disease is necessarily an incurable disease. This notion is as widespread as it is misleading. . . .

When a patient consults his doctor about some condition that troubles him, he does not ask the doctor to change his genes, which the doctor is utterly unable to do, only to advise him on how to change his environment in such a way that the genes with which he was born will be helped to produce a satisfactory phenotype.

Neither genetic health nor genetic disease can be defined except in terms of its opposite. And although "genetic" refers to something in the genotype, health and disease are observable only in the phenotype. If the environment in which people live were the same everywhere and could not be altered, one might conceivably select the one healthy genotype; all other genotypes would, by definition, contain genetic defects. Even then the selection would be difficult because it would involve a value judgment on which agreement would not be easy to reach. Is the optimum genotype the one giving to its carriers the greatest euphoria, or the highest working capacity, or the aesthetically most pleasing appearance, or the longest life?

In reality, the environments in which people live are not only infinitely variable, but man is able, through his technology, to devise new ones according to his specifications. A genetic endowment may give different phenotypes in different environments. Some of these phenotypes may be agreeable to their possessors and acceptable to the society in which they live; they will accordingly be regarded as healthy. Other phenotypes may give discomfort or suffering to their possessors and be socially unsatisfactory; they are unhealthy or defective. The same genotype may give health in some, but cause disease or discomfort in other environments. For example, most people do not need insulin injections, because their bodies manufacture their own insulin, but diabetics are better off in an insulin-supplying environment. Myopics need an environment that gives them glasses, or else the environment of a profession such as watch repairing, in which nearsightedness is less of an obstacle.

We have seen that no two persons, and probably no two individuals of any sexually reproducing species, have the same genotype. Excepting only identical twins, a given constellation of genes occurs only once, in only a single person. The evaluation of these genotypes is most simply made according to what phenotypes they produce in ordinary, widespread, customary environments in which most people live at present on this planet. A majority of these genotypes belong to what is described as *l'homme moyen sensuel*, the happy average, neither a giant nor a dwarf, neither completely resistant nor easily succumbing to the shocks unavoidable in customary environments, hence ordinarily enjoying a robust to good to fair health, neither a hero nor a villain, in short, the general public. This multitude of genotypes, and of phenotypes, constitutes the adaptive norm of mankind, the human species.

In addition to the adaptive norm, there is a minority of genotypes that are free of overt defects, and in addition can do in some respects much better than the norm; this is the genetic elite. And there is another minority that, in ordinary environments, develops weak or infirm bodies or minds, described as hereditary diseases, malformations, or constitutional weaknesses. This is the genetic load, or the genetic burden, which our species, like other living species, always carries. . . . To forestall a possible misunderstanding, it must, however, be stressed that the categories of the adaptive norm, genetic elite, and genetic load are not fixed or unchangeable forever. They depend on the environments and, in man, on social requirements. Let us be reminded that the curability or incurability of a disease has little to do with its genetic or nongenetic origin. At the present level of advancement of the medical arts there are, unfortunately, many as yet incurable environmental as well as genetic diseases. In principle, no disease need be incurable, but medicine is certainly far from having achieved a level of knowledge that would make all human suffering remediable.

EUTHENICS AND EUGENICS

If mankind succeeds in improving genetics, genetics may succeed in improving mankind. The ways to improve the human lot generally fall into two broad categories: environmental engineering, or euthenics, and genetic engineering, or eugenics. Methods of making better environments range from spiritual and cultural uplift, political and social reforms, improved education, and superior medical care to better nutrition and sanitation. Environments can be said to have been improved if the existing array of human genotypes reacts to these environments by developing healthier, happier, or in some other ways more desirable or admirable phenotypes. On the other hand, one may endeavor to improve the array of human genetic endowments. Measures may be taken to prevent the formation of genotypes that are environmentally refractory, that react badly to most existing or easily contrived environments. This is negative eugenics. Or else, one may seek to encourage the increase and spread of good genotypes, which react favorably to existing environments or to environments in the offing. This is positive eugenics.

Recent advances in genetics, in Mendelian organismic genetics, but particularly in biochemical or molecular genetics, are grounds for optimism. The work on heredity coded by means of the four genetic "letters" in the DNA chain molecules will surely abide as one of the greatest, if not the greatest, achievement of biology in our age. The hopes for new approaches to both euthenics and eugenics, based on the application of the discoveries in molecular genetics, go very far. We have seen that treatments have already been discovered, that is, artificial environments have been devised in which certain genotypes that in ordinary environments react by genetic disease can live fairly happily. Insulin supplied to diabetics, withholding galactose from infants with galactosemia genes, the drug that relieves acrodermatitis enteropathica, these and other examples show how much can be accomplished if the biochemical basis of the genetic defect is understood, and measures to correct it properly applied.

It is only reasonable to expect that the knowledge of the developmental mechanisms involved in other hereditary diseases and malformations will progressively increase. It may then become possible to manage their manifestation in the phenotype. Hopefully, the carriers of defective genes may be helped to live happy and useful lives. An even bolder approach is conceivable. This is to control the gene action itself at the level of the DNA-RNA-ribosome-enzyme chains of reactions. Evidence is rapidly growing that the genes in the chromosomes of a cell nucleus are not continuously active; the reactions that lead to formation of enzymes occur only at certain stages in the development of the body; different genes seem to be active in different cells and at different times. Cellular mechanisms exist that cause the action of this or that gene to be "switched on" or "off." My colleague A. E. Mirsky has pointed out the possibility of discovering methods to suppress the action of undesirable genes and perhaps to enhance the action of favorable ones.

Euthenics is, however, not alternative, but complementary to eugenics. Even if we had methods available to control or to suppress the action of undesirable genes, it is certainly better to have no need to resort to such operations. At the very best they are likely to be laborious and costly.

Selective Breeding

Man scans with scrupulous care the character and pedigree of his horses, cattle, and dogs before he matches them; but when he comes to his own marriage he rarely or never takes any such care. He is impelled by nearly the same motives as the lower animals, when they are left to their own free choice, though he is in so far superior to them that he highly values mental charms and virtues. On the other hand he is strongly attracted by mere wealth or rank. Yet he might by selection do something not only for the bodily constitution and frame of his offspring, but for their intellectual and moral qualities. Both sexes ought to refrain from marriage if they are in any marked degree inferior in body or mind; but such hopes are Utopian and will never be even partially realized until the laws of inheritance are thoroughly known. Every one does good service, who aids toward this end. When the principles of breeding and inheritance are better understood, we shall not hear ignorant members of our legislature rejecting with scorn a plan for ascertaining whether or not consanguineous marriages are injurious to man.

Charles Darwin
The Descent of Man (1871)

Superman *

by Friedrich Nietzsche (1883)

Zarathustra spake thus unto the people:
I teach you the Superman. Man is something that is to be surpassed. What have ye done to surpass man?

All beings hitherto have created something beyond themselves: and ye want to be the ebb of that great tide, and would rather go back to the beast than surpass man?

What is the ape to man? A laughing-stock, a thing of shame. And just the same shall man be to the Superman: a laughing-stock, a thing of shame.

Ye have made your way from the worm to man, and much within you is still worm. Once were ye apes, and even yet man is more of an ape than any of the apes.

* From Friedrich Nietzsche, "Zarathustra's Prologue," *Thus Spake Zarathustra,* translated by Thomas Common, 1883.

Even the wisest among you is only a disharmony and hybrid of plant and phantom. But do I bid you become phantoms or plants?

Lo, I teach you the Superman!

The Superman is the meaning of the earth. Let your will say: The Superman *shall be* the meaning of the earth! . . .

Man is a rope stretched between the animal and the Superman—a rope over an abyss.

A dangerous crossing, a dangerous wayfaring, a dangerous looking-back, a dangerous trembling and halting.

What is great in man is that he is a bridge and not a goal: what is lovable in man is that he is an *over-going* and a *down-going*.

I love those that know not how to live except as down-goers, for they are the over-goers.

I love the great despisers, because they are the great adorers, and arrows of longing for the other shore.

I love those who do not first seek a reason beyond the stars for going down and being sacrifices, but sacrifice themselves to the earth, that the earth of the Superman may hereafter arrive.

I love him who liveth in order to know, and seeketh to know in order that the Superman may hereafter live. Thus seeketh he his own down-going.

I love him who laboureth and inventeth, that he may build the house for the Superman, and prepare for him earth, animal, and plant: for thus seeketh he his own down-going. . . .

I love all who are like heavy drops falling one by one out of the dark cloud that lowereth over man: they herald the coming of the lightning, and succumb as heralds.

Lo, I am a herald of the lightning, and a heavy drop out of the cloud: the lightning, however, is the *Superman.*—

When Zarathustra had spoken these words, he again looked at the people, and was silent. "There they stand," said he to his heart; "there they laugh: they understand me not; I am not the mouth for these ears.

"Must one first batter their ears, that they may learn to hear with their eyes? Must one clatter like kettledrums and penitential preachers? Or do they only believe the stammerer?

"They have something whereof they are proud. What do they call it, that which maketh them proud? Culture, they call it; it distinguisheth them from the goatherds.

"They dislike, therefore, to hear of 'contempt' of themselves. So I will appeal to their pride.

"I will speak unto them of the most contemptible thing: that, however, is *the last man!*"

And thus spake Zarathustra unto the people:

It is time for man to fix his goal. It is time for man to plant the germ of his highest hope.

On Good Breeding, Property and Marriage *

by Bernard Shaw (1903)

The cry for the Superman did not begin with Nietzsche, nor will it end with his vogue. But it has always been silenced by the same question: what kind of person is this Superman to be? You ask, not for a super-apple, but for an eatable apple; not for a super-horse, but for a horse of greater draught or velocity. Neither is it of any use to ask for a Superman: you must furnish a specification of the sort of man you want. Unfortunately you do not know what sort of man you want. Some sort of good-looking philosopher-athlete, with a handsome healthy woman for his mate, perhaps.

Vague as this is, it is a great advance on the popular demand for a perfect gentleman and a perfect lady. And, after all, no market demand in the world takes the form of exact technical specification of the article required. Excellent poultry and potatoes are produced to satisfy the demand of housewives who do not know the technical differences between a tuber and a chicken. They will tell you that the proof of the pudding is in the eating; and they are right. The proof of the Superman will be in the living; and we shall find out how to produce him by the old method of trial and error, and not by waiting for a completely convincing prescription of his ingredients.

❋   ❋   ❋

What is really important in Man is the part of him that we do not yet understand. Of much of it we are not even conscious, just as we are not normally conscious of keeping up our circulation by our heart-pump, though if we neglect it we die. We are therefore driven to the conclusion that when we have carried selection as far as we can by rejecting from the list of eligible parents all persons who are uninteresting, unpromising, or blemished without any set-off, we shall still have to trust to the guidance of fancy (*alias* Voice of Nature), both in the breeders and the parents, for that superiority in the unconscious self which will be the true characteristic of the Superman.

At this point we perceive the importance of giving fancy the widest possible field. To cut humanity up into small cliques, and effectively limit the selection of the individual to his own clique, is to postpone the Superman for eons, if not forever. Not only should every person be nourished and trained as a possible parent, but there should be no possibility of such an obstacle to natural selection as the objection of a countess to a navvy or of a duke to a charwoman. Equality is essential to good breeding; and equality, as all economists know, is incompatible with property. . . .

One fact must be faced resolutely, in spite of the shrieks of the romantic. There is no evidence that the best citizens are the offspring of congenial marriages, or that a conflict of temperament is not a highly important part of what breeders call crossing. On the contrary, it is quite sufficiently probable that good results may be obtained from parents who would be extremely unsuitable companions and partners, to make it certain that the experiment of mating them will sooner or later be tried purposely almost as often as it is now tried accidentally. But mating such couples must clearly not involve marrying them. In conjugation two complementary persons may supply one another's deficiencies: in the domestic partnership of marriage they only feel them and suffer from them. Thus the son of a robust, cheerful, eupeptic British country squire, with the tastes and range of his class, and of a clever, imaginative, intellectual, highly civilized Jewess, might be very superior to both his parents; but it is not likely that the Jewess would find the squire an interesting companion, or his habits, his friends, his place and mode of life congenial to her. Therefore marriage, whilst it is made an indispensable condition of mating, will delay the advent of the Superman as effectually as Property, and will be modified by the impulse towards him just as effectually.

. . . It cannot be denied that one of the changes in public opinion demanded by the need for the Superman is a very unexpected one. It is nothing less than the dissolution of the present necessary association of marriage with conjugation, which most unmarried people regard as the very diagnostic of marriage. They are wrong, of course: it would be quite as near the truth to say that conjugation is the one purely accidental and incidental condition of marriage. Conjugation is essential to nothing but the propagation of the race; and the moment that paramount need is provided for otherwise than by marriage, conjugation, from Nature's creative point of view, ceases to be essential in marriage.

The Result of All Race-Crossing *
by Adolf Hitler (1925)

Any crossing between two beings of not quite the same high standard produces a medium between the standards of the parents. That means: the young one will probably be on a higher level than the racially lower parent, but not as high as the higher one. Consequently, it will succumb later on in the fight against the higher level. But such a mating contradicts Nature's will to breed life as a whole towards a higher level. The presumption for this does not lie in blending the superior with the inferior, but rather in a complete victory of the former. The stronger has to rule and he is not to amalgamate with the weaker one, that he may not sacrifice his own greatness. . . .

Just as little as Nature desires a mating between weaker individuals and stronger ones, far less she desires the mixing of a higher race with a lower one, as in this case her entire work of higher breeding, which has perhaps taken hundreds of thousands of years, would tumble at one blow.

Historical experience offers countless proofs of this. It shows with terrible clarity that with any mixing of the blood of the Aryan with lower races the result was the end of the culture-bearer. . . . The result of any crossing, in brief, is always the following:

(a) Lowering of the standard of the higher race,

(b) Physical and mental regression, and, with it, the beginning of a slowly but steadily progressive lingering illness.

To bring about such a development means nothing less than sinning against the will of the Eternal Creator.

This action, then, is also rewarded as a sin.

Man, by trying to resist this iron logic of nature, becomes entangled in a fight against the principles to which alone he, too, owes his existence as a human being. Thus his attack is bound to lead to his own doom.

* From Adolph Hitler, *Mein Kampf*, translated by Manheim, Reynal & Hitchcock, New York, Hurst & Blackett, London, 1939. Copyright 1925 & 1927 by Verlag Frz. Eher Nachf., G.m.b.H. Copyright 1939 by Houghton Mifflin Company. Reprinted by permission of Houghton Mifflin Company.

The Importance of Genetic Variety *

by George Gaylord Simpson (1949)

Eugenics has deservedly been given a bad name by many sober students in recent years because of the prematurity of some eugenical claims and the stupidity of some of the postulates and enthusiasms of what had nearly become a cult. We are also still far too familiar with some of the supposedly eugenical practices of the Nazis and their like. The assumption that biological superiority is correlated with color of skin, with religious belief, with social status, or with success in business is imbecile in theory and vicious in practice. The almost equally naïve, but less stupid and not especially vicious, idea that prevention of reproduction among persons with particular undesirable traits would quickly eradicate these traits in the population has also proved to be unfounded. The incidence of a few clearly harmful hereditary defects could be reduced by sterilization of the individuals possessing them, but they could not be wholly eliminated and in the light of present knowledge it is highly doubtful whether this means can produce any really noteworthy physical improvement in the human species as a whole.

Selection was, nevertheless, the means by which man arose and it is the means by which, if by any, his further organic evolution must be controlled. Man has so largely modified the impact of the sort of natural selection which produced him that desirable biological progression on this basis is not to be expected. There is no reason to believe that individuals with more desirable genetic characteristics now have more children than do those whose genetic factors are undersirable, and there is some reason to suspect the opposite. The present influence of natural selection on man is at least as likely to be retrogressive as progressive. Maintenance of something near the present biological level is probably about the best to be hoped for on this basis. The only proper possibility of progress seems to be in voluntary, positive social selection to produce in offspring new and improved genetic systems and to balance differential reproduction in favor of those having desirable genes and systems. As soon as we know what the desirable human genes and systems are and how to recognize them! The knowledge is now almost wholly lacking, but it seems practically certain that it is obtainable.

* From George Gaylord Simpson, *The Meaning of Evolution,* Yale University Press, New Haven. Copyright 1949 by Yale University Press. Reprinted by permission.

One thing that is definitely known now is that breeding for uniformity of type and for elimination of variability in the human species would be ethically, socially, and genetically bad and would not promote desirable evolution. This variability, with accompanying flexibility and capacity for individualization, is in itself ethically good, socially valuable, and evolutionarily desirable. It happens that the present human breeding structure is excellent for the promotion of adaptability and desirable variability and for control of evolution by selection. The theoretically ideal conditions in this respect involve a large population with wide genetic variety (reflected also in local polymorphism), divided into many relatively small, habitually interbreeding groups which are not, however, completely isolated but also have some gene interchange between them. This ideal is actually rather closely approached in the present breeding structure of the human species. Its continuance as a basis for effective selection and maintenance of desirable variability demands avoidance of both of two extremes. On one hand, a completely classless society or habitual general intermingling in marriage of all racial or other groups would be bad, from this point of view. On the other hand, effective segregation and prohibition of interbreeding between any two or more racial, religious, or other groups would be even worse.

Selection can in the long run make the most of the genetic factors now existing in the human species. Eventually its possibilities would be exhausted if there were not also new mutations. Mutations are known to occur in mankind at rates comparable to those in other animals and consistent with sustained evolution at moderate speed. Almost all of those known are disadvantageous and produce abnormalities definitely undesirable in present society and probably of no value for any desirable future development. We probably completely miss in study of human heredity the very small favorable mutations that are likely to exist and to be more frequent than these larger and generally unfavorable mutations. The ability to recognize these small mutations will be a necessary factor if man is to advance his own biological evolution by voluntary selection. At present we do not have the slightest idea as to how to produce to order the sort of mutation that may be needed or desired. We do not even know whether this is physically possible. If it is, and if man does discover the secret, then indeed evolution will pass fully into his control.

Now we cannot predict for sure whether the future course of human evolution will be upward or downward. We have, however, established the fact that it *can* be upward and we have a glimpse, although very far from full understanding, as to how to ensure this. It is our responsibility and that of our descendants to ensure that the future of the species is progressive and not retrogressive.

New Techniques of Positive Eugenics

Human Evolution by Voluntary Choice of Germ Plasm *
by H. J. Muller (1961)

For some decades the term *eugenics* has been in such disrepute, as a result of its spurious use in support of the atrocities committed by those with class and race prejudices, that few responsible students of evolution or genetics have dared to contaminate themselves by mentioning it, much less by dealing with the subject except in condemnation. However, it is now high time to take new stock of the situation. For the odious perversions of the subject should not blind us longer to a set of hard truths, and of genuine ethical values concerning human evolution, that cannot be permanently ignored or denied without ultimate disaster. On the other hand, if these truths are duly recognized and given expression in suitable policies, they may open the way to an immeasurable extension and enhancement of the potentialities of human existence.

In view of the signal defeat in World War II of the leading exponents of racism—a defeat which is still gathering momentum—and the declining prestige afforded in the Western world to the claims of aristocratic or bourgeois class differentiations, it at last becomes feasible to return, in a more reasonable spirit, to the theme of prospective human biological evolution. Moreover, for this job of re-examination we are now provided not only with a better understanding of genetic and evolutionary principles but also with a considerably reformed structure in most Western societies, liberalized mores, a heightened freedom of discussion, and a marked improvement in technologies, all of which combine to make possible approaches that earlier would have seemed out of the question.

It was Darwin who pointed out that modern culture is causing a relaxation and perhaps even a reversal of selection for socially desirable traits, and he expressed himself rather pessimistically about the matter, although in his time this process must have been much less pronounced than it is

* From H. J. Muller, "Human Evolution by Voluntary Choice of Germ Plasm," *Science*, September 8, 1961, 134: 3480. Copyright 1961 by the American Association for the Advancement of Science. Reprinted by permission.

nowadays. His cousin Galton, impressed by Darwin's arguments concerning evolution in general as well as by those pertaining to man, but unwilling to accept defeat or frustration for humanity on this score, proposed the idea that the trend might be counteracted consciously. For this course of action he coined the term *eugenics,* included within which he understood all measures calculated to affect the hereditary constitution in a favorable way. As he pointed out, these measures might be of very diverse kinds, lying not only in such fields as medicine but also in education, economics, public policy in general, and social customs, although he did not contemplate drastic changes from the mores of that Victorian age.

Unfortunately, although Galton realized to some extent the influence of the social and familial environment in the shaping of people's psychological traits, he was not sufficiently aware of the profundity of the environmental control. He therefore made the naive mistake, so widespread in his day, of looking upon the performances of different ethnic, national and social groups as indicative of their genetic capabilities and inclinations, although there were plenty of object lessons of the comparatively rapid transference of cultures that should have taught him better. Later, it was the madness of such out-and-out racists and so-called "social Darwinists" as Madison Grant, Lothrop Stoddard, Eugen Fischer, Lenz,[1] and the Hitlerites which, carrying these prejudices much further, brought such odium upon the whole concept of eugenics as to run it into the ground.

Meanwhile, a large group of psychologists, represented by the Watson school, and of other social scientists, social reformers, socialists, and communists, all of them persons of egalitarian sympathies, impressed by the enormous potency of educational and other cultural influences, and regarding all eugenics as a dangerous kind of reaction that threatened their own roads to progress, popularized the idea that differences in human faculties are of negligible consequence not only as between different peoples and social classes but even as between individuals of the same group. They held that genetics in man could be allowed to take care of itself. And even where some genetic defects were admitted to exist, it was maintained that improved medical, psychological, and other cultural ministrations would provide sufficient remedies for them. Moreover, added the many Lamarckians among these groups, the improvements thereby acquired would eventually pass into the hereditary constitution. In this way, not only would all men become equalized but they would rise to ever higher biological as well as cultural levels. By about 1936 it had become a dire heresy among the official communists to dispute this line of argument, and the word *eugenics* had become a favorite symbol of all that is vile.

1. H. J. Muller, *Birth Control Rev. 17,* 19 (1933).

It is no wonder that earnest students of evolution and genetics, confronted by the mighty currents of these contending movements, which had advanced into the area of power politics, seeing how intertwined truth and error had become, and aware that their own views would almost certainly be misconstrued, tended to withdraw into their ivory towers and to refuse to discuss seriously the possible applications of genetics to man. It is to their credit, however, that only a few floated with the current that happened to be around them. But even fewer tried to contend with that current by raising the voice of reason, for that way lay the path to martyrdom.

Perhaps the last attempt made, up until the past few months, to present an appraisal of eugenics undistorted by extremist politics was the drafting of the "Geneticists' Manifesto" [2] of 1939, signed by about a score of participants at the International Genetics Congress held in Edinburgh just as the curtain began to rise on World War II. In this document it was pointed out that by far the greatest causes of differences between human groups in regard to psychological traits were environmental, predominantly cultural, whereas in the causation of such differences between individuals within the same group, both environmental and genetic factors were very powerful, and often comparable in their potency of action. The need for far-reaching reforms—for affording more nearly equal opportunities to all groups as well as to all individuals and for removing biases —was stressed in this document, not only for the sake of persons directly concerned but also to provide a groundwork for the truer assessment of genetic differences, in the interest of more soundly based eugenics.

These reforms in society were also needed, it was pointed out, for the attainment by the population of a sounder set of values, applicable equally to eugenic and to cultural purposes: values by which active service and creativity would be regarded more highly than either passive submissiveness or self-aggrandizement. The "Geneticists' Manifesto," far from discarding the concept of eugenics per se, acknowledged that it afforded, when rightly used, a means of making far-reaching human progress of a kind that must complement purely cultural advancement. Beyond that, genetic improvement was even affirmed to be a right which future generations would consider those of the past who were aware of the situation as having been obligated to accord them, just as they in turn would consider themselves as being similarly obligated to their own successors. This obligation would not be regarded as a burden, however, but rather as a high privilege and a challenge to their creativity. At the same time, it was recognized that adequate implementation of eugenic policies also required a clearing away of the ancient heritage of super-

2. *J. Heredity* 30, 371 (1940); *Nature* 144, 521 (1939).

stition and taboos that hitherto had so obstinately enshackled human usages and preconceptions in matters of sex and reproduction.

It is true that our own world of today is still grievously beset with the old inequalities, prejudices, and mummeries. However, all peoples have by now seen the handwriting on the wall that spells the end of these irrationalities. For modern technologies have, on the one hand, made it too dangerous for the world to remain divided. They have, on the other hand, provided the means for achieving an unparalleled interdiffusion of techniques, ideas, personnel, education, and socioeconomic organization, and for raising standards of living. In the process, provincialisms are at last being ground down, though not without much friction. Opportunities, educational, economic, and social, are being extended ever more effectively to the more depressed social classes and ethnic groups in our own country and elsewhere. A real effort is being made to bring the viewpoint of science home to the general population. The battles that superstition is still winning take place ever closer and closer to its heartland, as was so well depicted in the moving picture on the Scopes trial, *Inherit the Wind.* And it is even becoming permissible to debate seriously matters that in the days before the nuclear weapons, space ships, Kinsey, and the Darwin centenary were taboo among all nice people.

Thus the scene has at last been shifted to such an extent as to make it fitting to re-examine even such a scandalous subject as eugenics, with a view to preparing the new forces now arising in the world to deal with it both realistically and humanistically. For, as we shall see, cultural progress of the kinds mentioned has already proceeded far enough here and there to make the beginnings of a new approach to the subject possible, and the attitudes now formed and the preparations now made may presently lead, when the time is riper, to more salutary developments in this field than could ever before have occurred.

CONTRADICTIONS IN THE TRADITIONAL EUGENIC METHODS

Let us first examine the methods by which it has hitherto been thought that eugenics might operate. These methods have taken their cue from the natural selection of the past. All evolution has had its direction determined in some way by the force of selection. Selection chooses among the materials available to it, namely, diverse mutations, which occur in a manner that is fortuitous so far as their adaptation to the needs of the organism in the given situation is concerned. Selection acts entirely through differential multiplication, but this process can be conceptually divided into two parts, namely, unequal rates of survival (or, conversely stated, of mortality), on the one hand, and unequal rates of reproduction of the survivors, that is, differential fertility (or, conversely stated, differential infertility), on the other hand.

Eugenists have therefore distinguished between two conceivable methods—differential control over mortality, and differential control over reproductive rate. The first method, however, although practiced by the Spartans and by primitive tribes who destroyed infants regarded as undesirable, is universally acknowledged to be inconsistent with the respect for human beings that forms an essential part of civilization. It might be contended that artificial abortion is an intermediate method, but everyone recognizes this also to be an undesirable means where other procedures are available. Essentially, then, this has left for eugenics the second alternative: that of a qualitatively differential control over reproduction prior to or at conception.

In Galton's time, before the advent of modern contraceptive techniques, it was indeed rather visionary to conceive of people's reproduction being governed in the interests of the progeny. For this could be done only by the drastic method of surgical sterilization, of a type that interfered with the sexual life, or by such consummate self-control as voluntary abstention from intercourse, or from its completion.

The development of techniques for cutting or ligating the tubes that conduct the mature reproductive cells afforded less objectionable means of sterilization, but this procedure was still usually regarded by most people—rightly or wrongly—as too irrevocable, except, perhaps, for persons who were mentally or morally hopelessly irresponsible. For them, enforced operations of this type were legalized in some regions, although it was rightly pointed out that there was grave danger of abuse of the practice unless it were confined to the most extreme cases. For attitudes that seem wrong in one place or setting may seem right elsewhere, and nonconformists may at times have moral standards superior, in a longer perspective, to those of the majority who condemn them. Thus, the amount of sterilization resulting from legal applications of an advisable kind would be so minute as to have very little eugenic influence.

However, the invention of fairly practicable artificial means of voluntary contraception opened up much wider possibilities for the control of reproduction in economically developed countries. As we all know, advantage has been taken of these techniques on a large scale, and they have become one of the indispensable procedures whereby the general standard of living has been so greatly raised. Still more practicable means of contraception seem at last to be on the way, thanks to the efforts of a handful of devoted scientists, and they cannot come too soon, for it is imperative to make similar benefits possible in the less developed regions.

But although contraception that is used for the enhancement of cultural benefits through the control of population quantity is at the same time a *potential* instrument for the improvement of genetic quality, such improvement does not occur unless the contraception is specifically aimed in this direction: To be sure, this purpose might be achieved if the individual couples concerned were to reach their decisions about how many children to have in a highly idealistic spirit, one guided by almost heroic self-criticism and wisdom. We shall presently consider whether or not it is realistic to expect this. A second proposal has been that of altering the economic and social system in such a way that people of higher gifts and greater natural warmth of fellow feeling—that is, the genetically more highly endowed—would be normally led into occupations and modes of life more conducive to having a large family. Conversely, the organization of society which this view would hold to be ideal would tend to lead persons less well endowed to choose, of their own accord, situations in life that would encourage them to expend their energies in other pursuits than reproduction, and would give them less inducement for raising families.

These two approaches, the individual and the societal, are, of course, not mutually exclusive, and most 20th-century eugenists have advocated a combination of the two. But let us examine each of them more closely. First, as regards the individual approach, which is supposedly to be adopted by people in general once they have been well educated in matters of evolution and genetics, it should be acknowledged that people in general can in fact be taught to take pride in making great sacrifices for what they recognize to be a great cause, especially when they win social approval thereby. This has often happened in times of war as well as after social revolution. However, it seems asking almost too much to expect those individuals who are *really* less well equipped than the average, in mentality or disposition, to acknowledge to themselves that they are genetically inferior to their neighbors in these respects, and then to publicly admit this low appraisal of themselves by raising no family at all or a smaller one than normal, especially since at the same time they would often be thwarting a natural urge to achieve the deep fulfillments, accorded to their neighbors, that go with having little ones to care for and bring up. Moreover, those with physical impairments would likewise tend to rationalize the situation, by thinking that they possessed some superior psychological qualities that more than compensated for their physical defects.

In fact, then, the ones most likely to comply with the idea of restraining their own reproduction would be those who had such strong social feelings, such a sense of duty, so high a standard of what is good, so little egotism, and such an urge for objectivity, as actually to lean over backward and so underrate themselves. Thus we would be likely to lose for the next generation much of what might have been its best material.

On the other hand, for many of the really gifted there are often unusual opportunities for achievement, for rich experiences, and for service along other lines than those of bringing up a large family. Hence it would be only human of them, even though they were in sympathy with eugenic ideals, to expend a larger share of their energies in these other ways than does the average man or woman, for whom the home is often both a refuge and the chief stage on which to express leadership. In view of these considerations, it is not at all surprising that eugenic practices of this intentional, personal type, that require a correlation between the size of one's family and one's realistically made appraisal of one's genetic endowments, have made so little headway, even where they were approved theoretically. Thus, even among eugenists themselves, one seldom finds much evidence that these principles are being acted upon.

What, then, about the proposal that our society should introduce features into its structure whereby the more gifted, the abler, and the more socially minded would find conditions more conducive to their raising a large family, while those less capable or relatively antisocial would tend automatically to be deflected from family life? Surely we would not want a dictatorship to institute such a system, for dictators are oftener wrong than right in their decisions. Moreover, their subjects are not able to become truly men, in the all-round sense, and those who shine under such circumstances are not likely to be the wise and the responsible.

Under a democracy, on the other hand, is it not likely that "the common man" will refuse to subject himself to such manipulations? Certainly if the proposal took some such crude form as a subsidy for the raising of children, allotted to those who already occupied better positions or who had scored higher on certain tests, it would rightly be resented and defeated as discriminatory by the great majority. And even if subtler forms of influence were used, such as special aids to family life for those in occupations requiring greater skills, responsibility, or sacrifice, there would soon be a clamor on all sides to have these advantages extended to every responsible citizen. No, we can hardly use democracy to support any kind of aristocracy. To be sure, ways might eventually be found to reduce the present strong negative correlation between educational or social achievement and size of family, and these would be all to the good. But no major formula is in sight for restoring the greater family size of the fitter, while retaining that most essential feature of our culture—the extension to all of mutual aid based on the most advanced technologies available.

It might seem to follow that we have now, as a result of our improved techniques for living, reached an inescapable genetic cul-de-sac. It might be concluded that we should therefore confine ourselves entirely to the immediate job on hand—the pressing and rewarding one of all social reformers and educators—that of making the best of human nature as it is, the while allowing it to slide genetically downhill, at an almost imperceptible pace in terms of our mortal time scale, hoping trustfully for some miracle in the future.

THE NEW APPROACH: GERM-CELL CHOICE

However, it is man who has made the greatest miracles of any species, and he has overcome difficulties arising from his technologies by means of still better basic science, issuing in still better technologies. And so in the case of the genetic cul-de-sac of the present day, he has even now possessed himself of the means of breaking through it. For he is no longer limited, like species of the past which had the family system, to the two original methods of genetic selection applying to them: that of differential death rate on the one hand, and differential birth rate or family size, on the other hand. He has now given himself, in addition, the possibility of exerting conscious selection by making his own choice of the source of the germ cells from which the children of his family are to be derived.[3] At present, this choice is confined to the male germ cells, but there are indications that with a comparatively small amount of research it might in some degree be extended to those of the female as well.[4]

It is pretty common knowledge nowadays that some tens of thousands of babies have already been born, in the United States alone, that were derived by "AID," that is, by artificial insemination with sperm obtained by the physician from a donor chosen by him, but whose identity was kept unknown to all others, including even the parents. In the great majority of these cases the husband had been found to be irremediably sterile. And although, in view of the prevalence of the traditional mores, the whole matter was kept secret from the children themselves, both members of the couple had in these cases been eager to avail themselves of this opportunity to have one or more children. This method has, of course, been allowed only to those likely to make good parents—or, shall we say, good "love parents," as distinguished from "gene parents"? Moreover, follow-up studies have shown that these parents did truly love their

3. H. J. Muller, *Out of the Night* (Vanguard, New York, 1935); also, H. Brewer, *Eugenics Rev.* 27, 121 (1935); *Lancet* 1, 265 (1939); and H. J. Muller, *Perspectives in Biol. and Med.* 3, 1 (1959); ———, in *Evolution after Darwin*, Sol Tax, Ed. (University of Chicago Press, Chicago, 1960), vol. 2, p. 423.

4. Research indicates that it will be possible to deep-freeze human ova which can at some later time be engrafted into "foster" mothers. Ed.

children, as the children did their parents. It is noteworthy that this proved to be as much the case for the father as for the mother, and that the marriage was strengthened thereby.

Here, then, we see repeated what is typically found in those cases of early adoption in which the children have been genuinely desired. However, this "pre-adoption" (as Julian Huxley has termed it) is likely to prove even more binding and satisfying than "post-adoption." And the method of pre-adoptional choice, despite its relative crudity at the present time, has demonstrated its capability of producing a superior lot of children. Thus, the couples that practice it have made a virtue of necessity by inducing the genesis of children of whom they can usually be even prouder than of the children they would have had if they had been free from reproductive infirmity.

Recently, as more people have become alive to matters of genetics, an increasing number of couples have resorted to AID when the husband, although not sterile, was afflicted with some probably genetic impairment, or likely to carry such an impairment that had been found in his immediate family. Similarly, those with incompatibilities in blood antigens have also made use of the method with good results. In these ways a beginning has already been made in the conscious selection of germinal material for the benefit of the progeny. It is to be expected that many more people will seek such advantages for their prospective children when there are available for creating them the germ cells of persons who are decidedly superior in endowment to the normally healthy and capable persons that are commonly sought as donors by the physicians of today. For there is no physical, legal, or moral reason why the sources of the germ cells used should not represent the germinal capital of the most truly outstanding and eminently worthy personalities known, those who have demonstrated exceptional endowments of the very types most highly regarded by the couple concerned, and whose relatives also have tended to show these traits to a higher-than-average degree. How happy and proud many couples would be to have in their own family, to love and bring up as their own, children with such built-in promise.

Today, of course, most people have such an egotistical, individualistic feeling of special proprietorship and prerogative attaching to the thought of their own genetic material as to be offended at the suggestion that they might engage in such a procedure. No one proposes that they do so as long as they feel this way about the matter. However, they should not try to prevent others who would welcome such an opportunity from ordering their lives in accord with their own ideals. And as the prejudice against the practice gradually dwindles, the manifest value of the results for those who had participated in it would appeal to an increasingly large portion of the population.

In this connection it is important to bear in mind that there is no such thing as a paternal instinct in the sense of an inherent pride in one's own genetic material or stirps. Some primitive peoples, even including a few still in existence in widely separate regions, have had, strange as it may seem to us, no knowledge that the male plays a role in the production of the child, much less have they had any conception of genetic material or genes. Thus, among some of them the mother's brother has effectively filled the role of father in regularly caring for mother and children. Moreover, among some peoples, such as the Hawaiians, children are rather freely adopted at an early age into other families, into whose bosom they are warmly, fully, and unambiguously accepted as equals in every way to the natural children. It is "second nature," but not "first nature," for us in our society to exalt our own stirps.

It is, however, "first nature" for men and women to be fond of children and to want to care for them, and more especially, those children with whom they have become closely associated and who are dependent on them. If the love of a man for his dog, and vice versa, can go to such happy lengths as it often does, how much stronger does the bond normally become between the older and younger generations of human beings who live together. And since, in the past, the children *have* usually been those of the parents' own stirps, it has been a natural mistake to suppose that these stirps, rather than the human associations of daily life, formed the chief basis of the psychological bonds that existed between parent and child. Yet, as our illustrations have shown, this view is incorrect, and a family life of deep fulfillment can just as well develop where it is realized that the genetic connection lies only in our common humanity.

The wider adoption of the method of having children of chosen genetic material rather than of the genetic material fate has chanced to confer on the parents themselves implies, of course, that material from outstanding sources become available by having it stored in suitable banks.[5] It would be preferable to have it in that glycerinized, deep-frozen condition developed by modern technology, in which it remains unchanged for an unlimited period without deterioration. It is true that research is badly needed for finding methods by which immature germinal tissue can, after the deep-freezing which it is known to survive, be restored to a state where it will multiply in vitro and subsequently produce an unlimited number of

5. H. Hoagland, *Sci. Monthly* 56, 56 (1943); also C. W. Kline, unpublished address before the Society for Scientific Study of Sex, New York (1960), and H. J. Muller, *Out of the Night*.

mature reproductive cells. Even without this further development, however, the way is already open, so far as purely biological considerations are concerned, for gathering and inexpensively storing copious reserves of that most precious of all treasures: the germinal material that has formed the biological basis of those human values that we hold in highest regard.

The high potential service to humanity represented by the pre-adoption of children should carry with it the privilege, for the parents, of having a major voice in choosing from what source their adopted material is to be derived. Surely, if they have ever had the right to produce, willy-nilly, the children that would fall to their lot as a result of natural circumstances, they should have the right of choice where they elect to depart from that haphazard method. They certainly would not wish knowingly to propagate manifest defectives, and, being idealistic enough to undertake this service at all, they would in most cases be glad to give serious consideration to the best available assessments of the genetic probabilities involved, as well as be open to advice regarding relative values and needs.

It would be made clear to parents that there is always an enormous amount of uncertainty concerning the outcome in the genetics of an organism so crossbreeding as man, especially since the most important traits of man are so greatly influenced by his cultural environment. Nevertheless, facing this, they would realize that the degree of promise was in any such case far greater than for those who followed the traditional course. It would be in full awareness of this situation that they would exercise their privilege of casting the *loaded* dice of their own choosing.

This kind of choice means that the physician can no longer be the sole arbiter of destiny in this matter. Clearly, if the couple are to accept their share of the responsibility and privilege here involved, the practice of keeping the donor unknown to them must be relinquished in these cases. Knowledge of the child's genetic lineage will also be needed later, so that sounder judgments may be reached concerning his genetic potentialities in the production of the generation to follow his own. This lifting of the veil of secrecy will become ever more practicable, and in fact even necessary, as the having of children by chosen genetic material becomes more widely accepted and therefore more frequent. Moreover, the attitude of others toward the couples who have employed this means of having their children will gradually become one of increasing acceptance and then of approbation and even honor.

Today's fear that knowledge by the mother of the identity of the gene father may lead to personal involvement between the two, to the detriment of normal family life, will recede when the gene source is remote in space or time, as when the germinal material has been kept in the deep-frozen state for decades. This procedure will also allow both the individual worth of those being considered as donors, and their latent genetic potentials, to be viewed in better perspective, and will reduce the danger that choices will be based on hasty judgments, swayed by the fads and fashions of the moment.

Let us see in what ways this method of reproduction from chosen material tends to avoid the difficulties that are encountered in attempting to reconcile traditional reproduction with the interests of genetic quality by somehow controlling the size of families. For one thing, as previously pointed out, most men would resist accepting and acting on the conclusion that they are below their next-door neighbor, or below average, in genetic quality. They would particularly resist the idea that they themselves are in that lowest fifth which would be required to refrain from having children if an equilibrium of genetic quality were to be maintained in the face of a 20-percent mutation rate. Yet most of these same people would willingly accept without resentment the idea that they are not among the truly exceptional who conform most closely to their own ideals. And so, when encouraged by the community mores, they would be glad and proud to have at least one of their children derived, by choices of their own, from among such sources. Thus they would continue to have families of a size more nearly conforming with their inclinations.

On the other hand, the worthy but humble, those who might otherwise, from overconscientiousness, limit their families unduly, would often be eager to serve as love parents. And although in that capacity they would tend to derive the germ cells from outside sources, they would be especially likely to have a well developed sense of values and so to choose sources even worthier than themselves. Finally, the really highly endowed but realistic would not be confronted with the sore dilemma of choosing between exercising their special gifts, on the one hand, or having the large family to which genetic duty seemed to obligate them, on the other hand. For their germinal material would tend to be sought by others, if not in their own generation, then later, and to a degree more or less in proportion to their achievements. Thus they would be freed to give their best services in whatever directions they elected.

In all these ways, the diverse obstacles encountered by a eugenics that tried to function by means of a consciously differential birth rate—that is, by adjustment of family size—would be avoided. Thereby, a salutary separation would be effected between three functions that often have conflicting needs today. These are, first, the choice of a conjugal partner; this should be determined primarily by sexual love, companionability, and compatible mentality and interests. Second, there is the determination of the size of the family; this should depend largely on the degree of parental love that the partners have, and on how successfully they can express it. Third, there is the promotion of genetic quality, both in general and in given particulars; these qualities are often very little connected with the first two kinds of specifications. By thus freeing these three major functions from each other, all of them can be far better fulfilled. Under these circumstances the conjugal partners need not be chosen by criteria in which a compromise is sought between the natural feelings and considerations of eugenics. Neither need the family size be restricted or expanded according to eugenic forebodings or feelings of duty. Yet at the same time there can be a far more effective differential multiplication of worthy genetic material than in any other humanly feasible way.

FURTHER PROSPECTS

It is likely that the avoidance of the effects of sterility will not be the only door through which such a change of mores will be approached. Facilities for keeping germ cells, suitably stored below ground in a deep-frozen condition, in areas relatively free from radiation and chemical mutagens, may well be provided in our generation for an increasing number of people.[6] Among these would be persons subject to the growing radiation hazards of industry, commerce, war, and space flight, and those exposed to the as yet unassessed hazards of the chemical mutagens of modern life. The same means would greatly retard the accumulation of spontaneous mutations which probably occurs during ordinary aging. Thus, wives may in time demand such facilities for the storage of their husbands' sperm. These facilities would be provided not only for the sake of reducing mutational damage but also as a kind of insurance in the event of the husband's death or sterility. In these ways great banks of germinal material would eventually become available. They would be increasingly used not only as originally intended but also for purposes of conscious choice. Moreover, some of these stocks might become recognized as especially worthy only after those who had supplied them had passed away.

The cost of storage is, relatively, so small that failure to make such a provision will eventually be considered gross negligence. As Calvin Kline[7] points out, this will be especially the case where (as in India today) voluntary vasectomy becomes more prevalent as the surest and, in

6. H. J. Muller, in *The Great Issues of Conscience in Modern Medicine* (Dartmouth Medical School, Hanover, N.H., 1960), p. 16.

7. C. W. Kline, unpublished address before the Society for Scientific Study of Sex, New York (1960).

the end, the cheapest means of birth control. For when vasectomy is complemented by stores of sperm kept in vitro, the process of procreation thereby achieves its highest degree of control—control not subject to the impulses of the moment but only to more considered decisions.

It is true that most people's values, in any existing society, are not yet well enough developed for them to be trusted to make wise decisions of the kind needed for raising themselves by their bootstraps, as it were.[8] But this type of genetic therapy, of "eutelegenesis," as Brewer termed it when he advocated it in 1935,[9] is certainly not going to spring into existence full fledged overnight. It will first be taken up by tiny groups of the most idealistic, humanistic, and at the same time realistic persons, who will tend to have especially well developed values. This mode of origination of such practices was pointed out by Weinstein in "Palamedes" in 1932.[10] These groups will tend to emphasize the most basic values that are distinctive of man, those that have raised him so far already, but which still may be enormously enhanced. Foremost among these are depth and scope of intelligence, curiosity, genuineness and warmth of fellow feeling, the feeling of oneness with others, joy in life and in achievement, keenness of appreciation, facility in expression, and creativity.

Those who follow these lodestars will blaze the trail, and others will follow and widen this trail as the results achieved provide the test of the correctness with which its direction was chosen. Meanwhile, the world in general, through its reorganizations of society and education, is moving in the same direction, by a de-emphasis of its provincialisms and a consequent recognition of the supreme worth of these basic human values. For after all, these values have been prominent in the major ethical systems of the whole world.

At the same time, plenty of diversity will inevitably be developed. For each especially interested group will naturally seek to enhance its particular proclivities, and this is all to the good. But on the whole, the major gifts of man have been found to be not antagonistic but correlated. Thus we may look forward to their eventual union with one another in a higher synthesis. And from each such synthesis in turn, divergent branches will always be budding out, to merge once more on ever higher levels.

It may be objected that we have next to no knowledge of the genes for those traits we value most, and that their effects are inextricably interwoven with those of environment. As was acknowledged earlier, this is quite true. However, it has also been true in all the natural selection of

8. H. J. Muller, *Sci. Monthly* 37, 40 (1933).
9. H. Brewer, *Eugenics Rev.* 27, 121 (1935); *Lancet* 1, 265 (1939).
10. A. Weinstein, *Am. Naturalist* 67, 222 (1933).

the past and in the great bulk of artificial selection. Yet these empirical procedures, based entirely on the accomplishment of the individuals concerned, did work amazingly well. We can do a good deal better by also taking advantage of the evidence from relatives and progeny. Yet that evidence also is furnished mainly by accomplishments or output. Where those were high, the environment was, to be sure, usually favorable, but so was the heredity. And as, in our human culture, social reform proceeds and opportunities become better distributed, our genetic judgments will become ever less obscured by environmental biases, while at the same time our knowledge of genes will improve.

Meanwhile, the efforts of educators and the lessons of world affairs will serve to emphasize the same values for us. And these attitudes we will take over for our genetic judgments also. In this connection, another consideration deserves mention here. The preference which most parents will inevitably have for the genes of persons of truly remarkable achievement and character, rather than for those of the merely eminent or powerful, will at the same time serve to direct the stream of genetic progress toward the factors underlying creativity, initiative, originality, and independence of thought, on the one hand, and toward genuineness of human relations and affections, on the other hand. Otherwise the genetic movement might, as so often happens in other affairs of men, become directed toward skill at conformity, showmanship, and the dignified hypocrisy that often brings mundane success and high position.[11] This would have been a far greater danger in the case of the old-style eugenics.

But, it may be objected, does all this really represent conscious control in an over-all sense? Is it not merely a type of floating along in a chaotic manner, each straw making its own little movement independently of the rest, without a general plan or goal or stream? The answer is that humanity is as yet too limited in knowledge and imagination, too underdeveloped in values, to see more than about one step ahead at a time. That step, however, can be discerned clearly enough, and by enough people, to give rise to a general trend in a salutary direction. And at the higher level to which each step taken will bring us we will be able to see an increasing measure of advance ahead. So we humans will achieve, not through dictation but through better general understanding and ever more clearly seen values, increasing mutual consent both concerning the means to be used and the aims toward which to orient. Thus an ever wider over-all view will emerge, and a surer, greater over-all plan, or rather, series of plans. To create them and to put them into effect will then enlist our willing efforts. And the very enjoyment of their fruits will bring us further for-

11. See Weinstein and Brewer, *op. cit.*

ward in our great common endeavor: that of consciously controlling
human evolution in the deeper interests of man himself.[12]

Genetic Control of the Human Stock *
by René Dubos (1962)

Hereditary changes in man have resulted so far from undirected processes, blind selective forces tending to favor the survival of the persons
best fitted to a given environment. A few geneticists believe, however,
that the time has now come to control scientifically the evolutionary
changes, in order to prevent the degenerative effects of civilization on the
human stock. Whereas mutation rates are increasing, our ways of life are
interfering with the elimination of undesirable genes; natural selection is
more and more embarrassed by social and medical practices. True
enough, the genetic processes of degeneration are so slow, that one is
tempted to shrug off their effects, because most of them take hundreds of
years to become evident. As in the case of erosion they come to be accepted as a part of the natural order. Social scientists and men of affairs
are no more interested in such remote problems than is the general public. The fact is, however, that the most destructive, as well as the most
creative operations of the living world, have been of a creeping, secular
character. In the long run they are probably the most influential determinants of the future of the human race.

In view of these facts, it would seem logical to take advantage of modern biological knowledge and attempt some measure of control over the
hereditary endowment of man. The celebrated American geneticist Hermann J. Muller has recently made concrete proposals to this effect. What
Professor Muller suggests is simply the widespread practice of artificial
insemination, using for this purpose preserved sperm obtained from human males whose life record is well known. In his view, this policy of

---

* From René Dubos, *The Torch of Life,* the Credo series, edited by Ruth Nanda
Anshen, Simon & Schuster, Inc. Copyright © 1962 by Pocket Books, Inc.; Also *Man
Adapting,* Yale University Press, New Haven. Copyright © 1965 by Yale University
Press. Reprinted by permission.

12. The Muller article was originally prepared as a lecture to be given on the in-
vitational program of the Society for the Study of Evolution at its annual meeting in
New York, 29 Dec. 1960, but adventitious circumstances prevented its delivery on
that occasion.

positive eugenics would facilitate the spread through the general population of a number of desirable genetic characters, including physical and intellectual endowments.

Professor Muller's program of positive eugenics has stirred up passionate controversies, and there is as yet no indication that it will be applied on any significant scale. Objections to it come from two very different angles. First is the fact that genetic science is not yet sufficiently developed to permit prediction of what will happen if steps are taken to control human evolution. Many geneticists take, in this regard, a position almost completely opposite to that of Professor Muller; they claim that far from trying to channel genetic make-up in certain directions, we should maintain as much genetic diversity as possible. It is important, according to Professor P. B. Medawar,[1] to "maintain a population versatile enough to cope with hazards that change from time to time and from place to place. A case can be made for saying that a genetical system that attaches great weight to genetic diversity is part of our heritage, and part of the heritage of most other free living and outbreeding organisms." Instead of trying to tailor man to a future which is at best dimly visualized, it is probably wiser at this stage to develop a culture fitted to man's needs.

The other fundamental objection to positive eugenics is that no one really knows what characters to breed in order to improve the human stock. Everyone agrees, of course, that it would be desirable to eliminate certain obviously objectionable traits, such as gross physical and mental defects—though even this limited program poses problems of judgment and of execution far more complex than is usually realized. But the selection of positive qualities raises questions of a much more subtle nature.

It is relatively easy to formulate a genetic program aimed at producing larger pigs, faster horses, better hunting dogs, or more friendly cats. But what is the ideal for human societies? The cave man, for all his strength and resourcefulness, would not get along well in a modern city. On the other hand, our present way of life may soon be antiquated, and future life may demand a kind of endurance undreamt of at the present time. For all we know, resistance to radiation, to noise, to intense light, to the monotony of continuous stimuli, and to the eternal repetition of boring activities, may become essential for biological success in future civilizations. Who knows, furthermore, whether mankind is better served by the gentleness of Saint Francis of Assisi and Fra Angelico or by the dynamism of the Industrial Revolution and modern art? Is the higher type of society one which prizes, above all, individual and self-development, or one which regards devotion to the common welfare as the highest standard of morality? Is it at all desirable to reproduce on a large scale a trait the ap-

1. See P. B. Medawar selection, Part 8, pp. 474–88. Ed.

peal of which is perhaps its uniqueness? How many Beethovens would it take to make his genius commonplace? Furthermore, would a human being endowed with Beethoven's genes have the kind of musical genius that would enable him to express the moods of the automation age?

The fundamental difficulty in formulating a program for the genetic improvement of man is that we do not know what we are nor where we are, what we want to become nor where we want to go. In fact, mankind has not yet developed the skill to think effectively about these problems. To a large extent we still hold to a static and mechanical view of the universe and of life. The discussions about evolutionary development rarely encompass a continuously emergent novelty, an open future, not predictable from what is known. Yet this is probably the most profound meaning of the evolutionary concept—a view of life as a continuous act of creation, in which man has become the most important actor.

✻   ✻   ✻

The practice of a eugenic policy would thus require great wisdom; but there is little indication that such wisdom is widespread today, or that we know how to recognize it where it exists. The likelihood is that if people really had the chance to choose the fathers of their children, they would be likely to choose more pronounced projections of their self-images. The disturbing diversity of opinion as to desirable human goals that came to light among a group of illustrious scientists, recently gathered to discuss the "future of man," made it clear that even the most learned and sophisticated human beings do not have the wisdom required to formulate long-term eugenic objectives.[2]

Many biologists believe that results even better than those which might follow selection from existing genotypes could be accomplished more rapidly by direct mutagenic operations on the genetic material. Others advocate that man could be rapidly improved by modifying the course of development and physiological processes. Still others are in favor of creating artificial contrivances capable of replacing or supplementing normal bodily structures.

The following quotations illustrate statements recently made with regard to these different possibilities.

Perhaps, following the current use of plastic heart valves and arterial walls, it will be possible to construct plastic prostheses of heart or kidney which would be accepted by the human body and remain functional in it.[3] It would be incredible if we did not soon have the basis of developmental engineering

2. Wolstenholme, G. [ed.]. 1963. *Man and his Future*. Boston, Little, Brown.
3. Kroprowski, H. 1963. Future of infectious and malignant diseases, pp. 196–216. *In* G. Wolstenholme [ed.], *Man and his Future*.

technique to regulate, for example, the size of the human brain by prenatal or early postnatal intervention. In fact, it is astonishing how little experimental work has been done to test some elementary questions on the hormonal regulation of brain size in laboratory animals or the functional interconnexion of supernumerary brains. Needless to say, "brain size" and "intelligence" should be read as euphemisms for whatever each of us projects as the ideal of human personality. . . . Only preliminary suggestions are possible, but even imperfect ones may help to illuminate the possibilities:

(1) Accelerated engineering development of artificial organs, e.g. hearts, which may relieve intolerable economic pressures on transplant sources.

(2) Development of industrial methodology for synthesis of specific proteins: hormones, enzymes, antigens, structural proteins. For example, large amounts of tissue antigens would furnish the most likely present answer to the homotransplantation problem and its possible extension to heterotransplantation from other species. Structural proteins may also play an important role in prosthetic organs.

(3) A vigorous eugenic programme, not on man, but on some non-human species, to produce genetically homogeneous material as sources for spare parts.[4]

Thus medical science is now contemplating the supplementation and modification of man by biotechnological procedures and mechanical contrivances that alter his very personality. The requirements for space travel may encourage even more drastic changes. Under these conditions, it has been suggested:

The human legs and much of the pelvis are not wanted. Men who had lost their legs by accident or mutation would be specially qualified as astronauts. If a drug is discovered with an action like that of thalidomide, but on the leg rudiments only, not the arms, it may be useful to prepare the crew of the first spaceship to the Alpha Centauri system, thus reducing not only their weight, but their food and oxygen requirements. A regressive mutation to the condition of our ancestors in the mid-pliocene, with prehensile feet, no appreciable heels and an ape-like pelvis, would be still better. There is no immediate prospect of men encountering high gravitational fields, as they will when they reach the solid or liquid surface of Jupiter. Presumably they should be short-legged or quadrupedal.[5]

Ultimately the question will arise as to the identification of the person modified by biotechnology. What is the moral, legal, or psychiatric identity of a human being so modified by medical manipulation that he has almost become an artificial chimera? Poets have long been aware of the loss of human quality symbolized by the "hollow man." Yet at a recent

4. Lederberg, J. 1963. Biological future of man, p. 263–74. *In* G. Wolstenholme [ed.], *Man and his Future.*

5. Haldane, J. B. S. 1963. Biological possibilities for the human species in the next ten thousand years, p. 337–83. *In* G. Wolstenholme [ed.], *Man and his Future.*

meeting a physician concerned with space medicine was willing to advocate the creation of an "optiman," in whom essential organs had been replaced by different mechanical contrivances more efficient for operation within the environment of a space ship than those provided by nature. The present interest in the creation of an "optiman" may have encouraged the use of "instant men," consisting of large bodies of plastic having the same sound-absorptive powers as a human audience, to test the acoustic qualities of the new Philharmonic Hall in Lincoln Center! The poverty of the results shows that musically as well as biologically "instant men" and "optimen" are poor substitutes for real men of flesh and bone, despite all their imperfections.

The intensity of efforts devoted to problems of human development makes it likely that techniques will soon become available to affect the human brain by prenatal or early postnatal intervention. Even at present, mental states can be influenced by many different techniques, from yoga to hypnosis and drugs. In man as well as in animals, electric stimulation of a particular area in the brain can produce a sense of well-being in the whole organism. Similar effects can be produced by drugs such as mescalin, lysergic acid, and psilocybin. This kind of knowledge is immensely exciting because it enlarges the understanding of the human mind, but by the same token, it is also frightening because, almost inevitably, knowledge is used for control. Electrodes have already been placed in the pleasure centers of the brain. Presumably they were the brains of hopeless psychotics, but it takes little imagination to realize how the power to manipulate human behavior could become an instrument of tyranny.

The moral aspect of the biomedical techniques that alter man's nature must be emphasized because it will often be difficult to recognize the real motivation of those who use them. At the least, a sharp distinction must be made between changing people's ideas by dialogue and by manipulation. The manipulator regards the other member of the interplay almost as an object; whereas in dialogue, there is a reciprocity of interaction and therefore a greater chance that human freedom and rights are respected.

Fifty years ago, a physician, like an engineer, could approach his tasks with the confidence that he was acting as a benefactor of humanity. In contrast, medical technology is now so powerful and often so indiscriminate that it can damage human personality even as it improves the functions of the body. The most cruel dilemma of modern medicine is to decide which aspects of man's nature can be ethically tampered with and which ones should be respected at all cost. There is no guide to resolve this dilemma, beyond Montaigne's admonition: "Science without conscience is but death of the soul."

Editor's Note. Rapid advances in biochemistry and genetics in recent years have introduced the possibility of controlling human heredity at the molecular level. Geneticists have been able to transplant DNA extracted from certain bacterial cells into the cells of other strains of bacteria. They hope in the near future to be able to extract, or even to synthesize, parts of the DNA chain molecules that act as desirable genes and to introduce these good genes into the chromosomes of human cells, replacing detrimental genes. These techniques of transforming one living cell into another are called *genetic alchemy* or *algeny.* Biologists are also exploring the possibility of causing undifferentiated living cells to divide and reproduce themselves in a vegetative manner without fertilization. Some of the revolutionary results that might be achieved by these techniques are described in the following selection.

Experimental Genetics and Human Evolution *
by Joshua Lederberg (1966)

Although the eugenic controversy started in the infancy of genetic science, the recent integration of experimental genetics with biochemistry has provoked a new line of speculation about more powerful techniques than the gradual shift of gene frequencies by selective breeding for the modification of man. . . .

VEGETATIVE PROPAGATION

Many plants spread almost entirely by asexual growth and reproduction—hence the very phrase "vegetative propagation." A colony of organisms derived from a single ancestor by vegetative propagation—without a sexual union or genic recombination as a step in the process—is called a "clone," from a Greek root, "cutting" which is related to "colony." (In a certain sense, the single organism is one clone derived from the original fertilized egg, but we refer here to communities of whole organisms.) That clonal reproduction is mainly confined to plants may be a mere accident of cell biology; nevertheless, a phylum that fell into this trap might be greatly impeded in its evolutionary experimentation toward creative innovation.

Vegetative—or clonal—reproduction has a certain interest as an investigative tool in human biology, and as an indispensable basis for any sys-

* From Joshua Lederberg, "Experimental Genetics and Human Evolution," *The American Naturalist.* Reprinted by permission of the author.

tematic algenics, but unlike other techniques of biological engineering there might be little delay between demonstration and use. Clonality outweighs algeny at a much earlier stage of scientific sophistication, primarily because it answers the technical specifications of the eugenicists in a way that Mendelian breeding does not. If a superior individual—and presumably, then, genotype—is identified, why not copy it directly, rather than suffer all the risks, including those of sex determination, involved in the disruptions of recombination. The same solace is accorded the carrier of genetic disease: why not be sure of an exact copy of yourself, another healthy carrier, rather than risk an overtly diseased offspring; at worst, copy your spouse and allow some degree of biological parenthood. . . .

Clonality as a way of life in the plant world is well understood as an evolutionary cul-de-sac, often associated with hybrid luxuriance. It can be an unexcelled means of multiplying a rigidly well-adapted genotype to fill a stationary niche. So long as the environment remains static, the members of the clone might congratulate themselves that they had outwitted the genetic load; and they have indeed won a short-term advantage. In the human context, it is at least debatable whether sufficient latent variability to allow for any future contingency would be preserved if the population were distributed among some millions of clones. From a strictly biological standpoint, tempered clonality could allow the best of both worlds: we would at least enjoy being able to observe the experiment of discovering whether a second Einstein would outdo the first one. How to temper the process and the accompanying social frictions is another problem.

CLONAL POSSIBILITIES

The internal properties of the clone open up new possibilities, for example, the free exchange of organ transplants with no concern for graft rejection. More uniquely human is the diversity of brains. How much of the difficulty of intimate communication between one human being and another, despite the function of common learned language, arises from the discrepancy in their genetically determined neurological hardware? Monozygotic twins are notoriously sympathetic, easily able to interpret one another's minimal gestures and brief words; I know, however, of no objective studies of their economy of communication. For further argument, I will assume that genetic identity confers neurological similarity, and that this eases communication. This has never been systematically exploited as between twins, though it might be singularly useful in stressed occupations—say a pair of astronauts, or a deep-sea diver and his pump-tender, or a surgical team. It would be relatively more important in the discourse between generations, where an elder would teach his infant copy. A systematic division of intellectual labor would allow efficient communicants to have something useful to say to one another.

The burden of this argument is that the cultural process poses contradictory requirements of uniformity, for communication, and of heterogeneity, for innovation. We have no idea where we stand on this scale. At least in certain areas—say the military—it is almost certain that clones would have a self-contained advantage partly independent of, partly accentuated by, the special characteristics of the genotype which is reproduced. This introverted and potentially narrow-minded advantage of a clonish group may be the chief threat to a pluralistically dedicated species.

Even when nuclear transplantation has succeeded in the mouse, there would remain formidable restraints on the way to human application, and one might even doubt the further investment of experimental effort. However, several lines are likely to become active. Animal husbandry, for prize cattle and racehorses, could not ignore the opportunity, just as it bore the brunt of the enterprises of artificial insemination and oval transplantation. The dormant storage of human germ plasm as sperm will be replaced by the freezing of somatic tissues to save potential donor nuclei. Experiments on the efficacy of human nuclear transplantation will continue on a somatic basis, and these tissue clones used progressively, in patchwork combinations, chimeras. Human nuclei, or individual chromosomes and genes, will also be recombined with those of other animal species; these experiments are now well under way in cell culture. Before long we are bound to hear of tests of the effect of dosage of the human twenty-first chromosome on the development of the brain of the mouse or the gorilla. Extracorporeal gestation would merely accelerate these experiments. As bizarre as they seem, they are direct translations to man of classical work in the fruit-fly and in many plants. They need no further advance in algeny; just a small step in cell biology.

My colleagues differ widely in their reaction to the idea that anyone could conscientiously risk the crucial experiment, the first attempt to clone a man. Perhaps this will not be attempted until gestation can be monitored closely to be sure the fetus meets expectations. The mingling of individual human chromosomes with those of other mammals assures a gradualistic enlargement of the field and lowers the threshold of optimism or arrogance, particularly if cloning in other mammals gives incompletely predictable results.

THE GOALS

What are the practical aims of this discussion? It might help to redirect energies now wasted on naive eugenics and to protect the community from a misapplication of genetic policy. It may sensitize students to recognize the significance of the fruition of experiments like nuclear transplantation. Most important, it may help to provoke more critical use of the lessons of history for the direction of our future. This will need a much wider participation in these concerns. It is hard enough to approach verifiable truth in experimental work; surely much wider criticism is needed for speculations whose scientific verifiability is elusive in proportion to their human relevance. Scientists are by no means the best qualified architects of social policy, but there are two functions no one can do for them: the understanding and interpretation of technical challenges, to expose them for political action; and forethought for the balance of scientific effort that may be needed to manage such challenges. Popular trends in scientific work toward effective responses to human needs move just as slowly as other social institutions, and good work will come only from a widespread identification of scientists with these needs.

The foundations of any policy must rest on some consideration of purpose. One test that may appeal to skeptical scientists is to ask what they admire in the trend of human history. Few will leave out the growing richness of man's inquiry about nature, about himself, and his purpose. As long as we insist that this inquiry remains open, we have a pragmatic basis for a humble appreciation of the value of innumerable different approaches to life and its questions, of respect for the dignity of human life, and of individuality; and we decry the arrogance that insists on an irrevocable answer to any of these questions of value. The same humility will keep open the options for human nature until their consequences for the legacy momentarily entrusted to us are fully understood. These concerns are entirely consistent with the rigorously mechanistic formulation of life which has been the systematic basis of recent progress in biological science. . . .

Algeny presupposes a number of scientific advances that have yet to be perfected, and its immediate application to human biology is, probably unrealistically, discounted as purely speculative. In this paper, I infer that the path to algeny already opens up two major diversions of human evolution: clonal reproduction, and introduction of genetic material from other spheres. Indeed the essential features of these techniques have already been demonstrated in vertebrates, namely, nuclear transplantation in amphibia, and somatic hybridization of a variety of cells in culture, including human.

Paradoxically, the issue of "subhuman" hybrids may arise first, just because of the touchiness of experimentation on obviously human material. Tissue and organ cultures and transplants are already in wide experimental or therapeutic use, but there would be widespread inhibitions about risky experiments leading to an object that could be labelled as a human or para-human infant. However, there is enormous scientific interest in organisms augmented by fragments of the human chromosome set, especially as we know so little in detail of man's biological and genetic homology with other primates. This is being and will be pushed in steps as far as biology will allow, to larger and larger proportions of human genome in intact animals, and to organ combinations and chimeras with varying proportions of human, subhuman, and hybrid tissue. Note that there have been efforts to transplant primate organs to man. The hybridization is likely to be somatic, and the elaboration of these steps will make full use of nuclear transplantation to test how well these assorted genotypes will support the full development of a zygote.

Other techniques may well be discovered as shortcuts, especially how to induce the differentiation of a competent egg directly from somatic tissue. This process has no experimental foundation at present, but plenty of precedent in natural history.

These are not the most congenial subjects for friendly conversation, especially if the conversants mistake comment for advocacy. If I differ from the consensus of my colleagues, it may be only in suggesting a time scale of a few years rather than decades. Indeed, we will then face two risks: (1) that our scientific position is extremely unbalanced from the standpoint of its human impact; and (2) that precedents affecting the long-term rationale of social policy will be set, not on the basis of well-debated principles, but on the accidents of the first advertised examples. The accidentals might be as capricious as the nationality, batting average, or public esteem of a clonont; the handsomeness of a para-human progeny; the private morality of the experimenters; or public awareness that man is part of the continuum of life.

Back of all the more radical human breeding proposals hovers the thought that we might some day be able to produce a race of "superior" people, or, in another direction, that we might breed distinct types of human beings for specific purposes, as we now breed lower animals. That this *could* be done— once we have determined the exact genes responsible for the manifold and complex human traits—is hardly doubted by any geneticist. What it would demand, however, is the assumption by the state of supreme power over all human reproduction—the decision as to who should be allowed to reproduce, who should mate with whom, which children should survive, and how each child should be reared, graded, trained, and bred in turn. An unwelcome idea, but not an impossible one. It virtually happened in Sparta long ago, and it came very close to happening in Nazi Germany. . . . a program of breeding human beings to order, might bring upon our heads a deluge of evils that might far outweigh the good to be derived.

Amram Scheinfeld
The New You and Heredity (1950)

The Future of Mankind *
by Pierre Teilhard de Chardin (1955)

It has been said that we might well blush comparing our own mankind, so full of misshapen subjects, with those animal societies in which, in a hundred thousand individuals, not one will be found lacking in a single antenna. In itself that geometrical perfection is not in the line of our evolution whose bent is towards suppleness and freedom. All the same, suitably subordinated to other values, it may well appear as an indication and a lesson. So far we have certainly allowed our race to develop at random, and we have given too little thought to the question of what medical and moral factors *must replace the crude forces of natural selection* should we suppress them. In the course of the coming centuries it is indispensable that a nobly human form of eugenics, on a standard worthy of our personalities, should be discovered and developed.

Eugenics applied to individuals leads to eugenics applied to society. It would be more convenient, and we would incline to think it safe, to leave the contours of that great body made of all our bodies to take shape on their own, influenced only by the automatic play of individual urges and whims. "Better not interfere with the forces of the world!" Once more we are up against the mirage of instinct, the so-called infallibility of nature. But is it not precisely the world itself which, culminating in thought, expects us to think out again the instinctive impulses of nature so as to perfect them? Reflective substance requires reflective treatment. If there is a future for mankind, it can only be imagined in terms of a harmonious conciliation of what is free with what is levelled up and totalised.

Suggestions for Further Reading

* Asimov, Isaac: *The Genetic Code,* Signet Science Library, The New American Library of World Literature, 1963. An excellent book for students who would like to understand more about the biochemical nature of heredity.

* Bonner, David M.: *Heredity,* Foundations of Modern Biology Series, Prentice-Hall, Inc., 1961. A detailed and technical account of the biochemistry of genetics.

* Hardin, Garrett: *Nature and Man's Fate,* A Mentor Book, The New American Library of World Literature, Inc., 1961. A thought-provoking book about heredity and evolution with special emphasis on the social, political, and ethical implications. The final chapter, "In Praise of Waste," presents arguments against "eugenic utopias."

* Medawar, P. B.: *The Future of Man,* A Mentor Book, The New American Library of World Literature, Inc., 1961. This is a collection of Reith Lectures all of which are relevant to these readings. The selections beginning on p. 474 were taken from these lectures.

Scheinfeld, Amram: *Your Heredity and Environment,* J. B. Lippincott Company, Philadelphia, 1965. An excellent introduction to heredity for the layman.

* Young, Louise B., ed.: *The Mystery of Matter,* Oxford University Press, 1965. Part 9 contains readings on the biochemical theory of heredity.

* Paperback edition

10

Can Science Lead Mankind into a Better World?

These readings discuss the extent to which scientific power and progress threaten human progress. How can a proper balance be maintained between organization and freedom? And how can science be used to further the development of human potentialities?

Can Science Lead Mankind into a Better World?

Introduction

THE SCIENTIST, says C. P. Snow, has progress in his bones.[1] This strong conviction grows out of science in general because each step is built upon the preceding one. The biologist in particular sees the whole universe as changing and developing with time. In this scientific age the belief in progress has permeated our whole culture and has become a dominant characteristic of modern thought. The very word "progress" applied to almost any project is sufficient to sell it, and no sentimental attachment to the past is allowed to interfere with its realization.

Progress, however, can be variously defined, and the pursuit of it can sometimes lead to injustices as well as to genuine improvement. Are faster airplanes, taller skyscrapers, and longer limousines examples of progress? Modern man, Loren Eiseley says, has confused progress with his mechanical gadgets. He has failed to see that the triumph of the machine, without an accompanying inner triumph, is not the progress which biologists visualize as the ultimate aim of human evolution.

The illusion of progress is often used to justify the perpetration of evils in the present on the grounds that some benefit in the distant future may be so achieved. The argument that the end justifies the means has been used by all modern dictators to gain public support for changes that would otherwise be unacceptable. The history of these movements shows that while the evils do indeed become fact, the promised good remains a "pie in the sky" that is never actually tasted. Once basic freedoms have been relinquished, the power that science has made available delivers the people entirely into the hands of those in authority. Revolt is impractical against such great concentrations of might.

These and other dangers of our technological civilization are vividly set forth by Aldous Huxley both in essay form and as science fiction in a selection from *Brave New World*. This imaginative anti-utopia, which is probably familiar to most readers, is particularly interesting when viewed in the light of the previous readings. We are now in a position to assess Huxley's imaginary society from the point of view of its effect on human evolution. Are Community, Identity, and Stability constructive aims for a human society? How would adaptability and genetic diversity be affected by such a system? The methods Huxley invents, such as the Bokanovsky Process, do not seem particularly fantastic after reading Lederberg's proposal of "cloning" a man.

1. From C. P. Snow, *The Two Cultures and the Scientific Revolution*, Cambridge University Press, 1959.

The theme of *Brave New World,* Huxley tells us, is the advancement of science as it affects human individuals. "The really revolutionary revolution is to be achieved, not in the external world, but in the souls and flesh of human beings." [2] Huxley dramatizes his fears concerning the direction of technological change and portrays a world which he hopes will not occur.

On the other hand, B. F. Skinner describes a social system which he believes would represent a real improvement over our present society. In this utopia, people are trained from infancy to be tolerant, self-sacrificing, and co-operative. The system, however, is based on the proposition that man is not a free agent. The average citizen has no important decisions to make. He takes no part in the government and no responsibility for himself or his society. His character and personality are entirely molded by his cultural training.

Skinner believes that to apply scientific method to human behavior it is necessary to assume that man is not free. Man, he declares, is a mechanism controlled entirely by forces outside himself. Just as in a mechanical system we can predict the position and motion of a body by measuring the forces acting upon it, so it is with man. But, as other readings in this series have shown, the extension of this mechanistic view to the whole physical universe is no longer supported by modern physics.[3] The behavioral and social scientists took over many ideas from the older sciences but have not caught up with the fact that rigidly causal determinism cannot give us a complete description of any physical system, let alone of a living organism as complex as man.

Loren Eiseley suggests that the mechanistic view of man, pioneered by Watson and carried on by Freud and other psychologists, has created modern man's "moral predicament, his loss of authority over himself." [4] Both Archibald MacLeish and Edmund Sinnott concur in this opinion that the feeling of frustration and despair so widespread in our time comes from man's loss of faith in his own ability. "We no longer believe in man," says MacLeish. We have been sold the judgment that man is an empty organism, not possessing any inner power of self-direction. Furthermore, this judgment is believed to be verified by scientific fact. Why is it, then, MacLeish goes on to ask, that the humanists support this judgment of man's nature more than the scientists do today? [5]

The readers of this book know that many eminent biologists do not be-

2. See Aldous Huxley selection, p. 584.
3. See *Exploring the Universe,* by Louise B. Young, ed., McGraw-Hill Book Co., 1962, Part 6; and *The Mystery of Matter,* by Louise B. Young, ed., Oxford University Press, 1965, Part 3.
4. See Loren Eiseley selection, p. 604.
5. See Archibald MacLeish selection, pp. 607–13.

lieve that man is an empty organism, that he is nothing but an animal, or that he is controlled by environmental factors and can take no part in molding his own destiny. Julian Huxley's article in this part restates the optimistic biologist's view of man's unique responsibility as the spearhead of the evolutionary process, and his hopeful conviction that we are entering a humanist era in which the discoveries of modern science will be used to further human potentialities.

However, there has also been a recurrent concern voiced throughout these readings that scientific technology applied to human affairs will increasingly curtail individual freedom. The possibility of controlling human reproduction, human personality, and human evolution is seen by some authors to spell the end of the self-directing responsible person. The pressure of expanding population may necessitate a more regimented and specialized society. Improvement of computer technology will make it possible to feed all the information about each human being into a machine, and what will come out at the other end will be a number rather than an individual. Will changes of this kind inevitably occur simply because we will be swept along by the rising tide of technology? Or will it occur, as one author suggests, because we have abdicated the responsibility for shaping our own future?

Dostoievsky once said, "Nothing has ever been more insupportable for a man and a human society than freedom." [6] And yet it is this very freedom to choose that has given man his unique place in the evolutionary process.

The shape of the future, whether it be a brave new world or a truly humanistic era, will depend to a large extent on what man understands and believes about himself. The sciences of life are constantly increasing our knowledge of the origin, the nature, and the goals of humanity. They are placing in our hands the means of reaching these goals. And yet, in human evolution the idea is a greater moving force than any physical means. The idea of Man is still torn with dissension and veiled with mystery. "The serious division in the world today," says Edmund Sinnott, "is not so much on social or economic or political issues as on this basic question of what man really is. . . . What man believes about *himself* is of the utmost moment for it will determine the kind of world that he will make and even his own fate." [7]

6. From Feodor Dostoievsky, *The Brothers Karamazov* (1881), The Modern Library, Book V, Chapter V.
7. See Edmund W. Sinnott selection, p. 603.

The Technological Society

———————

The Dangers of the Doctrine of Progress *

by Aldous Huxley (1946)

The continuous advance of science and technology has profoundly affected the prevailing mental climate. The basic postulates of thought have been changed, so that what to our fathers seemed obviously true and important strikes us as either false or negligible and beside the point. Let us consider a few of the more significant of these changes and their effects upon the social and political life of our times.

a) Unlike art, science is genuinely progressive. Achievement in the fields of research and technology is cumulative; each generation begins at the point where its predecessor left off. Furthermore, the results of disinterested research were from the first applied in such a way that the upper and middle classes of all industrialized societies found themselves becoming steadily richer and richer. It was, therefore, only to be expected that the professional thinkers who sprang from these classes, and who were familiar with the methods and achievements of science, should have based upon the facts of technological and economic progress a general theory of human life. The world, they affirmed, was becoming materially, intellectually and morally better and better, and this amelioration was in some way inevitable. The theory of progress—a theory that soon became a dogma, indeed an axiom of popular thought—was novel and, from an orthodox Christian point of view, heretical. For orthodoxy, man was a fallen being. Humanity if not actively deteriorating, was statically bad, with a badness which only grace in co-operation with the individual's free will could possibly mitigate. In illustration of this, let us consider how the thirteenth century was regarded by those who lived through it, and how it is regarded by modern historians. For the latter it seems one of the most glorious periods in European history; the former were unanimous (as Professor Coulton has shown) in regarding it as an age of peculiar wickedness and manifest degeneracy. Even in the age of Queen Elizabeth

* From Aldous Huxley, *Science, Liberty and Peace.* Copyright 1946 by Aldous Huxley. Reprinted by permission of Harper & Row, Publishers and Mrs. Laura Huxley.

thoughtful men were still ta̧lking of humanity's decline. It was not until the late seventeenth century (the age of the rise of modern science) that the note of bumptious self-congratulation began to be sounded, not until the eighteenth and nineteenth centuries that the dogma of inevitable progress became an unquestioned article of popular faith.

The belief in all-round progress is based upon the wishful dream that one can get something for nothing. Its underlying assumption is that gains in one field do not have to be paid for by losses in other fields. For the ancient Greeks, *hubris,* or overweening insolence, whether directed against the gods, or one's fellow men, or nature, was sure to be followed, sooner or later, in one way or another, by avenging Nemesis. Unlike the Greeks, we of the twentieth century believe that we can be insolent with impunity.

So intense is our faith in the dogma of inevitable progress that it has survived two world wars and still remains flourishing in spite of totalitarianism and the revival of slavery, concentration camps and saturation bombing.

Faith in progress has affected contemporary political life by reviving and popularizing, in an up-to-date, pseudo-scientific and this-worldly form, the old Jewish and Christian apocalypticism. A glorious destiny awaits mankind, a coming Golden Age, in which more ingenious gadgets, more grandiose plans and more elaborate social institutions, will somehow have created a race of better and brighter human beings. Man's Final End is not in the eternal timeless Now, but in a not too distant utopian future. In order to secure the peace and happiness of their great-great-grandchildren, the masses ought to accept and their rulers need feel no qualms in imposing, any amount of war and slavery, of suffering and moral evil, in the present. It is a highly significant fact that all modern dictators, whether of the Right or of the Left, talk incessantly about the golden Future, and justify the most atrocious acts here and now, on the ground that they are means to that glorious end. But the one thing we all know about the future is that we are completely ignorant of what is going to happen, and that what does in fact happen is often very different from what we anticipated. Consequently any faith based upon hypothetical occurrences a long time hence must always, in the very nature of things, be hopelessly unrealistic. In practice, faith in the bigger and better future is one of the most potent enemies to present liberty; for rulers feel themselves justified in imposing the most monstrous tyrannies on their subjects for the sake of the wholly imaginary fruits which these tyrannies are expected (only an implicit faith in progress can say why) to bear some time, let us say, in the twenty-first or twenty-second century.

b) As theory, pure science is concerned with the reduction of diversity to identity. As a praxis, scientific research proceeds by simplification. These habits of scientific thought and action have, to a certain extent, been carried over into the theory and practice of contemporary politics. Where a centralized authority undertakes to make plans for an entire society, it is compelled by the bewildering complexity of the given facts to follow the example of the scientific experimenter, who arbitrarily simplifies his problem in order to make it manageable. In the laboratory this is a sound and entirely justifiable procedure. But when applied to the problems of human society, the process of simplification is a process, inevitably, of restraint and regimentation, of curtailment of liberty and denial of individual rights. This reduction of human diversity to a military and quasi-mechanical identity is achieved by propaganda, by legal enactments and, if necessary, by brute force—by the imprisonment, exile or liquidation of those persons, or those classes, who persist in their perverse desire to remain themselves and are obstinate in their reluctance to conform to the pattern which the political and economic bosses find it, at the moment, most convenient to impose. Philosophically, this ironing out of individual idiosyncrasies is held to be respectable, because it is analogous to what is done by scientists, when they arbitrarily simplify an all too complex reality, so as to make nature comprehensible in terms of a few general laws. A highly organized and regimented society, whose members exhibit a minimum of personal peculiarities, and whose collective behavior is governed by a single master plan imposed from above, is felt by the planners and even (such is the power of propaganda) by the planees to be more "scientific," and therefore better, than a society of independent, freely cooperating and self-governing individuals.

c) The first step in this simplification of reality, without which (since human minds are finite and nature is infinite) scientific thought and action would be impossible, is a process of abstraction. Confronted by the data of experience, men of science begin by leaving out of account all those aspects of the facts which do not lend themselves to measurement and to explanation in terms of antecedent causes rather than of purpose, intention and values. Pragmatically they are justified in acting in this odd and extremely arbitrary way; for by concentrating exclusively on the measurable aspects of such elements of experience as can be explained in terms of a causal system they have been able to achieve a great and ever increasing control over the energies of nature. But power is not the same thing as insight and, as a representation of reality, the scientific picture of the world is inadequate, for the simple reason that science does not even profess to deal with experience as a whole, but only with certain aspects of it in certain contexts. All this is quite clearly understood by the more philosophically minded men of science. But unfortunately some scientists, many technicians and most consumers of gadgets have lacked the time and the inclination to examine the philosophical foundations and background of the sciences. Consequently they tend to accept the world picture implicit in the theories of science as a complete and exhaustive account of reality; they tend to regard those aspects of experience, which scientists leave out of account, because they are incompetent to deal with them, as being somehow less real than the aspects which science has arbitrarily chosen to abstract from out of the infinitely rich totality of given facts. Because of the prestige of science as a source of power, and because of the general neglect of philosophy, the popular *Weltanschauung* of our times contains a large element of what may be called "nothing-but" thinking. Human beings, it is more or less tacitly assumed, are nothing but bodies, animals, even machines; the only really real elements of reality are matter and energy in their measurable aspects; values are nothing but illusions that have somehow got themselves mixed up with our experience of the world; mental happenings are nothing but epiphenomena, produced by and entirely dependent upon physiology; spirituality is nothing but wish fulfillment and misdirected sex; and so on. The political consequences of this "nothing-but" philosophy are clearly apparent in that widespread indifference to the values of human personality and human life, which are so characteristic of the present age.

The Organization of Human Beings *
by Harrison Brown (1954)

Today we see about us on all sides a steady drift toward increased human organization. Governments are becoming more centralized and universal. In practically all areas of endeavor within industrial society—in our systems of production, in the fields of labor, capital, commerce, agriculture, science, education, and art—we see the emergence of new levels of organization designed to coordinate, integrate, bind, and regulate men's actions. The justifications for this increasing degree of organization to which man must accommodate himself are expressed in terms such as "stability," "security," and "efficiency." The end result of this rapid transition might well be the emergence of a universal, stable, efficient, industrial society within which, although all persons have complete personal

* From Harrison Brown, *The Challenge of Man's Future,* Viking Press, New York. Copyright 1954 by Harrison Brown. Reprinted by permission.

security, their actions are completely controlled. Should that time arrive, society will have become static, devoid of movement, fixed and permanent. History will have stopped.

Here we indeed find ourselves on the horns of the dilemma. To what purpose is industrialization if we end up by replacing rigid confinement of man's actions by nature with rigid confinement of man's actions by man? To what purpose is industrialization if the price we pay for longer life, material possessions, and personal security is regimentation, controlled thoughts, and controlled actions? Would the lives of well-fed, wealthy, but regimented human robots be better than the lives of their malnourished, poverty-stricken ancestors? At least the latter could look forward to the unexpected happening—to events and situations which previously had been outside the realm of their experiences.

In a modern industrial society the road toward totalitarianism is unidirectional. In days gone by men could revolt against despotism. People could arise against their governments in the absence of legal recourse, and with muskets, sticks, knives, and stones as their weapons they could often defeat the military forces of the central authorities. But today our science and our technology have placed in the hands of rulers of nations weapons and tools of control, persuasion, and coercion of unprecedented power. We have reached the point where, once totalitarian power is seized in a highly industrialized society, successful revolt becomes practically impossible. Totalitarian power, once it is gained, can be perpetuated almost indefinitely in the absence of outside forces, and can lead to progressively more rapid robotization of the individual.

Thus we see that, just as industrial society is fundamentally unstable and subject to reversion to agrarian existence, so within it the conditions which offer individual freedom are unstable in their ability to avoid the conditions which impose rigid organization and totalitarian control. Indeed, when we examine all of the foreseeable difficulties which threaten the survival of industrial civilization, it is difficult to see how the achievement of stability and the maintenance of individual liberty can be made compatible.

The view is widely held in our society that the powers of the machine will eventually free man from the burden of eking out an existence and will provide him with leisure time for the development of his creativity and enjoyment of the fruits of his creative efforts. Pleasant though this prospect may be, it is clear that such a state cannot come into existence automatically; the pressures forcing man into devising more highly organized institutions are too great to permit it. If he is to attain such an idyllic existence for more than a transitory period he must plan for that existence carefully, and in particular he must do everything within his power to reduce the pressures that are forcing him to become more highly organized.

One of the major pressures that give rise to the need for increasing numbers of laws, more elaborate organization, and more centralized government is increase of population. Increase of numbers of people and of population density results in greater complexities in day-to-day living and in decreased opportunities for personal expression concerning the activities of government. But even more important, as populations increase and as they press more heavily upon the available resources there arises the need for increased efficiency, and more elaborate organizations are required to produce sufficient food, to extract the necessary raw materials, and to fabricate and distribute the finished products. In the future we can expect that the greater the population density of an industrial society becomes, the more elaborate will be its organizational structure and the more regimented will be its people.

A second pressure, not unrelated to the first, results from the centralization of industrial and agricultural activity and from regional specialization in various aspects of those activities. One region produces textiles, another produces coal, another automobiles, another corn, and another wheat. Mammoth factories require mammoth local organizations. Centralized industries must be connected, and this requires elaborate transportation systems. Regional localization of industries gives rise to gigantic cities, which in turn give rise to elaborate organization for the purpose of providing the inhabitants with the necessary food, water, and services. All of these factors combine to produce vulnerability to disruption from the outside, increased local organization and regimentation, more highly centralized government, and increasing vulnerability to the evolution of totalitarianism.

A third pressure results from increasing individual specialization and the resultant need for "integration," "coordination," and "direction" of activities in practically all spheres of vocational and leisure activity. It results in the placing of unwarranted trust in "integrators," "coordinators," and "directors." Early specialization results in lack of broad interests, lessened ability to engage in creative activity during leisure hours, decreased interest in the creative activities of other individuals, and lessened abilities to interpret events and make sound judgments. All of these factors combine to pave the way for collectivization, the emergence of strong organization, and, with it, the great leader.

Strong arguments can be presented to the effect that collectivization of humanity is inevitable, that the drift toward an ultimate state of automatism cannot be halted, that existing human values such as freedom, love, and conscience must eventually disappear.[1] Certainly if we used the present trends in industrial society as our major premises, the conclusion would appear to be inescapable. Yet is it not possible that human beings, recognizing this threat to the canons of humanism, can devise ways and means of escaping the danger and at the same time manage to preserve those features of industrial civilization which can contribute to a rich, full life? Is it really axiomatic that the present trends must continue and that in the long run industrial civilization and human values are incompatible? Here, in truth, we are confronted with the gravest and most difficult of all human problems, for it is one that cannot be solved by mathematics or by machines, nor can it even be precisely defined. Solutions, if they exist, can arise only in the hearts and minds of individual men.

The machine has divorced man from the world of nature to which he belongs, and in the process he has lost in large measure the powers of contemplation with which he was endowed. A prerequisite for the preservation of the canons of humanism is a reestablishment of organic roots with our natural environment and, related to it, the evolution of ways of life which encourage contemplation and the search for truth and knowledge. The flower and vegetable garden, green grass, the fireplace, the primeval forest with its wondrous assemblage of living things, the uninhabited hilltop where one can silently look at the stars and wonder—all of these things and many others are necessary for the fulfillment of man's psychological and spiritual needs. To be sure, they are of no "practical value" and are seemingly unrelated to man's pressing need for food and living space. But they are as necessary to the preservation of humanism as food is necessary to the preservation of human life.

I can imagine a world within which machines function solely for man's benefit, turning out those goods which are necessary for his well-being, relieving him of the necessity for heavy physical labor and dull, routine, meaningless activity. The world I imagine is one in which people are well fed, well clothed, and well housed. Man, in this world, lives in balance with his environment, nourished by nature in harmony with the myriads of other life forms that are beneficial to him. He treats his land wisely, halts erosion and over-cropping, and returns all organic waste matter to the soil from which it sprung. He lives efficiently, yet minimizes artificiality. It is not an overcrowded world; people can, if they wish, isolate themselves in the silence of a mountaintop, or they can walk through primeval forests or across wooded plains. In the world of my imagination there is

1. These views have geen forcefully and eloquently expressed by Roderick Seidenberg in his book *Post-Historic Man* (Chapel Hill: University of North Carolina Press, 1950).

organization, but it is as decentralized as possible, compatible with the requirements for survival. There is a world government, but it exists solely for the purpose of preventing war and stabilizing population, and its powers are irrevocably restricted. The government exists for man rather than man for the government.

In the world of my imagination the various regions are self-sufficient, and the people are free to govern themselves as they choose and to establish their own cultural patterns. All people have a voice in the government, and individuals can move about when and where they please. It is a world where man's creativity is blended with the creativity of nature, and where a moderate degree of organization is blended with a moderate degree of anarchy.

Is such a world impossible of realization? Perhaps it is, but who among us can really say? At least if we try to create such a world there is a chance that we will succeed. But if we let the present trend continue it is all too clear that we will lose forever those qualities of mind and spirit which distinguish the human being from the automaton.

We have seen that population stabilization within a framework of low birth rates and low death rates is a major key to the avoidance of collectivization and robotization of humanity and to the perpetuation of machine civilization. . . .

If all babies were born from test tubes, as in Aldous Huxley's *Brave New World,* the solution would be fairly simple: The number of babies produced on the production line each year could be made to equal the number of deaths. In years of unusually high death rates, the production line could be speeded up; in years of low death rates, the line could be slowed down. Further, if we cared little for human emotions and were willing to introduce a procedure which most of us would consider to be reprehensible in the extreme, all excess children could be disposed of much as excess puppies and kittens are disposed of at the present time. But let us hope it will be a long time before a substantial number of our babies are born from test tubes.

Utopias and Anti-Utopias

Brave New World *

by Aldous Huxley (1932)

CHAPTER ONE

A squat grey building of only thirty-four stories. Over the main entrance the words, CENTRAL LONDON HATCHERY AND CONDITIONING CENTRE, and, in a shield, the World State's motto, COMMUNITY, IDENTITY, STABILITY.

The enormous room on the ground floor faced towards the north. Cold for all the summer beyond the panes, for all the tropical heat of the room itself, a harsh thin light glared through the windows, hungrily seeking some draped lay figure, some pallid shape of academic goose-flesh, but finding only the glass and nickel and bleakly shining porcelain of a laboratory. Wintriness responded to wintriness. The overalls of the workers were white, their hands gloved with a pale corpse-coloured rubber. The light was frozen, dead, a ghost. Only from the yellow barrels of the microscopes did it borrow a certain rich and living substance, lying along the polished tubes like butter, streak after luscious streak in long recession down the work tables.

"And this," said the Director opening the door, "is the Fertilizing Room."

Bent over their instruments, three hundred Fertilizers were plunged, as the Director of Hatcheries and Conditioning entered the room, in the scarcely breathing silence, the absent-minded, soliloquizing hum or whistle, of absorbed concentration. A troop of newly arrived students, very young, pink and callow, followed nervously, rather abjectly, at the Director's heels. Each of them carried a notebook, in which, whenever the great man spoke, he desperately scribbled. Straight from the horse's mouth. It was a rare privilege. The D.H.C. for Central London always made a point of personally conducting his new students round the various departments.

"Just to give you a general idea," he would explain to them. For of course some sort of general idea they must have, if they were to do their work intelligently—though as little of one, if they were to be good and happy members of society, as possible. For particulars, as every one knows, make for virtue and happiness; generalities are intellectually necessary evils. Not philosophers but fret-sawyers and stamp collectors compose the backbone of society.

"To-morrow," he would add, smiling at them with a slightly menacing geniality, "you'll be settling down to serious work. You won't have time for generalities. Meanwhile . . ."

Meanwhile, it was a privilege. Straight from the horse's mouth into the notebook. The boys scribbled like mad.

Tall and rather thin but upright, the Director advanced into the room. He had a long chin and big rather prominent teeth, just covered, when he was not talking, by his full, floridly curved lips. Old, young? Thirty? Fifty? Fifty-five? It was hard to say. And anyhow the question didn't arise; in this year of stability, A. F. 632, it didn't occur to you to ask it.

"I shall begin at the beginning," said the D.H.C. and the more zealous students recorded his intention in their notebooks: *Begin at the beginning.* "These," he waved his hand, "are the incubators." And opening an insulated door he showed them racks upon racks of numbered test-tubes. "The week's supply of ova. Kept," he explained, "at blood heat; whereas the male gametes," and here he opened another door, "they have to be kept at thirty-five instead of thirty-seven. Full blood heat sterilizes." Rams wrapped in theremogene beget no lambs.

Still leaning against the incubators he gave them, while the pencils scurried illegibly across the pages, a brief description of the modern fertilizing process; spoke first, of course, of its surgical introduction—"the operation undergone voluntarily for the good of Society, not to mention the fact that it carries a bonus amounting to six months' salary"; continued with some account of the technique for preserving the excised ovary alive and actively developing; passed on to a consideration of optimum temperature, salinity, viscosity; referred to the liquor in which the detached and ripened eggs were kept; and, leading his charges to the work tables, actually showed them how this liquor was drawn off from the test-tubes; how it was let out drop by drop onto the specially warmed slides of the microscopes; how the eggs which it contained were inspected for abnormalities, counted and transferred to a porous receptacle; how (and he now took them to watch the operation) this receptacle was immersed in a warm bouillon containing free-swimming spermatozoa—at a minimum concentration of one hundred thousand per cubic centimetre, he insisted; and how, after ten minutes, the container was lifted out of the liquor and its contents re-examined; how, if any of the eggs remained unfertilized, it was again immersed, and, if necessary, yet again; how the fertilized ova went back to the incubators; where the Alphas and Betas remained until definitely bottled; while the Gammas, Deltas and Epsilons were brought out again, after only thirty-six hours, to undergo Bokanovsky's Process.

"Bokanovsky's Process," repeated the Director, and the students underlined the words in their little notebooks.

One egg, one embryo, one adult—normality. But a bokanovskified egg will bud, will proliferate, will divide. From eight to ninety-six buds, and every bud will grow into a perfectly formed embryo, and every embryo into a full-sized adult. Making ninety-six human beings grow where only one grew before. Progress.

"Essentially," the D.H.C. concluded, "bokanovskification consists of a series of arrests of development. We check the normal growth and, paradoxically enough, the egg responds by budding."

Responds by budding. The pencils were busy.

He pointed. On a very slowly moving band a rack-full of test-tubes was entering a large metal box, another, rack-full was emerging. Machinery faintly purred. It took eight minutes for the tubes to go through, he told them. Eight minutes of hard X-rays being about as much as an egg can stand. A few died; of the rest, the least susceptible divided into two; most put out four buds; some eight; all were returned to the incubators, where the buds began to develop; then, after two days, were suddenly chilled, chilled and checked. Two, four, eight, the buds in their turn budded; and having budded were dosed almost to death with alcohol; consequently burgeoned again and having budded—bud out of bud out of bud—were thereafter—further arrest being generally fatal—left to develop in peace. By which time the original egg was in a fair way to becoming anything from eight to ninety-six embryos—a prodigious improvement, you will agree, on nature. Identical twins—but not in piddling twos and threes as in the old vivaparous days, when an egg would sometimes accidentally divide; actually by dozens, by scores at a time.

"Scores," the Director repeated and flung out his arms, as though he were distributing largesse. "Scores."

But one of the students was fool enough to ask where the advantage lay.

"My good boy!" The Director wheeled sharply round on him. "Can't you see? Can't you *see?*" He raised a hand; his expression was solemn. "Bokanovsky's Process is one of the major instruments of social stability!"

Major instruments of social stability.

Standard men and women; in uniform batches. The whole of a small factory staffed with the products of a single bokanovskified egg.

"Ninety-six identical twins working ninety-six identical machines!" The voice was almost tremulous with enthusiasm. "You really know where you are. For the first time in history." He quoted the planetary motto. "Community, Identity, Stability." Grand words. "If we could bokanovskify indefinitely the whole problem would be solved."

Solved by standard Gammas, unvarying Deltas, uniform Epsilons. Millions of identical twins. The principle of mass production at last applied to biology.

"But, alas," the Director shook his head, "we *can't* bokanovskify indefinitely."

Ninety-six seemed to be the limit; seventy-two a good average. From the same ovary and with gametes of the same male to manufacture as many batches of identical twins as possible—that was the best (sadly a second best) that they could do. And even that was difficult.

"For in nature it takes thirty years for two hundred eggs to reach maturity. But our business is to stabilize the population at this moment, here and now. Dribbling out twins over a quarter of a century—what would be the use of that?"

Obviously, no use at all. But Podsnap's Technique had immensely accelerated the process of ripening. They could make sure of at least a hundred and fifty mature eggs within two years. Fertilize and bokanovskify— in other words, multiply by seventy-two—and you get an average of nearly eleven thousand brothers and sisters in a hundred and fifty batches of identical twins, all within two years of the same age.

"And in exceptional cases we can make one ovary yield us over fifteen thousand adult individuals."

Beckoning to a fair-haired, ruddy young man who happened to be passing at the moment. "Mr. Foster," he called. The ruddy young man approached. "Can you tell us the record for a single ovary, Mr. Foster?"

"Sixteen thousand and twelve in this Centre," Mr. Foster replied without hesitation. He spoke very quickly, had a vivacious blue eye, and took an evident pleasure in quoting figures. "Sixteen thousand and twelve; in one hundred and eighty-nine batches of identicals. But of course they've done much better," he rattled on, "in some of the tropical Centres. Singapore has often produced over sixteen thousand five hundred; and Mombasa has actually touched the seventeen thousand mark. But then they have unfair advantages. You should see the way a negro ovary responds to pituitary! It's quite astonishing, when you're used to working with European material. Still," he added, with a laugh (but the light of combat was in his eyes and the lift of his chin was challenging), "still, we mean to beat them if we can. I'm working on a wonderful Delta-Minus ovary at this moment. Only just eighteen months old. Over twelve thousand seven hundred children already, either decanted or in embryo. And still going strong. We'll beat them yet."

"That's the spirit I like!" cried the Director, and clapped Mr. Foster on the shoulder. "Come along with us, and give these boys the benefit of your expert knowledge."

Mr. Foster smiled modestly. "With pleasure." They went.

In the Bottling Room all was harmonious bustle and ordered activity. Flaps of fresh sow's peritoneum ready cut to the proper size came shooting up in little lifts from the Organ Store in the sub-basement. Whizz and then, click! the lift-hatches flew open; the bottle-liner had only to reach out a hand, take the flap, insert, smooth-down, and before the lined bottle had had time to travel out of reach along the endless band, whizz, click! another flap of peritoneum had shot up from the depths, ready to be slipped into yet another bottle, the next of that slow interminable procession on the band.

Next to the Liners stood the Matriculators. The procession advanced; one by one the eggs were transferred from their test-tubes to the larger containers; deftly the peritoneal lining was slit, the morula dropped into place, the saline solution poured in . . . and already the bottle had passed, and it was the turn of the labellers. Heredity, date of fertilization, membership of Bokanovsky Group—details were transferred from test-tube to bottle. No longer anonymous, but named, identified, the procession marched slowly on; on through an opening in the wall, slowly on into the Social Predestination Room.

"Eighty-eight cubic metres of card-index," said Mr. Foster with relish, as they entered.

"Containing *all* the relevant information," added the Director.

"Brought up to date every morning."

"And co-ordinated every afternoon."

"On the basis of which they make their calculations."

"So many individuals, of such and such quality," said Mr. Foster.

"Distributed in such and such quantities."

"The optimum Decanting Rate at any given moment."

"Unforeseen wastages promptly made good."

"Promptly," repeated Mr. Foster. "If you knew the amount of overtime I had to put in after the last Japanese earthquake!" He laughed good-humouredly and shook his head.

"The Predestinators send in their figures to the Fertilizers."

"Who give them the embryos they ask for."

"And the bottles come in here to be predestined in detail."

"After which they are sent down to the Embryo Store."

"Where we now proceed ourselves."

And opening a door Mr. Foster led the way down a staircase into the basement.

The temperature was still tropical. They descended into a thickening twilight. Two doors and a passage with a double turn insured the cellar against any possible infiltration of the day.

"Embryos are like photograph film," said Mr. Foster waggishly, as he pushed open the second door. "They can only stand red light."

And in effect the sultry darkness into which the students now followed him was visible and crimson, like the darkness of closed eyes on a summer's afternoon. The bulging flanks of row on receding row and tier above tier of bottles glinted with innumerable rubies, and among the rubies moved the dim red spectres of men and women with purple eyes and all the symptoms of lupus. The hum and rattle of machinery faintly stirred the air.

"Give them a few figures, Mr. Foster," said the Director, who was tired of talking.

Mr. Foster was only too happy to give them a few figures.

Two hundred and twenty metres long, two hundred wide, ten high. He pointed upwards. Like chickens drinking, the students lifted their eyes towards the distant ceiling.

Three tiers of racks: ground floor level, first gallery, second gallery.

The spidery steel-work of gallery above gallery faded away in all directions into the dark. Near them three red ghosts were busily unloading demijohns from a moving staircase.

The escalator from the Social Predestination Room.

Each bottle could be placed on one of fifteen racks, each rack, though you couldn't see it, was a conveyor traveling at the rate of thirty-three and a third centimetres an hour. Two hundred and sixty-seven days at eight metres a day. Two thousand one hundred and thirty-six metres in all. One circuit of the cellar at ground level, one on the first gallery, half on the second, and on the two hundred and sixty-seventh morning, daylight in the Decanting Room. Independent existence—so called.

"But in the interval," Mr. Foster concluded, "we've managed to do a lot to them. Oh, a very great deal." His laugh was knowing and triumphant.

"That's the spirit I like," said the Director once more. "Let's walk around. You tell them everything, Mr. Foster."

Mr. Foster duly told them.

Told them of the growing embryo on its bed of peritoneum. Made them taste the rich blood surrogate on which it fed. Explained why it had to be stimulated with placentin and thyroxin. Told them of the *corpus luteum* extract. Showed them the jets through which at every twelfth metre from zero to 2040 it was automatically injected. Spoke of those gradually increasing doses of pituitary administered during the final ninety-six metres of their course. Described the artificial maternal circulation installed in every bottle at Metre 112; showed them the reservoir of blood-surrogate, the centrifugal pump that kept the liquid moving over the placenta and drove it through the synthetic lung and waste product filter. Referred to the embryo's troublesome tendency to anaemia, to the massive doses of hog's stomach extract and foetal foal's liver with which, in consequence, it had to be supplied.

Showed them the simple mechanism by means of which, during the last two metres out of every eight, all the embryos were simultaneously shaken into familiarity with movement. Hinted at the gravity of the so-called "trauma of decanting," and enumerated the precautions taken to minimize, by a suitable training of the bottled embryo, that dangerous shock. Told them of the test for sex carried out in the neighborhood of Metre 200. Explained the system of labelling—a T for the males, a circle for the females and for those who were destined to become freemartins a question mark, black on a white ground.

"For of course," said Mr. Foster, "in the vast majority of cases, fertility is merely a nuisance. One fertile ovary in twelve hundred—that would really be quite sufficient for our purposes. But we want to have a good choice. And of course one must always have an enormous margin of safety. So we allow as many as thirty per cent of the female embryos to develop normally. The others get a dose of male sex-hormone every twenty-four metres for the rest of the course. Result: they're decanted as freemartins—structurally quite normal (except," he had to admit, "that they *do* have the slightest tendency to grow beards), but sterile. Guaranteed sterile. Which brings us at last," continued Mr. Foster, "out of the realm of mere slavish imitation of nature into the much more interesting world of human invention."

He rubbed his hands. For of course, they didn't content themselves with merely hatching out embryos: any cow could do that.

"We also predestine and condition. We decant our babies as socialized human beings, as Alphas or Epsilons, as future sewage workers or future . . ." He was going to say "future World controllers," but correcting himself, said "future Directors of Hatcheries," instead.

The D.H.C. acknowledged the compliment with a smile.

They were passing Metre 320 on Rack 11. A young Beta-Minus mechanic was busy with screw-driver and spanner on the blood-surrogate pump of a passing bottle. The hum of the electric motor deepened by fractions of a tone as he turned the nuts. Down, down . . . A final twist, a glance at the revolution counter, and he was done. He moved two paces down the line and began the same process on the next pump.

"Reducing the number of revolutions per minute," Mr. Foster explained. "The surrogate goes round slower; therefore passes through the lung at longer intervals; therefore gives the embryo less oxygen. Nothing like oxygen-shortage for keeping an embryo below par." Again he rubbed his hands.

"But why do you want to keep the embryo below par?" asked an ingenuous student.

"Ass!" said the Director, breaking a long silence. "Hasn't it occurred to you that an Epsilon embryo must have an Epsilon environment as well as an Epsilon heredity?"

It evidently hadn't occurred to him. He was covered with confusion.

"The lower the caste," said Mr. Foster, "the shorter the oxygen." The first organ affected was the brain. After that the skeleton. At seventy per cent of normal oxygen you got dwarfs. At less than seventy eyeless monsters.

"Who are no use at all," concluded Mr. Foster.

Whereas (his voice became confidential and eager), if they could discover a technique for shortening the period of maturation what a triumph, what a benefaction to Society!

"Consider the horse."

They considered it.

Mature at six; the elephant at ten. While at thirteen a man is not yet sexually mature; and is only full-grown at twenty. Hence, of course, that fruit of delayed development, the human intelligence.

"But in Epsilons," said Mr. Foster very justly, "we don't need human intelligence."

Didn't need and didn't get it. But though the Epsilon mind was mature at ten, the Epsilon body was not fit to work till eighteen. Long years of superfluous and wasted immaturity. If the physical development could be speeded up till it was as quick, say, as a cow's, what an enormous saving to the Community!

"Enormous!" murmured the students. Mr. Foster's enthusiasm was infectious.

He became rather technical; spoke of the abnormal endocrine coordination which made men grow so slowly; postulated a germinal mutation to account for it. Could the effects of this germinal mutation be undone? Could the individual Epsilon embryo be made a revert, by a suitable technique, to the normality of dogs and cows? That was the problem. And it was all but solved.

Pilkington, at Mombasa, had produced individuals who were sexually mature at four and full-grown at six and a half. A scientific triumph. But socially useless. Six-year-old men and women were too stupid to do even

Epsilon work. And the process was an all-or-nothing one; either you failed to modify at all, or else you modified the whole way. They were still trying to find the ideal compromise between adults of twenty and adults of six. So far without success. Mr. Foster sighed and shook his head.

Their wanderings through the crimson twilight had brought them to the neighborhood of Metre 170 on Rack 9. From this point onwards Rack 9 was enclosed and the bottle performed the remainder of their journey in a kind of tunnel, interrupted here and there by openings two or three metres wide.

"Heat conditioning," said Mr. Foster.

Hot tunnels alternated with cool tunnels. Coolness was wedded to discomfort in the form of hard X-rays. By the time they were decanted the embryos had a horror of cold. They were predestined to emigrate to the tropics, to be miner and acetate silk spinners and steel workers. Later on their minds would be made to endorse the judgment of their bodies. "We condition them to thrive on heat," concluded Mr. Foster. "Our colleagues upstairs will teach them to love it."

"And that," put in the Director sententiously, "that is the secret of happiness and virtue—liking what you've *got* to do. All conditioning aims at that: making people like their unescapable social destiny."

In a gap between two tunnels, a nurse was delicately probing with a long fine syringe into the gelatinous contents of a passing bottle. The students and their guides stood watching her for a few moments in silence.

"Well, Lenina," said Mr. Foster, when at last she withdrew the syringe and straightened herself up.

The girl turned with a start. One could see that, for all the lupus and the purple eyes, she was uncommonly pretty.

"Henry!" Her smile flashed redly at him—a row of coral teeth.

"Charming, charming," murmured the Director and, giving her two or three little pats, received in exchange a rather deferential smile for himself.

"What are you giving them?" asked Mr. Foster, making his tone very professional.

"Oh, the usual typhoid and sleeping sickness."

"Tropical workers start being inoculated at Metre 150," Mr. Foster explained to the students. "The embryos still have gills. We immunize the fish against the future man's diseases." Then, turning back to Lenina, "Ten to five on the roof this afternoon," he said, "as usual."

"Charming," said the Director once more, and, with a final pat, moved away after the others.

On Rack 10 rows of next generation's chemical workers were being trained in the toleration of lead, caustic soda, tar, chlorine. The first of a batch of two hundred and fifty embryonic rocket-plane engineers was just passing the eleven hundred metre mark on Rack 3. A special mechanism kept their containers in constant rotation. "To improve their sense of balance," Mr. Foster explained. "Doing repairs on the outside of a rocket in mid-air is a ticklish job. We slacken off the circulation when they're right way up, so that they're half starved, and double the flow of surrogate when they're upside down. They learn to associate topsy-turvydom with well-being; in fact, they're only truly happy when they're standing on their heads.

"And now," Mr. Foster went on, "I'd like to show you some very interesting conditioning for Alpha Plus Intellectuals. We have a big batch of them on Rack 5. First Gallery level," he called to two boys who had started to go down to the ground floor.

"They're round about Metre 900," he explained. "You can't really do any useful intellectual conditioning till the foetuses have lost their tails. Follow me."

But the Director had looked at his watch. "Ten to three," he said. "No time for the intellectual embryos, I'm afraid. We must go up to the Nurseries before the children have finished their afternoon sleep."

Mr. Foster was disappointed. "At least one glance at the Decanting Room," he pleaded.

"Very well then." The Director smiled indulgently. "Just one glance."

CHAPTER TWO

Mr. Foster was left in the Decanting Room. The D.H.C. and his students stepped into the nearest lift and were carried up to the fifth floor.

INFANT NURSERIES. NEO-PAVLOVIAN CONDITIONING ROOMS, announced the notice board.

The Director opened a door. They were in a large bare room, very bright and sunny; for the whole of the southern wall was a single window. Half a dozen nurses, trousered and jacketed in the regulation white viscose-linen uniform, their hair aseptically hidden under white caps, were engaged in setting out bowls of roses in a long row across the floor. Big bowls, packed tight with blossom. Thousands of petals, ripe-blown and silkily smooth, like the cheeks of innumerable little cherubs, but of cherubs, in that bright light, not exclusively pink and Aryan, but also luminously Chinese, also Mexican, also apoplectic with too much blowing of celestial trumpets, also pale as death, pale with the posthumous whiteness of marble.

The nurses stiffened to attention as the D.H.C. came in.

"Set out the books," he said curtly.

In silence the nurses obeyed his command. Between the rose bowls the books were duly set out—a row of nursery quartos opened invitingly each at some gaily coloured image of beast or fish or bird.

"Now bring in the children."

They hurried out of the room and returned in a minute or two, each pushing a kind of tall dumb-waiter laden, on all its four wire-netted shelves, with eight-month-old babies, all exactly alike (a Bokanovsky Group, it was evident) and all (since their caste was Delta) dressed in khaki.

"Put them down on the floor."

The infants were unloaded.

"Now turn them so that they can see the flowers and books."

Turned, the babies at once fell silent, then began to crawl towards those clusters of sleek colours, those shapes so gay and brilliant on the white pages. As they approached, the sun came out of a momentary eclipse behind a cloud. The roses flamed up as though with a sudden passion from within; a new and profound significance seemed to suffuse the shining pages of the books. From the ranks of the crawling babies came little squeals of excitement, gurgles and twitterings of pleasure.

The Director rubbed his hands. "Excellent!" he said. "It might almost have been done on purpose."

The swiftest crawlers were already at their goal. Small hands reached out uncertainly, touched, grasped, unpetaling the transfigured roses, crumpling the illuminated pages of the books. The Director waited until all were happily busy. Then, "Watch carefully," he said. And, lifting his hand, he gave the signal.

The Head Nurse, who was standing by a switchboard at the other end of the room, pressed down a little lever.

There was a violent explosion. Shriller and ever shriller, a siren shrieked. Alarm bells maddeningly sounded.

The children started, screamed; their faces were distorted with terror.

"And now," the Director shouted (for the noise was deafening), "now we proceed to rub in the lesson with a mild electric shock."

He waved his hand again, and the Head Nurse pressed a second lever. The screaming of the babies suddenly changed its tone. There was something desperate, almost insane, about the sharp spasmodic yelps to which they now gave utterance. Their little bodies twitched and stiffened; their limbs moved jerkily as if to the tug of unseen wires.

"We can electrify that whole strip of floor," bawled the Director in explanation. "But that's enough," he signalled to the nurse.

The explosions ceased, the bells stopped ringing, the shriek of the siren died down from tone to tone into silence. The stiffly twitching bodies relaxed, and what had become the sob and yelp of infant maniacs broadened out once more into a normal howl of ordinary terror.

"Offer them the flowers and the books again."

The nurses obeyed; but at the approach of the roses, at the mere sight of those gaily-coloured images of pussy and cock-a-doodle-doo and baa-baa black sheep, the infants shrank away in horror; the volume of their howling suddenly increased.

"Observe," said the Director triumphantly, "observe."

Books and loud noises, flowers and electric shocks—already in the infant mind these couples were compromisingly linked; and after two hundred repetitions of the same or a similar lesson would be wedded indissolubly. What man has joined, nature is powerless to put asunder.

"They'll grow up with what the psychologists used to call an 'instinctive' hatred of books and flowers. Reflexes unalterably conditioned. They'll be safe from books and botany all their lives." The Director turned to his nurses. "Take them away again."

Still yelling, the khaki babies were loaded on to their dumb-waiters and wheeled out, leaving behind them the smell of sour milk and a most welcome silence.

One of the students held up his hand; and though he could see quite well why you couldn't have lower-caste people wasting the Community's time over books, and that there was always the risk of their reading something which might undesirably decondition one of their reflexes, yet . . . well, he couldn't understand about the flowers. Why go to the trouble of making it psychologically impossible for Deltas to like flowers?

Patiently the D.H.C. explained. If the children were made to scream at the sight of a rose, that was on grounds of high economic policy. Not so very long ago (a century or thereabouts), Gammas, Deltas, even Epsilons, had been conditioned to like flowers—flowers in particular and wild nature in general. The idea was to make them want to be going out into the country at every available opportunity, and so compel them to consume transport.

"And didn't they consume transport?" asked the student.

"Quite a lot," the D.H.C. replied. "But nothing else."

Primroses and landscapes, he pointed out, have one grave defect: they are gratuitous. A love of nature keeps no factories busy. It was decided to abolish the love of nature, at any rate among the lower classes; to abolish the love of nature, but *not* the tendency to consume transport. For of course it was essential that they should keep on going to the country, even though they hated it. The problem was to find an economically sounder reason for consuming transport than a mere affection for primroses and landscapes. It was duly found.

"We condition the masses to hate the country," concluded the Director. "But simultaneously we condition them to love all country sports. At the same time, we see to it that all country sports shall entail the use of elaborate apparatus. So that they consume manufactured articles as well as transport. Hence those electric shocks."

"I see," said the student, and was silent, lost in admiration. . . .

✿　　　✿　　　✿

Fifty yards of tiptoeing brought them to a door which the Director cautiously opened. They stepped over the threshold into the twilight of a shuttered dormitory. Eighty cots stood in a row against the wall. There was a sound of light regular breathing and a continuous murmur, as of very faint voices remotely whispering.

A nurse rose as they entered and came to attention before the Director.

"What's the lesson this afternoon?" he asked.

"We had Elementary Sex for the first forty minutes," she answered. "But now it's switched over to Elementary Class Consciousness."

The Director walked slowly down the long line of cots. Rosy and relaxed with sleep, eighty little boys and girls lay softly breathing. There was a whisper under every pillow. The D.H.C. halted and, bending over one of the little beds, listened attentively.

"Elementary Class Consciousness, did you say? Let's have it repeated a little louder by the trumpet."

At the end of the room a loud speaker projected from the wall. The Director walked up to it and pressed a switch.

". . . all wear green," said a soft but very distinct voice, beginning in the middle of a sentence, "and Delta Children wear khaki. Oh no, I don't want to play with Delta children. And Epsilons are still worse. They're too stupid to be able to read or write. Besides they wear black, which is such a beastly colour. I'm *so* glad I'm a Beta."

There was a pause; then the voice began again.

"Alpha children wear grey. They work much harder than we do, because they're so frightfully clever. I'm really awfully glad I'm a Beta, because I don't work so hard. And then we are much better than the Gammas and Deltas. Gammas are stupid. They all wear green, and Delta children wear khaki. Oh no, I *don't* want to play with Delta children. And Epsilons are still worse. They're too stupid to be able . . ."

The Director pushed back the switch. The voice was silent. Only its thin ghost continued to mutter from beneath the eighty pillows.

"They'll have that repeated forty or fifty times more before they wake; then again on Thursday, and again on Saturday. A hundred and twenty times three times a week for thirty months. After which they go on to a more advanced lesson."

Roses and electric shocks, the khaki of Deltas and a whiff of asafoetida —wedded indissolubly before the child can speak. But wordless conditioning is crude and wholesale; cannot bring home the finer distinctions, cannot inculcate the more complex courses of behaviour. For that there must be words, but words without reason. In brief, hypnopaedia.

"The greatest moralizing and socializing force of all time."

The students took it down in their little books. Straight from the horse's mouth.

Once more the Director touched the switch.

". . . so frightfully clever," the soft, insinuating, indefatigable voice was saying, "I'm really awfully glad I'm a Beta, because . . ."

Not so much like drops of water, though water, it is true, can wear holes in the hardest granite; rather, drops of liquid sealing-wax, drops that adhere, incrust, incorporate themselves with what they fall on, till finally the rock is all one scarlet blob.

"Till at last the child's mind *is* these suggestions, and the sum of the suggestions *is* the child's mind. And not the child's mind only. The adult's mind too—all his life long. The mind that judges and desires and decides —made up of these suggestions. But all these suggestions are *our* suggestions!" The Director almost shouted in his triumph. "Suggestions from the State." He banged the nearest table. "It therefore follows . . ."

A noise made him turn round.

"Oh, Ford!" he said in another tone, "I've gone and woken the children."

Comments on *Brave New World* Fifteen Years Later *

by Aldous Huxley (1946)

But *Brave New World* is a book about the future and, whatever its artistic or philosophical qualities, a book about the future can interest us only if its prophecies look as though they might conceivably come true. From our present vantage point, fifteen years further down the inclined plane of modern history, how plausible do its prognostications seem? What has happened in the painful interval to confirm or invalidate the forecasts of 1931?

* From the Foreword to *Brave New World,* Harper & Row, Publishers, New York. Copyright 1946 by Aldous Huxley. Reprinted by permission.

One vast and obvious failure of foresight is immediately apparent. *Brave New World* contains no reference to nuclear fission. That it does not is actually rather odd, for the possibilities of atomic energy had been a popular topic of conversation for years before the book was written. My old friend, Robert Nichols, had even written a successful play about the subject, and I recall that I myself had casually mentioned it in a novel published in the late twenties. So it seems, as I say, very odd that the rockets and helicopters of the seventh century of Our Ford should not have been powered by disintegrating nuclei. The oversight may not be excusable; but at least it can be easily explained. The theme of *Brave New World* is not the advancement of science as such; it is the advancement of science as it affects human individuals. The triumphs of physics, chemistry and engineering are tacitly taken for granted. The only scientific advances to be specifically described are those involving the application to human beings of the results of future research in biology, physiology and psychology. It is only by means of the sciences of life that the quality of life can be radically changed. The sciences of matter can be applied in such a way that they will destroy life or make the living of it impossibly complex and uncomfortable; but, unless used as instruments by the biologists and psychologists, they can do nothing to modify the natural forms and expressions of life itself. The release of atomic energy marks a great revolution in human history, but not (unless we blow ourselves to bits and so put an end to history) the final and most searching revolution.

This really revolutionary revolution is to be achieved, not in the external world, but in the souls and flesh of human beings. Living as he did in a revolutionary period, the Marquis de Sade very naturally made use of this theory of revolutions in order to rationalize his peculiar brand of insanity. Robespierre had achieved the most superficial kind of revolution, the political. Going a little deeper, Babeuf had attempted the economic revolution. Sade regarded himself as the apostle of the truly revolutionary revolution, beyond mere politics and economics—the revolution in individual men, women and children, whose bodies were henceforward to become the common sexual property of all and whose minds were to be purged of all the natural decencies, all the laboriously acquired inhibitions of traditional civilization. Between sadism and the really revolutionary revolution there is, of course, no necessary or inevitable connection. Sade was a lunatic and the more or less conscious goal of his revolution was universal chaos and destruction. The people who govern the *Brave New World* may not be sane (in what may be called the absolute sense of the word); but they are not madmen, and their aim is not anarchy but social stability. It is in order to achieve stability that they carry out, by scientific means, the ultimate, personal, really revolutionary revolution.

But meanwhile we are in the first phase of what is perhaps the penultimate revolution. Its next phase may be atomic warfare, in which case we do not have to bother with prophecies about the future. But it is conceivable . . . that we are capable of learning as much from Hiroshima as our forefathers learned from Magdeburg, [and that] we may look forward to a period, not indeed of peace, but of limited and only partially ruinous warfare. During that period it may be assumed that nuclear energy will be harnessed to industrial uses. The result, pretty obviously, will be a series of economic and social changes unprecedented in rapidity and completeness. All the existing patterns of human life will be disrupted and new patterns will have to be improvised to conform with the nonhuman fact of atomic power. Procrustes in modern dress, the nuclear scientist will prepare the bed on which mankind must lie; and if mankind doesn't fit—well, that will be just too bad for mankind. There will have to be some stretching and a bit of amputation—the same sort of stretching and amputations as have been going on ever since applied science really got into its stride, only this time they will be a good deal more drastic than in the past. These far from painless operations will be directed by highly centralized totalitarian governments. Inevitably so; for the immediate future is likely to resemble the immediate past, and in the immediate past rapid technological changes, taking place in a mass-producing economy and among a population predominantly propertyless, have always tended to produce economic and social confusion. To deal with confusion, power has been centralized and government control increased. It is probable that all the world's governments will be more or less completely totalitarian even before the harnessing of atomic energy; that they will be totalitarian during and after the harnessing seems almost certain. Only a large-scale popular movement toward decentralization and self-help can arrest the present tendency toward statism. At present there is no sign that such a movement will take place.

There is, of course, no reason why the new totalitarianisms should resemble the old. Government by clubs and firing squads, by artificial famine, mass imprisonment and mass deportation, is not merely inhumane (nobody cares much about that nowadays); it is demonstrably inefficient and in an age of advanced technology, inefficiency is the sin against the Holy Ghost. A really efficient totalitarian state would be one in which the all-powerful executive of political bosses and their army of managers control a population of slaves who do not have to be coerced, because they love their servitude. To make them love it is the task assigned, in present-

day totalitarian states, to ministries of propaganda, newspaper editors and
schoolteachers. But their methods are still crude and unscientific. The old
Jesuits' boast that, if they were given the schooling of the child, they
could answer for the man's religious opinions, was a product of wishful
thinking. And the modern pedagogue is probably rather less efficient at
conditioning his pupils' reflexes than were the reverend fathers who educated Voltaire. The greatest triumphs of propaganda have been accomplished, not by doing something, but by refraining from doing. Great is
truth, but still greater, from a practical point of view, is silence about
truth. By simply not mentioning certain subjects, by lowering what Mr.
Churchill calls an "iron curtain" between the masses and such facts or arguments as the local political bosses regard as undesirable, totalitarian
propagandists have influenced opinion much more effectively than they
could have done by the most eloquent denunciations, the most compelling
of logical rebuttals. But silence is not enough. If persecution, liquidation
and the other symptoms of social friction are to be avoided, the positive
sides of propaganda must be made as effective as the negative. The
most important Manhattan Projects of the future will be vast government-
sponsored enquiries into what the politicians and the participating scientists will call "the problem of happiness"—in other words, the problem of
making people love their servitude. Without economic security, the love
of servitude cannot possibly come into existence; for the sake of brevity, I
assume that the all-powerful executive and its managers will succeed in
solving the problem of permanent security. But security tends very
quickly to be taken for granted. Its achievement is merely a superficial,
external revolution. The love of servitude cannot be established except as
the result of a deep, personal revolution in human minds and bodies. To
bring about that revolution we require, among others, the following discoveries and inventions. First, a greatly improved technique of suggestion
—through infant conditioning and, later, with the aid of drugs, such as
scopolamine. Second, a fully developed science of human differences, enabling government managers to assign any given individual to his or her
proper place in the social and economic hierarchy. (Round pegs in square
holes tend to have dangerous thoughts about the social system and to in-
fect others with their discontents.) Third (since reality, however utopian,
is something from which people feel the need of taking pretty frequent
holidays), a substitute for alcohol and the other narcotics, something at
once less harmful and more pleasure-giving than gin or heroin. And
fourth (but this would be a long-term project, which it would take generations of totalitarian control to bring to a successful conclusion) a fool-
proof system of eugenics, designed to standardize the human product and
so to facilitate the task of the managers. In *Brave New World* this stan-

dardization of the human product has been pushed to fantastic, though not perhaps impossible, extremes. Technically and ideologically we are still a long way from bottled babies and Bokanovsky groups of semi-morons. But by A. F. 600, who knows what may not be happening? Meanwhile the other characteristic features of that happier and more stable world—the equivalents of soma and hypnopaedia and the scientific caste system—are probably not more than three or four generations away. Nor does the sexual promiscuity of *Brave New World* seem so very distant. There are already certain American cities in which the number of divorces is equal to the number of marriages. In a few years, no doubt, marriage licenses will be sold like dog licenses, good for a period of twelve months, with no law against changing dogs or keeping more than one animal at a time. As political and economic freedom diminishes, sexual freedom tends compensatingly to increase. And the dictator (unless he needs cannon fodder and families with which to colonize empty or conquered territories) will do well to encourage that freedom. In conjunction with the freedom to daydream under the influence of dope and movies and the radio, it will help to reconcile his subjects to the servitude which is their fate.

All things considered it looks as though Utopia were far closer to us than anyone, only fifteen years ago, could have imagined. Then, I projected it six hundred years into the future. Today it seems quite possible that the horror may be upon us within a single century. That is, if we refrain from blowing ourselves to smithereens in the interval. Indeed, unless we choose to decentralize and to use applied science, not as the end to which human beings are to be made the means, but as the means to producing a race of free individuals, we have only two alternatives to choose from: either a number of national, militarized totalitarianisms, having as their root the terror of the atomic bomb and as their consequence the destruction of civilization (or, if the warfare is limited, the perpetuation of militarism); or else one supra-national totalitarianism, called into existence by the social chaos resulting from rapid technological progress in general and the atomic revolution in particular, and developing, under the need for efficiency and stability, into the welfare-tyranny of Utopia. You pays your money and you takes your choice.

Editor's Note. Walden Two is an imaginary utopian community somewhere in the eastern United States. It was founded by an experimental psychologist named Frazier on the premise that "what was needed was a new conception of man, compatible with our scientific knowledge, which would lead to a philosophy of education bearing some relation to educational practices. But to achieve this, education would have to abandon the technical limitations which it had imposed upon itself and step forth into a broader sphere of human engineering. Nothing short of the complete revision of a culture would suffice." [1]

Walden Two *

by B. F. Skinner (1948)

"Mr. Castle," said Frazier very earnestly, "let me ask you a question. I warn you, it will be the most terrifying question of your life. *What would you do if you found yourself in possession of an effective science of behavior?* Suppose you suddenly found it possible to control the behavior of men as you wished. What would you do?". . .

"What would I do?" said Castle thoughtfully. "I think I would dump your science of behavior in the ocean."

"And deny men all the help you could otherwise give them?"

"And give them the freedom they would otherwise lose forever!"

"How could you give them freedom?"

"By refusing to control them!"

"But you would only be leaving the control in other hands."

"Whose?"

"The charlatan, the demagogue, the salesman, the ward heeler, the bully, the cheat, the educator, the priest—all who are now in possession of the techniques of behavioral engineering. . . . Many of its methods and techniques are really as old as the hills. Look at their frightful misuse in the hands of the Nazis! And what about the techniques of the psychological clinic? What about education? Or religion? Or practical politics? Or advertising and salesmanship? Bring them all together and you have a sort of rule-of-thumb technology of vast power. No, Mr. Castle, the science is there for the asking. But its techniques and methods are in the wrong hands—they are used for personal aggrandizement in a competi-

---

* From B. F. Skinner, *Walden Two,* The Macmillan Company, New York. Copyright 1948 by B. F. Skinner. Reprinted by permission.

1. B. F. Skinner, *Walden Two,* p. 312.

tive world or, in the case of the psychologist and educator, for futilely corrective purposes. My question is, have you the courage to take up and wield the science of behavior for the good of mankind? You answer that you would dump it in the ocean!"

"I'd want to take it out of the hands of the politicians and advertisers and salesmen, too."

"And the psychologists and educators? You see, Mr. Castle, you can't have that kind of cake. The fact is, we not only *can* control human behavior, we *must*. But who's to do it, and what's to be done?"

"So long as a trace of personal freedom survives, I'll stick to my position," said Castle, very much out of countenance.

"Isn't it time we talked about freedom?" I said. "We parted a day or so ago on an agreement to let the question of freedom ring. It's time to answer, don't you think?"

"My answer is simple enough," said Frazier. "I deny that freedom exists at all. I must deny it—or my program would be absurd. You can't have a science about a subject matter which hops capriciously about. Perhaps we can never *prove* that man isn't free; it's an assumption. But the increasing success of a science of behavior makes it more and more plausible. . . . We can achieve a sort of control under which the controlled, though they are following a code much more scrupulously than was ever the case under the old system, nevertheless *feel free*. They are doing what they want to do, not what they are forced to do. That's the source of the tremendous power of positive reinforcement—there's no restraint and no revolt. By a careful cultural design, we control not the final behavior, but the *inclination* to behave—the motives, the desires, the wishes.

"The curious thing is that in that case *the question of freedom never arises*. . . . The question of freedom arises when there is restraint— either physical or psychological.

"But restraint is only one sort of control, and absence of restraint isn't freedom. It's not control that's lacking when one feels 'free,' but the objectionable control of force. . . . When men strike for freedom, they strike against jails and the police, or the threat of them—against oppression. They never strike against forces which make them want to act the way they do. Yet, it seems to be understood that governments will operate only through force or the threat of force, and that all other principles of control will be left to education, religion, and commerce. If this continues to be the case, we may as well give up. A government can never create a free people with the techniques now allotted to it.

"The question is: Can men live in freedom and peace? And the answer is: Yes, if we can build a social structure which will satisfy the needs of everyone and in which everyone will want to observe the supporting code. But so far this has been achieved only in Walden Two. Your ruthless accusations to the contrary, Mr. Castle, this is the freest place on earth. And it is free precisely because we make no use of force or the threat of force. Every bit of our research, from the nursery through the psychological management of our adult membership, is directed toward that end—to exploit every alternative to forcible control. By skillful planning, by a wise choice of techniques we *increase* the feeling of freedom."

*   *   *

Frazier appeared just as we were finishing breakfast and announced that we were to spend the morning visiting the schools. . . .

A young woman in a white uniform met us in a small waiting room near the entrance [of the nursery]. Frazier addressed her as Mrs. Nash.

"I hope Mr. Frazier has warned you," she said with a smile, "that we're going to be rather impolite and give you only a glimpse of our babies. We try to protect them from infection during the first year. It's especially important when they are cared for as a group."

"What about the parents?" said Castle at once. "Don't parents see their babies?"

"Oh, yes, so long as they are in good health. Some parents work in the nursery. Others come around every day or so, for at least a few minutes. They take the baby out for some sunshine, or play with it in a play room." Mrs. Nash smiled at Frazier. "That's the way we build up the baby's resistance," she added.

She opened a door and allowed us to look into a small room, three walls of which were lined with cubicles, each with a large glass window. Behind the windows we could see babies of various ages. None of them wore more than a diaper, and there were no bedclothes. . . .

"This is a much more efficient way of keeping a baby warm than the usual practice of wrapping it in several layers of cloth," said Mrs. Nash, opening a safety-glass window to permit Barbara and Mary to look inside. "The newborn baby needs moist air at about 88 or 90 degrees. At six months, 80 is about right.". . .

"Clothing and blankets are really a great nuisance," said Mrs. Nash. "They keep the baby from exercising, they force it into uncomfortable postures—"

"When a baby graduates from our Lower Nursery," Frazier broke in, "it knows nothing of frustration, anxiety, or fear. It never cries except when sick, which is very seldom, and it has a lively interest in everything."

"But is it prepared for life?" said Castle. "Surely you can't continue to protect it from frustration or frightening situations forever."

"Of course not. But it can be prepared for them. We can build a tolerance for frustration by introducing obstacles gradually as the baby grows strong enough to handle them. But I'm getting ahead of our story. Have you any other point to make, Mrs. Nash?"

"I suppose you'd like to have them know how much work is saved. Since the air is filtered, we only bathe the babies once a week, and we never need to clean their nostrils or eyes. There are no beds to make, of course. And it's easy to prevent infection. The compartments are soundproofed, and the babies sleep well and don't disturb each other. We can keep them on different schedules, so the nursery runs smoothly. Let me see, is there anything else?"

"I think that's quite enough," said Frazier. "We have a lot of ground to cover this morning." . . .

The quarters for children from one to three consisted of several small playrooms with Lilliputian furniture, a child's lavatory, and a dressing and locker room. Several small sleeping rooms were operated on the same principle as the baby-cubicles. The temperature and the humidity were controlled so that clothes or bedclothing were not needed. The cots were double-decker arrangements of the plastic mattresses we had seen in the cubicles. The children slept unclothed, except for diapers. There were more beds than necessary, so that the children could be grouped according to developmental age or exposure to contagious diseases or need for supervision, or for educational purposes.

We followed Mrs. Nash to a large screened porch on the south side of the building, where several children were playing in sandboxes and on swings and climbing apparatuses. A few wore "training pants"; the rest were naked. Beyond the porch was a grassy play yard enclosed by closely trimmed hedges, where other children, similarly undressed, were at play. Some kind of marching game was in progress.

As we returned, we met two women carrying food hampers. They spoke to Mrs. Nash and followed her to the porch. In a moment five or six children came running into the playrooms and were soon using the lavatory and dressing themselves. Mrs. Nash explained that they were being taken on a picnic.

"What about the children who don't go?" said Castle. "What do you do about the green-eyed monster?"

Mrs. Nash was puzzled.

"Jealousy. Envy," Castle elaborated. "Don't the children who stay home ever feel unhappy about it?"

"I don't understand," said Mrs. Nash.

"And I hope you won't try," said Frazier, with a smile. "I'm afraid we must be moving along."

We said good-bye, and I made an effort to thank Mrs. Nash, but she seemed to be puzzled by that too, and Frazier frowned as if I had committed some breach of good taste.

"I think Mrs. Nash's puzzlement," said Frazier, as we left the building, "is proof enough that our children are seldom envious or jealous. . . . the emotions which breed unhappiness—are almost unknown here, like unhappiness itself. We don't need them any longer in our struggle for existence, and it's easier on our circulatory system, and certainly pleasanter, to dispense with them."

"If you've discovered how to do that, you are indeed a genius," said Castle. He seemed almost stunned as Frazier nodded assent. "We all know that emotions are useless and bad for our peace of mind and our blood pressure," he went on. "But how arrange things otherwise?"

"We arrange them otherwise here," said Frazier. He was showing a mildness of manner which I was coming to recognize as a sign of confidence.

"But emotions are—fun!" said Barbara. "Life wouldn't be worth living without them."

"Some of them, yes," said Frazier. "The productive and strengthening emotions—joy and love. But sorrow and hate—and the high voltage excitements of anger, fear, and rage—are out of proportion with the needs of modern life, and they're wasteful and dangerous. Mr. Castle has mentioned jealousy—a minor form of anger, I think we may call it. Naturally we avoid it. It has served its purpose in the evolution of man; we've no further use for it. If we allowed it to persist, it would only sap the life out of us. In a cooperative society there's no jealousy because there's no need for jealousy."

"That implies that you all get everything you want," said Castle. "But what about social possessions? Last night you mentioned the young man who chose a particular girl or profession. There's still a chance for jealousy there, isn't there?"

"It doesn't imply that we get everything we want," said Frazier. "Of course we don't. But jealousy wouldn't help. In a competitive world there's some point to it. It energizes one to attack a frustrating condition. The impulse and the added energy are an advantage. Indeed, in a competitive world emotions work all too well. Look at the singular lack of success of the complacent man. He enjoys a more serene life, but it's less likely to be a fruitful one. The world isn't ready for simple pacifism or Christian humility, to cite two cases in point. Before you can safely train out the destructive and wasteful emotions, you must make sure they're no longer needed."

"How do you make sure that jealousy isn't needed in Walden Two?" I said.

"In Walden Two problems can't be solved by attacking others," said Frazier with marked finality.

"That's not the same as eliminating jealousy, though," I said.

"Of course it's not. But when a particular emotion is no longer a useful part of a behavioral repertoire, we proceed to eliminate it."

"Yes, but how?"

"It's simply a matter of behavioral engineering," said Frazier.

"Behavioral engineering?" . . .

"The techniques have been available for centuries. We use them in education and in the psychological management of the community. . . .

"Each of us has interests which conflict with the interests of everybody else. That's our original sin, and it can't be helped. Now, 'everybody else' we call 'society.' It's a powerful opponent, and it always wins. Oh, here and there an individual prevails for a while and gets what he wants. Sometimes he storms the culture of a society and changes it slightly to his own advantage. But society wins in the long run, for it has the advantage of numbers and of age. Many prevail against one, and men against a baby. Society attacks early, when the individual is helpless. It enslaves him almost before he has tasted freedom. The 'ologies' will tell you how it's done. Theology calls it building a conscience or developing a spirit of selflessness. Psychology calls it the growth of the super-ego.

"Considering how long society has been at it, you'd expect a better job. But the campaigns have been badly planned and the victory has never been secure. The behavior of the individual has been shaped according to revelations of 'good conduct,' never as the result of experimental study. But why not experiment? The questions are simple enough. What's the best behavior for the individual so far as the group is concerned? And how can the individual be induced to behave in that way? Why not explore these questions in a scientific spirit?

"We could do just that in Walden Two. We had already worked out a code of conduct—subject, of course, to experimental modification. The code would keep things running smoothly if everybody lived up to it. Our job was to see that everybody did. Now, you can't get people to follow a useful code by making them into so many jacks-in-the-box. You can't foresee all future circumstances, and you can't specify adequate future conduct. You don't know what will be required. Instead you have to set up certain behavioral processes which will lead the individual to design his own 'good' conduct when the time comes. We call that sort of thing 'self-control.' But don't be misled, the control always rests in the last analysis in the hands of society.

"One of our Planners, a young man named Simmons, worked with me. It was the first time in history that the matter was approached in an experimental way. . . . Simmons and I began by studying the great works on morals and ethics—Plato, Aristotle, Confucius, the New Testament, the Puritan divines, Machiavelli, Chesterfield, Freud—there were scores of them. We were looking for any and every method of shaping human behavior by imparting techniques of self-control. Some techniques were obvious enough, for they had marked turning points in human history. 'Love your enemies' is an example—a psychological invention for easing the lot of an oppressed people. . . .

"When Simmons and I had collected our techniques of control, we had to discover how to teach them. That was more difficult. Current educational practices were of little value, and religious practices scarcely any better. Promising paradise or threatening hell-fire is, we assumed, generally admitted to be unproductive. It is based upon a fundamental fraud which, when discovered, turns the individual against society and nourishes the very thing it tries to stamp out. What Jesus offered in return for loving one's enemies was heaven *on earth*, better known as peace of mind.

"We found a few suggestions worth following in the practices of the clinical psychologist. We undertook to build a tolerance for annoying experiences. The sunshine of midday is extremely painful if you come from a dark room, but take it in easy stages and you can avoid pain altogether. The analogy can be misleading, but in much the same way it's possible to build a tolerance to painful or distasteful stimuli, or to frustration, or to situations which arouse fear, anger or rage. Society and nature throw these annoyances at the individual with no regard for the development of tolerances. Some achieve tolerances, most fail. Where would the science of immunization be if it followed a schedule of accidental dosages?

"Take the principle of 'Get thee behind me, Satan,' for example," Frazier continued. "It's a special case of self-control by altering the environment. Subclass A 3, I believe. We give each child a lollipop which has been dipped in powdered sugar so that a single touch of the tongue can be detected. We tell him he may eat the lollipop later in the day, provided it hasn't already been licked. Since the child is only three or four, it is a fairly diff——"

"Three or four!" Castle exclaimed.

"All our ethical training is completed by the age of six," said Frazier quietly. "A simple principle like putting temptation out of sight would be acquired before four. But at such an early age the problem of not licking the lollipop isn't easy. Now, what would you do, Mr. Castle, in a similar situation?"

"Put the lollipop out of sight as quickly as possible."

"Exactly. I can see you've been well trained. Or perhaps you discovered the principle for yourself. We're in favor of original inquiry wherever possible, but in this case we have a more important goal and we don't hesitate to give verbal help. First of all, the children are urged to examine their own behavior while looking at the lollipops. This helps them to recognize the need for self-control. Then the lollipops are concealed, and the children are asked to notice any gain in happiness or any reduction in tension. Then a strong distraction is arranged—say, an interesting game. Later the children are reminded of the candy and encouraged to examine their reaction. The value of the distraction is generally obvious. Well, need I go on? When the experiment is repeated a day or so later, the children all run with the lollipops to their lockers and do exactly what Mr. Castle would do—a sufficient indication of the success of our training."

"I wish to report an objective observation of my reaction to your story," said Castle, controlling his voice with great precision. "I find myself revolted by this display of sadistic tyranny."

"I don't wish to deny you the exercise of an emotion which you seem to find enjoyable," said Frazier. "So let me go on. Concealing a tempting but forbidden object is a crude solution. For one thing, it's not always feasible. We want a sort of psychological concealment—covering up the candy by paying no attention. In a later experiment the children wear their lollipops like crucifixes for a few hours."

" 'Instead of the cross, the lollipop,
About my neck was hung,' "

said Castle.

"I wish somebody had taught me that, though," said Rodge, with a glance at Barbara.

"Don't we all?" said Frazier. "Some of us learn control, more or less by accident. The rest of us go all our lives not even understanding how it is possible, and blaming our failure on being born the wrong way."

"How do you build up a tolerance to an annoying situation?" I said.

"Oh, for example, by having the children 'take' a more and more painful shock, or drink cocoa with less and less sugar in it until a bitter concoction can be savored without a bitter face."

"But jealousy or envy—you can't administer them in graded doses," I said.

"And why not? Remember, we control the social environment, too, at this age. That's why we get our ethical training in early. Take this case. A group of children arrive home after a long walk tired and hungry. They're expecting supper; they find, instead, that it's time for a lesson in self-control: they must stand for five minutes in front of steaming bowls of soup.

"The assignment is accepted like a problem in arithmetic. Any groaning or complaining is a wrong answer. Instead, the children begin at once to work upon themselves to avoid any unhappiness during the delay. One of them may make a joke of it. We encourage a sense of humor as a good way of not taking an annoyance seriously. The joke won't be much, according to adult standards—perhaps the child will simply pretend to empty the bowl of soup into his upturned mouth. Another may start a song with many verses. The rest join in at once, for they've learned that it's a good way to make time pass. . . .

"In a later stage we forbid all social devices. No songs, no jokes— merely silence. Each child is forced back upon his own resources—a very important step."

"I should think so," I said. "And how do you know it's successful? You might produce a lot of silently resentful children. It's certainly a dangerous stage."

"It is, and we follow each child carefully. If he hasn't picked up the necessary techniques, we start back a little. A still more advanced stage" —Frazier glanced again at Castle, who stirred uneasily—"brings me to my point. When it's time to sit down to the soup, the children count off— heads and tails. Then a coin is tossed and if it comes up heads, the 'heads' sit down and eat. The 'tails' remain standing for another five minutes."

Castle groaned.

"And you call that envy?" I asked.

"Perhaps not exactly," said Frazier. "At least there's seldom any aggression against the lucky ones. The emotion, if any is directed against Lady Luck herself, against the toss of the coin. That, in itself, is a lesson worth learning, for it's the only direction in which emotion has a surviving chance to be useful. And resentment toward things in general, while perhaps just as silly as personal aggression, is more easily controlled. Its expression is not socially objectionable." . . .

"May you not inadvertently teach your children some of the very emotions you're trying to eliminate?" I said. "What's the effect, for example, of finding the anticipation of a warm supper suddenly thwarted? Doesn't that eventually lead to feelings of uncertainty, or even anxiety?"

"It might. We had to discover how often our lessons could be safely administered. But all our schedules are worked out experimentally. We watch for undesired consequences just as any scientist watches for disrupting factors in his experiments.

"After all, it's a simple and sensible program," he went on in a tone of appeasement. "We set up a system of gradually increasing annoyances and frustrations against a background of complete serenity. An easy environment is made more and more difficult as the children acquire the capacity to adjust." . . .

"They don't sound happy or free to me, standing in front of bowls of Forbidden Soup," said Castle, answering me parenthetically while continuing to stare at Frazier.

"If I must spell it out," Frazier began with a deep sigh, "what they get is escape from the petty emotions which eat the heart out of the unprepared. They get the satisfaction of pleasant and profitable social relations on a scale almost undreamed of in the world at large. They get immeasurably increased efficiency, because they can stick to a job without suffering the aches and pains which soon beset most of us. They get new horizons, for they are spared the emotions characteristic of frustration and failure. They get—" His eyes searched the branches of the trees. "Is that enough?" he said at last.

"And the community must gain their loyalty," I said, "when they discover the fears and jealousies and diffidences in the world at large."

"I'm glad you put it that way," said Frazier. "You might have said that they must feel superior to the miserable products of our public schools. But we're at pains to keep any feeling of superiority or contempt under control, too. Having suffered most acutely from it myself, I put the subject first on our agenda. We carefully avoid any joy in a personal triumph which means the personal failure of somebody else. We take no pleasure in the sophistical, the disputative, the dialectical." He threw a vicious glance at Castle. "We don't use the motive of domination, because we are always thinking of the whole group. We could motivate a few geniuses that way—it was certainly my own motivation—but we'd sacrifice some of the happiness of everyone else. Triumph over nature and over oneself, yes. But over others, never."

"You've taken the mainspring out of the watch," said Castle flatly. . . .

"Are your techniques really so very new?" I said hurriedly. "What about the primitive practice of submitting a boy to various tortures before granting him a place among adults? What about the disciplinary techniques of Puritanism? Or of the modern school, for that matter?"

"In one sense you're right," said Frazier. "And I think you've nicely answered Mr. Castle's tender concern for our little ones. The unhappinesses we deliberately impose are far milder than the normal unhappinesses from which we offer protection. Even at the height of our ethical training, the unhappiness is ridiculously trivial—to the well-trained child.

"But there's a world of difference in the way we use these annoyances," he continued. "For one thing, we don't punish. We never administer an unpleasantness in the hope of repressing or eliminating undesirable behavior. But there's another difference. In most cultures the child meets up with annoyances and reverses of uncontrolled magnitude. Some are imposed in the name of discipline by persons in authority. Some, like hazings, are condoned though not authorized. Others are merely accidental. No one cares to, or is able to, prevent them.

"We all know what happens. A few hardy children emerge, particularly those who have got their unhappiness in doses that could be swallowed. They become brave men. Others become sadists or masochists of varying degrees of pathology. Not having conquered a painful environment, they become preoccupied with pain and make a devious art of it. Others submit—and hope to inherit the earth. The rest—the cravens, the cowards—live in fear for the rest of their lives. And that's only a single field—the reaction to pain. . . . The English public school of the nineteenth century produced brave men—by setting up almost insurmountable barriers and making the most of the few who came over. But selection isn't education. Its crops of brave men will always be small, and the waste enormous. Like all primitive principles, selection serves in place of education only through a profligate use of material. Multiply extravagantly and select with rigor. It's the philosophy of the 'big litter' as an alternative to good child hygiene.

"In Walden Two we have a different objective. We make every man a brave man. They all come over the barriers. Some require more preparation than others, but they all come over. The traditional use of adversity is to select the strong. We control adversity to build strength. And we do it deliberately, no matter how sadistic Mr. Castle may think us, in order to prepare for adversities which are beyond control. Our children eventually experience the 'heartache and the thousand natural shocks that flesh is heir to.' It would be the cruelest possible practice to protect them as long as possible, especially when we *could* protect them so well."

＊   ＊   ＊

Frazier began to pace back and forth, his hands still thrust in his pockets.

"My hunch is—and when I feel this way about a hunch, it's never wrong—that we shall eventually find out, not only what makes a child mathematical, but how to make better mathematicians! If we can't solve a problem, we can create men who can! And better artists! And better craftsmen!" He laughed and added quietly, "And better behaviorists, I suppose!

"And all the while we shall be improving upon our social and cultural design. We know almost nothing about the special capacities of the *group*. We all recognize that there are problems which can't be solved by an individual—not only because of limitations of time and energy but because the individual, no matter how extraordinary, can't master all the aspects, can't think thoughts big enough. Communal science is already a reality, but who knows how far it can go? Communal authorship, communal art, communal music—these are already exploited for commercial purposes, but who knows what might happen under freer conditions?

"The problem of efficient group structure alone is enough to absorb anyone's interest. An organization of a committee of scientists or a panel of script writers is far from what it could be. But we lack control in the world at large to investigate more efficient structures. Here, on the contrary—here we can begin to understand and build the Superorganism. We can construct groups of artists and scientists who will act as smoothly and efficiently as champion football teams.

"And all the while, Burris, we shall be increasing the net power of the community by leaps and bounds."

The Scientific View of Man

Comments on Walden Two *
by Floyd W. Matson (1964)

Here, then, in textbook and storybook, is the outline of a total science of human behavior—at once descriptive and prescriptive, pure and applied—which has emerged from the experiments and scientific premises of a leading contemporary exponent of behaviorism. Skinner's behavioral science is, if anything, more uncompromising in its abolition of mind and purpose—not to mention the aspirations of what a prescientific age was wont to call the human spirit—than the classical doctrine of Watson himself. It is an altogether explicit disavowal of the freedom and responsibility, as well as the moral and political primacy, of the human person; and it is, correspondingly, an avowal of the superior value and importance of "society." Skinner's projection of the scientific utopia—a culture designed and engineered entirely on the rational counsel of behavioral science—is an unflinching extrapolation from the data, the methods and the underlying world view of the scientific mechanist. Given these assumptions about human nature and, more specifically, about the involuntary springs of action; given this expert knowledge concerning the new techniques, whether coercive or subtly palatable, of human conditioning and control; given the urge to practice these skills upon humanity in the service of a vision of perfection and efficiency—given all this, it is hard to see how Skinner's prescriptive "design for a culture" might be improved upon.

❀   ❀   ❀

To be sure, such a scientific view "is distasteful to most of those who have been strongly affected by democratic philosophies." But there is no help for it; the old way of thinking is doomed. Behavioral science, it now appears, is by no means a neutral instrument to be used however we see fit; once its mechanism is wound up and set going, it points its own direction and follows its own daemon. The creature thus comes to dominate its creator, *and this is as it should be;* our deepest beliefs and oldest traditions are helpless to check or divert its forward march. It would seem that all that is left for modern man is to regard the movement of science—specifically, of behavioral science—not as a juggernaut but as a band-wagon, and to climb aboard while there is time.

Matter, Mind, and Man *

by Edmund W. Sinnott (1957)

[If man] is a physical system, however exquisitely complex, he must obey the laws of the physical universe and is as subject to their control as any machine. As to just how this control operates there are two somewhat different ideas. It may be thought of as coming chiefly from within, as in an automatic mechanism. This is the general view of physiology. Or the control may be thought of as primarily external, imposed on a rather neutral, plastic system by factors in its environment. This is a view commonly held by psychologists.

It is here that the science of behavior carries still further the arguments drawn from the physical sciences and biology and applies them more closely to man himself. If man is a mechanism it ought to be possible to deal with him as we do with any other machine and to determine his behavior as we wish. Many psychologists believe that this can be done and that the highest hope for man is in his being so wisely guided and controlled that he will become, so to speak, a custom-made saint who never does wrong nor ever wants to do so. Since his acts are determined not by himself but by forces outside himself, his leaders can take the control of human progress into their own hands. A man is an "empty" organism, and everything he is and does is the result of environmental factors working upon him.

One of the leaders of this school of thought, Professor B. F. Skinner, puts it thus: "Just as biographers and critics look for external influences to account for the traits and achievements of the men they study, so science ultimately explains behavior in terms of 'causes' or conditions which lie beyond the individual himself. . . . Every discovery of an event which has a part in shaping a man's behavior seems to leave so much the less to be credited to the man himself; and as such explanations become more and more comprehensive, the contribution which may be claimed by the individual himself appears to approach zero. Man's vaunted creative powers, his original accomplishments in art, science and morals, his capacity to choose and our right to hold him responsible for the consequences of his choice—none of these is conspicuous in this new self-portrait." [1]

The techniques which science will use to control behavior are those of "supplying information, presenting opportunities for action, pointing out logical relationships, appealing to reason or 'enlightened understanding,'" and presumably more subtle forms of conditioning made possible by science. These are as much methods for the control of behavior, says Professor Skinner, as the bully's threat of force. We have long recognized education and persuasion as means of getting people to do what we want them to do, though often with only partial success, but these new planners hope to go much further and make men virtuous and happy almost automatically.

This challenging proposal to make over human nature and solve the problems of a troubled world by a judicial application of the principles of psychology has a wide appeal. If men's minds could thus be molded to insure any desired pattern of behavior, this would surely result in a radical change in the lives of all of us, though whether or not it would usher in the millennium may perhaps be doubted. The idea presents some grave questions, however. If man is indeed the tractable and docile creature that the psychologist seems to think him, our ideas about him must be greatly changed. Where, we may ask, is now that freedom we thought he prized so highly? It would have to be surrendered; not, indeed, to physical determinism but to the edicts of a culture-planner or a brain-washer. And where goes moral responsibility, the concept of what "ought" to be? What about values? What is the significance of man's passionate love of beauty, of the poet's rhapsody, the mystic's vision? Can proper conditioning confer the ability to create a work of art? And who would decide what sort of behavior to inculcate? Such questions are not altogether frivolous. Whether or not man can be so readily controlled will depend on whether he is indeed an "empty" organism, a blank sheet of paper, or has instead an inner quality that makes him resist environmental influences.

If life is mechanical, as both physiologist and psychologist believe, and thus is rigidly controlled, we may ask what place there is in an organism for mind, for any immaterial agency to guide its activities. This question of the relation between body and mind is one of the oldest with which philosophy has to deal and still remains unsettled. An automatic machine presumably does not have a mind, and mind in us may perhaps be only an illusion, a sort of by-product of physical acts; an epiphenomenon, pro-

1. B. F. Skinner, "Freedom and the Control of Men," *American Scholar*, 25:47–65, 1955–56.

duced by them as inevitably as the picture on a television screen is formed by its inner mechanism. The instinctive and common-sense opinion that we have minds and that they control what we do is hard to reconcile with the mechanical view of man that science often takes. In trying to follow both these concepts today we are in danger of becoming hopelessly divided ourselves.

But if mind is hard to find a place for in the human organism, what of the soul? the spirit? human values generally? The challenge of physiology and psychology to the old ideas about man's nature is evidently far more serious than that of evolution. Man might come to think of himself as an ennobled animal, moving up the evolutionary stairway toward a divine nature, and thus regain some of his old assurance and sense of cosmic significance; but to think that he is simply "an eddy in a stream of energy," a mechanism buffeted by fate and circumstance and tossed at last into the rubbish heap—such a concept completely destroys his traditional ideas about himself.

Man's true nature is therefore the real problem—whether he is a child of God with an immortal soul and actually a part of the great spiritual power that rules all nature, or whether he is simply a clever brute, risen out of the primordial slime; a chemical mechanism that has evolved into a glorified calculating machine whose aspirations, seemingly so exalted, are nothing but motions among molecules, a puppet whose fate is no longer in his own hands. The serious division in the world today is not so much on social or economic or political issues as on this basic question of what man really is. Until we can agree on this we shall never be able to build a peaceful and satisfying social order for we shall have no common philosophy on which it can be founded. What man believes about *himself* is of the utmost moment for it will determine the kind of world that he will make and even his own fate.

Free Will *

by Loren Eiseley (1960, 1958)

DETERMINISM

The "within," man's subjective nature, and the things that come to him from without often bear a striking relationship. Man cannot be long studied as an object without his cleverly altering his inner defenses. He thus becomes a very difficult creature with which to deal. Let me illustrate this concretely.

Not long ago, hoping to find relief from the duties of my office, I sought refuge with my books on a campus bench. In a little while there sidled up to me a red-faced derelict whose parboiled features spoke eloquently of his particular weakness. I was resolved to resist all blandishments for a handout.

"Mac," he said, "I'm out of a job. I need help."

I remained stonily indifferent.

"Sir," he repeated.

"Uh huh," I said, unwillingly looking up from my book.

"I'm an alcoholic."

"Oh," I said. There didn't seem to be anything else to say. You can't berate a man for what he's already confessed to you.

"I'm an alcoholic," he repeated. "You know what that means? I'm a sick man. Not giving me alcohol is ill-treating a sick man. I'm a sick man. I'm an alcoholic. I have to have a drink. I'm telling you honest. It's a disease. I'm an alcoholic. I can't help myself."

"Okay," I said, "you're an alcoholic." Grudgingly I contributed a quarter to his disease and his increasing degradation. But the words stayed in my head. "I can't help myself." Let us face it. In one disastrous sense, he was probably right. At least at this point. But where had the point been reached, and when had he developed this clever neo-modern, post-Freudian panhandling lingo? From what judicious purloining of psychiatric or social-work literature and lectures had come these useful phrases?

And he had chosen his subject well. At a glance he had seen from my book and spectacles that I was susceptible to this approach. I was immersed in the modern dilemma. I could have listened, gazing into his mottled face, without an emotion if he had spoken of home and mother. But he was an alcoholic. He knew it and he guessed that I might be a scientist. He had to be helped from the outside. It was not a moral problem. He was ill.

I settled uncomfortably into my book once more, but the phrase stayed with me, "I can't help myself." The clever reversal. The outside judgment turned back and put to dubious, unethical use by the man inside.

"I can't help myself." It is the final exteriorization of man's moral predicament, of his loss of authority over himself. It is the phrase that, above all others, tortures the social scientist. In it is truth, but in it also is a dreadful, contrived folly. It is society, a genuinely sick society, saying to its social scientists, as it says to its engineers and doctors: "Help me. I'm rotten with hate and ignorance that I won't give up, but you are the doctor;

fix me." This, says society, is *our* duty. We are social scientists. Individuals, poor blighted specimens, cannot assume such responsibilities. "True, true," we mutter as we read the case histories. "Life is dreadful, and yet—"

Man on the inside is quick to accept scientific judgments and make use of them. He is conditioned to do this. This new judgment is an easy one; it deadens man's concern for himself. It makes the way into the whirlpool easier. In spite of our boasted vigor we wait for the next age to be brought to us by Madison Avenue and General Motors. We do not prepare to go there by means of the good inner life. We wait, and in the meantime it slowly becomes easier to mistake longer cars or brighter lights for progress. And yet—

Forty thousand years ago in the bleak uplands of southwestern Asia, a man, a Neanderthal man, once labeled by the Darwinian proponents of struggle as a ferocious ancestral beast—a man whose face might cause you some slight uneasiness if he sat beside you—a man of this sort existed with a fearful body handicap in that ice-age world. He had lost an arm. But still he lived and was cared for. Somebody, some group of human things, in a hard, violent and stony world, loved this maimed creature enough to cherish him.

And looking so, across the centuries and the millennia, toward the animal men of the past, one can see a faint light, like a patch of sunlight moving over the dark shadows on a forest floor. It shifts and widens, it winks out, it comes again, but it persists. It is the human spirit, the human soul, however transient, however faulty men may claim it to be. In its coming man had no part. It merely came, that curious light, and man, the animal, sought to be something that no animal had been before. Cruel he might be, vengeful he might be, but there had entered into his nature a curious wistful gentleness and courage. It seemed to have little to do with survival, for such men died over and over. They did not value life compared to what they saw in themselves—that strange inner light which has come from no man knows where, and which was not made by us. It has followed us all the way from the age of ice, from the dark borders of the ancient forest into which our footprints vanish.

❃   ❃   ❃

INDETERMINISM

Evolution, if it has taught us anything, has taught us that life is infinitely creative. Whether one accepts Henri Bergson's view of the process or not, one of the profoundest remarks he ever made was the statement that "the role of life is to insert some *indetermination* into matter." An advanced brain capable of multiple choices is represented on this planet only by man. He is a "reservoir of indetermination" containing infinite possibilities of good and evil. He is nature's greatest attempt to escape the blind subservience of the lower world to instinct and those evolutionary forces which, in all other forms of life, channel its various manifestations into constricted nooks and crannies of the environment. Wallace saw, and saw correctly, that with the rise of man the evolution of parts was to a marked degree outmoded, that mind was now the arbiter of human destiny.

❊   ❊   ❊

We have spoken of the brain of man as a sort of organ of indetermination. We have seen through Wallace its ability to escape from mechanical specialization, its creation of a freedom unknown to any other creature on the planet. Ironically enough, that freedom, that power of choice on the part of man, represents in a curious way the belated triumph of Erasmus Darwin and Lamarck.[1]

Here at last volition has taken its place in the world of nature. It was not perhaps quite the place these evolutionists had foreseen, but in the end its part in the cultural drama of man could not be gainsaid by their scientific successors. The mind of man, by indetermination, by the power of choice and cultural communication, by the great powers of thought, is on the verge of escape from the blind control of that deterministic world with which the Darwinists had unconsciously shackled man. The inborn characteristics laid upon him by the biological extremists have crumbled away. Man is many things—he is protean, elusive, capable of great good and appalling evil. He is what he is—a reservoir of indeterminism. He represents the genuine triumph of volition, life's near evasion of the forces that have molded it. In the West of our day only one anachronistic force threatens man with the ruin of that hope. It is his confusion of the word "progress" with the mechanical extensions which represent his triumph over the primeval wilderness of biological selection. This confusion represents, in a way, a reversion. It is a failure to see that the triumph of the machine without an accompanying inner triumph represents an atavistic return to the competition and extermination represented in the old biological evolution of "parts." . . . Transcendence of self is not to be sought in the outer world or in mechanical extensions. These are merely another version of specialized evolution. They can be used for human benefit if one recognizes them for what they are, but they must never be confused with that other interior kingdom in which man is forever free to be better

1. David Bidney, *Theoretical Anthropology*, Columbia University Press, p. 82.

than what he knows himself to be. It is there that the progress of which he dreams is at last to be found. It is the thing that his great moral teachers have been telling him since man was man. This is his true world; the other, the mechanical world which tickles his fancy, may be useful to good men but it is not in itself good. It takes its color from the minds behind it and this man has not learned. When he does so he will have achieved his final escape from the world which Darwin saw and pictured.

Our Altered Conception of Ourselves *
by Archibald MacLeish (1968)

We, as Americans, we perhaps as members of our generation on this earth, have somehow lost control of the management of our human affairs, of the direction of our lives, of what our ancestors would have called our destiny.

It is a sense we have had in one form or another for a long time now, but not as an explicit, a formulated fear until we found ourselves deep in the present century with its faceless slaughters, its mindless violence, its fabulous triumphs over space and time and matter ending in terrors space and time and matter never held. Before that there were only hints and intimations, but they were felt, they were recorded where all the hints and intimations are recorded—in poems, fictions, works of art. From the beginning of what we used to call the industrial revolution—what we see today more clearly as a sort of technological coup d'état—men and women, particularly men and women of imaginative sensibility, have seen that something was happening to the human role in the shaping of civilization.

A curious automatism, human in origin but not human in action, seemed to be taking over. Cities were being built and rebuilt not with human purposes in mind but with technological means at hand. It was no longer the neighborhood which fixed the shape and limits of the town but the communications system, the power grid. Technology, our grand-

* From Archibald MacLeish, "The Great American Frustration," *Saturday Review*, July 13, 1968. Copyright 1968 by Saturday Review, Inc. Reprinted by permission. Based on the first Milton Eisenhower Lecture delivered by Mr. MacLeish at the Johns Hopkins University.

fathers said, "advanced" and it was literally true: it was technology which was beating the tambours, leading the march. Buildings crowded into the air not because their occupants had any particular desire to lift them there, but because the invention of electric elevators and new methods of steel and glass construction made these ziggurats possible and the possibility presented itself as economic compulsion.

Wildness and silence disappeared from the countryside, sweetness fell from the air, not because anyone wished them to vanish or fall but because throughways had to floor the meadows with cement to carry the automobiles which advancing technology produced first by the thousands and then by the thousand thousands. Tropical beaches turned into high-priced slums where thousand-room hotels elbowed each other for glimpses of once-famous surf not because those who loved the beaches wanted them there but because enormous jets could bring a million tourists every year—and therefore did.

The result, seen in a glimpse here, a perception there, was a gradual change in our attitude toward ourselves as men, toward the part we play as men in the direction of our lives. It was a confused change. We were proud—in England, and even more in America, raucously proud—of our technological achievements, but we were aware also, even from the beginning, that these achievements were not altogether ours or, more precisely, not altogether ours to direct, to control—that the *process* had somehow taken over leaving the purpose to shift for itself so that we, the ostensible managers of the process, were merely its beneficiaries.

Not, of course, that we complained of that, at least in the beginning. A hundred years ago, with the rare exception of a Dickens or a Zola, we were amenable enough—amenable as children at a Christmas party. Inventions showered on our heads: steam engines and electric lights and telegraph messages and all the rest. We were up to our knees, to our necks, in Progress. And technology had made it all possible. Science was the giver of every good and perfect gift. If there were aspects of the new world which were not perfect—child labor for example—progress would take care of them. If the ugliness and filth and smoke of industrial cities offended us, we put up with them for the sake of the gas lights and the central heating. We were rich and growing richer.

But nevertheless the uneasiness remained and became more and more evident in our books, our paintings, our music—even the new directions of our medical sciences. Who were *we* in this strange new world? What part did *we* play in it? Someone had written a new equation somewhere, pushed the doors of ignorance back a little, entered the darkened room of knowledge by one more step. Someone else had found a way to make use of that new knowledge, put it to work. Our lives had changed but without *our* changing them, without our intending them to change. Improvements had appeared and we had accepted them. We had bought Mr. Ford's machines by the hundreds of thousands. We had ordered radios by the millions and then installed TVs. And now we took to the air, flew from city to city, from continent to continent, from climate to climate, following summer up and down the earth like birds. We were new men in a new life in a new world . . . but a world *we* had not made—had not, at least, intended to make.

And a new world, moreover, that we were increasingly unsure, as time went by, we would have wanted to make. We wanted its conveniences, yes. Its comforts, certainly. But the world as a world to live in? As a human world? It was already obvious by the beginning of this century that many of our artists and writers—those not so silent observers of the human world who sit in its windows and lurk in its doorways watching— were not precisely in love with the modern world, were, indeed, so little in love with it that they had turned against life itself, accepting absurdity and terror in its place and making of human hopelessness the only human hope. And there were other nearer, stranger witnesses. Before the century was two-thirds over numbers of our children—extraordinary numbers if you stop to think about it—were to reject, singly and secretly, or publicly in curious refugee encampments, the whole community of our modern lives, and most particularly those aspects of our lives which were most modern: their conveniences, their comforts . . . their affluence.

It was inevitable under these circumstances that some sort of confrontation should occur between the old idea of man as the liver of his own life, the shaper of his own existence, and the new idea of world, the newly autonomous world—world autonomous in its economic laws, as the Marxists hoped, or autonomous in its scientific surge, its technological compulsions, as some in the West began to fear. And, of course, the confrontation did occur: first in rather fatuous academic ructions in which science and the humanities were made to quarrel with each other in the universities, and then, in 1945, at Hiroshima. What happened at Hiroshima was not only that a scientific breakthrough—"breakthrough" in the almost literal sense—had occurred and that a great part of the population of a city had been burned to death, but that the problem of the relation of the triumphs of modern science to the human purposes of man had been explicitly defined and the whole question of the role of humanity in the modern science age had been exposed in terms not even the most unthinking could evade.

Prior to Hiroshima it had still been possible—increasingly difficult but still possible—to believe that science was by nature a human tool obedient to human wishes and that the world science and its technology could create would therefore be a human world reflecting our human needs, our human purposes. After Hiroshima it was obvious that the loyalty of science was not to humanity but to truth—its own truth—and that the law of science was not the law of the good—what humanity thinks of as good, meaning moral, decent, humane—but the law of the possible. What it is *possible* for science to know science must know. What it is possible for technology to do technology will have done. If it is possible to split the atom, then the atom must be split. Regardless. Regardless of . . . anything.

There was a time, just after Hiroshima, when we tried—we in the United States, at least—to escape from that haunting problem by blaming the scientists as individuals: the scientists, in particular, who had made the bomb—the mysterious workers in the cellars at Stagg Field and the laboratories of the Manhattan Project. And the scientists themselves, curious as it now may seem, cooperated; many of them, many of the best, assuming, or attempting to assume, burdens of personal guilt or struggling, somehow, anyhow, to undo what had been done.

I remember—more vividly perhaps than anything else which happened to me in those years—a late winter evening after Hiroshima in a study at the Institute at Princeton—Einstein's study, I think—when Niels Bohr, who was as great a man as he was a physicist, walked up and down for hours beside the rattling radiators urging me to go to President Truman, whom I did not know, to remind him that there had been an understanding between Mr. Roosevelt and the scientists about the future neutralization of the bomb. I guessed that Bohr, even as he talked that evening, realized there was nothing Mr. Truman or anyone on earth could do to unknow what was known. And yet he walked up and down the freezing study talking. Things, of course, *were* "done"—attempted anyway. In the brief time when we alone possessed what was called "the secret," the American Government offered to share it with the world (the Baruch Plan) for peaceful exploitation. What we proposed, though we did not put it in these words, was that humanity as a whole should assert its control of science, or at least of this particular branch of science, nuclear physics, limiting its pursuit of possibility to possibilities which served mankind. But the Russians, with their faith in the dialectics of matter, demurred. They preferred to put their trust in *things,* and within a few short months their trust was justified: they had the bomb themselves.

The immediate effect in the United States was of course, the soaring fear of Russia which fed the Cold War abroad and made the black plague of McCarthyism possible at home. But there was also a deeper and more enduring consequence. Our original American belief in our human capability, our human capacity to manage our affairs ourselves, "govern ourselves," faltered with our failure to control the greatest and most immediate of human dangers. We began to see science as a kind of absolute beyond our reach, beyond our understanding even, known, if it was known at all, through proxies who, like priests in other centuries, could not tell us what they knew.

In short, our belief in ourselves declined at the very moment when the Russian belief in the mechanics of the universe confirmed itself. No one talked any longer of a Baruch Plan, or even remembered that there had been one. The freedom of science to follow the laws of absolute possibility to whatever conclusions they might lead had been established, or so we thought, as the unchallengeable fixed assumption of our age, and the freedom of technology to invent whatever world it happened to invent was taken as the underlying law of modern life. It was enough for a manufacturer of automobiles to announce on TV that he had a better idea—any better idea: pop-open gas-tank covers or headlights that hide by day. No one thought any longer of asking whether his new idea matched a human purpose.

What was happening in those years, as the bitterly satirical fictions of the period never tired of pointing out, was that we were ceasing to think of ourselves as men, as self-governing men, as proudly self-governing makers of a new nation, and were becoming instead a society of consumers: recipients—grateful recipients—of the blessings of a technological civilization. We no longer talked in the old way of The American Proposition, either at home or abroad—particularly abroad. We talked instead of The American Way of Life. It never crossed our minds apparently—or if it did we turned our minds away—that a population of consumers, though it may constitute an affluent society, can never compose a nation in the great, the human, sense.

But the satirical novels, revealing as they were, missed the essential fact that we were becoming a population of consumers, an affluent society, not because we preferred to think of ourselves in this somewhat less than noble role but because we were no longer able to think of ourselves in that other role—the role our grandfathers had conceived for us two hundred years ago. We were not, and knew we were not, Whitman's Pioneers O Pioneers. . . .

Which means, of course, however we put it, that we no longer believe in man. And it is that fact which raises, in its turn, the most disturbing of all the swarming questions which surround us: how did we come to this defeated helplessness? How were we persuaded of our impotence as men? What convinced us that the fundamental law of a scientific age must be the scientific law of possibility and that our human part must be a passive part, a subservient part, the part of the recipient, the beneficiary . . . the victim? Have the scientists taught us this? A few months ago one of the greatest of living scientists told an international gathering composed of other scientists: "We must not ask where science and technology are taking us, but rather how we can manage science and technology so that they can help us get where we want to go." It is not reported that Dr. René Dubos was shouted down by his audience, and yet what he was asserting was precisely what we as a people seem to have dismissed as unthinkable: that "we," which apparently means mankind, must abandon our modern practice of asking where science and technology are "taking *us*," and must ask instead how *we* can "manage" science and technology so that they will help us to achieve *our* purposes—our purposes, that is to say, as men.

Dr. Dubos, it appears, scientist though he is and great scientist, believes rather more in man than we do. Why, then, do we believe so little? . . . Is it *our* education which has shaped the very different estimate of man we live by? In part, I think; in considerable part. Education, particularly higher education, has altered its relation to the idea of man in fundamental ways since Adams's day and Jefferson's. . . . As specialized, professional training, higher education in the United States today is often magnificent. . . . But the educated *man*, the man capable not of providing specialized answers, but of asking the great and liberating questions by which humanity makes its way through time, is not more frequently encountered than he was two hundred years ago. On the contrary, he is rarely discovered in public life at all.

I am not arguing—though I deeply believe—that the future of the Republic and the hope for a recovery of its old vitality and confidence depend on the university. I am confining myself to Dr. Dubos's admonition that we must give up the childishness of our present attitude toward science and technology, our constant question where *they* are taking *us*, and begin instead to ask how *we* can manage *them* "so that they can help us get where we want to go." "Where we want to go" depends, of course, on ourselves and, more particularly, on our conception of ourselves. If our conception of ourselves as the university teaches it or fails to teach it is the conception of the applicant preparing for his job, the professional preparing for his profession, then the question will not be answered because it will not be asked. But if our conception of ourselves as the university teaches it is that of men preparing to be men, to achieve themselves as men, then the question will be asked *and* answered because it cannot be avoided. Where do we want to go? Where men can be most themselves.

How should science and technology be managed? To help us to become what we can be.

There is no quarrel between the humanities and the sciences. There is only a need, common to them both, to put the idea of man back where it once stood, at the focus of our lives; to make the end of education the preparation of men to be men, and so to restore to mankind—and above all to this nation of mankind—a conception of humanity with which humanity can live.

The frustration—and it is a real and debasing frustration—in which we are mired today will not leave us until we believe in ourselves again, assume again the mastery of our lives, the management of our means.

The Future of Man *

by Julian Huxley (1959)

Science provides increased control over the forces of nature, and so gives us the means of realizing our aims in practice. But it also provides fuller understanding and a truer vision of natural reality. This is in the long run the more important, for our vision of reality helps to determine our aims.

HUMANIST ERA

By discovering how to control intra-atomic energy, science has launched us into the Atomic Era, with all its attendant hopes and fears. But by giving us fuller comprehension of nature as a whole, it has set us on the threshold of a greater and more revolutionary age, which I will call the Humanist Era. It is the era in which the evolutionary process, in the person of man, is becoming purposeful and conscious of itself.

Today, for the first time in man's long and strange history, science is revealing a comprehensive picture of the natural universe and of man's place and role in it—in a word, of his destiny.

Thanks to the patient labors of thousands of scientists—biologists and

* From Julian Huxley, "The Future of Man," *Bulletin of the Atomic Scientists,* December 1959. Copyright 1959 by The Educational Foundation for Nuclear Science. Reprinted by permission. Originally delivered as one of a series of presentations on "The Future of Science" at the Day of Science, Brussels Exposition, 1958.

astronomers, geologists and anthropologists, historians and physicists—
the universe of nature in which man lives is now revealed as a single process of evolution, vast in its scales of space and time. Man is part of this
universal evolving world-stuff. He is made of the same matter, operated
by the same energy, as all the stars in all the galaxies.

Most of the universe is lifeless and its portentously slow evolution has
produced only simple patterns of organization and little variety. But on
our earth (and doubtless on other planetary specks) conditions permitted
the appearance of the self-reproducing and self-varying type of matter we
call life. With this, natural selection could begin to act and the biological
phase of evolution was initiated.

CHANGE BECOMES MORE RAPID

Through natural selection, change—though still slow by human standards—could become much more rapid, and surprising new possibilities
could be realized by the world-stuff. From the uniformity and relative
simplicity of submicroscopic particles, there was generated the astonishingly rich variety of life, from sea-anemones and ants to cuttlefish and
lions, from bacteria and toadstools to daisies and giant trees; the astonishingly high organization of a beehive or a bird; and most astonishing of all,
the emergence of mind, living matter's increasing awareness of itself and
its surroundings.

But there are restrictions on what the blind forces of natural selection
can accomplish. A few million years ago, it now appears, living matter
had reached the limits of purely material and physiological achievement:
only the possibilities of mind remained largely unrealized.

By exploiting the possibilities of mental advance, man became the latest dominant type of life, and initiated a new phase of evolution, the
human or *psychosocial* phase, which operates much faster than biological
evolution, and produces new kinds of results. Man's capacity for reason
and imagination, coupled with his ability to communicate his ideas by
means of the verbal symbols of language, provided him with a new
mechanism for evolution, in the shape of cumulative tradition. Pre-human
life depended only on the transmission of material particles, the genes in
the chromosomes, from one generation to the next. But man can also
transmit experience and its results. With this, mind as well as matter
acquired the capacity for self-reproduction. Natural selection became
subordinate to psychosocial selection, and the human phase of evolution
could begin.

Science has also shown man his position in evolutionary time. Life has
been evolving on the earth for over two thousand million years. Man-like

creatures have existed for only about one million years, and human civilization, with all its achievements, for a bare five thousand. But evolving man can reasonably expect an immensity of future time—another two thousand million or more.

The psychosocial phase of evolution is thus in its infancy: man as dominant evolutionary type is absurdly young. I may adapt a simile of Sir James Jeans: If you represent the biological past by the height of St. Paul's cathedral, then the time since the beginning of agriculture and settled life equals one postage stamp flat on its top. And, unless man destroys himself by nuclear war or other follies, he can look forward to evolving through at the least the time-equivalent of another St. Paul's.

ROLE OF MAN

Man's place and role in nature is now clear. No other animal can now hope to challenge his dominant position. Only man is capable of further real advance, of major new evolutionay achievement. He and he alone is now responsible for the future of this planet and its inhabitants. In him evolution is at last becoming conscious of itself; his mind is the agency by which evolution can reach new levels of achievement. Man's destiny, we now perceive, is to be the agent of evolution on this earth, realizing richer and ampler possibilities for the evolutionary process and providing greater fulfillment for more human beings.

The revelation of fulfillment as man's most ultimate and comprehensive aim provides us with a criterion for assessing our own psychosocial evolution. Already in its brief course psychosocial evolution has produced real progress—increased expectation of life, less disease, more knowledge, better communications, increase of mechanical power and decrease of physical drudgery, more varied interest, and enrichment through creative achievement—in buildings and works of art, in music and spectacle, in discovery and ideas. But it has also produced poverty and crime and slavery and organized cruelty, and its course has been accompanied by constant exploitation, indignity, and slaughter.

In this new perspective, we see that what Père Teilhard de Chardin called the process of hominization—the better realization of man's intrinsic possibilities—has barely begun. Few human beings realize more than a tiny fraction of their capacities, or enjoy any but the most meager degree of possible satisfaction and self-fulfillment. The majority are still illiterate, undernourished and short-lived, and their existence is full of misery and indignity. Nor have human societies realized more than a fraction of their capacities. They provide inadequate opportunities for expression and enjoyment, they still produce more ugliness than beauty, more frustration than fulfillment; they can easily lead to the dehumanization of life instead of its enrichment.

What has all this to do with science? I would say a great deal. First let us remember that most of what we can properly call advance in psychosocial evolution has stemmed from new or better organized knowledge, whether in the form of traditional skills, sudden inventions, new scientific discoveries, technological improvements, or new insights into old problems.

Science is a particularly efficient method for obtaining, organizing, and applying knowledge. Though modern science is barely three centuries old, it has led to the most unexpected discoveries and the most spectacular practical results. Scientific method involves controlled observation of fact, rational interpretation by way of hypothesis, the publication and discussion of procedures and results, and the further checking of hypothesis against fact. The use of scientific method has proved to be the best way of obtaining fuller intellectual understanding and increased practical control, in all fields where it has been tried. It leads inevitably toward more and fuller truth, to an increasing body of more firmly established factual knowledge, and more coherent principles and ideas.

Science is often used to denote only the natural sciences; but this is a false restriction, which springs from the historical fact that scientific method could be more readily applied to simpler subjects, and so first became effectively applied in non-human fields. But it can be applied to all natural phenomena, however complex, provided that we take account of their special peculiarities and go to the trouble of devising appropriate methods for dealing with them.

PSYCHOSOCIAL SCIENCE

Today, the time has come to apply scientific method to man and all his works. We have made a piecemeal beginning, with psychology, economics, anthropology, linguistics, social science, and so forth. But we need a comprehensive approach to the human field as a whole. We already have physical science, chemical science, and biological science: to deal with man as a natural phenomenon, we must develop psychosocial science.

The primary job of psychosocial science will be to describe and analyze the course and mechanism of psychosocial evolution in scientific terms. It will also include a science of human possibilities. What are the possibilities of man and his nature, individually and collectively? How is their realization helped or hindered by different types of psychosocial environment? How can we estimate human fulfillment; in what ways and to what extent can it be promoted by changes in psychosocial organization? In particular, such a science will involve a radical re-thinking of man's systems of education, their aims, content, and techniques.

The value of such an approach and such criteria is clear when we look at concrete problems. Two new challenges have recently appeared on the evolutionary scene—the threat of over-population and the promise of excess leisure. The population problem obstinately resists solution in terms of power-politics, economics, or religion, but the criterion of greater fulfillment immediately lights it up and indicates the general lines of the policy we should pursue in reconciling quantity with quality of human life.

The new possibilities opened up by science are exerting two effects on psychosocial evolution. Increased scientific control over the forces of nature has produced a flood of new conveniences and comforts, and has led directly to death-control and the recent alarming increase of human numbers. But the knowledge that healthier and longer life is possible, and that technology can provide higher standards of living and enjoyment, has changed the attitude of the vast underprivileged majority: they are demanding that the new possibilities shall be more abundantly realized.

LIMITED POSSIBILITIES

The next step must be to grasp the fact that the quantitative possibilities are not unlimited. Unless present-day man controls the exploitation of natural resources, he will impoverish his descendants; unless he supplements death-control with birth-control, he will become the cancer of the planet, ruining his earthly habitation and himself with it.

The leisure problem is equally fundamental. Having to decide what we shall do with our leisure is inevitably forcing us to re-examine the purpose of human existence, and to ask what fulfillment really means. This, I repeat, involves a comprehensive survey of human possibilities and the methods of realizing them; it also implies a survey of the obstacles to their realization.

Let me summarize the new picture of human destiny from a slightly different angle.

Man is the latest dominant type of life, but he is also a very imperfect kind of being. He is equipped with a modicum of intelligence, but also with an array of conflicting passions and desires. He can be reasonable but is often extremely stupid. He has impulses to sympathy and love, but also to cruelty and hatred. He is capable of moral action but also has inevitable capacities for sin and error.

As a result, the course of psychosocial evolution has been erratic, wasteful, and full of imperfection. It is easy to take a pessimistic view of man's history in general, and of his present situation in particular, where force and fear have become magnified on a gigantic scale.

HOPEFUL PROCESS

But when we survey the process as a whole, it looks more hopeful. During its course, there has been progress. Progress has always been the result of the discovery, dissemination, or application of human knowledge, and human knowledge has shown a cumulative increase. Furthermore, the erratic course of past psychosocial evolution was largely due to man as a species being divided against himself, and not having discovered any single overriding aim.

There is now a dramatic change in process. The human world has become inextricably interlocked with itself; the separate parts of the psychosocial process are being forced to converge toward some sort of organized unity. We are at last able and indeed compelled to think in terms of a single aim for mankind, while our increasing knowledge is enabling us to define our aim in relation to reality instead of in terms of wish-fulfillment: our knowledge of our imperfections and limitations is helping to define the possibilities of our improvement.

This marks a critical point in history. We have discovered psychosocial evolution as a complex but natural phenomenon, to be explored and controlled like other natural phenomena. Up till now, it has operated in erratic and often undesirable fashion, with self-contradictory aims. We now see that it could be transformed into an orderly mechanism for securing desirable results.

The idea of greater fulfillment for all mankind could become a powerful motive force, capable of influencing the direction of future evolution, and of overriding the more obvious motives of immediate personal or national self-interest. But it can only do so if it and its implications are properly understood, and made comprehensible to the bulk of men, all over the world. For this we need not only an extension of science but a reorientation of education; not only more knowledge, but also a better expression and a wider dissemination of ideas.

We must not imagine that the fuller realization of possibilities will be accomplished without effort, conflict, or suffering. This is inherent in the nature of man and of the psychosocial process: but so is hope.

The individual human brain and mind is the most complicated and highly organized piece of machinery that has ever existed on this earth. So-called electronic brains can perform extraordinary tasks with super-human rapidity: but they have to be given their instructions by men. The human organism can give instructions to itself; and can perform tasks outside the range of any inanimate machine. Though at the outset it is a feeble instrument equipped with conflicting tendencies, it can in the course of its development achieve a high degree of integration and performance.

HOW SHALL WE USE IT?

It is up to us to make the best use of this marvellous piece of living machinery. Instead of taking it for granted, or ignorantly abusing it, we must cherish it, try to understand its development, and explore its capacities.

The collective human organism, embodied in the psychosocial process, is an equally extraordinary piece of machinery. It is the mechanism for realizing human destiny. It can discover new aims for itself, and devise new methods for realizing them; but it is still primitive and inefficient. It is up to us to improve it, as we have improved our inanimate machines. Our ignorance about its potentialities is profound; therefore our immediate task is to understand the first principles of its operation, and think out their consequences.

Thus the new vision we owe to science is one of real though tempered optimism. It gives us a measure of significance and rational hope in a world which appeared irrational and meaningless. It shows us man's place and role in the universe. He is the earth's reservoir of evolutionary possibility; the servant of evolution, but at the same time its youthful master. His destiny is to pursue greater fulfillment through a better ordering of the psychosocial process. That is his extraordinary privilege, and also his supreme duty.

Our new vision assures us that human life could gradually be transformed from a competitive struggle against blind fate into a great collective enterprise, consciously undertaken. We see that enterprise as one for greater fulfillment through the better realization of human possibilities.

It is for us to accept this new revelation given us by science, examine it, and explore all its implications, secure in the knowledge that ideas help to determine events, that more understanding leads to more appropriate action, that scientific truth is an indispensable weapon against stupidity and wickedness and the other enemies of fulfillment, and true vision the parent of progress.

Epilogue *
by George Gaylord Simpson (1949)

It has been a long journey down the corridors of time to this point where mankind looks with foreboding and yet with hope into the mists of the future. We have followed a billion years of the history of life; we have examined some of the processes by which life has changed and developed; and we have come at the end to consideration of ourselves, the only form of life that knows it has a history, the thinking, responsible, and ethical animals. . . .

The fact of responsibility and the ethic of knowledge have many ethical corollaries, among them that blind faith ("blind," or "unreasoning," to be emphasized) is morally wrong. In connection with high individualization, another human diagnostic character, the resulting ethics include the goodness of maintenance of this individualization and promotion of the integrity and dignity of the individual. Socialization, a necessary human process, may be good or bad. When ethically good, it is based on and in turn gives maximum *total* possibility for ethically good individualization. . . . Variability and flexibility are in themselves desirable, from the point of view both of ethics and of evolution, biological and social.

Beyond the subjects of this enquiry are mysteries deeper still, unplumbed by paleontology or any other science. The boundaries of what can be achieved by perception and by reason are respected. The present chaotic stage of humanity is not, as some wishfully maintain, caused by lack of faith but by too much unreasoning faith and too many conflicting faiths within these boundaries where such faith should have no place. The chaos is one that only responsible human knowledge can reduce to order.

It is another unique quality of man that he, for the first time in the history of life, has increasing power to choose his course and to influence his own future evolution. It would be rash, indeed, to attempt to predict his choice. The possibility of choice can be shown to exist. This makes rational the hope that choice may sometime lead to what is good and right for man. Responsibility for defining and for seeking that end belongs to all of us.

Suggestions for Further Reading

Berger T., T. Abel, and C. H. Page, eds: *Freedom and Control in Modern Society*, D. Van Nostrand, Inc., 1954. A collection of essays on the relation of the individual to society in a technological world.

* Conant, James: *Modern Science and Modern Man*, Anchor Books, Doubleday & Company, 1952. A series of lectures delivered in 1952 discussing the role of modern science and its relation to society.

Dubos, René: *Dreams of Reason*, Columbia University Press, 1961. A thought-provoking collection of lectures on the fallacies inherent in most utopias and the dangers of our scientific civilization.

* Huxley, Julian: *Knowledge, Morality, and Destiny*, A Mentor Book, The New American Library of World Literature, Inc., 1960. Man's place at the summit of evolution and his responsibility as the agent of selective evolution are discussed in this collection of essays.

* Krutch, Joseph Wood: *The Measure of Man*, Charter Books, Bobbs-Merrill Company, Inc., 1962. A naturalist examines the influence that science has had upon our view of ourselves and suggests that a new attitude is overdue in the social sciences.

Matson, Floyd W.: *The Broken Image: Man, Science and Society*, George Braziller, 1964. This book (from which a short selection appears in this Part) offers a comprehensive analysis of the philosophy of the social sciences and compares it with the underlying philosophy of the other sciences, especially the new non-materialistic ideas emerging from modern physics.

* Whitehead, Alfred North: *Science and the Modern World*, Mentor Book, The New American Library of World Literature, Inc., 1948. This famous work examines the relationship between science and philosophy throughout the last three centuries and discusses the impact of science on modern thought.

* Paperback edition

BIOGRAPHIES

ANDRADA, Edward Neville da Costa, was born in 1887. He was educated at University College in London, the University of Heidelberg, Cavendish Laboratory in Cambridge, and the University of Manchester, where he worked under Ernest Rutherford. From 1928 to 1950 he was Quain Professor of Physics at the University of London. In 1950 he was appointed director of the Royal Institution and, some years later, became senior research fellow of the department of metallurgy at the Imperial College of Science in London. Andrada has written several books for the layman on science and scientists, and many technical publications. He is also known as a humanist and poet, having at one time written a column on food and wine for a literary magazine and having published two volumes of poetry.

ARDREY, Robert, was born in Chicago in 1908. After graduation from the University of Chicago, where he majored in natural sciences, he became a successful playwright and screen writer. Two of his plays have received special awards. In 1955 he began his African travels and studies which led to the publication of several books and articles on evolutionary anthropology.

ASIMOV, Isaac, was born in Russia in 1920. Brought as a child to the United States, he attended Columbia University where he received his Ph.D. in 1948. He is now an associate professor of biochemistry at Boston University School of Medicine. Asimov has an impressive record of more than thirty-five successful books to date and is considered one of today's most imaginative interpreters of scientific subjects. In 1965 he received the James T. Grady award for science writing from the American Chemical Society.

BATES, Marston, was born in 1906. He was graduated from the University of Florida and took his Ph.D. at Harvard. He worked first as an entomologist with the United Fruit Company. Among other subsequent positions, he served on the staff of the International Health Division of the Rockefeller Foundation and as special assistant to the President (1950–52). Since 1952 he has been a professor of zoology at the University of Michigan. He is the author of about a dozen books on natural history and related subjects for the general reader.

BERGSON, Henri (1859–1941), was born in Paris. He taught philosophy at Clermont-Ferrand, obtained his doctorate from the University of Paris in 1889, and from 1900 taught philosophy at the Collège de France. He was elected to the French Academy, and in 1927 was awarded a Nobel Prize for his books on philosophy. In 1940 he refused the Vichy government's offer to exempt him from the anti-Semitic decrees. He resigned his honorary chair at the Collège de France. Eighty-one years old and feeble, he rose from a sick bed and stood in line, supported by his valet and nurse, for the required registration. The effort may have hastened his death on January 4, 1941.

BERTALANFFY, Ludwig von, was born in Austria in 1901. He studied at the University of Innsbruck and took his Ph.D. at the University of Vienna. For fourteen years (1934–48) he was a professor at the University of Vienna. Then in 1949 he became a professor and director of biological research at the University of Ottawa, and followed this with several years as a visiting professor. He is now a professor of theoretical biology at the University of Alberta. He has written a number of books and articles in the English, German, French, Spanish, and Japanese languages.

BEVERIDGE, William Ian, was born in 1908. He was educated at Cranbrook School, St. Paul's College, University of Sydney, and holds an M.A. from Cambridge. He is a professor of animal pathology at Cambridge and a fellow at Jesus College.

BROWN, Harrison, was born in 1917. He received his B.S. from the University of California and his Ph.D. from Johns Hopkins. After teaching chemistry at Johns Hopkins he became an assistant director of chemistry at Clinton Laboratories at Oak Ridge, and then a research associate at the plutonium project at the University of Chicago. From 1946 to 1948 he was an assistant professor of the Institute for Nuclear Studies (associate professor, 1948–51). Since 1951 he has been a professor of geochemistry at the California Institute of Technology. He is a member of the National Academy of Sciences, has received awards in chemistry and geology, and has written several successful books.

CLARK, Colin, was born in England in 1905. He was educated at Oxford and the University of Milan in Italy. He has held research and lecturing posts in England, Australia, and the United States. He was an economic adviser to the government of Queensland, State Under-Secretary for Labor and Industry (1938–52). Since 1953 he has been Director of the Agricultural Economics Research Institute at Oxford.

DARWIN, Charles (1809–1882), was a grandson of the poet-naturalist Erasmus Darwin. He was sent to Edinburgh University to study medicine and then to Cambridge to study for the ministry, but he showed no aptitude for, or interest in, either of these professions. Then, in 1831, at the age of twenty-two he was offered the post of naturalist on the ship *Beagle* and he persuaded his father to allow him to accept this opportunity. The voyage lasted five years and left Darwin with frail health but with an intense interest in biology which was to win him fame. He married his first cousin Emma Wedgwood and they lived in a country house in Essex. Here Darwin wrote *The Origin of Species* and *The Descent of Man*.

DAVIS, Kingsley, was born in 1908. He was educated at the University of Texas and Harvard where he received his Ph.D. in 1936. He taught sociology at Smith College (1934–36), Clark University (1936–37), Pennsylvania State University (1937–1944). Here he became head of sociology in 1942. Later he held professorial posts at Princeton, Columbia, and the University of California at Berkeley, where he has been a professor of sociology and Director of International Population and Urban Research since 1955. From 1954 to 1961 he was the U.S. Representative to the Population Commission of the United Nations.

DOBZHANSKY, Theodosius, was born in Russia in 1900. He was educated at the University of Kiev. For several years he taught zoology and genetics in Russia before coming to the Rockefeller Foundation in New York to work with Professor T. H. Morgan. In 1929 he joined the staff of the California Institute of Technology. Later he became DaCosta Professor of Zoology at Columbia and is now a professor at the Rockefeller University. He has been awarded many honors, one of the most recent being the National Medal of Science in 1965.

DUBOS, René Jules, was born in Saint-Brice, France, in 1901. He attended the Collège Chaptal and Institut National Agronomique in Paris. In 1924, after a few years with the International Institute of Agriculture in Rome, he went to Rutgers University, where he was awarded his Ph.D. in 1927. Since that time, except for the period 1942–44, when he was George Fabyan Professor of Comparative Pathology and a professor of tropical medicine at Harvard University Medical School, Dubos has been associated with the Rockefeller University, where he did pioneer work in the field of antibiotics. In 1951 he was elected president of the Harvey Society and the Society of American Bacteriologists. He is also a member of the National Academy of Sciences. In addition to his scientific achievements and honors, Dubos is the author of twelve books.

EISELEY, Loren, was born in Lincoln, Nebraska, in 1907. Having received his A.B. from the University of Nebraska, he completed graduate work in anthropology at the University of Pennsylvania. He returned to the Midwest for his first academic job, at the University of Kansas. He later became head of the Department of Sociology and Anthropology at Oberlin College in Ohio. In 1947 he returned to the University of Pennsylvania to head the Department of Anthropology. Since 1948 he has been Curator of Early Man in the University Museum. For a number of years he was active in the search for early postglacial man in the western United States; he has worked in the high plains, mountains, and deserts bordering the Rocky Mountains from Canada into Mexico. Eiseley has lectured at a number of universities and has published many books and articles.

GALTON, Francis (1822–1911), was a grandson of Erasmus Darwin and a first cousin of Charles Darwin. He was born in England and educated at Cambridge. He entered British civil service and made a study of meteorology. His *Meteorographica*, published in 1863, is the basis of modern weather maps. He is best known, however, for his studies of heredity which founded the science of eugenics. His works on this subject were published between 1869 and 1889. Galton also invented a system of fingerprint identification.

HALDANE, J. B. S. (1892–1965), came from a distinguished Scottish family of biologists. He was born in England and educated at Eton and Oxford. From 1937 until 1957 he was a professor of genetics at London University. Then he emigrated to India and was appointed professor of the Indian Statistical Institute at Calcutta. He published a number of books on biology and on the ethical and philosophical implications of modern scientific concepts. Haldane was one of the three scientists principally responsible for working out the quantitative mathematical treatment of selection and mutation theory. In 1953 he was awarded the Darwin Medal of the Royal Society.

HALL, Edward T., was born in 1914. He attended the University of Denver, the University of Arizona, and Columbia where he received a Ph.D. in anthropology. After conducting field work in Micronesia, he served for five years as director of the State Department's Point IV training program. Subsequently he taught at the University of Denver, Bennington College, the Harvard Business School, and the Illinois Institute of Technology. He is now professor of anthropology in the College of Arts and Sciences and professor of organization theory in the School of Business at Northwestern University.

HILL, A. V., was born in England in 1886. He received his M.A. and Sc.D. from Cambridge University. He was a professor of physiology at Manchester University and University College in London, and Foulerton Research Professor of the Royal Society from 1926 to 1951. He served as a member of Parliament (1940–45) and science adviser to the government of India (1943–44). He received the Nobel Prize for physiology and several other honors. Now retired, he lives in Cambridge, England.

HITLER, Adolph (1889–1945), was born in Austria. He worked for a while as an architect's draftsman, before moving to Munich and serving in World War I. In 1923 he organized an unsuccessful revolt in Munich, known as the "Beer Hall Putsch," and as a consequence spent nine months in prison. During this time he dictated *Mein Kampf* to his secretary Rudolf Hess. In 1933 Hitler was swept into power on the rising tide of German nationalism and named Chancellor. He established a new economic program, rearmed Germany, and broke the conditions of the Versailles treaty, thus precipating the Second World War. He died in 1945, presumably in a suicide pact with his mistress Eva Braun, when the Allies entered Berlin.

HOAGLAND, Hudson, was born in 1899. He received his A.B. from Columbia, his M.S. from Massachusetts Institute of Technology, and his Ph.D. from Harvard. He has taught biology and physiology in various capacities at Harvard College and Clark University in Worcester, Massachusetts. He was also a research professor of physiology at Tufts Medical School. Since 1944 he has been Co-Director of the Worcester Foundation for Experimental Biology, and has been a professor of physiology at Boston University since 1950. He was president of the American Academy of Arts and Sciences (1961–64) and received the Humanist of the Year award in 1965.

HUXLEY, Aldous Leonard, grandson of Thomas H. Huxley, was born in England, July 26, 1894. After attending a preparatory school, he went to Eton on a scholarship in 1908. He began to study medicine and contracted keratitis, which left him almost blind. At eighteen he wrote a novel which he was unable to read, and when his sight was partially restored, the manuscript had been lost. He did his graduate work in literature at Oxford University and in 1919 joined the staff of *Athenaeum*. His best known works include *Chrome Yellow, Antic Hay, Point Counter Point,* and *Brave New World.* He died in 1963.

HUXLEY, Julian, was born in London in 1887, a grandson of Thomas H. Huxley, an eminent biologist. He was educated at Eton and Oxford. He has taught zoology at Rice Institute in Houston, Texas, at Oxford, and at the University of London. He has made many lecture tours in America and for three years served as Director General of UNESCO. Among the many honors he has received are the Huxley Memorial Lecture and Medal of the Royal Anthropological Institute and the Darwin Medal of the Royal Society. His countless publications are well known for their imaginative scope, wisdom, and responsibility. He was knighted in 1958.

HUXLEY, Thomas Henry (1825–1895), was a distinguished English biologist and the head of an accomplished family. His eldest son, Leonard, became a classical scholar. One of his grandsons, Julian, became a famous biologist and philosopher; another, Aldous, became a well-known writer. After Darwin's *Origin of Species* was published in 1859, Huxley became an important defender of the work, and *Darwiniana* was published in 1863. The last twenty years of Huxley's life were devoted to public service more than to science, and his term on the London School Board deeply influenced the English national elementary school system. In 1890 he moved away from London because of his health and died at Eastbourne on June 29, 1895.

KETTLEWELL, H. B. D., was educated at the University of Cambridge and practiced medicine for fifteen years. He had also been interested in moths and butterflies since childhood. In 1948 he gave up medicine in order to conduct research work on evolution in moths. As Nuffield Research Fellow in Genetics at Oxford, his field work has taken him to many countries, the Belgian Congo, Kenya, Namaqualand, South Africa, Norway, and Canada.

KRECH, David, was born in Russia in 1909. He came to the United States in 1913, was educated at New York University and the University of California at Berkeley. He taught psychology at Swarthmore, the University of Colorado, and the University of California at Berkeley, where he has been a professor of psychology since 1947. He has also held a number of lecturing and visiting professorial posts here and abroad.

KRUTCH, Joseph Wood, was born in Knoxville, Tennessee, in 1893. He took his A.B. at the University of Tennessee, and his M.A. and Ph.D. at Columbia, where he became Brander Matthews Professor of Dramatic Literature. One of America's distinguished naturalists, he has been equally well-known as a teacher, drama and literary critic, biographer, editor and journalist, and public speaker. Now retired, he lives in Tucson, Arizona.

LAMARCK, Jean Baptiste (1744–1829), was born in France, studied medicine and then botany in Paris. He was appointed Royal Botanist and Custodian of the Royal Herbarium, and later became Professor of Zoology at the Jardin des Plantes (1793–1818). Lamarck published a number of works in which he expounded his own theory of evolution.

LEAKEY, Louis Seymour Bazett, was born in Africa in 1903. His parents were the first missionaries to the Kikuyu tribe. He was educated at Cambridge University and then led several archaeological expeditions to east Africa between 1926 and 1935. During World War II he served with the Criminal Investigation Department in Nairobi. From 1945 to 1961 he was Curator of the Coryndon Memorial Museum and since 1962 has been Director of the Natural Museum Center for Prehistory and Paleontology. The discoveries at Olduvai Gorge have won him and his wife, Mary, world-wide recognition. He is honorary fellow of St. John's College at Cambridge and fellow of the British Academy.

LEDERBERG, Joshua, was born in 1925. He took his A.B. at Columbia and his Ph.D. at Yale University. From 1947 to 1959 he taught at the University of Wisconsin. In 1958 he was awarded the Nobel Prize in medicine for research in the genetics of microbes. Since 1959 he has been at Stanford University, where he is professor and executive head of the Genetics Department as well as Director of the Kennedy Laboratories for Molecular Medicine.

LÜSCHER, Martin, was born in Basel in 1917, attended the university there and received his Ph.D. in 1944. After four years of postgraduate research in England and France he returned to the University of Basel to teach experimental zoology. Since 1965 he has been professor and Director of the Zoological Institute at the University of Bern.

LYNCH, Kevin, was born in 1918 in Chicago. He studied at Yale University and Rensselaer Polytechnic Institute, also under Frank Lloyd Wright, and at Massachusetts Institute of Technology. Since 1949 he has been Professor of Planning at the Massachusetts Institute of Technology, a consultant to the State of Rhode Island and a number of organizations. He has published several books on urban planning.

MacLEISH, Archibald, was born in 1892. He was graduated from Yale, and has received a number of honorary degrees. He was Librarian of Congress 1939–44. From 1949 to 1962 he was Boylston Professor at Harvard and Simpson lecturer at Amherst College (1963–67). In addition to his career as a writer, he has held a number of responsible government posts and has served as delegate for several UNESCO conferences. He received many awards for poetry and drama, including the Pulitzer Prize in 1952 and 1959.

McCONNELL, James V., was born in 1925. He was graduated from Louisiana State University and received his Ph.D. in psychology from the University of Texas. He went to teach at the University of Michigan in 1956 and has been a professor of psychology there since 1963. He is editor and publisher of *The Worm Runner's Digest.*

MATSON, Floyd, was born in Honolulu in 1921. He took his A.B. and Ph.D. at the University of California at Berkeley and later worked as a newspaperman and as an editor of technical publications. He spent several years in Japan, first as civilian press and propaganda analyst, then as instructor of speech in the Far East Program. Returning to Berkeley, he lectured in speech before accepting his present position as an associate professor of political science and American studies at the University of Hawaii.

MEDAWAR, P. B., was born in Brazil in 1915. He received his M.A. and D.Sc. from Oxford University. From 1938 to 1947 he lectured in zoology at Oxford. Then he became professor of zoology at the University of Birmingham (1947–51) and the University of London (1951–62). He has been Director of the National Institute of Medical Research in London since 1962. In addition he has held a number of posts as a visiting lecturer both in England and in this country (Harvard, Cornell, Berkeley). In 1959 he received the Medal of the Royal Society and in 1960 the Nobel Prize in medicine and physiology.

MONTAGU, Ashley, was born in London in 1905. He studied at the University of London and the University of Florence, and took his Ph.D. at Columbia University. He taught anatomy at New York University and Hahnemann Medical College in Philadelphia. In 1949 he became Chairman of the Department of Anthropology at Rutgers University, which post he held until 1955. After that time he held a number of positions as visiting lecturer and editor. Among other activities, he produced, financed, wrote, and directed a film *One World or None* in 1946. He has served as a consultant on race for UNESCO and has been the author of many books on anthropology, especially on the subject of race.

MOORE, Ruth, was born in St. Louis, Missouri, and received her A.B. and A.M. from Washington University there. She has worked as a reporter on the *St. Louis Star-Times,* as Washington reporter for the *Chicago Sun,* and as an assistant editor of the *Kiplinger Magazine.* She is now in Chicago on the *Sun-Times,* where science feature stories are one of her specialties. She has become well known as a popular science writer, having published five books on science for the layman. Her interest in housing and urban renewal has also made her a specialist on these questions. In May 1960 she received the first individual award given by the American Association of Planning Officials for her 1959 series on urban renewal in seven major cities.

MORRIS, Desmond, was born in England in 1928. He received a B.Sc. in zoology from Birmingham University and a D.Phil. in animal behavior from Oxford. After several years of post-doctoral research at Oxford, he joined the staff of the Zoological Society of London where he held the post of Curator of Mammals from 1959 to 1967. From 1967 to 1968 he was Director of the Institute of Contemporary Arts in London. He is now devoting his full time to writing.

MULLER, H. J., was born in New York City in 1890 and was educated at Columbia University, where he took a Ph.D. in zoology in 1916. From 1915 to 1936 he taught biology, first at the Rice Institute, then at the University of Texas. In the years 1933–37 he worked at the Institute of Genetics in Moscow as a senior geneticist, but he later became a fierce foe of the Soviet system.

After spending some years at the University of Edinburgh and at Amherst College, he joined Indiana University in 1945. Professor Muller discovered that the rate of mutation could be accelerated by X rays. This discovery, which he made in 1927, won him the Nobel Prize in physiology and medicine for 1946.

NIETZSCHE, Friedrich (1844–1900), was born in Saxony and educated at the Universities of Bonn and Leipzig. He was professor of classical philology at Basel 1869–79. He subsequently lived in Switzerland and Italy, devoting all his time to writing poetry and philosophy. He denounced all religion and glorified the concept of the perfectibility of man (*Übermensch*). He had a nervous breakdown in 1889 and was cared for by his mother and sister until his death in 1900. His philosophy is believed to have influenced Adolf Hitler and the Germans of the Third Reich.

OSBORN, Fairfield, was born in 1887 and was educated at Princeton and Cambridge. Since 1940 he has been president of the New York Zoological Society, president of the Conservation Foundation since 1948, and Chairman of the Board since 1962. A distinguished naturalist, he has been awarded many honors for his books on conservation.

PALEY, William (1743–1805), was an English theologian and philosopher. He was Archdeacon of Carlisle and Subdean of Lincoln Cathedral. His lectures and treatises on the subject of religious philosophy had a considerable influence on the thought of his time.

PEATTIE, Donald Culross (1898–1965), was graduated from Harvard in 1922 and worked for the Department of Agriculture before becoming a full-time author. Writing was a popular profession in Peattie's family. His mother and father were editors of the Omaha *World Herald* and later the *Chicago Tribune*. His older brother, Roderick, his wife, Louise, and his son, Noel, are also authors. Donald Culross Peattie is best known for his nature studies that convey an unusually perceptive and poetical awareness of natural phenomena.

RODWIN, Lloyd, was born in 1919. He attended the College of the City of New York and received advanced degrees from the New School of Social Research, New York Law School, the University of Wisconsin, and Harvard. After serving for several years on government housing programs, he became a research assistant at the University of Wisconsin. In 1946 he went to Massachusetts Institute of Technology where he has been professor of land economics, Department of City and Regional Planning, since 1959. He has also served as director of the special program for urban and regional studies of developing areas at the School of Architecture and Planning since 1967. Rodwin has written and edited several books on urban planning.

SARNOFF, David, was born in 1891 in Russia. He came to the United States in 1900 and was educated as an electrical engineer at Pratt Institute in Brooklyn. He started work as a radio inspector and engineer with the Marconi Company and served in various capacities with that company until it was taken over by the Radio Corporation of America in 1921. At that time he was elected general manager, then vice president (1922), and president in 1930, becoming chairman of the board in 1947. He holds honorary degrees from twenty-six colleges and universities.

SELYE, Hans, was born in Austria in 1907 and went to college in Hungary. He took his medical work at the University of Paris, University of Rome, and the German University at Prague. A Rockefeller research fellowship brought him to Johns Hopkins and later to McGill University where he continued to serve in various teaching positions until 1945, when he became Director of the Institute of Experimental Medicine and Surgery at the University of Montreal. He was an expert consultant to the Surgeon General of the U. S. Army, 1947–57, and has been awarded many medals and other honors.

SHAW, George Bernard (1856–1950), was born in Dublin but spent most of his life in England. He was an art, music, and drama critic for London journals and established his reputation by sympathetic criticisms of the impressionist painters, Wagnerian music, and the Ibsen school of drama. He also became prominent as a Socialist. About 1892 he began writing plays and soon became known as the leading British dramatist. In addition to his many successful plays he wrote several tracts on Socialism. He was awarded the Nobel Prize for literature in 1925.

SHERRINGTON, Sir Charles Scott (1861–1952), was graduated from Cambridge University and became a professor of physiology. He taught first at Liverpool, then at the Royal Institution of Great Britain, and finally at Oxford. In 1932 he shared the Nobel Prize in physiology with Edgar O. Adrian for their discoveries regarding the function of the neuron.

SIMPSON, George Gaylord, was born in Chicago in 1902. He was graduated from Yale in 1923 and received his Ph.D. three years later. From 1945 to 1959 he was a professor of vertebrate paleontology at Columbia and was associated with the Museum of Natural History in New York. He left there to become Alexander Agassiz Professor of Vertebrate Paleontology at the Museum of Comparative Zoology at Harvard. He is a member of the National Academy of Sciences and the recipient of many awards. He is also the author of more than 350 articles and several books.

SINNOTT, Edmund W., was born at Cambridge, Massachusetts, in 1888. Educated at Harvard, he taught botany there, then at the Connecticut Agricultural College, and at Columbia, before becoming a member of the faculty at Yale. Now emeritus, he was Sterling Professor of Botany, Director of the Sheffield Scientific School, and Dean of the graduate school. He was also president of the American Association for the Advancement of Science, the Botanical Society of America, and the American Society of Naturalists. In addition to numerous papers in scholarly journals, he is the author of *The Biology of the Spirit, Botany Principles and Problems, Cell and Psyche; Two Roads to Truth,* and *Plant Morphogenesis.*

SKINNER, B. F., was born in 1904. He was graduated from Hamilton College and took his Ph.D. at Harvard. After several years of research work at Harvard, he joined the staff of the University of Minnesota where he remained until 1945. At that time he was made a professor and chairman of the Department of Psychology at Indiana University. In 1948 he went to Harvard where, since 1958, he has been Edgar Pierce Professor of Psychology. He is a member of the Academy of Sciences and has received several awards in the field of psychology.

SPENCER, Herbert (1820–1903), was born in England, the only surviving child of nine offspring of a radical, eccentric schoolteacher and a gentle, unhappy mother. His formal education ended when he was sixteen years old. Thereafter, he served for a while as an assistant schoolteacher; later, he worked as a civil engineer and dabbled in mechanical invention, modeling, and phrenology. In 1848 he became sub-editor of the *Economist*. Here he met and became friends with George Eliot, Carlyle, Huxley, and Tyndall. In 1853 he resigned from his editorial position to devote his time to free-lance writing. Durng the next fifty years he planned and wrote an entire system of philosophy covering metaphysics, biology, psychology, sociology, and ethics. This philosophy based on a principle of organic development had great influence throughout Europe and America.

SPERRY, Roger W., was born in 1913. He was graduated from Oberlin College and took his Ph.D. at the University of Chicago. He did research work at Harvard and the Yerkes Laboratory, taught anatomy at the University of Chicago, and since 1954 has been Hixon Professor of Psychobiology at California Institute of Technology. He has been recognized for his research on the origin of the brain and neural mechanism behavior. He is a member of the National Academy of Sciences.

STORER, John Humphreys, was born in 1888. He was graduated from Harvard and spent 25 years farming. In 1927 the Massachusetts Department of Agriculture awarded him its gold medal for his contributions in poultry breeding. Since 1937 he has devoted his time to the study of conservation and ecology. He has lectured and prepared motion pictures both free lance and for organizations such as the National Audubon Society and the Conservation Foundation of New York.

SUSSMAN, Maurice, was born in 1922. He took his Ph.D. at the University of Minnesota. After teaching for eight years at Northwestern University, he became a professor of biology at Brandeis University. During the summer months he is an instructor in marine embryology at the Woods Hole Marine Biological Laboratories in Massachusetts.

TEILHARD DE CHARDIN, Pierre (1881–1955), the son of a landowner in Auvergne, was educated at a Jesuit school and became a Jesuit priest. For a while he lectured in science at the Jesuit College in Cairo and then in 1918 became a professor of geology at the Institut Catholique in Paris. Several paleontological expeditions took him to England, China, and central Asia. His work in Cenozoic geology and paleontology became known and he was awarded academic distinctions, including the Legion of Honor in 1946. However, his philosophical writings did not conform to orthodox Jesuit teachings, and he was forbidden by his religious superiors to teach and publish. His most important work, *The Phenomenon of Man*, was written in 1938 but was not published until after his death in 1955, along with his other works. Teilhard spent the last few years of his life in America.

VOGT, William, was born in 1902. He took his A.B. and Ph.D. degrees at Bard College in New York. He was a field naturalist and lecturer for the National Association of Audubon Societies, editor for *Bird-Lore Magazine*, consultant for the War Department, chief of the conservation section of the Pan-American Union from 1943 to 1949, and National Director of Planned Parenthood from 1951 to 1961. Since 1964 he has been Secretary of the Conservation Foundation in New York City.

WALLACE, Alfred Russel (1823–1913), was born in England. His father lost his fortune in unwise ventures and was unable to provide his son with formal schooling, so Alfred was apprenticed to a watchmaker. Later he worked as a surveyor with his brother, and for a short time served as a master in a collegiate school. Here he became interested in botany and entomology. In 1848 he left England to collect botanical specimens in South America and the Malay Archipelago where he discovered a strait (now known as "Wallace's Line"). West of it the flora and fauna are Oriental in character, east of it, Australian. His observations led to an interest in evolution. One day as he lay ill with a fever the theory of natural selection suddenly occurred to him. He wrote down his idea and sent it to Charles Darwin who had been working on the same theory for almost twenty years. However, Darwin saw that Wallace was given credit with him for the idea. In 1862 Wallace returned to England. He wrote a number of books on other subjects, particularly social and economic questions. He lived to be 90 years old.

WELLS, Herbert George (1866–1946), was born in Bromley, Kent, of lower middle-class parents. After winning a scholarship to the Royal College of Science, he received his B.S. from London University in 1888. He became a teacher and wrote a biology textbook, but tuberculosis forced him to abandon his work and leave London for a period of convalescence. Upon his return he began his career as novelist and journalist. A prolific writer, he produced science-fiction, realistic novels, and novels designed primarily as vehicles for his political beliefs, which were Socialist but anti-Marxist. In addition to his fiction, Wells also worked on two compendiums of knowledge, *The Outline of History* and *The Science of Life*, the second written in collaboration with his son G. P. Wells and Julian Huxley.

WHYTE, Lancelot Law, was born in Edinburgh in 1896 and educated at Cambridge. He first worked as a scientist in industry. From 1936 to 1941 he was chairman and managing director of Power Jets, Ltd. From 1941 to 1945 he worked for the Ministry of Supply in London. Since 1945 he has devoted his time to writing and lecturing.

WIENER, Norbert (1894–1964), was graduated from Tufts College at fifteen and took graduate studies at Harvard, Cornell, Cambridge (where he studied under Bertrand Russell), and Columbia. In 1932 he became a professor of mathematics at the Massachusetts Institute of Technology where he had lectured since 1919. He worked on guided missile systems during World War II and studied the handling of information by electronic devices, drawing analogies between these devices and mental processes. This study led to the founding of the science of cybernetics and the publication of several books on the subject. He also wrote detective stories under the pseudonym "W. Norbert."

WILSON, John Rowan, was born in England in 1919. He was educated at Leeds University and Medical School, becoming fellow of the Royal College of Surgeons in 1950. He worked as a surgeon in various hospitals until 1954 when he left surgery to become medical director of Lederle Laboratories in London. In 1961 he became assistant editor of the *British Medical Journal*. At the present time he is international editor of the British medical magazine *World Medicine* and a free-lance writer. He has published six novels and several scientific works.

INDEX